INDIA'S MIXED ECONOMY

INDIA'S MIXED ECONOMY

*The Role of Ideology and Interest
in its Development*

BALDEV RAJ NAYAR

BOMBAY
POPULAR PRAKASHAN

POPULAR PRAKASHAN PRIVATE LIMITED
35-C, Pandit Madan Mohan Malaviya Marg
Popular Press Bldg., Tardeo, Bombay 400 034

HC
435.2
.N348
1989

© 1989 by BALDEV RAJ NAYAR

First published in 1989

(3361)

ISBN 0 86132 217 7

PRINTED IN INDIA

By Gopsons Papers Pvt. Ltd., A-28, Sector IX, Noida
and Published by Ramdas Bhatkal for Popular Prakashan Pvt. Ltd.
35-C, Pandit Madan Mohan Malaviya Marg
Bombay 400 034,

To
My wife

NANCY

Preface

India has a mammoth public sector. How this public sector came to be created and expanded is the central concern of this study. The study is a work in political economy, that is, it is concerned with the interaction of economics and politics. Unfortunately, the neglect of economics and economic problems has been characteristic of political scientists; equally, the neglect of political aspects of economic issues has been characteristic of economists. However, this work is rooted in the belief that more is to be gained in understanding the social reality of India's public sector by examining the interaction of economics and politics.

This study takes an agnostic posture toward ideologies on the interaction between economics and politics. That the position taken in this study may at some points coincide with some version of Marxism, while at other points with some version of liberalism, is simply a result of contingent appreciation of their instrumental utility. But otherwise, by and large, the study may be considered to be a dialogue with various schools of thought. Although the subject matter of the study tends to be surrounded by passionate controversy, the present work essentially aims not to be controversial but a serious contribution to the ongoing debate on the role of ideology and interest in policy. It is the first study which consciously and analytically examines the role of ideology and interest in the creation of India's public sector over the entire stretch of India's post-independence development, and does so systematically against the larger canvas of theory. If in the process it has tended to be long, that is a result of attempting to bring empirical evidence to bear on the issues in a thorough manner, for the purpose is to advance, not opinion, but knowledge.

No rigorous theory is available on the political economy of the public sector. Accordingly, the study attempts to cast the theoretical net wide in Chapter I and embed the examination of the public sector in the historic confrontation of rival ideologies, the epochal process of modernization and development, and the nature of the state. That chapter deals comprehensively with questions both of causes and performance but the empirical part of the present study concerns itself only with *causes*. It leaves the question of a rigorous evaluation of performance aside for future research, perhaps by other scholars. Chapter II elaborately examines the nature of the Indian state. The baffling and bewildering variety of interpretations revealed therein leads compellingly to an empirical and historical examination of the conflict between the social and political forces arrayed for and against the creation and expansion of the public sector. In this empirical examination, Chapters III, IV and V relate to the Nehru period. Among these, Chapter III examines the role of class in the nationalist movement, the development of Nehru's ideological thought, and his relationship to the capitalist class; Chapter IV is an analysis of the Nehru model of socialism and how it came to be implemented; especially noteworthy is Chapter V, which looks at the contrasting posture of the Indian bourgeoisie and the communist movement towards Nehru's model of socialism.

In a further extension of the empirical examination, Chapters VI, VII and VIII cover the period of Mrs. Gandhi in office. They pay special attention to questions concerning the role of ideology: among them, Chapter VI, examines the break-up of the Congress party and its roots in and consequences for opposed conceptions of the economic order; Chapter VII is the most thorough analysis so far of the process of nationalization under Mrs. Gandhi; and Chapter VIII takes as its focus the retreat from the radical course amidst economic crisis in the mid-1970s and subsequently its continuancas in the early 1980s. Chapter IX discusses the consequences of the cumulative decisions for the size and status of the public sector; the chapter intends not to furnish the most uptodate data on the subject but to suggest the larger emergent pattern. Finally, Chapter X provides the overall conclusions of the study.

The manuscript was revised during the summer of 1986, but understandably individual chapters are likely to bear the imprint of the time when they were initially written. Chapters I and II were written during the summer of 1983, Chapters III-V during that of 1984, Chapters VI-VIII during that of 1985, and Chapters IX and X during the summer of 1986. To facilitate reading, each chapter is divided into several sections; each also carries a summary at the end. Since footnotes are bunched together at the end of the each chapter, no separate bibliography has been included; however, an author index has been provided.

I have benefited immensely from the painstaking and penetrating comments on different parts of the study by Professors Stephen Bornstein, Michael Brecher, Thomas Bruneau, Dipankar Gupta, Jagdish Handa, Frank Kunz, Samuel Noumoff and Donald von Eschen, and Mr. Ashok Nigam. To all of them, but particularly Michael Brecher and Donald von Eschen, I am extremely grateful. Besides, I am especially indebted for financial support for research to the Social Sciences and Humanities Research Council and to McGill's Social Sciences Research Grants Committee. Without the generous word-processing facilities of McGill University, it would have been difficult to complete this study. I owe special thanks to Mr. P.N. Malik, Administrative Director, Shastri Indo-Canadian Institute, New Delhi, for his generous assistance in so many most ways during the course of the research. A past debt owed to Mr. R.R. Gulati, Department of Science and Technology, Government of India, for his many kindnesses and selfless help is here gratefully placed on record. Although not named here, the many officials and non-officials in Indian economic and political life have continually educated me over the years by sharing their knowledge and wisdom; no research would ever have been possible without their generous cooperation and help.

The vagaries of publishing have considerably delayed the coming out of this

Preface

book until 1989. But it is just as well that it should be published in Nehru's centenary year since much of the book deals with Nehru and with his socialist project. The momentous changes that have taken place in regard to the economy in the Soviet Union under Gorbachev since the book was initially written, though not incorporated in the opening theoretical chapter, are consistent with, indeed confirm, the basic thrust of the present work. I hope to examine in another study the changes in India's posture toward economic policy and the public sector under the Rajiv Gandhi administration.

BALDEV RAJ NAYAR

January 7, 1989

Contents

Preface — vii

A. Theory

I. National Development and the Public Sector — 1

II. Contending Approaches to the State and Public Sector — 62

B. The Nehru Era

III. Class and Ideology in the Age of Nationalism — 128

IV. State Entrepreneurship in the Nehru Era : Ideology vs Necessity — 174

V. Nehru's Socialism and Group Interests — 213

C. The Era of Mrs. Gandhi

VI. Economic-Political Crisis and the Transition to Radicalism (1964-1969) — 248

VII. The Reign of Ideology : The Grand Era of Nationalization (1969-1973) — 282

VIII. Economic-Political Crisis and the Retreat from Radicalism (1974-1984) — 328

D. Summing Up

IX. The Hegemonic Position of the Public Sector — 360

X. Conclusions — 386

Select Bibliography — 395

Index — 411

Chapter I

National Development and the Public Sector

The rationale for -- and evaluation of -- the public sector in a country's economy and polity, especially in the Third World, is entwined with questions about both consummatory and instrumental values.[1] From the perspective of consummatory values, the presence or absence of the public sector is related to, indeed identified with, the vision of society to be built. It is, as a consequence, entangled with the conflict among the world's great political ideologies that has raged in the modern era, but especially since the Russian revolution.

As a phenomenon of the modern age and the rationality characteristic of it, ideology provides an explanation of man's social condition; further, it offers a this-worldly prescription about replacing that condition with a morally more desirable one but distinctively, as a secular successor to religion, it demands a thorough-going commitment to the explanation and to purposive action to follow through with that prescription.[2] In origin, modern ideologies may have been rooted in contemporary social compulsions but they often continue to evoke a religious-type allegiance against which there can be no appeal, even though the historical process as it unfolded may have brought into question their intellectual assumptions and social predictions. At times, the original ideological paradigms may themselves have undergone substantial modification or decay in the course of actual implementation in specific social contexts, but their advocates, especially elsewhere, often continue to take extreme positions on their truth value or validity.

From the perspective of instrumental values, the public sector is linked with its possible role in the efficient management of the economy. Especially in the Third World, it is tied up with the advancement of modernization or development, the historic summary goal for underdeveloped countries in the present epoch. With amazing prescience, Marx had stated more than a century ago that "the country that is more developed industrially only shows, to the less developed, the image of its own future."[3] Of course, in the pioneer industrial country of Great Britain, development may have initially taken place as a spontaneous process, but since then all countries that have developed have done so under the sponsorship of the state. It became the historic function of the state to sponsor development, with industrialization as its core, once it was accomplished in the pioneer country, but the question of the appropriate role for the state in this process still continues to be a contentious issue. Should the state be merely a regulator of the development process, relying on the private sector for managing the actual process, or should it be also entrepreneur, owner and manager of the means of production? Or, should it combine reliance on both the private and public sectors?

At issue is the relative efficacy of the public and private sectors as instruments in the development of economically backward societies. However, this issue cannot be completely de-linked from ideology, which may in fact be the governing consideration in many cases. To be sure, in actual life, consummatory and instrumental values, ideology and efficacy, may be difficult to disentangle; indeed, ideology itself may encompass the requirements of efficacy, but the two may also be in conflict, especially as time passes. Therefore, they need to be analytically distinguished in any examination of the impact they have on policy-making and the implications they hold for the continuation or alteration of settled policies. In this chapter, three aspects of the issue are considered at some length: (1) the ideological context in terms not only of the conflict between different paradigms but of the substantial modification that has actually come to characterize them; (2) the historical compulsions for economically backward nations to accord a high priority to development and a central role to the state in it; and (3) the relationship of social class to state power and, therefore, to the requirements of development. This comprehensive perspective is indicated by the absence of any rigorous theory on the political economy specifically of the public sector.[4]

1. The Ideological Context : The Historical Dialectic versus the Invisible Hand

The public sector refers to that set of the means of production which is deemed to be owned by society as a whole and is managed through the organized institutions of society, such as the state. As a mode of economic organization it is contrasted with the private ownership and management of the means of production in capitalism. Although states have run some economic enterprises in the past for economic and other reasons, in the modern era the public sector as a morally desirable order of things is associated with the ideology of socialism which, in turn, is related to the thought of Karl Marx. What Marx really said or meant is a contentious issue, for he wrote prolifically and over an extended period of time. Here, what he was believed to have said by the early Marxists, both orthodox and revisionist, is taken as the point of departure.

Marx : Social Ownership and Comprehensive Planning

Marx's analysis and critique of capitalism was set in a broader conception of the movement of history : the operation of a dialectic propelled mankind through a succession of qualitatively different but progressive stages of societal development. The motor force in this movement was property-based class conflict whose two critical characteristics, for Marx, were (1) its bipolar nature, pitting the class of exploiters against that of the exploited; and (2) its zero-sum nature where, in an exploitative relationship, the gain of one is necessarily the loss of the other.

In the contemporary stage of societal development, Marx thought that —beset by a series of crises resulting from the basic contradiction between the *social* organization of the forces of production and the *private* appropriation of economic surplus — capitalist society would, of necessity, become increasingly polarized between a narrow exploiting class of the bourgeoisie or capitalists owning the means of production and the increasingly exploited class of the proletariat or workers constituting the bulk of society. In counterpoint to Adam Smith, in whom the anarchy of the market would lead to hitch-free benign results through the invisible hand, Marx held that capitalism's inherent contradictions would result in its inevitable breakdown.[5] Eventually, the proletariat, fueled by its increasing misery and its rising class consciousness, would overthrow the capitalists through a revolutionary seizure of power. However, if, for Marx, the model was the earlier bourgeois revolution, there would seem to be no warrant for optimism on the score of the proletariat enacting a socialist revolution; the dynamic agent of that bourgeois revolution lay outside the feudal mode of production, while the proletariat lies within the capitalist mode of production. In actuality, it has turned out to be the case that the proletariat has not been the vanguard of revolutions made in its name.

In any case, the enactment of the proletarian revolution was conditional on capitalism having earlier exhausted the possibilities of further advance in the productive forces of society, that is, on capitalism having fully replaced the precapitalist modes of production and thus having made the proletariat into the large majority of the population. To enact a revolution before then would naturally be adventuristic. This stance, however, became a contentious issue in the Marxist movement in Russia, which split up between Mensheviks and Bolsheviks, with the latter insisting that it was unnecessary to wait until capitalism had fully replaced feudalism. At any rate, for Marx, the proletariat would, after the revolution, launch the new social and economic order of socialism, based not on private profit but directly on the social good.

As a result of the work of Marx, a powerful mystique developed around the notion of the proletariat. On the one hand, his discussion of capitalist exploitation in the pursuit of private profit evokes nothing but compassion for labour and therefore sympathy for its deliverance from misery and "universal suffering" through revolution. On the other hand, Marx argued that what is morally desirable is, indeed, scientifically inevitable, and that in the accomplishment of it the proletariat will assume the role of saviour of mankind, delivering humanity from the perennial scourge of exploitation of man by man. The result was an apotheosization of the proletariat. In the new classless society of communism that the proletariat was expected to bring about, there will be no private ownership of the means of production, while production and distribution will take place without the mediation of money and market; instead, there will be "comprehensive planning" to organize economic affairs.[6] Such a society will be established after a transitional period of the dictatorship of the proletariat in order to remove the vestiges of the old order. Once it is established, this society will entail the withering away of the state — the organized instrument of domination — and the cessation of the operation of the historical dialectic, with mankind finally coming to rest in the communist stage.

However, although such a society would represent a tremendous social achievement, it is clear that it would be so only in distribution, not at production.

In regard to the latter, it is manifest that communism was dependent on the dirty work of establishing the productive forces having already been accomplished by capitalism. Where violent revolutions have occurred in the name of Marxism, such as the Soviet Union and China, however, the state has not withered away but rather has become total in the scope of its activity. Nonetheless, insofar as these revolutions involved the abolition of capitalism, there has come into being state or public ownership of the means of production, with the state apparently assuring the social or public interest directly through a centrally planned economy.

In addition to the revolutionary route, Marx and Engels envisaged an alternative road to socialism by way of an increasing expansion of the public sector under pressure from the working class in political systems based on adult franchise. But this was strictly by way of exception as a possibility in the cases of England and America, not as a general expectation, nor as undermining their overall conclusion on, and call for, the revolutionary path. [7] Bernstein and Kautsky, however, made this alternative route the centrepiece of their political position. [8] Through this route socialism would emerge in a painless manner; the anarchy of capitalism would in the end be replaced in an evolutionary way by state ownership of the means of production and by central management of the economy on the basis of economic rationality, informed by social purpose rather than private profit.

Smith: The Self-Regulating Market and the Absentee State

In his advocacy of the public ownership of the means of production and the replacement of private profit by social good as the guiding motive in economic activity, Marx had prescribed a course that was in direct ideological opposition to that of Adam Smith, the supreme advocate of laissez faire capitalism. Smith had posited only a restricted role for the state in economic affairs. No doubt, he allotted important functions to the state which were critical for the efficient functioning of the economy, but basically his prescription was that of a hands-off policy for the state in relation to direct participation in economic activity. However, Smith was not insensitive to the public interest or social good; indeed, that interest was central to his intellectual framework, but the mechanism for its achievement was decisively different. Of course, Adam Smith had preceded Marx by about three-quarters of a century; his targets were the mercantilists, but his prescription could just as well have been addressed to Marxists of a later era with their commitment to the public ownership of the means of production.

Rather than locating the public good in activity directly motivated by concern for it, Smith saw its most optimal manifestation in the unintended but enormously benign consequence of the pursuit of self-interest. For him, self-interest was the surest way of achieving the public or social good. Smith acknowledged that, insofar as the businessman was concerned, the community's interests "never enter into his thoughts" and his sole concern is "his own private profit," but "By pursuing his own interest he frequently promotes that of the society more effectually than when he really intends to promote it. I have never known much good done by those who affected to trade for the public good." [9] For Smith, the self-seeking of the businessman is automatically converted into the social good through the mechanism of the market which comes into being naturally in society as a "consequence of a certain propensity in human nature...the propensity to truck, barter, and exchange one thing for another." Man engages in this activity not to promote the social good but his own interest. However, the end result is economic benefit for the community: "It is not from the benevolence of the butcher, the brewer, or the baker, that we expect our dinner, but from their regard to their own interest." [10] Self-seeking men are, as it were, led by an "invisible hand" through the operation of the market to serve the general welfare. Out of private vice thus comes public virtue.

Unoperated by any authority, the market is, for Smith, an amazing mechanism for coordinating the myriad economic decisions and activities of countless consumers and producers; sensitive to demand and supply, it produces automatically economic organization and social harmony out of their self-seeking pursuits. It is the "invisible hand" of the market -- with its "automatic adjustment", "perfect liberty" -- that assures optimal allocation of society's resources and therefore the maximum output and the greatest economic growth, thus promoting the welfare of the community. The businessman will, of course, seek to invest his capital in areas of the greatest demand and therefore of the greatest profit, but "the study of his own advantage naturally, or rather necessarily leads him to prefer that employment which is most advantageous to the society." The lower classes gain, too, from the allocative mechanism of the market, for the maximization of output provides more and better employment and advances their living standards.[11]

Thus, the market assures not a zero-sum but a positive-sum game, where everybody gains, even if not in the same proportion. This, of course, is not the result of anyone willing it to be so but because of the operation of a free market. Indeed, Smith was not naive about businessmen; as a philosopher, he had a natural distrust of "the mean rapacity, the monopolizing spirit of merchants and manufacturers," bent on restrictive practices with the intent of raising prices; he held that "their interest is, in this respect, directly opposite to that of the great body of the people" and warned that "people of the same trade seldom meet together -- but the conversation ends in a conspiracy against the public." However, Smith, like Marx later, was interested not in private motive but social result. Central to his system for assuring the social good was therefore not simply reliance on private enterprise but competition.

The public interest as a function of self-interest, the social good as a result of personal greed, achieved through the instrumentality of the market's invisible hand -- a conceptual equivalent to the dialectic of Marx -- was an extraordinarily profound but counter-intuitive idea. It was not unknown before Smith, but it was his genius to have systematized it and to have made it the foundation stone of his economic science. Those countries in Europe and North America which appropriated this idea as the basis of their economic systems, regardless of the many modifications subsequently introduced into them, came in due course to dominate the rest of the world economically, militarily, and politically, precisely because of the explosion in wealth and power that those economic systems made possible.

From the notion that the competitive market as a vast but self-adjusting machine was the ideal mechanism for assuring the public good -- the theory of a pre-determined social harmony [12] —flowed a strong case against state intervention. Reisman sees several reasons in Smith for reducing the role of the state in the economy :[13] (1) The state's ignorance in matters economic leads to the selection of wrong policy goals and policy instruments and to the neglect of the right ones, the resulting misallocation of resources harming producer, consumer and labour alike. State action is also likely to fall under the influence of special interests, thus harming the public interest. (2) The state is likely to waste capital, for its officials are given to extravagance: "They are themselves always, and without any exception, the greatest spendthrifts in the society. Let them look well after their own expense, and they may safely trust private people with theirs. If their own extravagance does not ruin the state, that of their subjects never will." (3) Since the state is dependent on a bureaucracy, its economic activity will therefore necessarily be characterized by inefficiency and mismanagement: "The agents of a prince regard the wealth of their master as inexhaustible; are careless at what price they buy; are careless at what price they sell." The reason for this is that there is no incentive to perform well since their reward or salary does not depend on performance: "Public services are never better performed than when their reward comes only in consequence of their being performed, and is proportioned to the diligence employed in performing them." Smith held the system, not individuals, as responsible for inefficiency and mismanagement: "It is the system of government, the situation in which they are placed, that I mean to censure; not the character of those who have acted in it. They acted as their situation naturally directed." (4) State activity that runs counter to or transcends the natural economic order will be either unenforceable or lead to perverse results. Thus the imposition of high import duties "presents such a temptation to smuggling, that all the rigour of the law cannot prevent it," while high taxation is likely to lead to evasion and outflow of capital.

However, it needs to be understood that the state in Smith was not bereft of all functions; he did not reject it nor envisage its withering away. He singled out three functions that were of particular importance: (1) defence; (2) administration of justice; and (3) public works (roads, bridges, canals, ports) which are "in the highest degree advantageous to a great society."[14] However, the burden of Smith's message to the state was laissez faire, let it be, and he demanded the elimination of the system of state controls and state-sponsored monopolies that were part of mercantilism in favour of reliance on the efforts of individuals in an atomistic but self-regulating market. The operative concepts in Smith, however, were the state and individuals;[15] he was less sensitive to the constraints on economic activity, and therefore welfare, that may come from the power of structures in between the two. It is with these latter in the form of corporations and trade unions that society would have to contend with subsequently under capitalism.

The Revenge of History

On the foundation of the assumptions, theses and predictions of Marx and Smith, there arose fundamentally opposed ideological movements and economic systems known as socialism and capitalism. However, the actual course of history has brought into question the constitutive principles of their respective doctrines, indeed a drastic decay or erosion of the basic paradigms constructed by them.

In Respect of Capitalism

As far back as 1858, Marx had proclaimed that "on the Continent the revolution is imminent and will immediately assume a socialist character." [16] The astounding fact, however, is that during the century and more that has passed since, there has been no revolution in any advanced capitalist society. The survival of capitalism over this period, regardless of the post hoc rationalizations provided by Marxists, is a development that stands in dramatic contradiction to the prognosis by Marx. Amazingly, capitalism developed structurally in such a manner as to undermine both the assumption of the bipolarization of social stratification and that of the zero-sum game.

Instead of social bipolarization there took place what Dahrendorf has called the decomposition of classes.[17] On the one hand, labour became stratified into unskilled and skilled strata. Consisting of workers with special skills in management and technology, and differentiated by their higher income and status, the skilled strata identified themselves more with the interests of the owners of the means of production than with a generalized working class. With the development of this "new middle class", society in advanced capitalist countries did not divide itself into two opposed segments, a cleavage which is, as Marx correctly saw, productive of societal breakdown, if not revolution; rather, conflict was mitigated by precluding division of society along a single central axis.

On the other hand, with the rise of the corporation as the characteristic type of business organization, there took place not only the diffusion of ownership of capital but, more importantly, a disassociation between ownership of capital and management control of the corporation ("late capitalism"). The displacement of the individual capitalist -- with his singleminded pursuit of profit and commitment to capital accumulation ("early capitalism") -- by the corporation also mitigated capital-labour conflict. Unlike the individual capitalist-owner, who directly and personally appropriated whatever could be squeezed out of labour, the manager did not directly and personally benefit from keeping labour's wages low, and consequently tended to be less negative in relation to labour's demands; after all, managers, too, are employees of the corporation and the future of all rests on the success of the corporation, thus the necessity for cooperation between management and labour which assumes the form of a periodic negotiated settlement. The relationship between labour and management therefore came to be transformed into a variable-sum rather than negative-sum game.

Beyond the decomposition of classes -- a structural development which had not only not been foreseen but one which directly contradicted the social prognosis proclaimed by Marx -- there was additionally the critical role of politics in moderating social conflict. Marx had simply neglected politics as an independent variable, assuming it to be merely an epiphenomenon of the economic base. But the actual course of history has demonstrated the autonomous strength of the political variable in modifying the operation of the forces of the capitalist market. This is decisively manifest in three respects: (1) the countervailing collective power of labour; (2) the welfare state; and (3) the "mixed economy" -- all made possible because the state came to have a broader social base than simply the capitalist class.

The rise of the corporation as the leading form of business organization was paralleled by the rise of the collective power of labour in the form of the trade union movement. This resulted, as political parties in representative systems sought the electoral support of the working class, in labour being able to achieve higher living standards, greater job security and better health, pension and unemployment benefits, and also to obtain state protection for its rights and interests. Marx had attributed to a pauperized proletariat the class interest of making a revolution for a new society, but labour has instead chosen to act in a reformist fashion to improve its position as an integral part and partner of capitalism, seeing mutual (though still unequal) advantage in cooperation. Indeed, management and labour in the "managed capitalism" of the "mature corporation" can, apart from countervailing against each other, as well act in collusion, with management buying social peace through high wage settlements in favour of labour and then passing the cost to the public. [18]

Some Marxist scholars have, of course, sought to rationalize this development by holding that "the bourgeoisies of the center learned through historical experience that a situation that allows the standard of living of the proletariat to rise over time (a stable rate of surplus value combined with rising productivity) is not only functional but even indispensable for the operation of the system as a whole." Further, they maintain that, "alarmed by the revolutions of 1848, the ruling classes of the advanced capitalist countries began to reconsider their strategy, responding to the struggles of the workers more flexibly and discovering in the process that the new course paid both political and economic dividends." [19] But the point is that this development was not provided for in Marxist theory and could not have been provided for, without bringing into question the whole conceptual edifice of the inevitability of revolution. Beyond protection for labour, but in part through labour's political organization and its electoral strength, there came to be created also the "welfare state" which provided a minimum living standard for all citizens through income transfers and common social services. [20] In the words of a Marxist scholar, "The rise of welfare state capitalism is an irreversible development....The recent advances of labour therefore constitute a historical conquest which cannot be undone by democratic means." Surprisingly, "in advanced capitalist countries today, between one fifth and one third of all household income derives from public revenue and not from property or labour." [21]

Howsoever much Adam Smith and his successors may have praised the benign functioning of the invisible hand, and howsoever momentous the contribution to human progress may have been as a result of it, the market nonetheless imposed significant costs on society and groups within it. These included the externalization of costs that resulted from unregulated competition among profit-maximizing firms in a capitalist economy, such as in the case of damage to ecology. Furthermore, public interest could be, and was, hurt through the emergence of monopolies exploiting economies of scale but dictating prices and barring entry to new firms. Again, to protect the larger public interest the state resorted to regulation of economic activity, as in the case of smoke emissions, restrictive trade practices, excessive market shares, and high rates for public utilities. However, as time went on, the state's role in the economy developed further than that. This was spurred by the social havoc inflicted by capitalism through its periodic crises, especially during the 1930s. Here, history did not uphold the virtues of the invisible hand for vast segments of the population; rather, for them the hidden hand, regardless of its long-term potentialities, was a veritable carrier of deprivation and doom. Again, the state was seen as the appropriate instrument for correcting and mitigating the ravages of the business cycle.

The theoretical rationale for state intervention in the economy to augment aggregate market demand in order to pull out of, and in future prevent, economic slumps was provided by Keynes, but in actual fact the state had already intervened in the 1930s in several economies, such as in Germany, for this purpose. Keynes, however, occupies a key place in the history of economic thought. Whereas Adam Smith thought (before industrialization was really a fact) that capitalism automatically would function harmoniously and while Marx envisaged that capitalism would inevitably confront crises and eventually breakdown, Keynes believed that crisis-ridden capitalism could be fixed -- through demand management by government. The Keynesian "revolution" made possible for a quarter of a century after World War II a reconciliation between business and labour, by assuring a continued profitable investment environment to the former and a high level of employment to the latter. Since then, state intervention through various fiscal and monetary instruments has become a routine aspect of the national management of the economy.

Some among the advanced capitalist countries, such as France, however, went beyond this to make such national management of the economy part and parcel of a more comprehensive central economic plan. Indeed, economic planning assumed considerable importance, becoming "the most characteristic expression of the new capitalism. It reflects the determination to take charge rather than be driven by economic events." [22] And as one scholar observes: "The French government surely plans more than the Yugoslav, and only a little less than the Hungarian."[23] Subsequently, even Germany, while avoiding the rhetoric, came to follow the French in economic planning.[24] Furthermore, beyond regulation and intervention, many advanced capitalist countries, but especially Italy, France and Great Britain, have gone in for extensive state ownership of economic enterprises through nationalization or otherwise, and have used this state ownership as an instrument for a more effective control of the economy. These enterprises extend beyond public utilities and at times cover whole industries, such as coal and steel in Great Britain. In the case of Austria, the public sector accounts for 20 per cent of GNP and 25 per cent of exports.[25] Ideology has, of course, played some role in the establishment of state ownership, as in Great Britain after World War II, but in many cases contingent causes have been at work in the creation of a "public sector".[26] In any case, a substantial public sector as an integral part of the national economy can no longer be said to be a function of socialism alone.

Through the cumulative impact of state regulation, intervention, planning and ownership, there has, for some, taken place a significant transformation of capitalism and what exists is not strictly a capitalist economy but a "mixed economy", with some of the malign features of capitalism allegedly removed. Indeed, modern capitalism and "mixed economy", incorporating both large public and private sectors, have become equivalent terms. A new model of capitalism seems to have evolved. Surprisingly, governments now handle 40 per cent of GNP on the average among the advanced industrial countries.[27] In one group of OECD countries in 1982, public expenditure constituted the dominant part of GNP (Sweden 64%, Denmark 60%, Netherlands 59%, Belgium 52%, and France 52%) while in another group it was somewhat below half (Canada, Austria, Germany, Ireland, Italy, Norway and UK).[28] Even the Marxist scholar Miliband, whose intent is to downplay the significance of the change, acknowledges that "the scale and pervasiveness of state intervention in contemporary capitalism is now immeasurably greater than ever before, and will undoubtedly continue to grow; and much the same is also true for the vast range of social services for which the state in these societies has come to assume direct or indirect responsibility."[29] Whether capitalism has, indeed, transcended itself and has evolved into "an altogether *different* system (and, needless to say, a much *better* one)" may be a debatable issue, but socialism, especially through revolutionary means, seems to provide little attraction for the vast majority of the masses of the advanced capitalist societies. No extant socialist society constitutes a model for the advanced capitalist societies outside scattered groups of university academics. Rather, it is the advanced capitalist societies that are the model for other industrialized countries as is demonstrated by the one-way movement of people out of socialist countries notwithstanding the stringent barriers against it.

The attractiveness of the modified capitalist model derives, first, from the extraordinary advance in productive forces, and therefore material welfare of the masses that it has made possible in the postwar period. Continuing a tradition first established by Marx and Engels in the Communist Manifesto with their unexcelled praise of the achievements of the bourgeoisie, Schumpeter about a hundred years later could not resist being dazzled by its accomplishments. Undeterred by the depression years, he wrote in 1942 that "if capitalism repeated its past performance for another half century starting with 1928" -- and he had no doubt that it would, provided anti-capitalist policies were not followed -- "this would do away with anything that according to present standards could be called poverty, even in the lowest strata of the population, pathological cases alone excepted".[30] Of course, standards change and the sense of poverty perhaps can never be removed, while experts would undoubtedly differ over definitions of poverty, but that capitalism has provided unprecedented levels of mass welfare in material terms -- whatever the social and psychological costs of urbanization and industrialization -- there can be no doubt about. If Schumpeter prophesied that eventually but inevitably capitalism would metamorphose into socialism, it was so because of its achievements, not because of its failures, with the corporation making capitalists superfluous because of the "evaporation of the substance of property." It is interesting that the Marxist scholar Brus also sees a strong "continuity in the relationship between capitalism and socialism," a progressive evolution in the development of productive forces from (1) the individual capitalist-entrepreneur, to (2) the modern large capitalist corporation, to (3) the public sector, to (4) socialism.[31]

For a quarter of a century after 1948 in the postwar era, modern capitalism provided a period of remarkable and sustained economic growth, unmarred by any serious depression unlike every decade in the century prior to World War II. This long boom, which Shonfield characterized as "the age of acceleration," [32] certainly marked a triumph for Keynesianism, but in the eyes of some it only postponed the inevitable crisis, indeed made it even more serious. [33] The economic crisis that emerged in the mid-1970s has been seen, depending often on the initial predisposition of the observer, as representing one or more of the following three possibilities: (1) a downswing in one of the recurrent business cycles which are characteristic of capitalist development but, given the right set of policies, would be followed by an upswing; (2) the downward half of the fifty-year Kondratieff long wave but which will be followed by economic rejuvenation; and (3) the inevitable structural breakdown which cannot be accommodated within the present capitalist system.[34] Some have, of course, projected the early demise of capitalism, [35] but the failed predictions of the past are a warning against underestimating the flexibility and self-rejuvenating capacity of capitalism; the projections of its demise are likely to turn out to be highly exaggerated. What is striking about capitalism over the last century is rather its extraordinary capacity at adaptation to the various challenges that have confronted it. [36]

Despite the economic reverse in the 1970s, the remarkably impressive feature of the economic arrangement called capitalism, compared to all historically known economic systems, is the extraordinary advance in productive forces and thus in human welfare. Not only that, secondly, nowhere else is the achievement of capitalism more dramatically manifest than in the sphere of technological innovation; capitalism revolutionized and continues to revolutionize social life through technological innovation. In reality, capitalism as an economic system is synonymous with technological change. Marx himself had emphasized the thrust to technological innovation in capitalism, seeing it as an objective necessity for capitalists in order to lower costs so as to escape the situation of equalization of profits. However, Oskar Lange thought that the reduction in competition as a result of the rise of monopolies (that is, corporations) would diminish technological innovation because, while able to influence prices and entry by new firms, these monopolies would not want to devalue their existing capital by new innovations. [37] Schumpeter and Galbraith, on the other hand, have shown that the superiority of the corporation over the individual capitalist lies precisely in the acceleration of technological change and that the "perennial gale of creative destruction" (Schumpeter) operates even more intensely under managed capitalism, one reason why the "technostructure" or technocracy (Galbraith) has acquired a position of pre-eminence in the transformed capitalism. [38] The acceleration of technological change after World War II is well known, but it is incredible that one cannot think of any significant technological innovation which has changed modern social and economic life -- such as the automobile, jet aircraft, television, computers -- that owes its origin to an economic system outside capitalism. Capitalism continues to cast other economic systems in the role of technology borrowers.

All this is certainly a tribute to capitalism, but it is not the capitalism that Adam Smith, as its father-philosopher, had envisaged. Rather, it is a transformed capitalism where not only giant corporations have replaced Smith's individual capitalists but where the "invisible hand" is no longer left to its own devices but is monitored, controlled, guided and manipulated by the state; indeed, the invisible hand today can only be a fanciful mystifying term for the overall resultant of quite visible state activity in the economic sphere. To the extent that such state intervention was necessitated by the failings of the hidden hand, and to the extent that such state intervention has been accompanied by unprecedented economic growth and prosperity, history seems to have upheld, not so much the efficacy of the unperturbed invisible hand as of one guided by the state -- a state that at the same time has by and large astutely limited its intervention in the sense of affecting the market but not substituting itself for it.[39] For its part, in adjusting to the changed role of the state, capitalism has demonstrated its tremendous capacity for flexibility, its regenerative power and its remarkable resilience.

This mutual adjustment between the state and capitalism has been associated with, perhaps has been made possible by, another extraordinary phenomenon -- political democracy. Among the advanced industrial economies, it is only in the market economies -- howsoever modified by state intervention -- that democracy exists. "Democracy, capitalism and the welfare state march hand in hand," as one scholar puts it.[40] The perfect correlation of democracy with market economies among the advanced industrial countries is striking. Some would even hold that democracy is compatible only with market economies and that it cannot exist where there is complete state or societal (as in Yugoslavia) ownership of the means of production. That remarkable feature alone constitutes a major attraction of the market economy model, whatever its other weaknesses. Indeed, the combination of capitalism's economic success and practice of political democracy constitutes a legitimacy formula of great potency.

Of course, democracy itself has come to be held responsible for the crisis of modern capitalism in the 1970s insofar as it fostered excessive and incessant claims on the economic surplus on the part of insistent lobbies, which politicians readily conceded because of electoral requirements. In the process, it has hindered capital accumulation through precluding the "rational" management of the economy by the state, but presumably insulated from conflict and politics, that Keynes had envisaged. In this fashion, democracy with its compulsion for state intervention and capitalism with its thrust for laissez faire in the maximization of profit and capital accumulation are seen to be incompatible, with the Keynesian revolution having only postponed the inevitable clash rather than reconciled their competing requirements.[41] The solution may eventually lie in some form of corporate state.[42] But, in the meantime, even the very achievement of pluralist democracy has come to be qualified by its earlier votaries because of the perennial class hegemony of business, making such democracy relevant, in Lindblom's terms, "only on secondary issues, not on primary issues." Be that as it may, the deficiencies of democracy in the basically market economies are highlighted against its own ideals, not against other more attractive extant models; particularly apt is Dahl's comment that "one of the ironies of pluralism, it seems, is that it looks more attractive to people who are denied it than to those who possess it."[43]

Notwithstanding the achievements and virtues of capitalism, socialism as an ideal still continues to exercise a tremendous appeal, especially to sensitive intellects. Apart from socialism being a natural alternative as social discontent rises when capitalism is in crisis, particularly in terms of unemployment, that appeal stems from two vulnerable features of capitalism: one, the inequalities in income, no matter how high the floor for the lower incomes; and, two, the motive of private profit, rather than social good, as the basis of economic organization. Both these aspects have been found to be galling to idealistic minds, regardless of the rationale of the possibly superior benign outcomes from their existence. Economic inequality is considered a cultural affront, arousing revulsion against capitalism, while the profit principle is believed to run counter to man's inner nature and moral being.[44] Thus Galbraith, who in his earlier work had underlined the virtues of capitalism, finally opted for -- using precisely that taboo word -- "socialism."[45] Partly, it is because, as he says, "on the whole I'm committed to the idea of a much greater equality of income as an independent social good."[46]

Similarly, Lord Kaldor expressed his disdain for capitalism since it is "not only highly wasteful in terms of exhaustible resources, but creates a socially restless and basically frustrated competitive society which fosters a scale of values that moralists and religions throughout human history have regarded as reprehensible." He opposed the concentration of economic power, even if in the hands of managers, and held: "I cannot see that there is any long-run future for a society where some men are extremely powerful in relation to their fellow men because they possess a great deal of money or have the power of dismissal or patronage over a large number of employees..... It is the power conferred by the possession of wealth, far more than the inequalities of living standards occasioned by it, which makes modern capitalism so unsatisfactory as a method of organisation of human societies." He was also against "the making of investment decisions by a small minority, for their own benefit, which is not necessarily the social benefit." He thought it inconceivable that capitalism should continue and predicted its replacement by something else as inevitable, but sadly noted: "in the light of the history of the present century it is far less clear than it appeared at the end of the nineteenth century what that something else is going to be."[47]

It should be noted that in their criticism of capitalism, both presuppose an existing capitalist society from which to move to socialism -- a central idea in Marx. [48] As Galbraith has said, in order to attack capitalism for its economic inadequacies and moral shortcomings you have to have capitalism first: "The choice between capitalism and communism emerges only after there is capitalism."[49] In other words, regardless of all the morally appealing qualities of socialism — in abstract though — it can be decidedly premature to wish for them before capitalism has already created the productive forces for their achievement.

Notwithstanding that pertinent advice, it is among the less developed countries that Marxian socialism has come to exercise an enormous appeal as an ideology, if not always in practice.[50] That is supremely ironic because Marxism was, in origin, a prescription for the social predicament of already industrialized countries; it was of little relevance to countries that had much of neither bourgeoisie nor proletariat. The decisive factor in its popularity among the less developed countries is related, on the one hand, to their existential situation of being victims -- direct or indirect, past or present -- of an imperialism which issued out of countries whose economies were founded on capitalism and, on the other, to the reworking and revision of Marxism by Lenin in a fashion that spoke to that situation.

For one thing, Lenin made imperialist domination and exploitation comprehensible to its victims in terms of the inner expansionist logic of capitalism, thus calling forth among many a rejection of capitalism. Again, he held that deliverance of the less developed countries from their backward condition did not have to wait until the prolonged stage of capitalism had been completed, but that they could skip that stage entirely through combining the bourgeois and socialist revolutions into a single one. And, appropriate to the backward state of these countries, lacking the necessary proletarian base for a revolutionary seizure of power, he urged the formation of a coalition of workers and peasants under the auspices of a communist party to overcome that handicap. This was especially attractive to the political elites of the Third World because of the vanguard role Lenin posited for them in the transformation of their societies. More significantly, Lenin demonstrated the practical efficacy of his doctrinal position by actually leading a successful revolution in Russia, then the most advanced of the backward societies. Most critically, his successor Stalin made socialism almost irresistible for many Third World elites, not by attaining the utopian society, but by elevating the Soviet Union through rapid industrialization to the position of a first-rate industrial and military power, regardless of what some may consider to be the transformation or deformation of ideology in the process.

In Respect of Socialism

Public ownership of the means of production and comprehensive planning have been the central tenets of Marxism for the post-capitalist stage of socialism. There is therefore (1) in place of the ownership of the means of production by individual capitalists jointly or severally, ownership by society as a whole, thus removing the distinction between owners and non-owners in a classless social arrangement; and (2) in place of the individual calculus of the capitalist market, where capitalists pursue private profit, the economy is managed through a central allocative mechanism with the aim of achieving the interests of society as a whole. In brief, there is here the conception of the total productive economy constituting a single socially-owned factory, as it were, in which investments, materials and production quotas among its different branches are allocated by a central planning organization on the basis of social criteria determined by it, rather than by individual economic enterprises in response to market signals. Although Marx did not fully elaborate his vision of the post-capitalist society, public ownership and comprehensive economic planning are believed to have been the central components of his future society where man would be rid of exploitation and alienation. [51] In any case, these two features were turned into the heart of the economic design in the Soviet Union and became, in due course, identified with socialism. Subsequently, these two features were adopted in the countries of Eastern Europe that became part of the Soviet bloc after World War II, and resulted in a transformation of the economies of these countries. Regardless of the remarkable achievements in economic transformation, both features have come under attack as being either inadequate or inefficient.

Socialism's accomplishments in the reduction of economic and other inequalities are momentous, even though undeniably inequalities persist. [52] And undoubtedly, full employment, job security and public ownership have important social and psychological benefits for the populace. [53] Notwithstanding that, however, some consider it meaningless, from the viewpoint of the worker, whether the means of production are owned by private capitalists or by the state in behalf of society or the public, for the functional logic of central economic planning leads to a bureaucratic hierarchy whose top stratum administratively determines the consistency and coherence of the plan, resolves conflicts over allocation of resources, and issues directives for its implementation. [54] Bettelheim, after decades of support extended to the Soviet economic model, finally characterised it as state capitalism where "under cover of state ownership, relations of exploitation exist today in the USSR which are similar to those existing in the other capitalist countries, so that it is only the *form* of relations that is distinctive there." Complementing the proletariat, there exists a state bourgeoisie under this sub-type of capitalism:

> The factories are run by managers whose relations with "their" workers are relations of command, and who are responsible only to their superiors. Agricultural enterprises are run in practically similar ways. In general, the direct producers have no right to express themselves -- or rather, they can do so only when ritually called upon to approve decisions or "proposals" worked out independently of them in the "higher circles" of the state and the party.
>
> The rules governing the management of Soviet enterprises are to an increasing degree copied from those of the "advanced" capitalist countries...The producers are still wage earners working to valorize the means of production, with the latter functioning as collective capital managed by a state bourgeoisie. This bourgeoisie forms, like any other capitalist class, the corps of "functionaries of capital", to use Marx's definition of the capitalist class. The party in power offers to the working people only an indefinite renewal of these social relations. It is, in practice, the party of the "functionaries of capital", acting as such on both the national and international relations. [55]

Similarly, Poulantzas dismissed Soviet claims to socialism on the basis of formal juridical ownership of the means of production belonging to the state, "the people's state", holding: "but real control (economic ownership) certainly does not belong to the workers themselves...but to the directors of enterprises and to the members of the party apparatus." He went on to argue that "the form of collective juridical ownership conceals a new form of economic 'private' ownership; and hence that one should speak of a new 'bourgeoisie' in the USSR." [56]

Again, for Brus, public ownership, like private ownership, involves relations of dominance and subordination insofar as it vests control over the disposition of the means of production in "an established group holding state power, which in turn means deprivation of the majority of society of this right of disposition." The alienation of labour is thus not avoided by public ownership. Indeed, the domination over labour under public onwership can be far worse than under private ownership, "since (1) the state, gathering in one centre disposition over all -- or almost all -- the places and conditions of work, has in its hands an instrument of *economic coercion* the scale of which cannot be equalled by individual capitalists and corporations; (2) the state can directly link economic coercion with *political coercion*, in particular in a totalitarian system which actually liquidates political rights, the right of assembly and freedom of speech." [57] Much earlier, Djilas had come to the conclusion that social ownership was simply a cover for real ownership by a "new class" of political bureaucrats who made all the critical decisions in relation to economic development, the distribution of the economic product, and wages; he held that ownership meant the right to control, and this had come into the possession of the new class in place of the capitalists. [58]

For many Marxists now, public ownership by itself no longer constitutes socialism. For some, like Bettelheim and Poulantzas, it has to be supplemented by real, and not formal or fictional, control by workers. For others like Brus, control by workers at the enterprise level is not the right answer for the predicament, rather the solution lies in making public ownership really *social ownership* and the mechanism for that is *political democratism*, whereby those who have disposition over the means of production are made responsible to society. [59] However, no existing socialist country meets this criterion of social ownership. Everywhere in the socialist world, economic and political power is held by a self-perpetuating small group of leading party bureaucrats, recruitment to which is based on co-optation by existing incumbents.

The absence of democracy in socialist countries is as extraordinary a phenomenon as its presence in the advanced capitalist countries. It seems to suggest, perhaps wrongly, a basic incompatibility between socialism and democracy. Partly this is a function of the fact that everywhere socialist regimes in the Soviet bloc arose as a consequence of the seizure of power by a narrow group of dedicated revolutionaries representing a minority of the population. Such a premature installation of socialism necessarily resulted in a fundamental distortion of the ideal of socialism. On the other hand, precisely because they were not responsive to a democratic will, such groups were able to accomplish capital accumulation on an extraordinary scale through the exercise of totalitarian control. As one scholar so aptly points out, "Communist societies can hold down consumption not because they have abolished capitalism, but because they have abolished the multi-party system, the free press, and the right to strike. They have established political conditions for capital accumulation not dissimilar from those which existed in the nineteenth century." [60]

Whatever its economic accomplishments, however, socialism's association with the absence of democracy not only drastically deviates from Marx's benign vision but also dramatically diminishes its attraction for many groups outside the socialist bloc, especially in the advanced capitalist countries -- which are already ahead in development and material welfare even if not in equality and job security -- because of the fear of permanent and irreversible totalitarian control. At the same time, for the socialist countries themselves, their very success in industrialization combined with the lack of democracy creates a restless population, at times on the verge of a political explosion as in Eastern Europe. At any rate, the experience of socialism demonstrates that an ideology which has the social good, rather than private profit, as its central motivation is no assurance of the achievement of that social good; the nobility or purity of intentions is no guarantee of the desired results.

Besides the wide gap between public and social ownership which socialist countries have not even dared to bridge through democracy, central economic planning through administrative management rather than the market has come under increasing attack for its enormous inefficiency as an allocative mechanism, especially in an industrialized setting. Indeed, the inefficiency of socialist or authority-managed economies is perhaps comparable to the alleged waste of the anarchic relations and periodic crises of capitalist economies, except that the latter waste is the necessary price for organizational and technological innovation in what have been the world's pace-setter economies.

The management of an industrial economy, with all its complexity, as if it were a single factory is apparently beyond the capacity of a single centre. Alec Nove has sharply underlined the weaknesses of the Soviet economy that stem from the

> diseconomies of centralized scale. In a model which Brus has called 'etatist socialism', i.e. in which the state, acting on behalf of society, determines what is needed, what should be produced, how and by whom, the volume of decision making and information processing *enormously* exceeds the capacity of the centre. No amount of reorganization can change this basic fact... the system's functional logic requires an impossible centralization. Most of the weaknesses are to be found in the micro-economic realm; the transmission of user demand, quality, innovation, initiative, long-term responsibility, misleading prices, lack of objective criteria for choice, all these are problem areas....They are 'systemic', deeply embedded in the essence of the centralized 'directive planning' system, highly resistant to partial reform, or to any cure which does not tackle the nature of the system itself, including its political organization.....The West's superiority in the general area of micro-economics seems beyond dispute. Centralization leads to grave diseconomies of scale.[61]

The sheer amount of information to be processed at the centre leads to information overload and indigestion. [62] The number of prices alone required to be fixed administratively in the Soviet Union is mind-boggling; Lindblom estimates that there are some 20 million such prices involved. [63]

Objections to central economic planning on grounds of its inefficiency as an allocative mechanism in the absence of a market had been raised earlier on a theoretical basis before World War II, by von Mises and Hayek, and had been polemically countered by Lange. [64] However, after the political uprisings in the 1950s in several countries of Eastern Europe, voices for reform of the economic mechanism began to be raised, especially during the mid-1960s. The intent of the suggested reforms was to move towards reliance on the market mechanism in making decisions on production and resource allocations. Several experiments were tried but in the event only two countries -- Yugoslavia and Hungary -- have adopted the market model.

However, some feel that economic decentralization along with greater use of the market mechanism will ultimately prevail all over Eastern Europe. The reason for that is seen in the higher economic efficiency that the market model makes possible. Brus notes in this connection "the presence of economic needs -- the need to make production more flexible, reduce material content, stimulate the growth of labour productivity, etc. -- especially in the face of the necessity of solving the problems of further growth and technical progress ..."[65] Economic efficiency as a social requirement may be dismissed by some in favour of some other social good, but it would seem that it is a necessity for nation-states. In the past, civilizations came to an end -- or, more correctly, were put to an end -- because they were or became less efficient compared to other civilizations. History extracts a terrible price from those who lag behind in efficiency. In the international arena, economic efficiency matters, not just in absolute terms, but especially relative to other nations.

In moving away from the etatist model, Yugoslavia opted for the "self-management model" where workers in economic enterprises control those enterprises in the context of a market economy. Indeed, the Yugoslav economy has been held to correspond more to the nineteenth century small-producer market capitalism than any other modern economy, [66] and to have less central economic planning than France. Perhaps, as a consequence, it has been subject to balance of payments problems, inflation and income differentiation among workers in different enterprises and regions. Brus, who does not himself favour centralized economic planning, however, is critical of the Yugoslav model for its neglect of overall societal interests; besides, through this economic concession to workers, the single-party state elites have avoided the more basic question of political democratization. As he says: "The Yugoslav political system does not admit opposition, does not give the opportunity for presentation of an alternative solution....As a type of system, the Yugoslav political system belongs to the same category as the Soviet system. This therefore means that it does not meet the demands of the criterion of socialisation we have adopted." [67]

The origins of the Yugoslav model seemingly lay in the country's attempt to carve out a distinctive ideological position after having split from the Soviet bloc in 1948. Curiously, the Yugoslavs then suddenly discovered that the etatist model had all along rested on a misunderstanding of Marx and Engels, something that had not been apparent prior to the split. Socialism was now perceived as inconsistent with so-called state capitalism -- considered to be the inevitable result of state ownership combined with centralized economic planning -- leading to exploitation of workers by the new ruling class controlling the state. Accordingly, the new model replaced state ownership of the means of production with ownership by the direct producers through workers councils and also "a centrally managed economy by a self-governing economy." [68] Despite some adverse economic consequences associated with the Yugoslav market model, it has been politically functional for an ethnically-divided political system insofar as it shifted economic decision-making to the local nationalities even though the political system still cannot avoid the problem of regional imbalances.

The more interesting economic reform is that of Hungary since it has been carried out by a member of the Soviet bloc and also because reformers in other countries within the bloc find it attractive. Brus characterizes it as "the model of a planned economy with a regulated market mechanism" [69] and favours it over the Yugoslav model, for it combines central economic planning from the viewpoint of society-wide interests with the independence of enterprises responsive to market signals, while sidestepping the question of self-management by workers. Although Hungary did better economically than it would have without the reforms, its success was modest. At the same time, the reforms have not been greeted with particular enthusiasm by workers since managerial ruthlessness in the cause of economic efficiency accompanying the reforms has resulted in some unemployment, has made necessary geographic mobility for labour, and created a sense of job insecurity. However, the Hungarian model has not been accompanied by any political democratization, which fact, along with the lack of any self-management by workers, has perhaps made the reforms acceptable to the Soviet Union.

It may well be that Soviet accommodation to the reforms in Hungary was impelled by the political explosions in that country and at the same time facilitated by Hungary swearing loyalty to the Soviet bloc. Perhaps the reforms can be confined to the economic realm, for the threat of Soviet intervention places limits on their social consequences spilling over into the political arena. However, it would be a different matter if they were to be implemented in the Soviet Union. Apart from constituting an admission of national and ideological failure, reforms may well result in unraveling the political system through the creation of new autonomous centres of power, undermining the position of the present ruling class. Several scholars hold the fear of such possible political consequences as responsible for making attempts at economic reform come to nought. [70] Interestingly, China moved to the market model in the late 1970s in the context of the overthrow of an earlier leadership discredited for its alleged ideological excesses, and it did so under a leadership that had been purged and persecuted before.

The historical experience of the socialist countries suggests, on the one hand, that socialism today is no longer necessarily consistent with the central tenet of the philosophical position of Marx and Engels on the elimination of the market, leave aside commodity relations; witness the cases of Yugoslavia and Hungary. On the other hand, to the extent that the market is excluded in the degree that it is, it is no longer because of any belief in the superiority of centralized economic planning over the market; the reverse is the case. It is rather because of the potential political consequences for the state bourgeoisie, whose very existence, incidentally, stands in contradiction to the Marxist vision of a socialist society. Brus accurately concludes for the country in the vanguard of socialism that "the fundamental features of the production relations making up the etatist model of socialism remained undisturbed, in particular in the sector which we recognised as crucial, namely, in the political system, the real structure of power. Neither the whole of society nor the working class, whose dictatorship is supposed to be the essence of the socialist state, held power in the post-Stalinist period, and thus neither did they become the owners of the nationalised means of production."[72]

In the advanced industrial societies, both capitalist and socialist, the extent to which the public sector and market are employed seems then to be a function now more of interests, group and national, than of ideology; it is the instrumental value of each that is critical rather than their consummatory value. And it seems misleading for the followers of either Smith or Marx to make a case on grounds of the inherent moral superiority of their respective ideological paradigms without taking due note of the disintegration of the paradigms that has actually occurred in practice. At the same time, the fate of the two paradigms suggests that the question of the roles to be accorded to the public sector and the market mechanism needs to be examined more from the viewpoint of their actual utility as instruments in the service of national goals or social interests rather than uncritically on the basis of their ideological claims.

Notwithstanding how reality has emerged in consequence of attempting to achieve socialism in accordance with the system envisaged by Marx, this says nothing about the political function of the ideology of socialism in the modernization and development of underdeveloped or pre-industrial societies and the legitimacy that particular ideology accords to the role of the state in such modernization and development.

2. The Historical Compulsions of Economic Backwardness: The State as Development Agent

If the historical evolution of the advanced industrial societies testifies to the disintegration of the liberal and Marxist paradigms, a similar outcome is obvious in the intellectual treatment of the economically backward or underdeveloped societies constituting the Third World, though the heated rhetoric, even if one-sided, employed by the votaries of the respective paradigms would seem to hide this fact.

Diffusionism Versus Dependency

For a quarter century after World War II, the reigning paradigm in social science was "diffusionism" (or developmentalism, modernization theory). Diagnosing the problem of Third World countries as one of an endogenous low-level equilibrium trap, this paradigm recommended and Open Door policy on their part towards the inflow of foreign capital, technology and management from the developed countries, whose infusion would then break the low-level equilibrium trap and foster an ongoing process of development. The spokesmen of this specific manifestation of the liberal paradigm in relation to the Third World included W.W. Rostow, Bert Hoselitz, David McClelland and Harry Johnson among others. They were not all united in their prescriptions but were lumped together in the attack by the new "dependency" paradigm which emerged near the end of the 1960s.

According to the dependency paradigm, as most forcefully represented by Andre Gunder Frank, it was precisely the subjection of the Third World to the domination of western capital, technology and management that created and perpetuated the underdevelopment of Third World countries in the first place through structuring their inner social processes so as to facilitate the transfer of economic surplus to the developed capitalist countries. The prescription for a genuine development was accordingly the removal of this constraining dependency condition through revolution and disengagement from the world capitalist system. However, underdevelopment remained a curiously ambiguous term in the dependency vocabulary. There was no explanation of how development and dependence could manage to go together in Canada, Brazil or South Korea. [73] In the ultimate analysis, underdevelopment meant the absence of a socialist revolution, thus precluding development in a capitalist form. That was understandable. As neo-Marxists, many of the key spokesmen of the dependency school (Frank, Wallerstein, Amin) drew inspiration from the socialist tradition. They attacked the diffusionist scholars as bourgeois accomplices in the designs of American imperialism, facilitating capitalist exploitation of the Third World. Under their attack, the diffusionist school seemed to wither, conceding defeat without a fight, while the dependency school emerged dominant in the social sciences by the 1970s.

Significantly, however, the most devastating attacks against dependency theorists came from other Marxists, among them Warren (who took them to be simply ideologies for Third World nationalists), and Laclau and Brenner (who criticized them for their non-class understanding of capitalism). Nothing else better illustrated the present-day disintegration of the Marxian paradigm than this particular controversy among Marxists over development in the Third World. Ironically, the dependency theorists were, after Smith, dubbed Smithian. However, such controversy among Marxists hid the fundamental fact that the liberal paradigm with its individualistic calculus and the Marxian paradigm with its class-based calculus shared common assumptions in relation to the Third World despite the attempt by their votaries to counterpoise them as diametrically opposed ideologies.

First, both liberal and classical Marxian thought were characterized by optimism about the human fate, sharing the common belief of unilinear progress toward material abundance, with the path being marked by a series of stages in societal development, no matter how differently these stages were categorised. Regardless of their theoretical differences over the possible relationship between man and man under capitalism and socialism, both conceptualized the relationship between man and nature in the same terms.[74] Furthermore, both recognized capitalism as having made possible extraordinary economic advance. Indeed, whatever his vision of the desirable ultimate society, Marx held the capitalist stage to be a precondition and prerequisite for it, a stage that could not be skipped. For their part, liberals regarded the capitalist order as the ultimate form which allowed within-system progress towards a fuller development of individual and society.

Second, both regarded the dynamic agent for the development of capitalism to be individual capitalists or the bourgeoisie, moved by the motive of profit. Marx did not recognize any agent other than the bourgeoisie for the purpose of bringing about capitalism or, by extension, industrialization; the state as standing above the bourgeoisie had no role in this process. Furthermore, capitalism was not only a self-sustaining but also, unless deliberately thwarted, a self-expanding process destined to replace as a form of economic organization and behaviour all other forms, which dissolved before its onward march, not only nationally but globally.

Third, capitalism was a uniquely European achievement and other societies that today are said to constitute the Third World did not have the potentiality for generating capitalism on their own because of either their values or their social structure, which in turn may be the result of their particular natural environment. Marx referred to such societies as belonging to the Asiatic mode of production. Fourth, since endogenous economic development was not possible in the Third World, the source of this growth would have to be exogenous through the diffusion of western capital, technology and management; what is therefore required in Third World countries is, indeed, the Open Door and free trade. But since the door may not open automatically, it was the historic responsibility of the developed world to spread the benefits of European civilization everywhere, because of the white man's burden, or to force open the door for capitalism, because of the necessity to make the Third World part of universal history. Thus Marx himself assumed the role of a legitimizer of imperialism in India and elsewhere. Following this tradition in the contemporary era, the Marxist scholar Bill Warren appropriately chose to title his book against the dependency school as *Imperialism: Pioneer of Capitalism..*

Missing strikingly in both liberalism and classical Marxism was an independent and central role for the nation-state as against that exclusively accorded to individual capitalists or the bourgeoisie across the universe. There was no conception of capitalism (or economic development, industrialization) as a national project, sponsored by the state and not just left to spontaneous capitalist forces, and instituted as part of the perennial rivalry among nation-states. The prophet of that conception in the age of capitalism was the German economist, Friedrich List.

List and the Mercantilist Prescription

Born in 1789, List's intellectual achievement chronologically lay in between that of Smith's and Marx's; his major work *The National System of Political Economy* was published in German in 1841, seven years before the publication of the Communist Manifesto. Although this work had as its principal thrust an attack on the position of Adam Smith, it reflects more accurately what nation-states have actually done in the cause of economic development than the thought of either Smith or Marx even though they may have done so under the ideological banner of either liberalism or socialism. List opposed to Smith's individualistic calculus, in the context of universal free trade, the interest of the nation. He characterized Smith's work as representing "the shopkeeper's point of view" in emphasizing wealth and exchange values rather than the capacity to create wealth and the development of the productive forces of the nation. List attacked him for "this entire nullification of nationality and of State power, this exaltation of individualism" -- "for him no *nation* exists, but merely a community, i.e. a number of individuals dwelling together" -- finding his doctrine inadequate in that "it ignores the very nature of nationalities, seeks almost entirely to exclude politics and the power of the State, presupposes the existence of a state of perpetual peace and of universal union, underrates the value of a national manufacturing power, and the means of obtaining it, and demands absolute freedom of trade."[75]

List particularly objected to the doctrine's exclusion of the state from the economy: "The establishment of powers of production, it leaves to chance, to nature, or to the providence of God (whichever you please), only the State must have nothing at all to do with it, nor must politics venture to meddle with the business of accumulating exchangeable values.....it sets no value on the increase of the productive power, which results from the establishment of native manufactories, or on the foreign trade and national power which arise out of that increase. What may become of the entire nation in the future, is to it a matter of perfect indifference, so long as private individuals can gain wealth."[76] For his part, List believed not in the invisible hand but in purposive action by the state, urging in the name of history "the necessity for the intervention of legislative power and administration" and remonstrating that "private industry can only lay claim to unrestricted action so long as the latter consists [sic] with the well-being of the nation"; on the one hand, he recommended that, where necessary, private industry be supported with "the whole power of the nation" while on the other it ought "for the sake of its own interests to submit to legal restrictions."

The centrepiece of List's thought was the nation and national interest in the context of an anarchical international system: "But so long as other nations subordinate the interests of the human race as a whole to their national interests, it is folly to speak of free competition among the individuals of various nations.....Every great nation, therefore, must endeavour to form an aggregate within itself, which will enter into commercial intercourse with other similar aggregates so far only as that intercourse is suitable to the interests of its own special community. These interests of the community are, however, infinitely different from the private interests of all the separate individuals of the nation." For List, local industry was essential for both the wealth and power of the nation, and he strenuously objected to other nations serving merely as suppliers of raw materials while England would remain the world's industrial workshop through the mechanism of free trade.

The key instrument for the national development of industry was, for List, effective though not excessive protection against foreign competition through the imposition of tariff duties. He held England's contemporary advocacy of free trade as not only opportunistic -- following as it did an earlier policy of long duration of stringent restrictions on foreign imports, which enabled England to become an industrial power -- but also as typical English cant and self-interested ideological camouflage "to conceal the true policy of England under the cosmopolitical expressions and arguments which Adam Smith had discovered, in order to induce foreign nations not to imitate that policy." Quite the contrary, he recommended precisely the same path that England had followed to achieve its high position, holding that "more than one nation is qualified to strive to attain the highest degree of civilisation, wealth and power." Ironically, however, his anti-Smithian mercantilist prescription of industrialization as a national project for various nations through purposeful state action stopped at the shores of the North Atlantic, for he envisaged the tropical countries as not only destined by their moral condition for European political domination but also ordained by nature to serve as mere suppliers of raw materials in exchange for European manufactures;[77] his vision of the world economy was one in which the Third World had a role subordinate to, not coordinate with, Europe.

The State and Delayed Industrialization

It is List's mercantilistic prescription that came to dominate state policy in Europe, North America and Japan in the achievement of economic development, and such policy reached its zenith in the posture of economic autarky and state commandism in the Soviet Union after the revolution. Once development had taken place in the pioneer industrial country of Great Britain, industrialization among the major states everywhere else had to be and was state-sponsored. The reason for that is not that statesmen have been necessarily aware of List's prescription. Rather it lay in the fact that they operated, and continue to do so, in an international environment where interstate conflict is not only an invariant phenomenon but whose resolution, in a characteristically self-help system, depends in the ultimate analysis on war. In this context, capitalist industrialization added not only to economic capabilities, and thus indirectly to military capabilities, but also directly to the latter through superior, that is, more lethal, weaponry. Consequently, capitalist industrialization in Great Britain, even if accomplished largely through an autonomous and evolutionary process, propelled other major states to industrialize for the sake of national security, if for nothing else. Industrialization was thus an "imperative"[78]; it could be neglected only at the cost of political independence of the state. Significantly, "the first thrust of the liberal argument was that to free private initiative in a free market would increase the power of the nation-state."[79]

The goal of industrialization has therefore been built into the very historical process where states are embedded in an international system with one or more of them having become industrialized; it is thus historically pre-determined for the economically backward states; it is as teleological a phenomenon as one can find in history whose essence was captured in the significant comment by Marx that the image of the future of underdeveloped countries was embodied in the shape of the industrially developed countries. It is from this situation that there issues forth universally the necessity for the maximization of capital accumulation for less developed countries, regardless of the nature of regime and regardless of the presence or absence of capitalists. But both liberalism and Marxism, including dependency theory, are limited in perspective insofar as they concentrate solely on economics and neglect the significance of the security dimension. Less developed countries have acted in defiance of the advice of liberalism that economic growth is a sphere of the individual and of dependency theory that it is hopeless as a national project in the context of the world capitalist system.

While the goal of industrialization has been historically determined by a conflictual interstate system, the means for its accomplishment have varied among the late industrializers but fundamentally everywhere they have involved purposive state protection and encouragement of local industry. This is necessarily so because of the drastic difference in the situational context of early and late industrializers, in no small measure centered on the very existence of the early industrializers. First and foremost, the very attempt to industrialize among latecomers is perceived by the early industrializers -- unless tied into a dependent relationship to them -- as a threat, for they see such industrialization, and correctly so, as a zero-sum development in view of the impact it will have on their relative economic and military capabilities, and consequently undertake measures, overt or covert, to thwart it.

Second, even if no such measures were undertaken by design by the already industrialized countries, structurally they have the advantage of head start, by way of external economies in the form of industrial and communications infrastructure and trained and skilled labour for their industrial enterprises, all of which the late industrializers lack; local industry cannot therefore develop among the late industrializers without effective state protection. Third, the early industrializers as a result of their head start come to control world markets and colonies which, as resources for economic development, are no longer available to the late industrializers. Fourth, the radical discontinuity between the social structure of late industrializers and the advanced capital-intensive technologies available from early industrializers create severe adaptation problems in industrialization.

Fifth, the very reliance on import of foreign technology from the early industrializers places control of the industrialization process elsewhere than in their hands, while not availing of such technology retards the process among the late industrializers. Besides, the advanced manufacturing technologies require large blocks of capital which are not available in backward economies while their economies of scale are not easily adaptable to the limited markets available in the Third World. Sixth, technological and ideological developments as a result of the successful achievement of industrialization among the early industrializers serve to inhibit the industrialization process among the late industrializers by creating tremendous economic and political pressures; new medical technologies suppress death rates while earlier birth rates persist, creating serious demographic problems, while new ideas of welfare, political participation and human rights create other problems which the early industrializers did not have to encounter.[80]

Confronted with their particular historical situation, the late-comer states have had therefore to resort to a set of policies combining incentives and barriers in order to encourage local industrial development. In the process, as the economic historian Gerschenkron has pointed out, they developed new institutional instruments, such as industrial banks in Germany and the state in France -- "for which there was little or no counterpart in an established, industrial country" -- as substitutions for the diffuse social patterns, associated with the more gradual development among the early industrializers, precisely in order to compensate for their backwardness and developmental lag.[81] Such substitutions, however, extended beyond institutions into the realm of mobilization of new ideologies -- for example, Saint-Simonian socialism under Napoleon III in France and nationalism in Germany -- for "faith" was perceived to be essential to the accomplishment of the monumental task of "catch-up" industrialization.

The most dramatic combination of the ultimate in state sponsorship of industrialization through complete public ownership of the means of production and messianic ideology was achieved in the USSR. Whatever the original objectives of the Bolshevik revolutionaries, eventually the economic experiment turned necessarily, because of the dictates of the international system, into a gigantic mercantilistic undertaking to assure the survival of the Soviet state by the expansion of economic and military capabilities through forced-march industrialization, otherwise described as the full development of the forces of production. In doing so, the state was simply a surrogate for the bourgeoisie, transforming a feudal society into an industrial one. The universal consummatory ideology of communism served objectively as an instrument, perhaps as a camouflage, in the mobilization of support to build up through rapid industrialization the national power of the Soviet Union, summed up in the slogan of "socialism in one country", even if the original revolutionaries may not have so intended. In the case of China, even the very commitment to ideology may well have been rooted from the very beginning in a conception of its instrumental value for national reconstruction. Italian fascism, forthrightly nationalistic to begin with, has also been seen, in retrospect, as an ideology of delayed industrialization. [82]

The Public Sector in National Development

Both Marxist and liberal economic thought accepted an activist role for the state in the development of Third World countries in the postwar period. That was to be expected in the case of Marxism after the visible success of the Soviet experiment, but no less so in the case of liberal economic thought after the Keynesian revolution. W.W. Rostow, Paul Rosenstein-Rodan and Albert Hirschman, for example, in their various theories, assumed implicitly or explicitly an activist role for the state. But this stance was more by way of overall economic management by the state and its bringing about institutional change that would foster economic development which ought, however, to continue to be the responsibility of private entrepreneurs. Nonetheless, in actual practice, in the four decades of the postwar period the state's economic role in the Third World has extended beyond such management so as to encompass a considerable state ownership of the means of production.

The fundamental reasons for the resort to state ownership of the means of production are analytically either (1) consummatory or (2) instrumental, or a combination of both. In the first case, public ownership is considered a morally desirable end in itself as part and parcel of the preferred model of economic organization to which the political leadership is committed. This perhaps has been true of China and Cuba. On the other hand, their commitment to the particular model itself may have arisen from a perception of the functionality of the model to the goals of national power and welfare, especially given the rise of the Soviet Union to the status of a superpower on the basis of that model. If so, then in such cases, it may well be that, if public ownership of the means of production -- accompanied usually by the rejection of the market model of economic management in favour of a state-managed economy -- is not sufficiently productive in terms of the goals of national power and welfare, there would be an erosion in the rigour of the commitment. Such a change is obvious in China; illuminating in this regard is a statement by Huan Xiang, Senior Adviser, Chinese Academy of Social

Sciences, in 1983:

> We have given up our former 'Left' dogmatic position. We want to change the Chinese economic system in some very important ways. First, we have to change the view that only the public sector which is controlled by the State and by the collectives, should exist and operate and that the other sectors of the economy are useless. That was a 'Left' dogmatic position, and I think every communist economy has suffered because of that. We learnt that we were right in insisting on public ownership, both state and collective, but we were wrong in discarding all the other economic factors. Now we use the state economy and the collective as the backbone of the people's economy, and encourage private initiative at the same time. This is a very important change....We encourage private initiative in distribution, in agriculture, and even in the production of materials such as machines and equipment.....Another difficulty with the 'Left' dogma was that everything had to be 'planned'. How can you 'plan' everything? Life is very complicated and many-sided. That is almost impossible. For thirty years we 'planned' and suffered a great deal on account of that. Everybody was required to fulfill his quota for the state, but no one worried about the quality and content of performance. Ten thousand tons of steel must be produced, but what kind of steel? The law of demand and supply was disregarded as a capitalist law, but the state could not supervise the quality of individual performance. The result is that our society is a very backward society. You have to allow the law of demand and supply to spend its force. It has its force and if you disregard it, you might produce a million tons of steel but you will not find any demand for it.....There are also many activities which are now left to the market. All this has helped to break the 'Left' dogma that everything must be 'planned'.[83]

What this illustrates is that the higher requirements of development in the context of the economic backwardness of Third World countries are compelling and make for modification of ideology. The tension between the requirements of ideology and industrialization may reveal itself over a period of time rather than suddenly, and analysis therefore requires a time perspective. This is not meant in any way to denigrate the commitment of the Chinese leadership to ideology but rather to pay tribute to its appreciation of realities. The message underlined for Third World states would seem to be ideological flexibility in the face of the requirements of rapid industrialization.

On the other hand, it is equally significant that states that are not just ideologically capitalist but also rightist in orientation and are foreign policy allies or satellites of the United States, such as Brazil and South Korea, have large public sectors. In the case of South Korea, over the decade of the 1970s, about 30 per cent of total industrial investment went to public enterprises.[84] In Brazil, the state was responsible for 60 per cent of fixed investment in 1969.[85] The contrast between ideological rhetoric and practice is quite glaring. The reasons for the public sector in the case of Brazil and South Korea obviously relate to market imperfections or failures,[86] which in the context of economic backwardness are so compelling that even ideologically capitalist states seek recourse in the alternative to the public sector.

The public sector has been seen as instrumental in economic development for a variety of reasons: building the economic infrastructure and basic industries in an era of capital-intensive technologies and massive economies of scale, because private entrepreneurs are unlikely to undertake these on account of indivisibilities and externalities; advancing national independence, because the choice is seen as really between a national public sector, even if inefficient, and domination by multinational corporations; accelerating capital accumulation, because private entrepreneurs are likely to divert profits to consumption whereas the public sector makes them available for investment; facilitating better transfer of technology, because of stronger bargaining power; preventing excessive concentration of economic power in private hands; and assuring a balanced regional location of industry. [87]

Of course, some of these reasons may serve as rationalizations where there is already present, on account of other factors, a disposition toward the establishment of a public sector; they do not necessitate public ownership since other mechanisms -- such as risk insurance, loan guarantees, tax concessions and incentive subsidies -- are available to the state for accomplishing such ends. Indeed, "government failures" are as much a possibility as market failures; revisionist literature has come to question the market failure orthodoxy and to demonstrate that political biases may undermine expected gains of political solutions and may aggravate market failures.[88] Nonetheless, it is impressive that states with anti-socialist and pro-capitalist ideologies have felt compelled to resort to public ownership of the means of production in order to advance industrialization. The retreat from ideology is no less manifest here than it is in the case of socialist states such as China. Of course, the public sector may be seen by some as fortifying capitalism. On the other hand, a certain ambiguity characterizes the public sector, which will be seen by others in different contexts as facilitating a transition to socialism.

In looking at the instrumental role of the public sector, however, it is necessary to take a broader view than one merely restricted to economic development, for the state in the backward societies of the Third World has, as any state does, multiple goals. W.W. Rostow underlines three universal goals that all political systems attempt to achieve: (1) security; (2) welfare and growth; and (3) an appropriate constitutional order.[89] There is, of course, conflict among these goals, such as between defence and welfare, as well as tension within each category, such as between economic growth and welfare, and consequently statesmen must make trade-offs among them. More concretely, for Third World states, Organski outlines three major goals that have to be achieved almost simultaneously whereas they were accomplished sequentially in the states of the North Atlantic area: (1) primitive unification, i.e., national integration; (2) industrialisation; and (3) national welfare. [90] Almond lists state-building, nation-building, participation and distribution as characteristic goals for Third World states, but is insensitive to both national independence and industrialization. [91] The same is true of Dankwart Rustow -- who points to identity, authority and equality as national goals -- as it is of other mainstream political scientists.[92]

It is likely that such endeavours at goal specification for Third World states will be perceived as a peculiar enterprise of liberal scholarship, but the tradition may well have its roots in Marx and Engels who, in listing the achievements of the bourgeoisie while at the same time holding the industrially developed country to be the model for the underdeveloped country, forecast with amazing accuracy the future goals of Third World states. Almost all the goals articulated by mainstream social science are to be found in the Communist Manifesto of 1848:

> The bourgeoisie, historically, has played a most revolutionary part. The bourgeoisie, wherever it has got the upper hand, has put an end to all feudal, patriarchal, idyllic relations...The bourgeoisie cannot exist without constantly revolutionizing the instruments of production, and thereby the relations of production, and with them the whole relations of society...The bourgeoisie has subjected the country to the rule of the towns. It has created enormous cities, has greatly increased the urban population as compared with the rural, and has thus rescued a considerable part of the population from the idiocy of rural life....The bourgeoisie keeps more and more doing away with the scattered state of the population, of the means of production, and of property. It has agglomerated population, centralized means of production, and has concentrated property in a few hands. The necessary consequence of this was political centralization. Independent or but loosely connected provinces with separate interests, laws, governments and systems of taxation, became lumped together into one nation, with one government, one code of laws, one national class interest, one frontier and one customs-tariff. The bourgeoisie, during its rule of scarce one hundred years, has created more massive and more colossal productive forces than have all preceding generations together.[93]

What was consequence in the case of the activities of the bourgeoisie has, in considerable measure, become the intention of the state as the surrogate for the bourgeoisie in the Third World. At the same time, it would seem that the various functions that the bourgeoisie performed are intimately tied with each other and, in the Third World of today, are to be seen as constituting mutual prerequisites. It is not possible, for example, to have economic development without some measure of societal integration and political institutionalization. On the other hand, economic development in turn is necessary for integration and institution-building, because of the additional resources it provides. It was the genius of the Soviet leaders to have developed an organizational weapon in the form of the Communist party, which served as the focus of authority and an instrument of national integration, as also to have reformulated an ideology which served to unite an ethnically divided population on a universal basis and to legitimize forced-draft industrialization.

It would seem to follow that any adequate evaluation of the public sector requires consideration of the contribution it makes, or does not make, not only to (1) economic development and relatedly to national independence, but also more generally to (2) society in terms of integration and equality, and to (3) polity in terms of institution-building. Indeed, it may well be that the public sector as a mechanism for development is chosen for reasons pertaining to society and polity rather than economy. The choice of the public sector may have, in some cases, more to do with state-building or system-maintenance because of the concentration of resources in the hands of the state that it provides.

However, the assumption that the public sector is so chosen and "shapes economic and political life in modern states"[94] does not necessarily imply, any more than it does for economic development, that it will prove functional in practice. There is no reason to assume that purpose will guarantee performance or that the public sector is the appropriate instrument for it. In the case of the economy alone, an earlier enthusiast of the public sector, Richard Pryke, has after a decade of further research and reflection come to the conclusion that public sector industry, for example in Great Britain, is inherently inefficient and will remain so.[95] If so, even if functional in the short run for certain social and political purposes, the public sector may be subversive of long-term goals in relation to society and polity because of its failure at capital accumulation, the consequences of which will ultimately reverberate throughout the national system. That really underscores the necessity for judicious balancing among the various national goals in policy choices. Indeed, it may be advisable to restrict the role of the state, even simply to law and order, if it does not have adequate resources; concentration rather than dispersal and dissipation of resources should be the requirement.

Although a more adequate analysis of the public sector understandably requires an examination of its relationship to the broader requirements of society and polity in Third World countries, there is some hazard in seeing it only as functional for some generalized supra-class national interests, such as economic development. Quite the contrary, the public sector may be more narrowly instrumental -- and may perhaps have been designed to be so -- in terms of the interests directly of the power-wielders in the political system or indirectly of the class or class coalition they represent. One would thus have to supplement the "rational choice model," in which the public sector is seen as instrumental for larger developmental goals of the society as a whole, with a "politics model" in which it is seen as a function of political interests and power relationships. Indeed, the key problem of the public sector has been held to be its vulnerability to political pressures on behalf of different interests,[96] with the consequent inability to control costs, rendering it an inefficient and dysfunctional instrument for capital accumulation. All that simply underlines the importance of the relationship of the state to classes and groups in society.

3. The State and Society in the Third World

With the assumption by the state in the contemporary era of an elemental role in economic development, including the fostering of an active public sector, the question arises: Whose interests does the state represent in this endeavour? Does it represent a class-transcendent or multi-class national interest? Or, does it represent the interests of a specific class or class coalition? At issue here is the relationship between state and society. Two basic but contrasting models (also variously referred to as paradigms, perspectives or approaches) have dominated the social science literature: liberal-democratic (also known as pluralist and liberal-pluralist) and Marxist.[97] Interestingly, despite the antagonism between their respective advocates, both models share the common notion of the state as epiphenomenal.

The State as Epiphenomenal

The Liberal-Democratic Model

In the liberal-democratic model,[98] the underlying concept is of a class-neutral state, which is not the instrument of any particular class. An unstated but nonetheless essential assumption of the model is a prior successful industrialization and modernization which has brought about the integration of society on a new basis: primordial ties, if any, have been eroded and have given way to modern civil values; the pyramid-like stratification system characteristic of an agrarian society has been replaced by a new diamond-like stratification system with a large (actual or putative) middle class and, while inequalities exist, they tend to be dispersed rather than cumulative; enhanced economic resources that go with an industrial economy enable society to provide mobility opportunities across class lines on an achievement rather than ascriptive basis; instead of a few sharply divided cleavages, there exists, because of the social complexity resulting from industrialization, a multiplicity of diverse interest groups, with overlapping memberships and cross-cutting ties, and they compete on a roughly equal basis.

As a consequence, what prevails in the society is not a life and death struggle among social groups over fundamental values but rather a generalized consensus over basic norms of the system and division of labour which is considered a variable-sum, rather than a zero-sum, phenomenon. Within such a consensus, different sets of political elites engage in open competition in the political "market-place" for state power on the basis of aggregation of various interest groups -- the building blocks of the model -- through mediation and compromise. Questions concerning group domination are not relevant, for the diversity and complexity of interest groups militate against it and assure a shifting balance of forces.

As a complex of institutions for society-wide collective decision-making, the state in the liberal-democratic model is not autonomous; it is without any independent needs or compulsions of its own. Indeed, in the liberal-democratic conception, the state is simply a political arena where bargaining among interest groups takes place; at the most, it is a neutral arbiter among the interest groups. Politically neutral, it is at the same time highly accessible. As a result of its accessibility, neutrality and lack of autonomy, the state tends to reflect the changing parallelogram of competing social forces; it simply mirrors society, almost literally in the fashion of the cash register which indicates what is fed into it. The state is thus the servant of society, not its master; it stands in an instrumental relationship to society. Notwithstanding that, the state is not directly reducible to the economic division of labour because of the complexity of the interest group structure and the intervening mediatory role of politics. Furthermore, with industrialization an accomplished fact, the system as a whole is largely in equilibrium while change is basically incremental; state policy has to do with the management of a going concern, not with the reconstruction of society.

Although the liberal-democratic model reflects, to a greater or lesser extent, social and political reality in a considerable part of the world, yet it is deficient as an approach to understanding that reality. First, the assumption of the political neutrality of the state and its extensive political accessibility performs an ideological function in avoiding the question of which classes or groups have more or less power, which have critical control over the state, and which are excluded from power altogether. No doubt, the question can still be raised but the model itself provides no clue in this regard, which only underlines the fact that it is incomplete as a representation of reality insofar as inequalities in access are a fact of life. Indeed, it is the burden of the revisionist work of the pluralist scholar Lindblom that there exists a permanent class hegemony of the bourgeoisie in liberal-democratic societies.[99]

Second, in its assumption of the lack of autonomy of the state, the model ignores the fact that the state, by virtue of the resources -- managerial and economic -- at its command in an industrial society, may itself dominate over society and manipulate social forces to perpetuate its own interests rather than simply reflect those of society. The existence of state autonomy has come to receive belated scholarly recognition even in respect of democratic regimes.[100] Earlier, the "elite" approach to the state had more sharply pointed out how an interlocking set of power elites in command of huge organizations with powerful resources dominated over the masses in advanced industrial societies.[101] Third, the model is inadequate insofar as it sheds no light on the interaction of state and economy, which is so central to Marxist scholars. This may initially have been a result of the assumption that the state presides over a self-regulating economy, but the Keynesian revolution and the rise of the welfare state and the "mixed economy" seem to have had little impact on the conventional articulation of the model, perhaps a consequence of the separation of political science and economics in universities.

Fourth, the model is historically, culturally and socially specific rather than one that has general application. Most fundamentally, it is basically relevant to advanced capitalist societies where industrialization has already been achieved and where incremental, not fundamental, change is the issue. The attempt to transfer the model, in theory and in practice, to Third World countries can only have grievous consequences insofar as it enormously underestimates two important facts: (1) the immensity of the task for the state to transform an underdeveloped economy into a developed one through industrialization, and hence the need for the autonomy of the state rather than its being reduced to a reflection of society, whose own transformation is what is at issue; (2) the crises or "contradictions" that confront, and indeed overwhelm, the state through the leviathan of demands in behalf of groups of all kinds -- ethnic, religious, linguistic and economic -- as part of the process of social mobilization that inevitably accompanies industrialization and modernization, especially when an open political system with wide accessibility under the liberal-democratic model not only facilitates but fosters the articulation of such demands. The result is likely to be political and economic immobilisme, and system breakdown.

However, all this did not deter the transfer of the model -- often transmuted as the "systems", "consensual" or "structural-functional" model or approach -- to the analysis of underdeveloped societies. The initial installation by many post-colonial countries of liberal-democratic regimes, resting on the base of a small new middle class and a largely immobilized society, made the application of the model to the underdeveloped societies quite plausible. However, the "erosion of democracy", under the stresses and strains of increased social mobilization of the masses, and the rise of coercive regimes have called into question the model in both practical and theoretical terms.

The Marxist Model

The question of social and economic transformation and the relationship of the state to it is, of course, central to Marxism. However, although Marxism is often presented as a unified doctrine with an assured certainty and claim to universality, there is, in fact, such ambiguity, flexibility and diversity in it as to enable it to appropriate and encompass the most contradictory positions as its own, perhaps not simultaneously but certainly successively; indeed, the most contradictory hypotheses are today compatible with the corpus of Marxist "theory". Marxism has been constantly confronted by surprises, with historical developments undermining many of its central social predictions, but this has only led to revisionist formulations which at times have resulted in establishing new orthodoxies.

Notwithstanding the diversity within Marxism, the primary (or classical, or orthodox) Marxist model of the state [102] is rooted in a society whose basic characteristic is not consensus, but conflict and struggle; not diversity of social groups with overlapping memberships, but social polarization between the dominant class and the dominated class organized around the ownership of the means of production; not the sharing of power on the basis of rough equality, but coercion and exploitation. As Engels expressed it, the history of man is "a history of class struggles, contests between exploiting and exploited, ruling and oppressed classes."[103] In line with the basic proposition of Marxism that the superstructure of law and politics is essentially determined by the economic base, the state is not class neutral but an instrument of the dominant class for the exploitation of the dominated class. The Communist Manifesto had declared, "political power, properly so classed is merely the organized power of one class for oppressing another," while Lenin had underlined: "The state.....is a class concept. The state is an organ or instrument of violence exercised by one class against another."[104] Change in this situation is not incremental but (after considerable gestation) cataclysmic, erupting into a revolution when the dominant class is overthrown and a new class emerges in control of the state.

Given the centrality in Marxism of the notion of the bipolar division of society into two classes and of a zero-sum relationship between them, one class must necessarily dominate and exploit the other. Historically, the feudal state in Europe, representing the domination of the landed aristocracy over the serfs, was overthrown in the course of societal development and replaced by the capitalist state. In the context of a society sharply divided between capitalists and workers, the capitalist state is nothing but a "tool" of the capitalist class, the executive committee of the bourgeoisie. Marxists hold as false the liberal claim about the neutrality of the contemporary state. For them, it represents strictly the interests of the capitalist class; there is no mediation or compromise possible on behalf of the interests of the workers, and any apparent concessions in that regard are merely a sham. The capitalists assure their domination and control of the state, says Miliband under a Marxist approach that is referred to as "instrumentalist", through placing in strategic positions in the state apparatus personnel who are tied to them by their social background and ideological inclinations.[105] The political form associated with the bourgeois or capitalist state -- liberal democracy -- allegedly only solidifies the domination by the capitalist class; as Lenin said: "a democratic republic is the best possible shell for capitalism, and, therefore, once capital has gained control over this very best shell....it establishes a power so securely, so firmly, that *no* change, whether of persons, of institutions, or of parties in the bourgeois-democratic republic can shake it."[106]

The Marxist model certainly encompasses more of the social reality of Third World countries insofar as it underlines the sharp disparities in economic privilege and the domination of the privileged few over the deprived many. That, indeed, has been recognized as a key source of the actual appeal of Marxism in underdeveloped societies,[107] and therefore in the application of the model by intellectuals to the analysis of Third World societies. However, beyond that general appeal, the Marxist model would require much revision to be relevant to Third World societies, for it envisages a single mode of production and a single class holding state power whereas Third World societies manifest multiple modes of production without hegemonic social classes. One of the critical questions debated among Marxists in the Third World, and one productive of much controversy, has been precisely the nature of the mode of production in which they are situated and the consequences of that for political action.[108]

Application to Third World societies aside, there are other inadequacies in the primary Marxist model of the state. As in the liberal-democratic model, the state here also is epiphenomenal, derivative from society; it stands only in an instrumental relationship, not a dialectical one, to society. Strictly speaking, there is only a Marxist theory of society, for the state is held to be simply a reflection of society. On that account, Marxists in the period since World War II have recognized the lack of an adequate theory of the state as a serious shortcoming in Marxism and have worked to develop one. Like the liberal-democratic model, the primary Marxist model is noteworthy for its denial of any autonomy to the state, precluding its use as an instrument of social transformation. At the same time, however, Marxism has had to countenance two essential features of contemporary advanced capitalist societies: One, the state has legislated concessions to the working class and, even though intellectuals may dismiss them as sham, the working class has no intention of giving them up but rather struggles to hold on to them. Nor could the concessions, given their long duration, be denigrated by describing them as short term. Two, these concessions have often come to the working class against the strong opposition of the capitalist class, supposedly in control of the state.

Confronted with these material facts, many Marxist scholars have been engaged, with considerable intellectual ingenuity, in the rediscovery, reinterpretation and revision of Marxism, to advance the proposition of "the relative autonomy of the state", which is by now fairly well established in Marxist scholarship.[109] Of course, Marx and Engels had recognized that the state may in exceptional circumstances, such as the Bonapartist state (which is what has come to serve as the basis for the new models), have autonomy from social classes dominant in society and, indeed, lord it over them. But for contemporary scholars the issue concerns a more institutionalized feature of the capitalist state rather than a temporary aberration.

The State as "Relatively Autonomous"

Miliband blames "economist" misinterpretations of Marxism for the neglect of "the central importance of the concept of the relative autonomy of the political in Marxist theory". Invoking the classic statement on the state as executive committee of the bourgeoisie in the Communist Manifesto, he offers a revisionist interpretation which somehow had remained dormant and had eluded scholars and practitioners alike for more than a century:

> This has regularly been taken to mean not only that the state acts *on behalf* of the dominant or 'ruling' class, which is one thing, but that it acts *at the behest* of that class, which is an altogether different assertion and, as I would argue, a vulgar deformation of the thought of Marx and Engels. For what they are saying is that 'the modern state is but a committee for managing the *common* affairs of the *whole* bourgeoisie': the notion of common affairs assumes the existence of particular ones; and the notion of the whole bourgeoisie implies the existence of separate elements which make up that whole. This being the case, there is an obvious need for an institution of the kind they refer to, namely the state; and the state *cannot* meet this need without enjoying a certain degree of autonomy. In other words, the notion of autonomy is embedded in the definition itself, is an intrinsic part of it.[110]

Miliband admits that a major characteristic of capitalist regimes has been reform and that this reform has "generally been strongly and even bitterly opposed by one or another fraction of the `ruling class', or by most of it." But he sees such reform as solely intended to perpetuate capitalism and views the relative autonomy of the state as eminently useful for this purpose, for "what to concede and when to concede -- the two being closely related -- are matters of some delicacy, which a ruling class, with its eyes fixed on immediate interests and demands, cannot be expected to handle properly."[111] Even though operating autonomously, the state still acts in the long-term and real interests of the bourgeoisie because of -- at least in an earlier formulation by Miliband identified as "instrumentalist" -- the ties of common social background and values that bind the two together.

However, this linkage through personnel between the state and the bourgeoisie is not essential for Nicos Poulantzas, whose proposition on the "relative autonomy of the state" is embedded in a systematic theoretical treatment, which is often characterized as "structuralist" in contrast to the "instrumentalist" approach associated with Miliband. He dismisses the notion of the state as a simple instrument to be personally held by the hegemonic class or fraction in a physical manner for protecting and furthering its interests.[112] Departing from that perspective, Poulantzas recognizes the autonomy of the political from the economic, and finds that the capitalistic state "does not *directly* represent the dominant classes' economic interests, but their *political interests*: it is the dominant classes' political power centre, as the organising agent of their political struggle."[113] For him, the relative autonomy of the state from the *economic* interests of the capitalist class is determined structurally by the objective interests of the capitalist system as a whole:[114] "the capitalist State best serves the interests of the capitalist only when the members of this class do not participate in the State apparatus, that is to say when the *ruling class* is not the *politically governing class.*" The autonomy of the state is in relation to the capitalist class or its fractions, but not to the capitalist system as a whole which, in the ultimate analysis, determines the constitution of the state and its operation.

The relatively autonomous state performs two important functions for the capitalist order:[115] One, it serves to confuse and divide the working class through economic concessions that divert it from political struggle to economism. Two, the state serves to overcome the divisions and fractions within the bourgeoisie -- which necessarily follow from the competitive nature of capitalism -- and provide the needed cohesion and coherence to the capitalist order to implement measures, including concessions to the working class, for the long-run overall interests of capitalism, even if individual fractions of the bourgeoisie, or the bourgeoisie as a whole, are unable to see them as such and even oppose them. Says Poulantzas:[116] "It takes charge, as it were, of the bourgeoisie's political interests and realizes the function of political hegemony which the bourgeoisie is unable to achieve. But *in order to do this, the capitalist state assumes a relative autonomy with regard to the bourgeoisie.*" Neither of the two functions of organizing the bourgeoisie and disorganizing the working class could, on this understanding, be performed well if the state were a pliable tool of the bourgeoisie or its fractions.

Poulantzas sees the capitalist state as possessing an inherent structural flexibility to make concessions to the economic interests of the working class though, of course, "within the limits of the system". He goes on to acknowledge that "this state, by its very structure, gives to the economic interests of certain dominated classes guarantees which may even be contrary to the short-term economic interests of the dominant classes" but which, he nonetheless adds, "are compatible with their political interests and their hegemonic domination." He even admits that "it is true that the *political and economic struggles of the dominated classes* impose this on the capitalist state" but qualifies the admission by saying that it "cannot be seen *per se* as a restraint on the *political power* of the dominant classes" and that "in making this guarantee, the state aims precisely at the political disorganization of the dominated classes."[117]

The structuralist theory of the state has been criticised, and rightly so, for being highly abstract insofar as it fails to specify the precise social mechanisms that translate the interests of the bourgeoisie into the policies of the state. The linkage between capitalist interests and state actions remains at a rather metaphysical level.[118] In line with functionalism, the state here is destined to be capitalist simply by virtue of its insertion in a "capitalist" mode of production; the latter's needs structurally constrain it to serve nothing but the interests of the capitalist class; the very existence of capitalism apparently transforms structurally everything, including the state -- through what Miliband terms "another form of determinism"[119] -- into being functional for capitalism. There is an all or nothing posture here; everything the state does, in the absence of a proletarian revolution, by definition serves the interests of the capitalist class, no matter how state policy deals with them; only the complete destruction of that class and its system can assure that this is no longer the case. Poulantzas is quite clear on this point:[120] "So this 'social policy', though it may happen to contain real economic sacrifices *imposed on* the dominant class *by the struggle of the dominated classes*, cannot under any circumstances call into question the capitalist type of state, so long as it operates within these *limits.*"

Such a posture stems from pre-determined notions of what the "objective", "true", or "long-term" interests of the different classes are, but as another Marxist scholar has pointed out "Marxists who have employed the notion of class interest have encountered great difficulty in giving it a precise empirical meaning.....the notion seems to provide a spurious objectivity to essentially ideological evaluations."[121] Furthermore, Poulantzas' theory is not theory in the conventional sense that it is empirically verifiable; with the destruction of the capitalist order as the only test, it has to be taken on faith.

If, according to Poulantzas, the capitalist class as the ruling class is no longer the politically governing class -- and, indeed, it should not be -- and the state provides economic concessions to the working class which the capitalist class opposes, then "the relative autonomy of the state" theory comes perilously close to the liberal-democratic model. Insofar as the liberal model posits a sharing of political power among different groups, it underlines a fundamental aspect of social reality in advanced capitalist societies, which Marxism did not provide for. What is obvious is that attempts by Marxists to come to grips with that reality end up sounding similar to the model that they attack.

This is quite manifest in the work of the Marxist scholar Hamza Alavi, who adopts the position of the "relative autonomy" school in the Third World context with some variations. While paying continual homage to class struggle and domination, he accepts the position of "the state as an arena of class struggle among rival fundamental classes. Likewise, within much narrower limits subordinate classes may also find it possible to establish positions within the political system and the state and achieve a degree of class representation in order to achieve some limited gains -- such policies can exist in conditions of electoral politics, where rival parties need to rally some degree of mass support....[thus] the orderly incorporation of the subordinate classes into the framework of state power.[122] Nothing could make more obvious than this that the model of "the relative autonomy of the state", even if only "a cosmetic modification of Marxism's tendency to reduce state power to class power,"[123] certainly represents a major retreat from the primary, classical Marxist model of the capitalist state and can be said to be a half-way house between that and the liberal-democratic model. For their part, proponents of the liberal model are likely to find little to object to in the Marxist reference to the policies of the state as strengthening the capitalist order, especially at the cost of the capitalist class and to the benefit of the working class, if that is understood to mean simply the general economic or social order. The same comment would basically apply in relation to the analysis of the state by James O'Connor in his work on the fiscal crisis in advanced capitalist societies.

O'Connor states that the capitalist state must perform two fundamental but usually contradictory functions -- accumulation and legitimization.[124] Without capital accumulation, the source of the state's own power is in danger of evaporating, therefore it must assist the accumulation process undertaken by capitalists. This would seem to correspond to Lindblom's stance that government officials need economic growth to stay in office and this, rather than any conspiracy theory or common origins, explains their according business a privileged position.[125] On the other hand, for O'Connor "a capitalist state that openly uses its coercive forces to help one class accumulate capital at the expense of other classes loses its legitimacy and hence undermines the basis of its loyalty and support." Therefore, "the state must try to maintain or create the conditions for social harmony."

With this intellectual concession, of course, nothing remains of the view of the capitalist state as an instrument of class oppression, since the business of that state is not social harmony but class exploitation. With its assumption of a zero-sum relationship, Marxism had not, and could not have, provided for this contingency. Moreover, with this qualification, nothing that the capitalist state does could ever be inconsistent with the interests of the capitalist class, for it would have simply been required by its legitimacy needs. Of course, it would matter little from the viewpoint of the working class that the state adopted benign and beneficial policies toward it, not out of some morally virtuous intentions, but out of the supposedly morally-tainted need for legitimacy to assure long-term accumulation. And, in any case, would accumulation necessarily be contrary to the interests of labour? Could it not be that both capital and labour have a common interest in accumulation? O'Connor himself has to acknowledge:[126]

> First, monopoly capital and organized labour have supported the growth of state-financed social investments [urban renewal, highway construction, higher education]. From the standpoint of monopoly capital the greater the socialization of social investment the greater the profits. From the standpoint of organized labor the greater the socialization of these outlays, the greater the rise in productivity and wages....Second, monopoly capital and labor also have favored socializing social consumption expenditures such as medical costs and workers' retirement income...Third, monopoly capital and labor have advocated increased social expense outlays [defence, welfare]....Finally, monopoly capital and the unions have collaborated in the introduction of labor-saving technology......

Furthermore, the dual but contradictory functions of the state in respect of accumulation and legitimacy are generic to contemporary states, not limited to the capitalist state; the dilemmas, or "contradictions", they pose may disappear in utopia, but in the existential world they are present everywhere. Again, despite the apparent Marxist flavour to the terms employed, the functional requirements they represent have long been familiar to conventional social scientists.[127]

Notwithstanding the recognition of concessions to the working class, even against opposition by the capitalist class, the refrain among Marxist scholars is that the capitalist state adopts them in order to maintain the interests of capitalism. But some Marxist scholars have even broken away from this refrain to forthrightly acknowledge that the state may also act counter to the fundamental interests as well of the capitalist class and that it does so out of its own needs and compulsions that lie outside the realm of economics. Thus Skocpol states:[128]

> Indeed, a state's involvement in an international network of states is a basis for potential autonomy of action over and against groups and economic arrangements within its jurisdiction -- even including the dominant class and existing relations of production. For international military pressures and opportunities can prompt state rulers to attempt policies that conflict with, and even in extreme instances contradict, the fundamental interests of a dominant class. State rulers may, for example, undertake military adventures abroad that drain resources from economic development at home, or that have the immediate or ultimate effect of undermining the position of dominant socioeconomic interests. And, to give a different example, rulers may respond to foreign military competition or threats of conquest by attempting to impose fundamental socioeconomic reforms or by trying to reorient the course of national economic development through state intervention.

What this leaves of the primary Marxist model of the capitalist state can be well imagined. To meet the challenge of such a fundamental shift in perspective, Miliband makes a further intellectual concession: rather than conceptualizing the state any more as a committee of the bourgeoisie, he now states -- incorporating notions of both the self-interests of the power-wielders or "state bourgeoisie" and the national interest -- that "an accurate and realistic 'model' of the relationship between the dominant class in advanced capitalist societies and the state is one of *partnership between two different, separate forces,* linked to each other by many threads, yet each having its own separate sphere of concerns."[129] Similarly, earlier, another Marxist scholar, Fred Block, rejected the notion of "a conscious, politically directive, ruling class," and posited the capitalist class, the managers of the state apparatus, and the working class as separate sets of agents with different sets of interests which may converge or conflict, depending on circumstances.[130]

The notion of state autonomy from society, indeed of state domination over society, even if some Marxists would find the concept objectionable, is of course not altogether inconsistent with Marxism as a whole, nor is it limited, despite protestations to the contrary, to exceptional circumstances. Notwithstanding the primary or classical model, Marxism provides for two situations in which the state dominates over society.[131] One is the non-hegemonic situation, where there is a balance among classes, in contrast to the position that the state, as Engels declared, "in all typical periods is exclusively the state of the ruling class." However, such a non-hegemonic situation is seemingly neither untypical nor exceptional, when it is conceptualised to cover vast territories and to stretch over a couple of centuries; the exception may well be the rule. Note, for example, the instances of the non-hegemonic situation cited by Engels:[132]

> By way of exception, however, periods occur in which the warring classes balance each other so nearly that the state power, as ostensible mediator, acquires, for the moment, a certain degree of independence of both. Such was the absolute monarchy of the seventeenth and eighteenth centuries, which held the balance between the nobility and the class of the burghers; such was the Bonapartism of the First, and still more of the Second French Empire, which played off the proletariat against the bourgeoisie and the bourgeoisie against the proletariat. The latest performance of this kind, in which ruler and ruled appear equally ridiculous, is the new German Empire of the Bismarck nation; here capitalists and workers are balanced against each other and equally cheated for the benefit of the impoverished Prussian junkers.

And this does not even include the Asiatic mode of production, where the state dominated over society as literally a permanent phenomenon, even as Miliband admits that "it departs considerably from the classical theory of the state associated with Marx and Engels. In the formulation above, the state not only acquires a very high degree of independence, albeit by way of exception, but this independence appears to free it from its character as a class state; it seems to have become what might be described as a 'state for itself'."[133] The conclusion, as Stepan correctly points out, is obvious: the position of a class in relation to the state is not a matter of *a priori* assumption or assertion but of empirical investigation, especially in Third World societies where the national bourgeoisie has failed to achieve hegemony and a hegemonic proletariat has not come into existence.

The second situation which may result in the state dominating over society pertains to the rather permanent tendency toward administrative parasitism associated with the bureaucratic apparatus of the state. This general feature results from the increasing expansion of the state's repressive machinery as a necessary accompaniment of the intensification of class conflict, with the state machinery assuming then a life of its own rather than being a reflection of society. Engels saw this "transformation of the state and the organs of the state from servants of society into masters of society" as "an inevitable transformation in all previous states." Marx had also noted that the state in nineteenth century France "constantly maintains an immense mass of interests and livelihoods in the most absolute dependence; where the state enmeshes, controls, regulates, superintends and tutors civil society." Stepan, quite legitimately, adds that the problem ought to be even more severe in the Third World where the state is precisely what Marx describes but the civil society is far less developed.[134] And this is so, indeed. States are not simply relatively autonomous from society here but are dominant over it. Of course, such domination is never absolute; states are always Janus-faced, they have to be responsive to society in some measure, for the sake of their own stability. But state domination over society as the encompassing principle is the characteristic pattern in the Third World. Meanwhile, to notions of the democratic state and capitalist state has been added that of the dominant state (described as the "managerial state" or corporate state), or that of the "statist" approach.[135]

Classes and the State in the Third World

A strong tendency toward authoritarianism has been characteristic of Third World states, but surprisingly there is substantial agreement, by way of explanation of this tendency, between conventional analyses of "imbalanced political development" and more recent Marxist analyses of the "overdeveloped state". In the former, the colonial power is taken to have fostered authoritative, output, governmental structures (civil and military bureaucracy) and neglected, opposed and suppressed the development of non-authoritative, input, political infrastructures (political parties and interest groups), with the resulting imbalance in development between output and input structures, combined with increased social mobilization, manifesting itself in the eclipse and collapse of participatory institutions.[136]

In the analysis by the Marxist scholar Hamza Alavi, the metropolitan bourgeoisie is considered to have replicated in the colonial state a political superstructure, corresponding to the one it had established under its auspices in the advanced home country, in order to "exercise dominion over *all* the indigenous social classes." Accordingly, the state in the colony is, firstly, strong and "overdeveloped" compared to society and, secondly, it is not organically linked to society but dominates over it. The postcolonial society thus inherits "a powerful bureaucratic-military apparatus with governmental mechanisms that enable it, through routine operations, to subordinate the native social classes."[137] Elsewhere, in following the "relative autonomy" approach, Alavi maintains that "it is not the subordinate classes alone that confront the state as an alien force. The fundamental classes also have to do the same -- and here it is mainly the administrative arm of the state that is deployed against them....there is instead a very considerable accretion of powers of control and regulation over the 'dominant' fundamental classes in the hands of a powerful and centralized state; here the fundamental classes do not have any *direct* control over the state."[138]

Critics have pointed out that the same pattern of bureaucratic-military domination characterizes Third World societies which were never colonized. The reason for that, however, as Toynbee would have responded,[139] is that it is easier to adopt administrative technique (as in bureaucracy) than political values (as in democracy). What is manifest is that the state is strong and autonomous in Third World countries, but this is only relative to the weakness of society with its absence of hegemonic social classes; often the state is, in reality, weak in respect of capabilities to accomplish national tasks relating to economy, society and polity.

From scholars on Latin America has come a more dynamic model which relates the rise of the "bureaucratic-authoritarian" state to the changing requirements of economic development. In the "bureaucratic-authoritarian" model articulated by Guillermo O'Donnell, a popularly-based coalition of urban-industrial interests during the easy phase of "consumer goods" imports-substitution industrialization breaks down and gives way to an authoritarian regime as a deeper "capital goods" industrialization is attempted to cope with, in large part, the balance of payments and indebtedness problems resulting from the earlier phase.[140] The deeper phase, in turn, requires large blocks of capital from foreign corporations and international agencies, which are made available only on condition of assured economic and political stability, thus calling forth a repressive state under high-level civil and military technocrats (allied in this case with foreign capital and the local economic elite, thus constituting the "triple alliance".[141] However, it has been suggested that the economic causes in the model are perhaps exaggerated, since some industrially more advanced countries in Latin America have not succumbed to bureaucratic-authoritarianism, and that political causes have been more important in the rise of the bureaucratic-authoritarian state.[142] Be that as it may, the authoritarian state has become the modal state in the Third World.

Even if civil or military bureaucrats are in power in most Third World countries, it is still a relevant question as to who the state represents or whose interests the state serves. In other words: What class or class coalition governs? Who controls the state? In addressing these questions, Marxist scholars -- who are the ones, rather than conventional scholars, sensitive to these issues -- have distinguished, following Lenin, between two aspects of the state: state apparatus and state power. Apart from the ordinary meaning associated with the term as a particular set of structures, Poulantzas tells us: [143] by *state apparatus* Lenin means *"the personnel of the state, the ranks of the administration, bureaucracy, army, etc., whereas by state power, Lenin means the social class or fraction of a class which holds power."* For Poulantzas this meaning of the state apparatus "relates at one and the same time to the problem of how the classes holding power is related to the personnel who are 'in charge' of the state", but he resolves the problem by his overall structuralist position that the nature of state personnel is largely immaterial, for the capitalist state must ultimately always serve the interests of the capitalist class.

Therborn takes the issue somewhat further, however. For him, the state apparatus is "a material condensation of class relations" while state power essentially means the nature of the policy outputs of the state: "state power is a *relation* between social class forces expressed in the content of state policies. The class character of these policies may be seen in their *direct effects* upon the forces and relations of production, upon the ideological superstructure, and upon the state apparatus." Again, he says "the class character of state power is thus defined by the *effects* of state measures on class positions in these three spheres."[144] But then he goes on to reify what was identified as a relational element, stating that "the central question must concern the class character of state power, since the ruling class is defined as such by its exercise of that power".[145] However, it would seem that one cannot just assume that, because state policies favour a certain class, therefore automatically that class holds state power, especially when it may not even be in charge of the state apparatus; state policies may be simply the reflection of the ideological belief of decision-makers that favouring a particular class, whether yet in existence or not, is essential for other interests. Otherwise, there is only circular reasoning instead of explanation. Marxist scholars are particularly susceptible to this failing. From effects of state policies they infer what class has state power, and then proclaim that the state follows these policies because this particular class has state power. There has to be independent empirical evidence on who has state power. In the absence of independent empirical evidence, all that one can say is that state policies have a certain impact; the evaluation of policies does not tell us who the ruling class is, especially if it is not in charge of the state apparatus.

Nonetheless, the distinction between state apparatus and state power is of some advantage in that it alerts us to the two sets of considerations that may be employed in the analysis of what classes or groups control the state: (1) the social composition of the personnel occupying strategic positions in the state apparatus; and (2) the classes and groups that benefit from state policies. This distinction may be said to correspond in a general way, though without their larger doctrinal baggage, to the "instrumentalist" (identified, perhaps wrongly, with Miliband) and "structuralist" (identified with Poulantzas) approaches. At the same time, notwithstanding the distinction, it seems that Marxist scholars, perhaps persuaded that there is lack of especially strong evidence that the instrumentalist approach demands, implicitly adopt the structuralist approach with its inherent inadequacies. However, it would seem that only by taking the instrumentalist position seriously, and examining the class background of those who occupy the strategic positions in the state apparatus, can the fallacy of circular reasoning be countered.

What are the different models relating social classes to the state in the Third World? Models abound, of course, in the literature, and one can at best touch on only a few of them, oriented towards the distinctive insights that they provide. These can be distinguished here according to whether they are based on social class or social strata; the class-based models can be further distinguished according to whether they involve a single class or a class coalition. An initial point that hardly needs emphasis is that the class structure of Third World societies is exceedingly complex, incorporating elements of both agricultural (feudal) and industrial (capitalist) societies, while at the same time entangled in ethnic heterogeneity.[146] Although social reality seems recalcitrant to attempts at the simplification inherent in model-building, the latter seems essential as a first step in a more adequate comprehension of the world.

· Social Class as Basis of State Power

Because of the impact of imperialism on Third World societies over a long period of time, it is rare to find the argument now that these are feudal societies with the landed aristocracy as the hegemonic class, even though some communist parties, following the Maoist model, do characterize them as semi-colonial and semi-feudal. It is more likely that, if a single class has to be underlined as holding state power, the state will be characterized as capitalist. That, indeed, is the position of two Marxist scholars, Patankar and Omvedt, who assert the existence of "a capitalist state" in peripheral state formations.[147] However, they make this claim not by any objective criteria but by what they take to be "the driving force in society," stating that "this means that the state is bourgeois state when this process is only beginning" and that "African states such as Kenya could be called bourgeois states even when the bourgeoisie is in the process of forming itself." Such crude procedure, of course, absolves them of all obligation to prove anything but nonetheless makes mandatory the call for a proletarian revolution even in the absence of a significant proletariat. It is pertinent further to ask whether the category of the "capitalist state" is meaningful at all in scholarly analysis, if the state in both the industrially advanced First World and the non-industrial Third World is called capitalist.

This becomes readily apparent in the analysis by two other neo-Marxist scholars, Duvall and Freeman, whose characterization of the state in the periphery and semi-periphery as capitalist -- simply because it functions in the context of dependency in a "world capitalist system"[148] -- provides no clue as to why the state behaves as it does. The thrust of their analysis is that Third World states are building a dependent capitalism, "largely a reflection of the dynamic in the center", but they are unable to relate this phenomenon to any particular class, or to explain why such dependent capitalist states pursue counter-dependent strategies when they are supposed to be dependent on the international capitalist class, or why they follow non-capitalist, even anti-capitalist, methods when they are supposed to be states of the capitalist class. The solution to this dilemma is sought through scholastic definitional jugglery so that the characterization of the state as dependent and capitalist, no matter what the state behaviour, remains inviolate. Since what the states do cannot be defined as building socialism, they take them necessarily to be capitalist.

Though Duvall and Freeman tend toward the notion of the "triple alliance" -- the national bourgeoisie, the international capitalist class and the techno-bureaucratic state elite -- the all-encompassing nature of the class basis of their capitalist state is truly breathtaking: "it means that the social foundations of the state must be in capitalist classes (that is, a bourgeoisie -- national or international -- or petit bourgeoisie) and/or in capitalist quasi-classes (for example, a techno-bureaucratic elite), rather than in pre-capitalist classes (the landed aristocracy)."[149] One must wonder whether there is any state in the world which is not covered by this class description. But they immediately add in what is obviously a tautology: "Simply stated, only capitalist states are committed to capitalist development." In the final analysis, it is not so much the class basis that explains what the state does but its values: "Simply put, the entrepreneurial state is distinguished from the socialist state in that it is guided by capitalist ideology and acts in accordance with a capitalist model of development."[150] The analysis, then, is descriptive, not explanatory.

Why does the capitalist state resort to non-capitalist methods such as state ownership? That, the authors respond, "represents an effort to correct or overcome inadequacies in the functioning of the private sector": "agencies of a capitalist state take on direct responsibility for the production of goods and services, in large part, because it is believed the private sector is unwilling or unable to do so, or because in doing so the private sector is generating some highly undesired patterns of development (such as imbalance, vulnerability, or loss of national control)". What should have been obvious from this analysis is that there is an overall national posture to which both the commitment to capitalism and the dependence on the international bourgeoisie are subordinate -- and that is one of national industrial development -- for, if the state were truly capitalist and really dependent, it is inconceivable that there is any industrial project that could not be accomplished through international capitalist enterprise. Even Duvall and Freeman do acknowledge at one place that "state ownership and operation of productive enterprise thus is a *nationalist* response designed to achieve advanced forms of industrialization in the face of deeply rooted obstacles to that development through the private sector."[151] But they refuse to draw the appropriate implications of this insight for their analysis, which remains an uncritical prisoner of the generalized "world capitalist system" framework.

Class Coalition as Basis of State Power

Classical Marxism envisaged an inescapable and basic contradiction between feudalism and capitalism as modes of production, and therefore antagonism and struggle between their respective class structures. However, in the mid-1950s, Paul Baran put forward a model which went against this central grain of Marxist theory by positing "a political and social coalition of wealthy compradors, powerful monopolists, and large landowners dedicated to the defense of the existing feudal-mercantile order." Not only that, this model proffers a world very different from that of Duvall and Freeman who, at least, see the dependent capitalist state as very busy with capital accumulation and industrialization, removing structural obstacles to development and resorting, if necessary, to state ownership toward that end, despite or perhaps because of the dependency context. For Baran, too, the coalition existed under the umbrella of foreign capital, but that foreign capital was intent on stifling development because of the threat such development would pose to its interests. Nor was the class coalition any more eager for development; Baran thus saw no prospect for economic growth under its auspices: "ruling the realm by no matter what political means -- as a monarchy, as a military-fascist dictatorship, or as a republic of the Kuomintang variety -- this coalition has nothing to hope for from the rise of industrial capitalism which would dislodge it from its positions of privilege and power. Blocking all economic and social progress in its country, this regime has no real political basis in city or village, lives in continual fear of the starving and restive popular masses, and relies for its stability on praetorian guards of relatively well kept mercenaries."[152] Even so, such regimes would have been swept away but for the aid and support given to them by Western powers.

With his "stagnationist" thesis, Baran became the mentor for the dependency school that developed later, but his thesis did not turn out to be accurate as many Third World countries made major strides in economic development over the subsequent decades. The dependency school, however, tended to dismiss such development as either being superficial, or dependent and distorted, or simply not socialist.

Alavi also posited a class-coalition model of the state that is rather close to Baran's but one that was not opposed to development, albeit capitalist development. In this model, the civil and military bureaucracy is in charge of "the apparatus of the state in postcolonial societies", but Alavi attempts to relate it, though quite ambiguously and unconvincingly, to social classes. The class basis of the state in his model is a complex one, with three propertied classes with competing interests rather than one as in the classical model -- a non-ascendant indigenous bourgeoisie, neo-colonial metropolitan bourgeoisies, and the landed classes -- so that "the state in the postcolonial society is not the instrument of a single class", none of the three classes "exclusively command the state apparatus" "nor do they command it collectively."[153]

On the other hand, both the civil and military bureaucracy "are highly developed in comparison with their indigenous class bases" and therefore relatively autonomous. The postcolonial state assumes also "a new and relatively autonomous *economic* role, which is not paralleled in the classical bourgeois state, because the state in the postcolonial society directly appropriates a very large part of the economic surplus and deploys it in bureaucratically directed economic activity in the name of promoting economic development. These are conditions which differentiate the postcolonial state fundamentally from the state as analyzed in classical Marxist theory." More specifically in the case of Pakistan, which started out literally without a capitalist class, "capitalist development in Pakistan has taken place under the corrupt patronage and close control by the bureaucracy"; in other words, the state has been a class creator rather than a class representative.

Still, Alavi insists, in what seems to be an understatement of the state's conception and perception of its role in Pakistan, that the "historically specific role" of the postcolonial state is to mediate and arbitrate among the three propertied classes. His own claim to originality in the model lies in this state acting on behalf of several classes unlike the analysis of the Bonapartist state by Marx, where the state served to organize the hegemony of a single class divided into diverse fractions. However, Alavi does not entertain the possibility that the state headed by "the military-bureaucratic oligarchy" -- which others would even consider as constituting a class in its own right -- may be more than relatively autonomous, that is, hegemonic over all classes, regardless of the historical role it may perform for capitalism in the long run.

Set within orthodox Marxism, the preceding models take the top class, whether in the feudal or capitalist mode of production or both, to be the ruling class and therefore holding state power. In a sharp departure from this practice, but "attempting to use Marx's own method of analysis on problems that have come up since his day",[154] Michal Kalecki, the profound Marxist economist from Poland, offers a new model of "intermediate regimes" based on the lower-middle class and the rich and medium-rich peasantry.[155] For that reason, it may be anathema to Marxists, especially since it is not a reincarnation of the Marxist model of the Bonapartist state, where the petty bourgeoisie and the small-holding peasantry were merely the support-base of the state without any share in political power.[156] Kalecki instead took the middle or intermediate sectors of the stratification system, consisting of the lower-middle class and "the corresponding strata of the peasantry", to constitute the ruling class and hold state power. For him, writing in 1964, India and Egypt were the archetypes of the intermediate regime.

Kalecki does not tell us precisely what the lower-middle class consists of. But he says that the upper-middle class or big business, allied with foreign capital, and feudal landowners constitute the antagonists from above to this intermediate group. On the other hand, the antagonists from below are the poor peasants and landless labour in rural areas and workers in small factories and the unemployed in urban areas. However, he regards white-collar workers and the small labour force in large factories, especially state enterprises, as "allies" of the lower-middle class since they are "in a privileged position" compared to the rural and urban poor. Kalecki is not clear on why the lower-middle class should have become the ruling class but his references to "the lower-middle class is very numerous", or to "the numerous ruling class", would indicate that he implicitly regards numbers, not just property, as particularly relevant to achievement of political power in the contemporary era.[157] Undoubtedly, its emergence to state power is related to the nationalist movement, but his reference to this fact is brief: "in the process of political emancipation -- especially if this is not accompanied by armed struggle -- representatives of the lower-middle class rise in a way naturally to power". Actually, this is so because the landed aristocracy had been compromised through its alliance with the colonial power while big industry was usually under the control of foreign capital,[158] leaving the lower-middle class to be the natural leader, given its resources in property, social status and numbers.

After the achievement of independence, the lower-middle class as the new ruling class adopts a programme of land reform that reduces the power of the landed aristocracy and -- recognizing "the weakness of the native upper-middle class and its inability to perform the role of 'dynamic entrepreneurs' on a large-scale"-- decides that "the basic investment for economic development must therefore be carried out by the state, which leads directly to the pattern of amalgamation of the interests of the lower-middle class with state capitalism." Expanding on the last point, but without explaining why the lower middle class should be interested in economic development at all, Kalecki adds:

> this system is highly advantageous to the lower-middle class and the rich peasants; state capitalism concentrates investment on the expansion of the productive potential of the country. Thus there is no danger of forcing the small firms out of business, which is a characteristic feature of the early stage of industrialization under *laissez faire*. Next, the rapid development of state enterprises creates executive and technical openings for ambitious young men of the numerous ruling class. Finally, the land reform, which is not preceded by an agrarian revolution, is conducted in such a way that the middle class which directly exploits the poor peasants -- i.e. the moneylenders and merchants -- maintains its position, while the rich peasantry achieves considerable gains in the process.

In regard to the upper-middle class, the regime's policies may "range from far-reaching nationalization (usually with compensation) to a mere limitation of the scope of private investment coupled with attempts, as a rule rather ineffective, to adjust its structure to the general goals of development."

As a Marxist, Kalecki was quite sensitive to the treatment of communists by intermediate regimes. He, of course, took the lower classes to be unhappy with their economic position but did not consider them as constituting a serious threat to the regime for some time since they were controlled by "some form of a local oligarchy comprised of the petty bourgeoisie (merchants and moneylenders), the richer peasants and smaller landlords". But nonetheless "the lower-middle class is quite rightly afraid of the political activisation" of the lower classes and therefore bore hard with repression on the communists who were their potential spokesmen.

Kalecki made an insightful correlation between the internal and foreign policies of such regimes that stood, as it were, intermediate between the two ideological power blocs: "the internal position of the ruling lower-middle class finds its counterpart in the policy of neutrality between the two blocs; an alliance with any of the blocs would strengthen the corresponding antagonist at home." Also, this policy enhances their bargaining power to extract foreign credits to meet balance of payments problems: "the intermediate regimes are the proverbial clever calves that suck two cows; each bloc gives them financial aid competing with the other. Thus has been made possible the 'miracle' of getting out of the USA some credits with no strings attached as to internal economic policy."

Basically, Kalecki believed that the intermediate regime was an unstable and ineffective political arrangement: "History has shown that lower-middle class and rich peasantry are rather unlikely to perform the role of the ruling class." But there seems to be no evidence, as Joan Robinson seems to think, that "he was too optimistic in supposing that it might give birth to a viable socialist alternative."[159] Quite the contrary, he felt that the representatives of these classes "invariably served the interests of big business (often allied with the remnants of the feudal system)", though he did not specify the precise mechanisms through which this takes place, nor how the antagonists from above are transformed into beneficiaries.

Social Strata as Basis of State Power

John Kautsky, a non-Marxist scholar who employs the concept of class in his work, takes the position that Third World states share a great many features in common, regardless of the differences among them, including the distinction between communist and non-communist regimes. The centrepiece of his model is not class but a stratum of the new middle class which results from the impact of colonialism -- the intelligentsia. Essentially, those who rose to power in Third World states, at least initially, came at the head of nationalist anti-imperialist movements, which are movements "led by modernizers, that is, mostly intellectuals"; "their leadership is almost invariably in the hands of men with a modern higher education."

For Kautsky, these leaders had largely a similar social support-base inasmuch as they "sought support from members of all classes." They were alike also in their goals: "rapid industrialization and opposition to the native aristocracy, traditionalism and colonialism, with nationalization of industry and land reform designed to serve these goals." Moreover, the states that these leaders came to head developed certain common political processes: single-party systems; centralized economic planning; government control or ownership of industry; mass terror, mass regimentation and mass persuasion to counter opposition. To the extent that such regimes are successful in their goal of industrialization, "a new stratum of managerial modernizers develops and may come into conflict with the revolutionary modernizers."[160]

In stating all this, Kautsky seems to have underlined that differences among Third World states are not so much related to the social background of their leadership, which is generally drawn from the intelligentsia, as perhaps to other factors, which may include: the destruction of the country's social fabric in war, the economic advance already made, the rigour with which ideological beliefs are held, and the nature of party organization at the disposal of the leadership. Under the leadership of a largely similar stratum -- referred to as intellectuals, modernizers, modernizing elites -- Third World states have been building different socio-economic systems variously described as: capitalism, state capitalism, national capitalism, dependent capitalism, dependent state capitalism, and socialism. Indirectly perhaps, Kautsky underscored the failure of class as an explanation of state power, at least in its "instrumentalist" version.

Quite close to the Kalecki model of the "intermediate regime", but focusing on institutionally-determined strata rather than social classes, is the national state-capitalist model by the Marxist scholar James Petras.[161] Here, the state incorporates socialist forms -- both political (such as the one-party state and radical slogans) and economic (such as state ownership and central planning) -- in order to achieve the capitalist ends of surplus expropriation in a class society. The state thus manifests collectivism but without redistribution. It is, in the words of another Marxist scholar, "a capitalist economy run by noncapitalists".[162] The key stratum presiding over this national political-economic project directed against neo-colonialism consists of "the state sector employees, civil and/or military", who have arrived at this position through different routes, such as political evolution, military coups or popular revolt. Such a stratum does not have an independent socio-economic base of any significance of its own; its narrow base therefore necessitates military or single-party rule, or both.

Holding that this social stratum "does not fit any of the classes described by Marx in the development of capitalism in Europe", Petras dismisses attempts, "through definitional acrobatics, to redefine this stratum so that it can be accounted for within the classical schema." It is neither bourgeois nor petty bourgeois since it does not own the means of production, nor is it proletarian because it is not directly involved in the process of production. Rather it stands in between property owners and workers as a propertyless "intermediary stratum" (comprising professionals, employees, military, university groups). What marks it off as a class-conscious stratum is the disparity in political power that goes with its status -- "their key weapon is political capacity: their ability to take hold of the state machinery, alter the distribution of social power, and reorganize economy."

Petras' analysis diverges critically from the model of dependent state capitalism offered by Duvall and Freeman. The bureaucratic milieu having shaped its vision, the intermediary stratum sees the state, in the context of the feeble development of society and its domination by foreign enterprise, as the only potent instrument for economic development. In brief, "the military-national-state-capitalist regime attempts to substitute itself for the absence of a coherent capitalist class, and through the state it attempts to perform the tasks of the bourgeois revolution." Imbued with economic nationalism, the intermediary stratum characteristically focuses on industrialization through state enterprises, development of a domestic market through agrarian reform, and national control of the economy through nationalization of foreign firms. However, since state enterprises function in a market context, there is only a shift in the method of exploitation of labour, not its elimination; besides, enterprises owned by the domestic bourgeoisie are not nationalized. Operating thus within a capitalist framework, the regime must inevitably face contradictions and crises.

For Petras, the state-capitalist regime is an unstable and transient phenomenon. Its national project with a multi-class appeal is bound to erode when threatened by foreign capital and the working class, forcing it on a zig-zag course in the search for new allies, given its narrow base. The regime is thus destined to evolve into a "coercive" state to suppress workers while simultaneously submitting to re-integration into the international capitalist system through compromise with foreign capital.[163]

It should be noted that the "intermediary stratum" central to Petras' model is precisely what some Marxist scholars have pointedly underlined as a social class, especially in relation to Africa, calling it "the new petty bourgeoisie", the "state bourgeoisie", or the "bureaucratic bourgeoisie";[164] it is not without significance that at one point Petras himself refers to "the surplus that accrues to the state bourgeoisie." The state here then does not in any sense "represent" the dominant class or class coalition as commonly understood in either the instrumentalist or structuralist version of Marxism; rather, it is the dominant class. But this class and its project are not lacking in a more positive appreciation and prognosis in the eyes of some analysts; rather than considering state capitalism to be a passing phase, in which its progressive features wither away, they believe it may transcend and transform itself into a non-capitalist future.[165] In any case, it is instructive to note here that in respect of a set of Third World states corresponding to what Petras has discussed, Miliband is driven to acknowledge:[166]

> in such societies, the state must be taken mainly to 'represent' itself, in the sense that those people who occupy the leading positions in the state system will use their power, *inter alia*, to advance their own economic interests, and the economic interests of their families, friends, and followers, or clients....In such cases the relation between economic and political power has been inverted; it is not economic power which results in the wielding of political power and influence and which shapes political decision making.' It is rather political power (which also means here administrative and military power) which creates the possibilities of enrichment and which provides the basis for the formation of an economically powerful class, which may in due course become an economically dominant one. The state is here the source of economic power as well as an instrument of it: state power is a major 'means of production'.
>
> It is an instrument of economic power, not in the sense that those who hold state power serve the interests of an economically dominant class separate from these power-holders and located in society at large; but that those who hold state power *use* it for their own economic purposes and the economic purposes of whoever they choose. This use of state power assumes many different forms, including of course the suppression of any challenge to the supremacy of what turns in effect into an *economically and politically dominant class*.

Summary and Conclusions

The public sector has mostly been studied in terms of ministerial control by political scientists and in terms of economic efficiency by economists. But its quite substantial role in the Third World calls for studying the public sector in a broader and more comprehensive framework that looks at both its origins and at its functions in relation to a number of important societal interests.

In respect of its origins as well as prospects of its continuance, the public sector has necessarily been implicated in the controversy relating to the competing ideologies of capitalism and socialism. That would seem to be unfortunate. For, even while ideological ballast often prevents social scientists from taking adequate account of it, there is obvious a substantial decay and disintegration in the respective ideological paradigms; equally, there is manifest an increasing pragmatism on this issue in countries that are identified with one or another of the opposed ideologies.

Although at one time the public sector was understood to stand in contradiction to the very principles of capitalism it has, along with economic planning, become today an integral part of capitalism in many advanced industrial societies. Indeed, the opponents of capitalism now interpret it as fortifying capitalism rather than modifying it. However, this much is certain that capitalism in the advanced industrial societies, whether modified or fortified by economic planning and the public sector, has gone hand in hand with technological innovation and therefore material welfare at unprecedentedly high levels compared to all other economic systems thus far known to man. Equally, confronted with problems of economic efficiency in the management of a centralized state-administered economy, socialist regimes have been willing to incorporate the market as an allocative mechanism in the economic system even while not allowing private ownership of the means of production. In some cases, socialist regimes have actually gone in for "market socialism" while in other cases its introduction has been resisted largely because of the adverse political consequences it would have for the ruling class. The lesson for the Third World from all this would seem to be that the determination of whether or not to have, or whether or not to continue to have, a public sector ought to be based less on considerations of ideology than those of its instrumentality in terms of national interests.

For the Third World, the overall comprehensive national goal in the present epoch is industrialization. Not only that, it is an "imperative". It is so because industrialization is a prerequisite for national independence, precisely as a result of the difference it makes to economic and military capabilities in the context of an international system whose key characteristic is the struggle for power. In this light, the industrialization already achieved by the developed countries constitutes literally a threat to the very independence of Third World countries and compels the latter to match it, no matter how daunting the task. The "mercantilist" prescription for industrialization, with the state assuming a critical role, is thus a historical compulsion for them in their condition of economic backwardness. If state sponsorship of industrialization is a historical compulsion, equally compelling is the resort to the public sector, given the absence of an adequate economic infrastructure as also the underdeveloped nature of entrepreneurial capabilities in their societies.

Since the public sector often originates in a historical compulsion, even if at times facilitated by ideology, its evaluation ought to be in terms of its adequacy in response to that compulsion; in other words, the test ought to be efficacy rather than ideology. However, for Third World countries the test of efficacy cannot be conceived of in narrow terms relating to economic criteria alone. Given the need for overall societal development, partly as a prerequisite for industrialization and partly compelled by the social and political process associated with industrialization, Third World countries often have a whole spectrum of national goals pertaining to economy, society and polity -- economic growth and welfare; equality and integration; participation and institution-building. The public sector may, indeed, often originate in -- even as its continuity may be rooted in -- goals other than simply that of industrialization narrowly understood. All that underlines the need for evaluating the public sector against the entire spectrum of national goals.

While it is essential for a more adequate understanding of the public sector that it be evaluated against the whole spectrum of national goals, there is considerable hazard as well in such a procedure. This is so because the multiplicity of goals prevents the application of stringent criteria in evaluation; the public sector's weakness in one sector is likely to be rationalized by its critical, but vague, role in some other sector. However, this ought to underscore the necessity for a more hardheaded and critical perspective rather than to detract from the need for a more comprehensive framework, since that corresponds closer to reality, in any evaluation of the public sector.

One noteworthy aspect deserving of attention in any study of the origins and role of the public sector as a national project, arising out of the historical compulsion for development of economically backward countries, is the public sector's relationship to more narrow interests of those who control it rather than the larger interests of economy, society and polity as a whole. This raises the important question of the relationship of the state to society, a subject of special interest in Marxist scholarship. At one time, Marxism presented a definite and unified position on this issue, but today it is characterized by diversity, confusion and disarray. Indeed, some Marxist reinterpretations bring its model of the state quite close to the liberal-democratic model. In regard to the Third World, beyond the recognition that the state here is more or less autonomous in relation to society, there is a whole variety of models pertaining to the nature of the class or class coalition that commands state power.

The diversity of models of the ruling class really testifies to the complexity of the Third World and to the futility of any attempt to develop a single comprehensive model for it. Although the Third World does share a common cluster of problems, there has in the four decades since World War II taken place a considerable decomposition of the Third World over and above the earlier division between large and small size countries -- oil exporting and oil importing countries, capital surplus and capital short countries, new industrializing and largely agricultural countries. It is therefore necessary to view the Third World in a more disaggregated manner than the overall popular label implies. Considered in that light, there is no one correct model among the different class models of the state. Rather, they are all correct, depending on the particular set of Third World countries chosen for examination. But it is significant that in many of them the role of the state is not always reducible to the economic base or to social classes. That point needs to be underlined, for there is a tendency among scholars to revert to old assumptions of the epiphenomenal nature of politics even after having acknowledged the primacy of politics in the Third World.

In summary, then, the preeminent role of the public sector in the Third World demands its study in terms of (1) its origins, and (2) its functions. In respect of both aspects, it raises the question of its relation to ideology and interests. The latter, of course, relate not only to a whole set of national goals for Third World countries issuing out of the bifurcation of the world into industrialized countries and economically backward countries, but also to the class or class coalition in control of the state.

NOTES

1. On the distinction, see David E. Apter, *The Politics of Modernization* (Chicago: University of Chicago Press, 1966), ch.3. The distinction is, of course, related to and rooted in an older tradition that separates *gemeinschaft* (based on natural will) from *gesellschaft* (based on rational will) as in Toennies.
2. See, for example, Willard A. Mullins, "On the Concept of Ideology in Political Science," *American Political Science Review*, vol. 66, no. 2 (June 1972), pp. 498-510.
3. Karl Marx, *Capital*, vol. I (Moscow: Progress Publishers, 1954), p. 18.
4. I do not take into account here notions such as Wagner's Law of Increasing State Activity or the Displacement-Effect Hypothesis (relating to resistance to downward movement of revenues once they are increased in the context of social upheavals, such as war), for these relate to growth in government expenditures, rather than productive enterprises of the state, and also because they focus on developed countries.
5. On the various theories and the debate and divisions within the Marxist camp on the breakdown of capitalism, see F.R. Hansen, *The Breakdown of Capitalism: A History of the Idea in Western Marxism, 1883-1983* (London: Routledge & Kegan Paul, 1985).
6. Branko Horvat, *The Yugoslav Economic System* (White Plains, N.Y.: International Arts and Sciences Press, 1976), p. 10.
7. Sherman H.M. Chang, *The Marxian Theory of the State* (Philadelphia: 1931), pp. 74-75.
8. See Sidney Hook, *Marx and the Marxists* (New York: D. Van Nostrand, 1955), pp. 43, 50-57, 66-70.
9. D. A. Reisman, *Adam Smith's Sociological Economics* (London: Croom Helm, 1976), pp. 220-21.
10. Adam Smith, *An Inquiry into the Nature and Causes of the Wealth of Nations* (London: Methuen & Co. 1925), pp. 15, 16.
11. Reisman, pp. 146-47, 220-21.
12. Eli F. Heckscher, *Mercantilism* (London: George Allen & Unwin, 1955), vol. II, p. 327.
13. Reisman, pp. 211-22.
14. *Ibid.*, pp. 223-26.
15. Heckscher, p. 327.
16. Letter dated October 8, 1858 from Marx to Engels, in Shlomo Avineri (ed.), *Karl Marx on Colonialism and Modernization* Garden City, N.Y.: Anchor Books, 1969), pp. 463-65.

17. Ralf Dahrendorf, *Class and Class Conflict in Industrial Society* (London: Stanford, Calif.: Stanford University Press, 1959), ch. 2. For the view that working class divisions are a product not only of technological and economic change but also of a class-based strategy of labour segmentation by capital, see David M. Gordon, Richard Edwards, and Michael Reich, *Segmented Work, Divided Workers: The Historical Transformation of Labor in the United States* (Cambridge, UK: Cambridge University Press, 1982). For a different statement on the growing homogenization of the working class, see Harry Braverman, *Labor and Monopoly Capital: : The Degradation of Work in the Twentieth Century* (New York: Monthly Review Press, 1974).
18. John Kenneth Galbraith, *American Capitalism: The Concept of Countervailing Power* (Boston: Houghton Mifflin, 1956).
19. Paul M. Sweezy, "Center, Periphery, and Crisis," Hamza Alavi and Teodor Shanin (eds.), *Introduction to the Sociology of "Developing" Societies* (New York: Monthly Review Press, 1982), p. 215.
20. For a pioneering analysis of the welfare state, see Asa Briggs, "The Welfare State in Historical Perspective," *European Journal of Sociology*, II, no.2. (1961), pp. 221-58. See also Peter Flora and Arnold J. Heidenheimer (eds.), *The Development of Welfare States in Europe and America* (New Brunswick, N.J.: Transaction Books, 1981). For the view that the various measures of the "welfare state" constitute not the adaptability of capitalism but an achievement of labour and as well a transition to socialism, largely consistent with Marx, see John D. Stephens, *The Transition from Capitalism to Socialism* (Atlantic Highlands, N.J.: Humanities Press, 1980).
21. Goran Therborn, "The Prospects of Labour and the Transformation of Advanced Capitalism," *New Left Review*, No. 145 (1984), 5-38. This is an iconoclastic article challenging Marxist presuppositions about labour advances and welfare state capitalism.
22. Andrew Shonfield, *Modern Capitalism: The Changing Balance of Public and Private Power* (London: Oxford University Press, 1965), p. 121.
23. Jack Hayward, "Introduction," in Jack Hayward and Olga A. Narkiewicz, *Planning in Europe* (London: Croom Helm, 1978), p. 21. For a critical analysis, see Stephen S. Cohen, *Modern Capitalist Planning: The French Model* (Berkeley: University of California Press, 1977).
24. Andrew Shonfield, *In Defence of the Mixed Economy* (Oxford: Oxford University Press, 1984), pp. 38-39.
25. William J. Baumol (ed.), *Public and Private Enterprise in a Mixed Economy* (New York: St. Martin's Press, 1980), p.3.
26. David Coombes, *State Enterprise: Business or Politics?* (London: George Allen and Unwin, 1971), ch. 2.
27. Andrew Shonfield, *The Use of Public Power* (Oxford: Oxford University Press, 1982), pp. v, xviii-xix.
28. Therborn, "The Prospects," pp. 26-27.
29. Ralph Miliband, *The State in Capitalist Society* (New York: Basic Books, 1969), p. 9.
30. Joseph A. Schumpeter, *Capitalism, Socialism and Democracy* (London: George Allen & Unwin, 1950), p. 66.
31. Wlodzimierz Brus, *Socialist Ownership and Political Systems* (London: Routledge and Kegan Paul, 1975), pp. 7-11.
32. Shonfield, *The Use of Public Power*, ch. 1.

33. See Robert Skidelsky (ed.), *The End of the Keynesian Era: Essays on the Disintegration of the Keynesian Political Economy* (New York: Holmes & Meier Publishers, 1977), chs. 5 and 6.
34. See Alan Wolfe, "Has Social Democracy a Future?", *Comparative Politics*, XI (October 1978), p. 119.
35. See Samir Amin, et al., *Dynamics of Global Crisis* (New York: Monthly Review Press, 1982), pp. 12, 50.
36. See Sayre P. Schatz, "Socializing Adaptation: A Perspective on World Capitalism," *World Development* XI, no. 1 (1983), pp. 1-10.
37. Oskar Lange and Fred M. Taylor, *On the Economic Theory of Socialism* (New York: McGraw-Hill, 1964), pp. 25-27.
38. Schumpeter, chs. VII-VIII, and John Kenneth Galbraith, *Economics and the Public Purpose* (New York: Houghton Mifflin, 1973) and *The New Industrial State* (Boston: Houghton Mifflin, 1978).
39. Note the comment: "State activity has emerged not as a substitute for but as a complement to the market....The success story of postwar economies has been a story of the market mechanism complemented by state regulation and intervention." S.K. Kuipers and G.J. Lanjouw (ed.), *Prospects of Economic Growth* (Amsterdam: North-Holland, 1980), p. 15.
40. Ira Katznelson, "Considerations on Social Democracy in the United States," *Comparative Politics*, XI (October 1978), p. 80.
41. See Skidelsky (ed.), *The End of the Keynesian Era* and "The Decline of Keynesian Politics," in Colin Crouch (ed.) *State and Economy in Contemporary Capitalism* (New York: St. Martin's Press, 1979); James M. Buchanan and Richard E. Wagner, *Democracy in Deficit: The Political Legacy of Lord Keynes* (New York: Academic Press, 1977); Fred Hirsch, *Social Limits to Growth* (Cambridge: Harvard University Press, 1976); Fred Hirsch and John H. Goldthorpe (eds.), The *Political Economy of Inflation* (Cambridge: Harvard University Press, 1978); and Leon N. Lindberg, et al., (ed.), *Stress and Contradiction in Modern Capitalism: Public Policy and the Theory of the State* (Lexington, Mass.: Lexington Books, 1975).
42. Among others, see Robert A Dahl, *Dilemmas of Pluralist Democracy* (New Haven: Yale University Press, 1982), and Shonfield, *The Use of Public Power*, pp. 109-111.
43. See John F. Manley, "Neo-Pluralism: A Class Analysis of Pluralism I and Pluralism II," *American Political Science Review*, vol. 77, no. 2 (June 1983), pp. 368-83, and communications by Charles E. Lindblom and Robert Dahl, *ibid.*, pp. 384-89.
44. John Dunn, *The Politics of Socialism: An Essay in Political Theory* (Cambridge, UK: Cambridge University Press, 1984), pp. 18, 53, 58.
45. See "The Socialist Imperative," John Kenneth Galbraith, *Economics and the Public Purpose*, ch. XXVII.
46. Interview, in Myron E. Sharpe, *John Kenneth Galbraith and the Lower Economics* (White Plains, N.Y.: International Arts and Sciences Press, 1974), p. 114.
47. Nicholas Kaldor, in Baumol (ed.), pp. 10, 136.
48. For a controversy among Marxists on this point, see Andrew Levine and Erik Olin Wright, "Rationality and Class Struggle," *New Left Review*, No. 123 (1980), pp. 47-68.
49. John Kenneth Galbraith, *The Voice of the Poor: Essays in Economic and Political Persuasion* (Cambridge, Mass.: Harvard University Press, 1983), p. 41.
50. On the subject, see M. George Zaninovich, "Socialist Models and Developing Nations," in

Wilard A. Beling and George O. Totten (ed.), *Developing Nations ; Quest for a Model* (New York: Van Nostrand Reinhold Company, 1970), pp. 116-51.
51. Horvat, p. 10.
52. John M. Echols III, "Does Socialism Mean Greater Equality? A Comparision of East and West Along Several Major Dimensions," *American Journal of Political Science*, vol. 25, no. 1 (1981), pp. 1-26. For a more extended treatment, see David Lane, *The End of Social Inequality? Class, Status and Power under State Socialism* (London: George Allen & Unwin, 1982). On a theoretical level, see John E. Roemer, *A General Theory of Exploitation and Class* (Cambridge, Mass.: Harvard University Press, 1982), esp. pp. 260-63.
53. See Ernest Mandel, "On the Nature of the Soviet State," *New Left Review*, No. 108 (1978), 23-45.
54. Alec Nove, *Political Economy and Soviet Socialism* (London: George Allen & Unwin, 1979), pp. 200-201.
55. Charles Bettelheim, *Class Struggles in the USSR: First Period: 1917-1923* (New York: Monthly Review Press, 1976), vol. I, pp. 17, 44.
56. Nicos Poulantzas, "On Social Classes," *New Left Review*, No. 78 (1973). For a summary discussion of the various characterizations of the Soviet state ("degenerate workers' state", "state capitalism", "a new mode of production issuing out of the Asiatic mode of production"), see Lane, ch. 5. For an attack on the application of the notion of state capitalism to the Soviet Union, or for that matter to any system, see Alex Dupuy and Barry Truchil, "Problems in the Theory of State Capitalism," *Theory and Society*, VIII, no. 1 (1979), 1-38.
57. Brus, pp. 17-18.
58. See Nove, p. 205. For a strong dissent on the bureaucracy as a ruling class, see Mandel, pp. 23-45.
59. Brus, pp. 18, 30.
60. Skidelsky, "The Decline of Keynesian Politics," p. 78.
61. Alec Nove, *The Soviet Economic System* (London: George Allen & Unwin, 1977), pp. 371, 377.
62. Nove, *Political Economy*, p. 155.
63. Charles E. Lindblom, *Politics and Markets: The World's Political-Economic Systems* (New York: Basic Books, 1977), p. 304.
64. Lange and Taylor, *On the Economic Theory of Socialism*.
65. Brus, p. 202.
66. David Granick, *Enterprise Guidance in Eastern Europe* (Princeton: Princeton University Press, 1975), pp. 25, 468.
67. Brus, p. 94.
68. See Horvat, pp. 10-13, 58; see also Nove, *The Soviet Economic System*, pp. 299-303.
69. Brus, pp. 74, 167-69; see also Nove, *The Soviet Economic System*, pp. 290-98, and Rezso Nyers and Marton Tardos, "Enterprises in Hungary Before and After the Economic Reform," in Baumol (ed.), ch. 10.
70. Lindblom, pp. 304-308; Nove, *The Soviet Economic System*, pp. 310-316; and Brus, p. 153. For an analysis of the possibility of implementing different types of models of reforms and their implications, see Joseph S. Berliner, "Planning and Management," in Abram Bergson and Herbert S. Levine (ed.), *The Soviet Economy: Toward the Year 2000* (London: George Allen & Unwin, 1983), 350-90.
71. See Edmund Lee, "Economic Reform in Post-Mao China: An Insider's View," *Bulletin of*

Concerned Asian Scholars, XV, no. 1. (1983), pp. 16-25; Peter Van Ness and Satish Raichur, "Dilemmas of Socialist Development: An Analysis of Strategic Lines in China, 1949-1981," *ibid.*, pp. 2-15; and Peter Moller Christensen, "Plan, Market or Cultural Revolution in China," *Economic and Political Weekly*, XVIII, no. 16-17 (April 16-23, 1983), pp. 848-54.

72. Brus, p. 171.
73. For the challenge posed by these countries to dependency theory, see John Browett, "The Newly Industrializing Countries and Radical Theories of Development," *World Development*, vol. 13, no. 7 (1985), 789-803; David Booth, "Marxism and Development Sociology: Interpreting the Impasse," *ibid.*, 761-787; and Aidan Foster-Carter, "Korea and Dependency Theory," *Monthly Review*, vol. 37 (October 1985), 27-34.
74. Richard Barnet, cited in Peter Dale Scott, "Peace, Power and Revolution," *Alternatives*, IX (1983-84), 351-72.
75. Friedrich List, *The National System of Political Economy* (New York: Augustus M. Kelley, 1966), pp. 347-49.
76. *Ibid.*, pp. 350-51.
77. *Ibid.*, pp. 172-73, 366, 368, 419-20.
78. See Baldev Raj Nayar, "Political Mainsprings of Economic Planning in the New Nations: The Modernization Imperative versus Social Mobilization," *Comparative Politics*, VI, no.3 (April 1974); reprinted in Norman W. Provizer (ed.), *Analyzing the Third World* (Cambridge, Mass.: Schenkman Publishing Company, 1978), pp. 460-85. See also Baldev Raj Nayar, *The Modernization Imperative and Indian Planning* (New Delhi: Vikas, 1972).
79. Robert A. Solo, *The Political Authority and the Market System* (Cincinnati: South-Western Publishing Co., 1974), p 29.
80. R.P. Dore, discussed in Ogura Mitsuo, "The Sociology of Development and Issues Surrounding Late Development," *International Studies Quarterly*, XXVI, no. 4 (1982), p. 610.
81. Alexander Gerschenkron, *Economic Backwardness in Historical Perspective* (Cambridge, Mass.: Belknap Press, 1966), pp. 529.
82. A James Gregor, *Interpretations of Fascism* (Morristown, N.J.: General Learning Press, 1974).
83. *Third World Quarterly*, V, no. 3 (1983), pp. 541-52.
84. Leroy P. Jones and Il Sakong, *Government, Business, and Entrepreneurship in Economic Development: The Korean Case* (Cambridge, Mass.: Harvard University Press, 1980), preface.
85. Peter Evans, *Dependent Development: The Alliance of Multinational, State, and Local Capital in Brazil* (Princetion: Princeton University Press, 1979), p. 220.
86. Jones and Sakong, p. 11.
87. On this, see Yash Ghai (ed.), *Law in the Political Economy of Public Enterprise: African Perspectives* (Uppsala, Sweden: Scandinavian Institute of African Studies, 1977), pp. 18-20; and Baumol, pp. 41-42, 44, 49. More broadly, Jones and Mason consolidate the reasons for establishing public enterprises into four groups: (1) ideological predilection; (2) acquisition or consolidation of political or economic power; (3) historical heritage and inertia; (4) pragmatic response to economic problems. See Leroy P. Jones (ed.), *Public Enterprise in Less Developed Countries* (Cambridge, UK: Cambridge University Press, 1982), ch. 2. Paul Streeten provides "Twenty-One Arguments for Public Enterprise," in Khadija Haq (ed.), *Global Development: : Issues and Choices* (Washington, D.C.: North South Roundtable, 1983), ch. 15. V.V. Ramanadham (ed.), *Public Enterprise and the Developing*

World (London: Croom Helm, 1984), p. 1, singles out "the development context" of Third World countries for attention in respect of the public sector. For a summary listing of goals of public sector, see W.T. Stanbury and Fred Thompson (ed.), *Managing Public Enterprises* (New York: Praeger, 1982), chs. 1 and 2.
88. Kenneth Shepsle and Barry R. Weingast, "Political Solutions to Market Problems," *American Political Science Review*, vol. 78 (1984), 417-33.
89. W.W. Rostow, *Politics and the Stages of Growth* (Cambridge: Cambridge University Press, 1971), pp. 11-12.
90. A.F.K. Organski, *The Stages of Political Development* (New York: Alfred A. Knopf, 1965).
91. Gabriel A. Almond and G. Bingham Powell, Jr., *Comparative Politics : A Developmental Approach* (Boston: Little, Brown, 1966), p. 35.
92. Dankwart A. Rustow, *A World of Nations: Problems of Political Modernization* (Washington, D.C.: Brookings Institution, 1967), pp. 35-36. Leonard Binder, et al., *Crises and Sequences in Political Development* (Princeton: Princeton University Press, 1971) lists five crises: identity, legitimacy, participation, distribution and penetration.
93. Karl Marx and Frederick Engels, *Manifesto of the Communist Party* (Moscow: Progress Publishers, 1975), pp. 44-48.
94. Allan Tupper, "The State in Business," *Canadian Public Administration*, XX, no. 1 (1981), pp. 124-50.
95. See Baumol (ed.), p. 230, and also Richard Pryke, *The Nationalised Industries: Policies & Performance Since 1968* (Oxford, England: Martin Robertson, 1981).
96. Baumol (ed.), pp. 42, 230-31.
97. Although many refinements and variations exist, these seem to be the two basic models. See Robert R. Alford, "Paradigms of Relations Between State and Society," in Leon N. Lindberg, et al., *Stress and Contradiction in Modern Capitalism: Public Policy and the Theory of the State* (Lexington, Mass.: Lexington Books, 1975), pp. 145-60; Colin Crouch (ed.), *State and Society in Contemporary Capitalism* (New York: St. Martin's Press, 1979); Stephen D. Krasner, *Defending the National Interest: Raw Materials and U.S. Foreign Policy* (Princeton: Princeton University Press, 1978); Richard Scase (ed.), *The State in Western Europe* (New York: St. Martin's Press, 1980); Alfred Stepan, *The State and Society: Peru in Comparative Perspective* (Princeton: Princeton University Press, 1978); and Albert Szymanski, *The Capitalist State and the Politics of Class* (Cambridge, Mass.: Winthrop Publishers, 1978).
98. Among other works, see David B. Truman, *The Governmental Process: Political Interests and Public Opinion* (New York: Knopf, 1951); Robert A. Dahl, *Who Governs? Democracy and Power in an American City* (New Haven: Yale University Press, 1961); and David Easton, *The Political System* (New York: Alfred A. Knopf, 1971).
99. Charles Lindblom, *Politics and Markets*.
100. Eric A. Nordlinger, *On the Autonomy of the Democratic State* (Cambridge, Mass.: Harvard University Press, 1981); see also Krasner, *Defending the National Interest*.
101. C. Wright Mills, *The Power Elite* (New York: Oxford University Press, 1957).
102. One scholar points to six different Marxist approaches to the state; see Bob Jessop, "Recent Theories of the Capitalist State," *Cambridge Journal of Economics*, I, no. 4 (1977), pp. 353-73.
103. Frederick Engels, "Preface," in Karl Marx, *The Communist Manifesto* (Chicago: Henry Regnery Company, 1965), p. 7.

104. Szymanski, p. 21.
105. Ralph Miliband, *The State in Capitalist Society* (New York: Basic Books, 1969), p. 146.
106. Cited in Alan Wolfe, "New Directions in the Marxist Theory of Politics," *Politics and Society*, IV, no. 2 (1973), pp. 131-59.
107. Robert Tucker, *The Marxian Revolutionary Idea* (New York: W.W. Norton, 1969), ch. 4.
108. See Anthony Brewer, *Marxist Theories of Imprialism: A Critical Survey* (London: Routledge & Kegan Paul, 1980), chs. 8, 11.
109. David A. Gold, Clarence Y.H. Lo, and Erik Olin Wright, "Recent Developments in Marxist Theories of the Capitalist State," *Monthly Review*, XXVII, no. 5 (October 1975), pp. 29-43, and no. 6 (November 1975), pp. 36-51.
110. Ralph Miliband, "Poulantzas and the Capitalist State," *New Left Review*, No. 82 (November-December 1973), pp. 83-92; emphasis in the original. Lest it be understood that the point was made in the context of a polemic with Poulantzas, it should be noted that it is repeated essentially in the same language in his larger work, *Marxism and Politics* (London: Oxford University Press, 1977), pp. 66-68, 74.
111. Miliband, *Marxism*, pp. 87-88.
112. Nicos Poulantzas, "On Social Classes," *New Left Review*, No. 78 (March-April 1973).
113. Nicos Poulantzas, *Political Power and the Social Classes* (London: New Left Books, 1973), p. 190.
114. Nicos Poulantzas, "The Problem of the Capitalist State," *New Left Review*, No. 58 (November-December 1969), pp. 67-78.
115. Poulantzas, *Political Power*, pp. 188-89.
116. *Ibid.*, pp. 284-85.
117. *Ibid.*, pp. 190-91.
118. To the extent that others (such as Alavi and Block) have specified the mechanisms, they unwittingly demonstrate that the capitalist state is no different from any other modern state in respect of responsiveness to the requirements of accumulation.
119. Miliband, *Marxism*, p. 73.
120. Poulantzas, *Political Power*, p. 194.
121. Goran Therborn, *What Does the Ruling Class Do When It Rules?* (London: New Left Books, 1978), p. 147.
122. Hamza Alavi, "State and Class under Peripheral Capitalism," in Hamza Alavi and Teodor Shanin (eds.), *Introduction to the Sociology of "Developing Societies"* (New York: Monthly Review Press, 1982), pp. 289-307.
123. Fred Block, "Beyond Relative Autonomy: State Managers as Historical Subjects," in Ralph Miliband and John Saville (ed.), *The Socialist Register 1980* (London: Merlin Press, 1980), pp. 227-42.
124. James O'Connor, *The Fiscal Crisis of the State* (New York: St. Martin's Press, 1973), p.6.
125. Lindblom, p. 175.
126. O'Connor, pp. 41-42, and ch. 6.
127. See the notions of "effectiveness" and "legitimacy" in Seymour Martin Lipset, "Conditions of Stable Democracy," in Harry Eckstein and David E. Apter (eds.), *Comparative Politics: A Reader* (London: The Free Press of Glencoe, 1963), p. 208; and of "instrumental legitimacy" and "consummatory legitimacy" in David E. Apter, *The Politics of Modernization* (Chicago: University of Chicago Press, 1966), pp. 236-37.

128. Theda Skocpol, *States and Social Revolutions* (Cambridge: Cambridge University Press, 1979), p.31.
129. Ralph Miliband, "State Power and Class Interests", *New Left Review*, No. 138 (1983), pp. 57-68; emphasis in the original.
130. Block, *op. cit.*, and Fred Block, "The Ruling Class Does Not Rule: Notes on the Marxist Theory of the State," *Socialist Revolution*, VII, No. 3 (1977), pp. 6-28.
131. Stepan, p. 22.
132. *Ibid.*
133. Miliband, *Marxism*, p. 87.
134. Stepan, p. 25.
135. See Robert R. Alford and Roger Friedland, *Powers of Theory: Capitalism, the State and Democracy* (Cambridge, UK: Cambridge University Press, 1985), and Krasner, *Defending the National Interest.*.
136. Samuel P. Huntington, *Political Order in Changing Societies* (New Haven: Yale University Press, 1968), and Fred W. Riggs, "Bureaucrats and Political Development: A Paradoxical View," in Joseph L. LaPalombara (ed.), *Bureaucracy and Political Development* (Princeton: Princeton University Press, 1963), p.p. 120-67.
137. Hamza Alavi, "The State in Postcolonial Societies: Pakistan and Bangladesh," in Kathleen Gough and Hari P. Sharma (eds.). *Imperialism and Revolution in South Asia* (New York: Monthly Review Press, 1973), pp. 145-73; emphasis added.
138. Alavi, "State and Class," p. 301.
139. Riggs, p. 124.
140. David Collier (ed.), *The New Authoritarianism in Latin America* (Princeton: Princeton University Press, 1979), ch. 1.
141. Peter Evans, *Dependent Development: The Alliance of Multinational, State and Local Capital in Brazil* (Princeton: Princeton University Press, 1978).
142. Collier, pp. 7-8.
143. Poulantzas, *Political Power*, pp. 16-17.
144. Therborn, *What Does the Ruling Class Do*, pp. 34-35, 151, 161; emphasis added.
145. *Ibid.*, p. 144.
146. See Ian Roxborough, *Theories of Underdevelopment* (Atlantic Highlands, N.J.: Humanities Press, 1979), ch. 6.
147. Bharat Patankar and Gail Omvedt, "The Bourgeois State in Post-Colonial Formations," *Insurgent Sociologist*, IX, no. 4 (Spring 1980), pp. 23-38.
148. Raymond D. Duvall and John R. Freeman, "The State and Dependent Capitalism," *International Studies Quarterly*, XXV, no. 1 (March 1981), pp. 99-118.
149. *Ibid.*, p. 112. Note, on the other hand, the comment by Poulantzas that the petit-bourgeoisie is non-capitalist; see Poulantzas, "The Problem of the Capitalist State," p. 71.
150. Duvall and Freeman, p. 104.
151. *Ibid.*, p. 113; emphasis added. See also John R. Freeman, "State Entrepreneurship and Dependent Development," *American Journal of Political Science*, vol. 26. no. 1 (1982), 90-112.
152. Paul A. Baran, "A Morphology of Backwardness," reprinted from *The Political Economy of Growth* (1957), in Alavi and Shanin (eds.), pp. 195-204; see also his "On the Political Economy of Backwardness," *The Manchester School* (January 1952), reprinted in Charles K. Wilber (ed.), *The Political Economy of Development and Underdevelopment* (2nd ed.;

New York: Random House, 1979), pp. 91-102.
153. Alavi, "The State in Postcolonial Societies," pp. 145-73.
154. Joan Robinson, "Introduction," to Michal Kalecki, *Essays on Developing Economies* (Hassocks, Sussex: Harvester Press, 1976), p. 11.
155. The essay "Observations on Social and Economic Aspects of 'Intermediate Regimes'", ch. 4 in *ibid.* was originally published in Poland in 1964 and, with some additions, appeared in *Coexistence*, IV (1), 1967, pp. 1-5.
156. Poulantzas, *Political Power*, pp. 283, 286.
157. On numbers as a political resource, see Gerhard Lenski, *Power and Privilege: A Theory of Social Stratification* (New York: McGraw-Hill, 1966), pp. 84, 318.
158. See John H. Kautsky, *The Political Consequences of Modernization* (New York: John Wiley, 1972), ch. 4; and Barrington Moore, Jr., *Social Origins of Dictatorship and Democracy* (Boston: Beacon Press, 1966), pp. 353-78.
159. Robinson, p. 11.
160. Kautsky, *The Political Consequences of Modernization*, pp. 237-51. This work builds on his earlier long essay in John H. Kautsky (ed.), *Political Change in Underdeveloped Countries: Nationalism and Communism* (New York: John Wiley, 1962), pp. 3-119.
161. "State Capitalism and the Third World," in James Petras, *Critical Perspectives on Imperialism and Social Class in the Third World* (New York: Monthly Review Press, 1978), pp. 84 - 102.
162. Teodor Shanin, in Alavi and Shanin, p. 319.
163. See also Berch Berberoglu, "The Nature and Contradictions of State Capitalism in the Third World," *Social and Economic Studies*, vol. 28, no. 3 (June 1979), pp. 341-63.
164. Yash Ghai (ed.), *Law in the Political Economy of Public Enterprise: African Perspectives* (Uppsala: Scandinavian Institute of African Studies, 1977), p. 22; Issa G. Shivji, *Class Struggles in Tanzania* (London: Heinemann, 1976); and John S. Saul, "The State in Post-Colonial Societies: Tanzania," in Ralph Miliband and John Saville (eds.), *Socialist Register 1974* (London: The Merlin Press, 1974), pp. 349-72. See also Poulantzas, *Political Power*, p. 334, who says: "A good example is the case of the *state* bourgeoisie in certain developing countries: the bureaucracy may, through the state, establish a specific place for itself in the existing relations of production, or even in the not-yet-existing relations of production. But in that case it does not constitute a class by virtue of being the bureaucracy, but by virtue of being an effective class."
165. Bjorn Beckman, "Public Enterprise and State Capitalism," in Ghai (ed.), pp. 127-36, and Archie Mafeje, *Science, Ideology and Development: Three Essays on Development Theory* (Uppsala: Scandinavian Institute of African Studies, 1978), ch. 2.
166. Miliband, *Marxism*, pp. 108-109. Similarly, Saul states in respect of Eastern Africa: "Indeed, in the absence of any indigenous economically dominant class -- bourgeoisie or landed aristocracy -- anchored in the production process, the state can be said to have a particularly *central* role in the economy and society and a role that is not paralleled in more fully developed capitalist systems." See John S. Saul, *The State and Revolution in Eastern Africa* (New York: Monthly Review Press, 1979), p. 5. Note, too, the statement: "Class relations, at bottom, are determined by relations of power, not production." Richard L. Sklar, "The Nature of Class Domination in Africa," *Journal of Modern African Studies*, vol. 17, no. 4 (1979). 531-52.

Chapter II

Contending Approaches to the State and Public Sector

India occupies an ambiguous position in the Third World. It is the archetypal poor country with a per capita income of less than $300, ranking near the bottom of the list, lower than Pakistan, among the less developed countries. At the same time, it is a vast and diverse country, with some of its constituent units rivalling major powers in size. Notwithstanding the poverty, India is, moreover, by virtue of the size of its economy, industrially and technologically one of the more advanced countries of the Third World and has perhaps the most developed entrepreneurial class.

Apart from the contrast between its poverty and its large and advanced industrial sector, what makes the country a real enigma in the Third World is that, regardless of how the future may unfold, it has for some four decades sustained an open competitive political system with periodic elections, accompanied by substantial peaceful transfer of power in the states and even some at the centre. Few other Third World states compare with India in this combination of underdevelopment and political competition. The installation of a democratic political system and its operation for four decades has accompanied, if not preceded, the attempt at industrialization. This certainly represents a departure, if not a reversal, of the historical pattern of economic and political development in Europe. Under these circumstances, what is the relationship between state and society? Is the state autonomous from society and dominant over it, or is it a reflection of society? What mutual changes occur as a result of the interaction between state and society which flows out of periodic elections? More particularly, who controls state power, and what are the implications of this for economic policy, especially in relation to the public sector? Again, does the public sector serve the interests of particular social classes, or is it an instrument of class-transcendent national development?

These questions have been contentious in India, and the answers have varied bewilderingly, while individuals and groups providing answers have also changed their positions over time, partly in response to a changing reality and partly because of a changing perception of reality. The purpose of this chapter is not to set out still another interpretation of state power, but rather the more prosaic one of providing a near-comprehensive inventory (thus its length) of the more significant contending approaches to these questions. Here, four different approaches to the analysis of the Indian state are considered: (1) liberal-democratic; (2) orthodox Marxist; (3) bureaucratic; and (4) "intermediate regime".

1. The Liberal-Democratic Model : Elites and Masses

Although many scholars have used the structural-functional (S-F) approach in the analysis of Indian politics, they have done so in relation to limited parts of the political system for limited periods of time. Rajni Kothari, the dean of political scientists in India, is one of the few who has systematically examined the political system as a whole and has done so over a considerable period of time. A keen and sensitive intellect, with an abiding commitment to liberal-democratic values, Kothari rose to prominence among analysts of politics in India with a series of articles in 1961 and, more especially, with a highly original article on "the Congress System" in 1964. There then followed a period of active cooperation and collaboration with some of the most eminent political scientists of the time in the U.S. employing the S-F approach under the leadership of Gabriel Almond. This culminated in a major book, *Politics in India*, in 1970, which used the tools of the S-F school though somewhat modified to suit the analysis of Third World countries.[1] Soon, however, he turned away both from this school and its proponents, apparently dissatisfied with its tools and the intellectual problems they focused on.

S-F analysis looks at social reality from the perspective of the system as a whole, rather than from the viewpoint of groups within it, and it basically asks two questions: What are the functions that need to be performed to ensure maintenance and adaptation of the system? And, what are the structures that perform these functions, and how effectively? What the approach does is to provide a framework for analysis; there is no theory here, least of all a dynamic theory of change. Though not necessarily inherent in it, a bias in favour of stability or maintenance of the system has marked the approach. Apart from a tendency toward viewing political reality from the perspective of those who manage the system, the approach has tended to lead to concentration on political epiphenomena rather than relating it to its social and economic base. In *Politics in India*, which is one of the most original though complex and difficult books on Indian politics, taking a holistic view of its subject matter, Kothari supplemented the S-F approach with an emphasis on the autonomy of the political process, institution-building and system performance but this attempt was in no way inconsistent with the overall approach.

The Indian State as Autonomous, Benign and Democratic

Though there is no reference to Kautsky's work, central to Kothari's analysis was the emergence of a new political elite in India, which was not only distinct from economic and social elites but had autonomy in relation to them. Kothari forcefully underlined the autonomy of the political elite and the political process, saying that "in the case of India it will not do to look at political institutions as some kind of superstructure that presides over more basic relationships in society and economy, or to look at elites as simple recipients of inputs from society to which they respond....Instead the whole process starts here...." Or again, he averred: "India seems to us to be the clearest refutation of the reductionist viewpoint which takes politics and government as phenomena whose explanation must properly be sought in social and economic spheres....To no small degree, the state has become the arbiter of society."[2] The political elite was drawn from the new English-educated urban middle class that had sprung out of the colonial impact, but that social origin was incidental to its role, for class analysis was simply marginal to the S-F approach. Policy was seen to derive from the values of the elite rather than its social connections.

During the course of the nationalist movement which it led against colonial rule, the small, homogeneous, upper-caste political elite developed a broad normative consensus centering around the values of national independence, liberal democracy and socialism. This benign and enlightened elite, functioning within an ethical code, was "enormously innovative and creative" after it inherited political power; it became "the dynamic agent of change" and proceeded to push forward simultaneously, rather than sequentially, on several fronts, adopting: an institutional strategy that founded and fostered an open liberal-democratic system as the ordering mechanism of the national state; an integrative strategy of nation-building through consultation, accommodation and consensus; an economic strategy organized around central planning, rapid industrialization, an expanding public sector, moderate reform and "socialism"; and a foreign policy strategy based on national autonomy and non-alignment. Kothari discerned that the elite's focus during the first two decades had necessarily to be on institution-building and national integration but he thought and urged that, sequentially, the thrust of the next stage would be and should be performance in regard to equity and distribution for the sake of, if nothing else, stability of the system. The public sector did not surface as a particularly salient ideological or practical issue in Kothari's analysis of the Indian state. The assumption apparently was that it was part and parcel of the total package for nation-building on the part of the elite.

The establishment of the new political centre under the elite's auspices brought about the interpenetration and interaction of centre and periphery, of modernity and tradition. The keystone of the polity, as also of Kothari's analysis of it, was the "Congress system", where the multi-class Congress party -- as literally the political embodiment of the nation which it had in fact created -- provided extensive representation within itself of interests from across the entire spectrum of social and political diversity in the nation. Almost the entire political process took place within the Congress party, with the opposition parties, having no expectation of overthrowing it, acting as parties of pressure influencing the balance of forces inside it. The Congress party thus became the historic instrument of legitimacy for the national elite, of consensus-building and, very importantly, of national integration, which constituted its greatest achievement.

Kothari's *Politics in India* displayed considerable pride in the Indian model of development; he was impressed with the model's emphasis on political participation and consensus-building even as it pushed through an "incremental revolution." He dismissed those who engaged in "a romantic disenchantment with any 'Western-type' political system," while he marvelled at the ability of the national elite to assure stability of the system:

> Although most of the measures taken were aimed at securing social and economic development, in the process they neutralized some of the important

cleavages that could have developed into major sources of instability in the country. Labor's potential for trouble, for instance, was greatly reduced by the host of labor laws that were passed, as well as by the direct interest taken by Congressmen in trade union activity. In the same way, measures enacted to give concessions to the scheduled castes and tribes and the backward classes in matters of admission to educational institutions and employment, along with other measures to raise their social status, prevented bitterness against the high castes from taking an ugly form. The abolition of feudalism and the redistribution of land rights in the rural areas removed another important root of social cleavage.

Undoubtedly, Kothari identified himself with the goals and policies of the national elite and with the political system it had built. He underlined the sturdiness of the system in withstanding the impact of crises of several wars and droughts, and believed that a world role for India in international affairs was a justified and normal aspiration, as was the building of a nuclear and satellite capability. At times, he bent over backwards to see functional aspects in less desirable phenomena, such as crises and violent conflict. However, he thought that the elite had been too doctrinaire on heavy industry and cooperative farming, but was enthused about the new agricultural strategy, especially its reliance on the entrepreneurship of the new kulak class. Many of his reservations about the system, particularly concerning the impact that lack of performance in respect of welfare and distribution would have on its future stability and integrity, came at the end in a concluding chapter but were not integral to the analysis as a whole. He nonetheless urged a dialectical shift in priorities for the next phase from institution-building to substantive outputs. But he felt that "the tasks that face the elite are not really that insurmountable," or again that: "There is, of course, no need to exaggerate these problems. What is said above is relevant only if governmental performance is not adequately generated. The institutional foundations and experiential grounds for such performance -- as well as the relevant technological and intellectual infrastructures -- are all there."[3] There was no notion here that class interests may act as constraints in this regard. He believed that policy changes were already under way or would emerge in due course in response to political pressure to cope with emergent problems.

The S-F approach as it was employed by Kothari seemed eminently appropriate for the period that was under examination, that is, about the first two decades after independence. After all, during this period the outstanding feature of India's development was, indeed, the establishment of an institutional framework and there could be no question that this was the handiwork of a national elite that was largely autonomous, and so the attention to the top rather than the bottom was apparently justified. Equally, that period was one of enormous achievement politically and economically; no scholar could but be impressed by India's ability to sustain through many crises a liberal-democratic framework in one of the poorest Third World countries, and consequently make it the focus for examination; and Pangloss was a term that was applied to many students of Indian politics of the time. However, the preoccupation with these elements led to a neglect of other aspects in the situation which became more consequential with the passage of time.

For one thing, elite autonomy was rather exaggerated. A deeper reflection on several policy issues, such as land reform, the new agricultural strategy, states reorganization, and language policy, would have shown that the elite was not so autonomous; rather, entrenched interests placed severe constraints on policy-making. A greater sensitivity to society would have warned that even the adoption of the basic political framework was not entirely a matter of choice for the political elite or a function of its goals, but that the configuration of societal forces, at the time at least, made it a necessity. It is quite intriguing, in this context, that an enlightened national elite would adopt a democratic constitution but not engage in redistribution.

Equally, it would seem that Kothari's hope of redistributive performance as following almost naturally and sequentially -- after institution-building and national integration had been accomplished -- was misplaced. A premature installation of democracy prior to redistribution could not but help allow powerful interests to pre-empt the political arena and thus prevent redistribution. There could be no surer prescription for ending the autonomy of a reformist political elite than to have a democratic system. A neglect of this aspect flowed from taking a narrow and static view of the autonomy of the national elite, rather than seeing it in dynamic interaction with society under democracy. Even when Kothari discussed state-society interaction, it was more in terms of tradition and modernity, periphery and center, rather than class and class interests. One could also ask if the situational context of economic poverty, deprivation and disparity itself would not corrupt and subvert the entire process of democracy once society became fully mobilized, a process inherent in a democratic framework. In this connection, it is apparent that the fact that the "Congress system" itself could function as effectively as it did was a contingent result of (1) a largely unmobilized population, (2) the non cumulation of demands on the centre because of adjudicative mechanisms at the local level, and (3) an initial consensus on goals with which the national elite came to power. In other words, the Congress was a house without much of a foundation, and Kothari was aware of this. Furthermore, the extensive representation of interests within the Congress could not but lead to political and economic immobilisme which would then require drastic measures for policy breakthroughs.

More fundamentally, Kothari did not ask, and still does not ask, how a national elite can assure development and industrialization under a democratic framework. A basic task for Third World states is not simply governance of ongoing systems but economic transformation. However, development requires huge resources which in turn demand tremendous sacrifices from the population, while democracy requires consent from that same population. This is quite apart from the question of social and economic dislocations that are inherent in development. There is thus limited compatibility between the requirements of development and democracy. But Kothari's analysis was flawed to the extent it delinked institution-building from both social mobilization and requirements of development. There was therefore a gross underestimation of the structurally-rooted challenges of development in relation to both elite and mass, as there was of those of operating a democratic system in a situation of poverty and scarcity; the result could only be, as proved to be the case, a sense of betrayal by elites when there was erosion of political institutions under the impact of those very challenges.

Politics in India represented a celebration of the Indian model of political development just as that model was about to be launched on a major course change. The Congress party split in 1969 and the system was beset with major upheavals in the 1970s. This whole period saw Kothari become increasingly disenchanted with Mrs. Gandhi and her politics; he was critical of the Emergency and then turned into an active opponent of Mrs. Gandhi and an intellectual spokesman for the Janata party. With her return to power in 1980, Kothari became a vehemently bitter critic of the state under her. It could well be said that it was the political reality that had changed, but his basic approach changed as well. He moved away from cooperation with Almond and his colleagues to collaborate now with the more radical "world models" project under the leadership of Saul Mendlowitz, and became the founder-editor of the project's journal whose very name *Alternatives* signified a shift away from the reigning paradigm in development theory. He now repudiated the idea that his approach was liberal, or Marxist, or any other,[4] but in fact his radical attack[5] drew on liberalism, Marxism and Gandhism, the common intellectual mix in India but susceptible to assuming diverse combinations and manifestations.[6]

The Indian State as Collusive, Corrupt and Fascist

In the radical analysis of the early 1980s, Kothari no more held the national elite as autonomous but linked it to powerful segments of society. The Indian state, especially the public sector, was no longer an instrument for general development in the hands of a political leadership guided by an ethical code, but had been transformed under a malign and corrupt second-generation elite into "a source of munificence and plunder, indeed, sacrilege all around." There had, for him, occurred "a consolidation of entrenched interests, a growing exploitation for private ends of the hegemonical view of the State which was inherent in the 'socialist' doctrine (according to which the State was to be the main source of patronage and power), and a general tendency for the State to become an instrument of a narrow class (which was of course the opposite of all socialist pretensions)." There thus now existed "a close linkage between financial, bureaucratic and political elites,"[7] a "systematic and institutionalized relationship between the Indian state and the sources of big money"; indeed, the state had entered into "collusion with big business on the one hand and lumpen criminality on the other."[8] It should be noted that Kothari does not say that business controls the state, only that it has links or colludes with the state apparatus under political elites.

Along with these changes, the elites had brought about an erosion of democratic institutions and of national autonomy, which together threatened the very existence of the state and the nation. "It is an illusion," maintained Kothari, "to think that it is any longer a democracy." More critically, he added: "It was neither a functioning democracy nor a functioning dictatorship but a tottering state structure." He discerned that: "We are fast moving to the model of a totalitarian democracy based on a centralized and increasingly brutal state apparatus, backed by world capitalism on the one hand and sophisticated military hardware on the other."

In his view, Mrs. Gandhi had brought about an enormous institutional decay in the political system with legislatures reduced to noise-making bodies, the Congress party to a personal instrument, the federal structure to a subordinate arm of an all-powerful central government, and elections to a contest in "muscle power and organized terror," while "the politics of persuasion has given place to the politics of manipulation, coercion and intimidation." This was the result of her resort to a single-person-centered plebiscitary democracy, populist rhetoric, personality cult, and fascist ideology. In the end, "what is left, then, is not only a totally non-functioning and non-performing system but one that is rotten to the core." The destruction of the institutional corpus by Mrs. Gandhi threatened not just democracy but the Indian state itself, for the absence of local intermediary institutions now resulted in the accumulation of demands at the national level.

Kothari now showed a new sensitivity to the condition of the poor and the downtrodden. The transformation of the Indian state had been accompanied by "an unprecedented increase in repression in the countryside." In elaboration, he continued:

> The horrendous waves of criminal assault on the rural poor and the landless at the hands of the land-owning castes are increasingly being backed by the law and order machinery...a virtual breakdown of the state at the lower levels...an increasing loss of authority of the institutions of adjudication and enforcement of law and a growing tendency of bypassing these institutions....a fear of one's environment and of the state....Almost everywhere any organised effort to mobilize the poor is leading to a violent backlash from the state apparatus...the 'hard state' that India has already become...the normal political process has been grossly distorted and camouflaged so that while formally the parliament, an independent judiciary and a free press are still there -- and they do help in exposing fragments of the larger reality -- in effect we have already moved into a harsh and oppressive state structure.[9]

Besides, corruption, gangsterism and mafia rule had become pervasive throughout the Indian state. Criminal elements and toughs had now entered the political mainstream in the respectable garb of legislators and ministers.

What had brought about this change in the Indian state? Whereas earlier in examining the moderate Indian state Kothari had looked at the elite at the top and its policies, now he turned to the masses at the bottom for an explanation of the change that had occurred. The earlier state with democracy as its key feature had resulted, as was expected, through the very operation of a competitive system, in the generation of demands on the part of traditionally deprived communities for a share in wealth, status and power. This threatened the existing privileged upper-caste groups not only in the economy and society but also in the state.[10] However, Kothari does not provide evidence whether the demands were from traditionally deprived communities or from privileged communities whose expectations had risen immensely but whose economic interests were not advancing sufficiently because of economic stagnation. In any case, the country's "westernized elite" was incapable of fostering the requisite structural changes and adopting appropriate policies specifically oriented to the politically-conscious deprived communities. Those policies that it did adopt became instead "instruments of privilege and concentration of power rather than of equity and broad-based participation."

At fault in all of this was the elite's basic economic strategy itself which concentrated on "achieving overall growth rates through the laying out of a considerable infrastructure for development and the building of a modern sector." While the elite placed reliance on the trickle-down effect in respect of the deprived communities, "no systematic effort was made to ensure that this would in fact happen. Distributive justice was not built into the nation-building design and the development model." Kothari had not raised these questions earlier but now, in retrospect, he underlined:

> Such a model of economic development was persuasive for groups that stood to benefit from it and identified national prosperity with their own. These groups included not only the commercial, industrial, finance and managerial segments of the capitalist structure and the upper peasantry in the rural areas but also a very wide spectrum of lower middle classes which were accommodated through a vast expansion of the middle and lower rungs of the state apparatus.

Interestingly, this description of state power is not very different, except for the doctrinal terminology, from those offered by some Marxist parties in India.

The failure at performance in response to the demands of the deprived communities, Kothari held, was sought to be compensated instead by a politics of postures, "a purposely diffuse populist rhetoric aimed at the poor and the dispossessed, dramatic overtures to socialism (which boils down to nationalisation and state ownership) and an avid assertion of developmentalism as the principal raison d'etre -- in short, a new genre of statism according to which the fate of the socially deprived and the destitute rests securely in the hands of the state and a strong central authority." This is how the new centralized state with its malignant features had come into existence.

However, Kothari's attack on the Indian elite's development strategy reflected a wider disenchantment with the very notions of development and the state. As against the earlier positive orientation toward modernization and toward the state as an instrument of modernization and social justice, he was now critical of these elements because of their perverse effects and their origins. Modernization was now perceived as constituting a "threat to cultural identity and civilizational values of Asia and Africa," "the political challenge involved in such a cultural encounter is therefore total and calls for a comprehensive corrective to the ideological framework based on the doctrine of progress, modernization and statism which originated in the West a long time back and whose latest incarnation is the theory of development of developing countries." Furthermore, humanity was threatened by "the all-encompassing and totalistic impact of the modern state," which was considered to be inherent in the progressivist creed encompassing welfarism for developed countries and "developmentalism" for developing countries; the progressivist creed had turned the state into a powerful centralized technocratic instrument and "an agency for exploitation, control and subjugation — nationally and internationally." Indeed, this outcome had been inherent in the very idea and institution of the state itself, only it had been held in check in some countries "largely through the consolidation of democratic institutions."

If this stance was representative of the classical liberal position, so was the strategy for rescuing India, which is noteworthy for its call on the very classes at the top which seem to have been responsible for the country's predicament in the first place:

> Much will depend on how effectively those among the middle classes — the bourgeoisie — who feel committed to the values of a liberal democracy and a just social order will throw their weight behind the forces struggling for an alternative political order and in the process save the country from both internal atrophy and eventual disintegration. This has been the main proposition of the liberal intellectual tradition of India — a convergence of interests between a liberal elite and a democratic mass, each moderating the other and the two together ensuring the broadest possible consensus for democratic nation-building.[11]

Kothari's searing indictment of the Indian state under Mrs. Gandhi was really a testimony to his integrity as an intellectual and a liberal; he stood by his values even at the cost of personal sacrifice. But it would seem that his critique suffered, apart from its partisan tone, from several weaknesses. First, it tended to hold Mrs. Gandhi overly responsible personally for whatever dose of authoritarianism had been introduced in India's political system. When most of the Third World is under authoritarian regimes, including several countries in South Asia, it does not seem appropriate to look for causes in the personal styles and motives of leaders rather than in the structural conditions that characterize Third World countries. One would need to ask whether there are some special factors that make India particularly immune from authoritarianism and that therefore particular individual leaders have to be held personally responsible for its political fate. Kothari's animus against Mrs. Gandhi, even as he denied it, persuaded him to make light even of external threats to national security; indeed, he held her responsible for them, which showed rather poor understanding of how the superpowers manipulate India's strategic environment. Again, he held Mrs. Gandhi personally responsible for not understanding the minds of specific linguistic and religious groups without any recognition of how these very groups in the past gave her father an equally tough time in far more favourable circumstances.

Second, Kothari is still unable to appreciate the difficulties of elites to manage a democratic system and at the same time meet the challenges of the international system and economic transformation, and all this in the context of immense poverty. Furthermore, it is an odd criticism to make of "developmentalism" that it is of Western origin and then to press for liberalism and democracy as if they were not. Third, decentralization as a generalized prescription against centralization may be worse than the disease. It has an instinctive appeal because of Mahatma Gandhi's emphasis on it, but unless adopted in a deliberately measured manner it may simply lead to the tyranny of the privileged which tends to be more effective in smaller political units than larger ones. If the condition of the underprivileged is the heart of the issue then Kothari needs to carry the analysis to its logical conclusion and ask for a revolution, which is likely to bring into being not a moderate but a real totalitarian state.

Fourth, the contrast between the two phases of the Indian state is decidedly overdrawn, though the reality of the second phase is unpleasant enough. There is no doubt about the political change that has occurred, but does the change make the Indian situation the equivalent of that in, say, Pakistan and Bangladesh? If so, how does one explain the ability of parties to come to power against the opposition of Mrs. Gandhi and her party? If this is an illusion, opposition groups in those countries would rather be its possessor. Again, was the condition of the underprivileged in the face of local repression and ruthlessness that much better earlier? Or is it that their condition did not receive much public notice by the middle classes because the poor were not mobilized and their tensions had not overflowed into the urban areas? In another era these very developments may have been interpreted more positively as signs of progress in contrast to age-old stagnation. The neglect earlier of the condition of the underprivileged, flowing from an intellectual framework which concentrated attention on the elite at the top, makes their present situation look that much worse. However, more activist proponents of class-conflict paradigms were able to focus on the condition of deprived groups at a far earlier stage in independent India's development. But, functioning in a doctrinaire fashion, they were not able to secure the support of these very groups.

2. Orthodox Marxist Interpretations of the Indian State

A striking feature of Marxist analyses of the Indian state is that the field is dominated by formulations by practitioners of politics in the Communist parties than by scholars. The role that scholars seem to have played in Marxist analyses is more by way of rhetorical repetition, rationalization and elaboration of party formulations. This is not to say that there is not some reflective and empirical work on the state by scholars but that the field as such is dominated by party theoreticians; the role of scholars seems to be secondary.

However, this ought not to detract other scholars from paying serious attention to the work of party theoreticians. For one thing, Communist parties represent a commitment to the theoretical work of one of the foremost intellects of the modern era and there is an obsession with theory within such parties. Because the Marxian perspective integrates theory and practice, those intellectuals and scholars who are attracted to Marxist ideology enter or sympathize with Communist parties, and their work and thinking appears in the political tracts of these parties. Within the Communist parties, it is not enough to advocate a political position, rather it must be theoretically situated and cogently argued. It is striking that party theoreticians are experts at writing long tomes the size of Ph.D. dissertations, especially at those turning points when there is a shift in political strategy; there is, at that time, particular attention directed to the exegesis of inherited texts, the relation of these to the present epoch and situation, the definition of terms, the nuances of words, and the stating of alternative hypotheses. Evidence is marshalled for and against different hypotheses, though not always on an objective or impartial basis. All this is attended to more seriously than by scholars, even if in a doctrinaire and one-sided fashion. In the Indian case, arguments are made in excellent English, a testimony to the superior educational calibre of intellectuals at work in Communist parties.

One particularly superior aspect of party analyses over academic scholarly analyses stems from the existential requirement that too much violence should not be done to facts, because the cost of that can be disastrous for parties and party members. Analysis by and in the party must lead to action, and action can generate reaction from other political forces. Unlike the work of conventional scholars, the intellectual work of party theoreticians has to meet the test of practice. Theory can literally kill. Yet it is testimony to the strong hold of ideology that there is selective perception of reality, which indeed has resulted in particularly grave results for the party. There is also the further hazard that party formulations will be left purposely vague and broad in order to stave off reaction from hostile forces or to mobilize the largest possible support within the party as well as without.

Although considerable serious intellectual activity takes place in the Communist parties and by Marxist scholars on the question of state power, indeed it is at the heart of Marxism, yet there is no special tool that Marxism has discovered to divine the nature of state power. One has only to note the considerable diversity of opinion among the Marxists, the vehemence with which they attack each other, and the dramatic shifts in positions even when reality has not changed, to appreciate the absence of any special skills. The usual procedure in determining which classes hold state power is, within the general framework of assumptions and theory of Marxism, to stake out a position on the basis of a selective analysis of the policy outputs of the state. Not that much reliance is placed on the social background of those in charge of the state apparatus, perhaps because it is either not likely to support the position taken or because the communist parties themselves are vulnerable on this point, drawing as they do heavily on the urban middle classes for their leadership and their cadres, a situation which may not be very different from that of other parties. In party analyses of state power, there is nonetheless the basic assumption of the lack of relative political autonomy of the state; the state is taken to be acting in the interests of the dominant class or class coalition. If and when the social and political consequences predicted do not follow then one of two paths is likely to be taken: (1) change the position to correspond to the reality, or (2) save the hypothesis by stating that the reality is not what it seems, or that the hypothesis pertains to the long term.

One of the central propositions in the Marxist position on the Indian bourgeoisie is that it lacks the dynamism to bring about economic and social transformation, that it is incapable, that it is, in other words, a lumpen bourgeoisie. What the Indian bourgeoisie's capabilities are for the future transformation of India can only be a matter of conjecture, but strangely some scholars have commented on the lack of original Marxist thinking in India. In his work on the communist movement in West Bengal, Franda states that India has not produced any creative or original

Marxist theoretician.[12] Along the same lines, an eminent Marxist scholar, P.C. Joshi, uses the near-derisive term "pamphlet Marxism" to characterize the Marxist literature in India, and asks:

> Why is the intellectual failure of Indian Marxism so pronounced both before and after independence? Why is it that India failed to create outstanding Marxist thinkers, and a body of Marxist thought and theory suited to Indian conditions? Why is it that Indian Marxism has been more derivative than original, more theological than scientific, more assertive than receptive, and more negative than positive? To raise these questions is to draw attention to the new properties and characteristics alien to its original character which Marxism acquired under conditions of colonial and semi-feudal backwardness...The critical, activist and creative tendencies in Marxism have often been overwhelmed in India by conformist, fatalist and mechanistic revision to Marxism itself. Marxists became hostile to questioning and independent thinking which is so necessary for the construction of a *Marxist* perspective on Indian problems.[13]

One of the problems with Marxist thinking in India on India has been to fit it into the mould of classical Marxist theory. This may well be appropriate but it is ironic that Marx himself found that India, at least of the past, fell outside the universal schema that he had developed on the movement of world history. Largely basing himself on the views of Hegel and the reports of British administrators, Marx thought of India as an unchanging society, lacking any internal dynamic or dialectic for development, largely because of the lack of private property in land. His perception of a stagnant economic and social structure led him to posit a special societal category labelled as the Asiatic mode of production, which lay outside his otherwise universalistic framework.

Some Marxist scholars in India have done a really impressive job in showing how poor Marx's understanding of India was,[14] but the point is that whatever Marx's failings he did not automatically apply his universal schema to India, rather he invented a new category. There is thus no more inherent reason that India of today should fall in any of Marx's universal categories than it did in the past.

Illustrative of the problems that Marxist theory has had in coming to terms with Indian reality is the shifting position of the Comintern, from the early 1920s on until the achievement of Indian independence, on the role of the Indian bourgeoisie which allegedly led the nationalist movement or the Congress party. To be sure, the Comintern stance may have been influenced by the foreign policy and other interests of the Soviet Union or groups within it, but two generations of thinking, perhaps unthinking, Indian communists supported whatever line the Comintern advocated.

What is significant about the nationalist movement in India was the completely marginal role of the Indian communist party in that movement, and the question that was central for the Comintern was what posture to adopt towards the Congress party. Following its assessment of the general weakness of the communists in colonial countries, the Comintern in the early 1920s decided on their entry into "bourgeois" nationalist parties in order to strengthen support for themselves. This was done in India and China. But after 1927 -- when Chiang Kai-shek had in a rightward shift attacked the Chinese Communist party -- the Comintern and its Indian allies for nearly a decade repeatedly declared (without taking into account the specific Indian situation where no similar rightward shift had taken place) the Congress party to be "counter revolutionary" and its work as "treachery" and "betrayal"; they characterized Mahatma Gandhi as "an agent of imperialism", or as "a police agent of British imperialism in India."[15]

The "Draft Platform of Action" of the Communist Party of India in 1930 declared that "the greatest threat to the victory of the Indian revolution is the fact that great masses of our people still harbour illusions about the National Congress and have not realized that it represents a class organisation of the capitalists working against the fundamental interests of the toiling mass of our country." It felt that "the capitalist class and the National Congress, in their search for a compromise with imperialism are betraying the interest not only of the workers and peasants but also of wide sections of the town petty bourgeoisie."[16] There was no explanation of why the Communists had earlier expected it to behave otherwise. What is more the Comintern and local communists declared that the Indian bourgeoisie was not really interested in independence, for its interests were tied up with foreign capital and imperialism. Even if it were to be interested, it was thought to be impotent. As Indian communists declared in the Meerut conspiracy case: "The reactionary policy of imperialism in relation to industry....all go to determine that the policy of the bourgeois class must be one of hostility to imperialism....Nevertheless, we consider that the Indian bourgeoisie is not objectively capable of pursuing a revolutionary policy.... It is too weak and its interests are bound up too closely with both British imperialism and Indian feudalism."[17] Any activity on the part of the Congress party that contradicted this stance was simply dismissed, in Banaji's words, as "empty gestures, maneuvers of hypocrisy." Yet it was the Congress that was bearing the brunt of the brutal attacks of imperialism.

The result of the opposition to the nationalist movement headed by the Congress was, of course, the isolation of the Communists themselves. But after having attacked the Congress party for many years, the Communists, in a somersault in the mid-1930s, following the "popular front" strategy of the Comintern, began to see virtue in the Congress and now urged support for it. Communist spokesmen now recognized that: "Our comrades in India have suffered for a long time from left sectarian errors; they did not participate in all the mass demonstrations organized by the National Congress and organizations affiliated with it,"[18] or that the Congress represented "the united front of the Indian people in the national struggle.... It is even possible that the National Congress, by the further transformation of its organisation and programme, may become the form of realisation of the Anti-Imperialism People's Front."[19] Why this potentiality in an organization led by the bourgeoisie, allegedly tied to imperialism, was not obvious earlier remained unexplained. Still later even Gandhi was seen to have a "progressive role." Then the outbreak of World War II saw the Communists opposing it on the ground of it being an imperialist war. But soon after it became "the people's war", following the Soviet Union's entry into it. The Communists then launched on a policy of collaboration with British imperialism along with determined opposition to any strike action by workers even as the Congress party was being ruthlessly crushed by British imperialism and its leaders were thrown into jail for the duration of the war. During the war years the Communist party declared the Congress party to be a Hindu body and extended support to the Muslim League and its demand for the partition of India. For its part, the Congress regarded all this as treacherous and traitorous, and swore never to forget it.

Three Waves of Analyses of the Indian State

The years after independence proved no less difficult for the Communist party in terms of figuring out what class or set of classes held state power in India. Its position changed dramatically several times within the span of a few years.

a. 1947: The Joshi Line -- Reformist Proletarian Cooperation with the Anti-Imperialist State of the National Bourgeoisie: The war years had seen the Communist party acquire an anti-nationalist character in the eyes of most politically aware Indians, and the party was isolated from the nationalist movement. As independence approached, the party had to come to terms with this isolation. In September 1946 the Congress party had been inducted into power as an interim arrangement, with Nehru as de facto head of the government. On June 3, 1947, the British government and the key Indian parties came to an agreement, known as the Mountbatten plan, to partition the country into the two states of India and Pakistan. The critical question that confronted the Communist Party of India (CPI) was as to how genuine this independence was and what class or class coalition did the new state represent.

In the same month, the party's Central Committee under the leadership of Secretary-General P.C. Joshi passed a resolution, often referred to as the Mountbatten resolution, in which it considered the agreement to represent both a compromise on the part of the nationalist leadership with British imperialism and at the same time a retreat imposed on imperialism by the popular national revolt on the part of the Indian masses. The resolution recognized that a genuine transfer of power was taking place to a new independent state, and that the plan, even though it did not concede complete independence, provided "important concessions" and "new opportunities for national advance."[20] However, it warned that, through its exploitation of various economic and political difficulties and reactionary forces such as the princes and the feudal classes and of its control of the economy, imperialism was intent on reducing the independence to being merely formal.

The resolution's special significance lay in the fact that it marked another departure in the party's posture toward the Congress. The CPI did not consider the new government to be an imperialist government or a satellite government on behalf of imperialism but a national independent government which nonetheless was confronted by an imperialist conspiracy to undermine it. Beyond that, in fact, it believed the new government to be a potent weapon for crushing the imperialist conspiracy. Basically, its position was that the political change constituted an advance for the Indian people, a concession to the nationalist movement, and a retreat by imperialism, but it did not mean complete independence.

Apparently, in an effort to politically disarm the nationalist critics of its anti-nationalist past, the CPI resolution extended support to the new government which was believed to represent the interests of the national bourgeoisie.[21] The party characterized the Congress as "the main national democratic organization" and pledged to "fully cooperate with the national leadership in the proud task of building the Indian Republic on democratic foundations".[22] In effect, the resolution sought to discourage any anti-government struggle, for it believed the main task to be not anti-bourgeois but anti-imperialist and anti-feudal, precisely the agenda attributed to the national bourgeoisie, allegedly represented by the new government. This non-class national posture would have led CPI on a course of either collaboration with the Congress, or becoming a democratic opposition. In many ways the CPI leadership of the time could be taken to be highly astute and uncannily prescient about the changed circumstances in deciding on its strategy, for CPI would have to return to essentially this path after a disastrous attempt at making a revolution. But for now, it was a path difficult to stomach for a group ideologically hardened as revolutionaries.

b. *1948-1950: The "Political Thesis" of B.T. Ranadive-Proletarian Revolution Against the Collaborationist State of the National Bourgeoisie (the "Russian model")*: The CPI could not sustain the stance of democratic opposition for long. The attack against it was led by B.T. Ranadive who called for a militant revolutionary strategy. Six months after the Mountbatten resolution, the party's Central Committee revolted in December 1947 to accept Ranadive's position; his faction's victory was consolidated at the Second Party Congress in February 1948, where Joshi was dismissed from office and his policies labeled reformist-deviationist and revisionist. The party Congress adopted the new militant line in a massive 118-page document titled "Political Thesis,"[23] which marked a dramatic shift from the analysis of the state in the Mountbatten resolution.

According to this "political thesis," the new state was not independent, rather it was a satellite state in which the collaborationist bourgeois class shared power as a junior partner. The document dealt at great length as to how this had come about. Confronted by a menacing revolutionary wave, which threatened both imperialism and the Indian bourgeoisie, "imperialism struck a deal with the bourgeoisie and proclaimed it as independence and freedom." In other words, what had been given to the nation was "not real but fake independence," so that in effect "Britain's domination has not ended, but the form of domination has changed." What Great Britain had done was to give the bourgeoisie, for long denied a share in state-power and desperately eager for it, "an important share of state-power, subservient to itself". This grant of the position of "a permanent junior partner in operating the state" had been done with the aim of installing "a reactionary government of vested interests in power which, while protecting the imperialist order, would screen imperialist designs," and "henceforth the bourgeoisie will guard the colonial order." This new arrangement by imperialism was based "on a new class -- the national bourgeoisie, whose leaders had placed themselves at the head of the national movement and who were immensely useful in beating down the revolutionary wave." [24]

The "political thesis" elaborated the relationship among bourgeoisie, state and imperialism. The national leaders, including Nehru and Patel, who now headed the government, represented "the class interests of the national bourgeoisie, the industrial bourgeoisie," "the interests of the capitalist class." But the state that the bourgeoisie had won was "dependent on imperialism and is a satellite state," where "the national bourgeoisie shares power with imperialism, with the latter still dominant." While earlier the bourgeoisie had depended on the masses, now it was considered to have entered into a collaboration with imperialism. The "thesis" rejected any attempt to distinguish between Nehru and Patel, stating "it must be clearly understood that Nehru is as much a representative of the bourgeoisie as Patel is." Further, it added:

> In fact all shades of difference within the bourgeois camp (such as those between Nehru and Patel) are entirely subordinated to the new basic realignment of the class as a whole, viz its role of collaboration with imperialism. Both Nehru and Patel represent this collaborationist class, and all differences between them are being and will be solved within the fundamental framework of the collaborationist policy of that class as a whole. [25]

The "political thesis" considered the national bourgeoisie to be the leading force of "the imperialist-bourgeois-feudal combine," that is, its most active partner, but it failed to make any differentiation within it.

The emergence to state power of the national bourgeoisie was detrimental to the cause of revolution, in the eyes of the "political thesis". The collaboration with imperialism not only revealed "the narrow and antinational character of its intentions" but it also "means continuation of feudal exploitation, low wages, no industrial revolution, but continued poverty, unemployment, crisis and famine-- the price of tying India to the capitalist order, of collaboration and joint exploitation. That is where the Indian bourgeoisie, and the national leadership which represents it, are taking India -- to economic dependence on Anglo-America, subservience to them and growing poverty for the people." The collaboration was "directed against the agrarian revolution, against the nationalisation of industries, a living wage and planning, and against the widespread industrial expansion which can only be realised on the basis of nationalisation." It was allegedly the fear of agrarian revolution that had made the bourgeoisie depend on the imperialist-colonial order, but the thesis held that on this basis "not only expansion of industries is not possible but even retention of the present production level is becoming impossible." [26]

Not only economic policy but also "the foreign policy of the government follows the class interests it represents," according to the "political thesis". Of course, over time, many different kinds of foreign policy lines have been attributed to the class interests of the bourgeoisie, but for now the thesis, asserting that "there can be no neutrality in the world struggle," took the position: "From the very beginning Pandit Nehru adopted a line of forming a socalled third bloc -- a line which represents the interests of big business inasmuch as it kept India away from the democratic camp and opened the way to the imperialist camp." For it, the Indian leadership, as an ally of imperialism for the purpose of "crushing the Indian revolution", "hides its subservience to the Anglo-American bloc in world politics under the cover of 'neutrality' between opposing camps, of frank opportunism to realise Indian bourgeois interests." It saw the Indian bourgeoisie as playing "the role of chief agent of the imperialists" in allegedly forming an anti-Communist bloc in South Asia. [27]

The party sensed that there existed a massive "revolutionary upsurge" against the "imperialist-bourgeois-feudal combine" and decided on a strategy of proletarian revolution through a new alliance from below of "the working class, the peasantry and the progressive intelligentsia" but under the hegemony of the working class. The underlying assumption was that the Indian situation corresponded to the Russian pattern, with Nehru standing for Kerensky and August 15 for the February revolution; consequently, further assuming that an insurrection was already maturing in India, the next logical step was to push for a revolution in which the democratic and socialist stages would be intertwined. However, the party's adventurist thrust at urban insurrection failed, not only because of the strong reaction of the coercive agencies but basically because, even besieged by overwhelming problems, the state commanded legitimacy. The Communist leaders gave calls for strikes and clashes with police and army but, as party historians noted in retrospect, "there was nobody to hear them, and the working class just turned its back thinking that the leaders had gone delirious." [28]

c. 1950-1951: Rajeshwar Rao and the "Andhra Thesis" Peasant - Revolution Against the State of "Imperialist-Bigbusiness-Feudal" Combine (the "Chinese model"): With the strategy following the Russian model in ruins -- later therefore to be characterized as "left sectarianism" -- Ranadive came under severe attack, not from the supporters of the moderate line advocated earlier by Joshi, but from a group in Andhra Pradesh (where an agrarian armed struggle was being successfully waged in Telengana) under the leadership of Rajeshwar Rao, who advocated the Chinese model. The basic assumption now was of Nehru standing for Chiang Kai-shek and August 15 for 1927 and that the heart of the revolutionary struggle lay in the countryside, and therefore the new revolutionary strategy was to be a protracted guerrilla war based in rural areas bringing about liberation of cities and then establishing a new democracy. In a document known as the "Andhra thesis", it was stated: "India like China is semi-colonial and semi-feudal in character.... The present stage of our revolution essentially, though not exactly, is similar to that of the present stage of Chinese revolution, the stage that opened since 1927 bourgeois offensive against communists and working class." Though Ranadive came under attack from the Andhra group from June 1948 onwards, he hung on to power until May 1950 when he was overthrown and replaced by Rao as Secretary-General. The new line advocated by the Andhra group became official party policy at the beginning of June 1950. [29]

The "Andhra thesis", as the new line was referred to, represented a continuity in some respects with the earlier line of revolution, but in certain others it was a departure, providing a more elaborated view of class, emphasizing the importance of differentiation within classes, and attributing distinctive patterns of behaviour to the differentiated strata. The Andhra thesis attacked the earlier line for assuming not only that the Indian economy was already a capitalist economy but an independent one at that, a position that had led to a faulty revolutionary strategy, for "it obliterated the differentiation between the revolutions in colonial and semi-colonial countries and in independent, capitalist, imperialist countries -- differentiation which is an accepted Marxist dictum." [30] For the Andhra thesis, on the other hand, India was a semi-colonial, semi-feudal society.

The Andhra group further attacked the earlier revolutionary line for its failure to see the differentiation within classes. More specifically, it attacked the failure to differentiate the big bourgeoisie from the middle bourgeoisie and to take adequate account of the comprador character of the collaborationist big bourgeoisie. Similarly, it attacked the earlier line for lumping the rich peasantry as an enemy of the revolution, taking it to be functioning in a capitalist economy. From these theoretical assumptions, the Andhra group articulated a different view of the class character of the state. It was now a state of imperialists, big bourgeoisie (rather than bourgeoisie as a whole), feudal princes and landlords. The principal form of revolutionary struggle for its overthrow and establishment of new democracy was therefore armed struggle. The revolutionary struggle would be based on the widest possible united national front from below, centered on the proletariat, the poor peasantry and agricultural laborers, who will constitute the immediate reserves of the revolutionary struggle, but with middle peasants and urban petty bourgeoisie as allies, while middle bourgeoisie and rich peasantry, who are likely to vacillate, would be neutralized. [31]

The Andhra group's attempt to enact a rural revolution against the "imperialist-bigbusiness-feudal combine" was no more successful than the earlier line of urban insurrection. Opposition arose then within the Central Committee which was reflected in the "Three P's Document," chastising the Andhra leadership for simply continuing the earlier discredited revolutionary strategy, for bringing about the isolation of the working class, and for leading the party to liquidation. The opposition attacked the Andhra group for living in a world of unrealism and make-believe, for not correctly studying the Indian situation, for making a false analogy between India and China, for finding short-cuts to revolution, for carrying illusions about the revolutionary potential of the peasantry, for exaggerating the strength of the Communist party, for neglecting the role of the working class, and for underestimating the strength of the Indian state, the prestige of the Congress party and the astuteness of the Indian bourgeoisie. [32] With the party decimated because of its adventurism, Rao resigned as Secretary-General and was in due course replaced by Ajoy Ghosh.

Adaptation, Reinterpretation of State Power, and Party Break-up

The failure of the attempt to stage a revolution discredited the revolutionary line inaugurated by the Second Party Congress in 1948, and the party had to come to terms with Indian realities rather than plans based on wishful thinking. An active part in this readjustment was played by the foreign policy requirements of Moscow. But for now the party was too seriously divided to come to a positive agreement on its future course; however, it developed in 1951 a "Programme" and a "Statement of Policy", as a basis for building unity within the party.[33] These documents were not terribly different from the positions articulated in the preceding phase of the Andhra thesis. They described the government as a "government of landlords and princes and the reactionary big bourgeoisie, collaborating with the British imperialists," and "tied to the chariot-wheels of British capital"; it was considered both antidemocratic and unpopular, thus making it opportune to remove it. Subservient to British imperialism, the government was held to be, in the words of a party historian, both "incapable and unwilling to take up real industrialisation, incapable of carrying land reforms."[34] Even as late as 1955, the CPI maintained that: "Allied with landlords and compromising with imperialism, the Indian bourgeoisie can not complete the bourgeois democratic tasks that our country has to fulfil in the present state."[35]

Even though the Cominform was urging a favourable view of Nehru's policy of peace and nonalignment, CPI found it hard to stomach the idea that a government headed by the big bourgeoisie could do anything other than be a lackey and stooge of imperialism; it condemned the government for carrying out essentially "the foreign policy of British imperialism," for "flirting with the USA" thus "facilitating the struggle of aggressors against peace loving countries," while describing the foreign policy of nonalignment as a "spurious" and "suspicious" play between the two camps.[36] The new Programme and Policy Statement projected the character of the revolution as anti-imperialist and anti-feudal, leading to the establishment of a government of people's democracy to replace the Nehru government, while it ruled out as premature at the present stage the establishment of socialism, which was now to be achieved over a long and indeterminate period in the future. The strategy for the achievement of people's democracy was to focus on the formation of a People's Democratic Front comprising the four-class bloc of the proletariat, the peasantry, the middle classes and the national bourgeoisie rather than the earlier three-class bloc of the former three.

In these respects the new line was not much of a departure from the Andhra thesis or even from the Second Party Congress thesis except for the differentiation of the big bourgeoisie from the national bourgeoisie as a whole. Its novelty rather lay in its putting aside the controversy over whether the ideal model for India was the Russian path with the general strike of industrial workers as its chief weapon or the Chinese path with peasant guerrilla warfare as the key weapon, and also in its commitment to working out "a path of Leninism applied to Indian conditions." "After long discussions, running for several months," declared the Statement of Policy, "the party has now arrived at a new understanding of the correct path for attaining the freedom of the country and the happiness of the people, a path which we do not and cannot name as either Russian or Chinese." It sharply underlined the differences in the Indian situation from that of China's in respect of India's superior communications system and larger working class. The statement turned the party to the task of building a mass movement which would make full use of India's democratic system, and called on it to fight parliamentary elections. In that context, its sobriety in evaluating the Indian situation was remarkable: "But it would be gross exaggeration to say that the country is already on the eve of armed insurrection or revolution, or that civil war is already raging in the country. If we were to read the situation so wrongly, it would lead us into adventurism and giving slogans to the masses out of keeping with the degree of their understanding and consciousness and their preparedness and the Government's isolation. Such slogans would isolate us from the people and hand over the masses to reformist disruptors."[37]

However, once launched on this course, even the party's analysis of state power and the alliance to be ranged against it was also fated to undergo changes, with disruptive results for the organization as a whole. But these changes did not come as a sudden development, rather the party was led "step by step" to them over a decade. What is impressive is that class analysis did not help the party to predict how the particular class or class coalition would act in government, but rather the policy behaviour of the government led the party to keep revising its assessment of the class or class coalition holding state power.

The Draft Programme and the Statement of Policy were adopted, after considerable debate, at the Third Party Congress at Madurai (1953). But no sooner had this been done than there occurred a tilt in India's foreign policy toward the Sino-Soviet bloc, inspired chiefly by the U.S. policy of arming Pakistan against India. At the same time, the Congress party proclaimed as its goal the achievement of a "socialistic pattern of society" and determined to push forward with the Second Five Year Plan envisaging rapid industrialization, with special emphasis on heavy industry.

The year 1955 was a turning point. In that year, some CPI leaders went to the extent of suggesting a policy of collaboration with the Nehru government in the cause of peace and freedom as well as national independence, [38a] while the Soviet Union urged a more cooperative attitude on CPI's part toward the Nehru government. A tortuous and painful debate developed within CPI, which now found that: "the understanding of the Programme was at crass variance with the reality. We began to change our understanding step by step, pragmatically and empirically. Some of the positions of the Programme regarding the role of the national bourgeoisie and its leadership in India were incorrect even at the time they were made." [39] The process of revision was not easy, for earlier formulations had "for some comrades acquired the sanctity of dogmas which could not be violated in any case. Any departure from these formulations was looked upon as repudiation of Marxism." [40] The process was painful, for

> this meant certain old positions which we held axiomatic had to be given up because they no more corresponded to reality; new paths and slogans of struggle corresponding with those new positions had to be found to move forward. There was a resistance to move out of old outmoded positions. There was often a lag in finding new paths and slogans of struggle corresponding to these new positions. There were charges and countercharges of dogmatism and revisionism.

Critical to revisions in the assessment of the Indian state and associated matters was the notion of the world situation. For the CPI, the changed world situation in the epoch of the crisis of colonialism -- with the socialist system apparently its determining force -- now opened up the possibility, even for states headed by the national bourgeoisie, both for consolidation of political independence and for advance towards economic independence. As CPI Secretary-General Ghosh acknowledged: "Such things were inconceivable in the past, but they are happening today." These new-found possibilities "meant giving up certain old theoretical positions which no longer corresponded with practice and reality.... But making this basic shift meant a break with certain sectarian positions of the past."[41] These changes in the party's theoretical position came after the Third Party Congress in Madurai (1953) and were ratified at the Fourth Party Congress at Palghat (1956) and Fifth Party Congress at Amritsar (1958).

The party now returned, in effect, to the assumption held under Joshi that there had, indeed, occurred in 1947 a transfer of power to the national bourgeoisie which, even though it had compromised, intended to use state power to overthrow the colonial order, not to preserve it. A similar re-evaluation occurred in relation to the class character of the state, but it is interesting to note how this re-evaluation followed, rather than precede, changes in state policy:

> When at Madurai we began to see that the foreign policy was becoming one of peace and later in 1954-55 the trend got confirmed and further when in the context of the formulation of the Second Five-Year Plan progressive economic policy began to be formulated, then we began to re-examine the class character of the government.[42]

The new assessment of the state's class character was noteworthy for its exclusion of the critical assumption that had been a consistent characteristic of all party formulations since 1948 -- collaboration of the bourgeoisie with imperialism, with its equally consistent and explicit logical implication of lack of capacity to consolidate independence, to pursue an independent foreign policy, to conceive and implement a programme of industrialization and agrarian reform. Not only that, the new assessment endowed the bourgeoisie with a positive urge for independence. It vested state power now in the national bourgeoisie as a whole, not just in big business, and stood it in contradiction to imperialism.

In line with the new assessment, CPI Secretary-General Ajoy Ghosh remarked: "No *section* of the bourgeoisie could be said to have *gone over* to imperialism though individuals might have. Contradiction between imperialism and the bourgeoisie as a whole remained....in that sense no section of the Indian bourgeoisie is interested in the *preservation* of the colonial order."[43] The report of the Central Committee to the party Congress at Palghat (1956) also noted: "This period has seen an immense growth of the forces of peace, freedom and democracy in our country. It has seen increasing assertion of independence by India and the sharpening of the conflict between the entire Indian people, including the government, on the one hand and the imperialist camp on the other. It has seen increasing assertion of freedom by India...the foreign policy of the Indian Government has steadily undergone a radical change -- a change of far-reaching significance.... Today, despite the vacillations and inconsistencies that still persist to some extent, it is essentially an independent policy, a policy of peace." [44] CPI's conception of the class character of the government also underwent change formally. No longer was the big or monopoly bourgeoisie the leading force, nor was it collaborating with imperialism. The political resolution adopted at the historic Palghat Congress (1956) stated simply: "It is a bourgeois-landlord government in which the bourgeoisie is the leading force. Its policies are motivated by the desire to develop India along independent capitalist lines." [45] More specifically, it noted that the state under the Congress party was actively attempting to reduce the role of British capital and feudalism in the Indian economy and to advance the public sector, policies which led to conflict with both imperialism and feudalism.

An interesting conceptual innovation, of Leninist origin but now coming secondhand through Mao, was employed as an explanatory device for the bourgeoisie's behaviour. This was the "dual role, duality or dualism of the national bourgeoisie, which struggled against as well as compromised with imperialism, in order to strengthen its class position both against imperialism abroad and against popular forces at home. [46] Since then, this concept has become a central explanatory mechanism in Marxist rhetoric about state policies in India as invariably strengthening the position of the bourgeoisie even when they seem to run counter to its interests. But in essence it is no explanatory device at all, for it is by definition always true; if the bourgeois state does one thing it is in its interests because it is directed against imperialism, but if it does the opposite that too is in its interest because it is directed against the toiling sections of the masses. In reality, the concept is simply descriptive, for the explanatory aspect -- that of serving the interests of the bourgeoisie -- never changes. But this did not prevent Marxists in India from supporting certain state policies, ostensibly because they ran against imperialism but in reality because they constrained the position of the bourgeoisie. The precise posture of Communist groups in India in this regard was determined by their perception of what would facilitate their own accession to power.

One of the issues in the reassessment by CPI of the class character of the state was the appraisal of economic planning and development. At Madurai in 1953, CPI had taken the position that the First Five Year Plan would result in no change but would eventuate in a series of crises, and that already the country was in the midst of "a maturing economic crisis and the initial stages of a political crisis"; this was a consequence to be expected, given the party's perception that the state was controlled by imperialists and landlords. However, at Palghat in 1956 the CPI came, in the words of an authoritative spokesman, "to the realistic appraisal that the government was pursuing policies of developing India's economic independence on capitalist lines",[47] and accordingly supported its plans for industrialisation and land reforms. No doubt, it was at the same time critical of the capitalist path of development since it would impose sacrifices on the population, foster the growth of monopolies, and in general result in conflicts and contradictions. But what was noteworthy was its positive appreciation of state policy in respect of economic transformation albeit capitalist. Even earlier in 1955, in a resolution, the Central Committee had concluded that the proposals for the Second Five Year Plan "if implemented, would reduce the dependence of India on foreign countries in respect of capital goods, strengthen the relative position of industry inside India and strengthen our economic position and national independence. The Party therefore supports these proposals, and also the proposal that these industries should be mainly developed in the public sector."[48] And in a report in 1956 to the Palghat Congress, the Central Committee underlined how what the bourgeoisie was now doing was based on national rather than class interests and therefore provided a basis for unity with the bourgeoisie:

> Today the aim of industrialization, of defence of peace, of independence and strengthening of freedom is the common task before the people. These are national tasks which constitute the basis of unity with the bourgeoisie..... These common features of our tasks in the past and in the present are not accidental. They arise from the fact that the democratic revolution has yet to be completed. They arise from the fact that our tasks, at the present stage, are national tasks.

More fully, the Central Committee explained:

> The Government of India is bourgeois-landlord Government in which the bourgeoisie is the leading force. Its policies are motivated by the desire to develop India along independent capitalist lines. The Government today defends the freedom against imperialist pressure, opposes the drive towards war and builds friendly relations with Socialist States. The Government strives to weaken the position of British capital in our economy. It strives to control and gradually eliminate feudal forms of exploitation, transforming feudal landlords into capitalist landlords and create a stratum of rich peasantry that can act as the social base of bourgeois rule in the countryside. It strives to extend and develop the State sector which in the existing situation is essential for the development of capitalism itself. These aims and the measures resulting therefrom inevitably bring the Government into conflict with imperialism, with feudalism and sometimes with the narrow interests of sections of the bourgeoisie as was seen in the nationalization of life insurance....
> The Communist Party is vitally interested in such developments and strives to strengthen them, for they help in strengthening the democratic movement and in strengthening and extending the democratic front. Every step that is taken by the Government for strengthening national freedom and national economy, against imperialists, feudal and monopoly interests, will receive our most energetic and unstinted support.

Of course, unity was not the only thing CPI stressed, rather it suggested a policy of simultaneous unity and struggle: "Unity for the aim of defence and strengthening of national freedom and for support to all measures that achieve this, even to a partial extent. Struggle against policies and methods that hamper these tasks and prevent rapid national advance."[49]

By the time of the Amritsar Congress (1958), however, CPI was in a dilemma as to the extent of support it should extend to government planning. The reason for this was the differential emphasis by one group on the rightward shift in government policies and by another on the rise of "extreme reactionary forces" outside which attacked economic planning and insisted on reducing the role of the public sector, opposing agrarian reform and adopting an open door policy for U.S. investment. Caught between "on the one hand, a national bourgeois government pursuing in general a policy of non-alignment, of economic independence and of ensuring some democracy, while pursuing anti-people and anti-democratic methods, developing capitalism and facilitating the growth of reaction; and on the other hand we have the rise of reactionary trend which is seeking to subvert the national policies and the democracy that exists," the CPI decided on the formula of "simultaneous struggle" against both. [50]

However, differences sharpened at the Sixth Party Congress at Vijayawada (1961), especially under the impact of the gathering crisis in Sino-Indian relations. For the time being, they were papered over by a unanimously-accepted resolution on the establishment of a state of the "national democratic front" as a route to power rather than simply opposing the existing government. The class composition and the programme of this state were the same as that of the state of "people's democracy" but in it "the leadership is shared between the national bourgeoisie and the proletariat." This proposition was based on the assumption of a growing differentiation in the national bourgeoisie -- "between the reactionary monopoly sections of the bourgeoisie and other feudal anti-national groups and elements on the one hand and patriotic national bourgeoisie on the other." The idea then was to detach, through a policy of unity and struggle, the patriotic progressive wing from the ruling class of the bourgeoisie, and to attach it to a wider front headed by the proletariat, in order to complete the tasks of the democratic stage. More precisely, the purpose was to isolate and defeat the reactionary forces attempting to subvert the "national policies -- of non-alignment, growth of public sector, radical agrarian reforms and democracy" -- and to "consistently implement national policies and advance forward to the non-capitalist path." [51] It is interesting to note the references to "national" policies without any explanation as to how they came to be adopted by a bourgeois government; only a liberal-pluralist model of the state, not a Marxist model, could provide an explanation here. At times, it is proclaimed by party spokesmen that these "national" policies have been adopted in response to the struggles of the masses, which can only imply that the masses have a share in state power rather than that state power is exclusively held by the dominant economic classes.

However, the idea of the "national democratic front" was unacceptable to many who characterized it and its advocates as revisionist, declaring that its result will be tailism in relation to the bourgeoisie; they, in turn, were met with cries of dogmatism. A central issue in this conflict between the "revisionists" who constituted the majority, and the "dogmatists", who were in the minority, related to the public sector. Their dispute over the public sector has special intellectual significance in that very contrasting hypotheses or conclusions are drawn from the same reality by leaders owing allegiance to the same larger ideological framework. The majority took the position at Vijayawada (1961) that the public sector, even though built by a state under the control of the bourgeoisie, was an instrument of the national bourgeoisie directed at undercutting the attempt of the monopolist section to acquire complete control of the economy and the state:

> The turn towards industrialization, the creation of the state sector, certain limitations on the activities of foreign private capital in key branches of industry, measures to channel and control investments of private capital -- all these marked a change, signifying the failure of the bid which monopolistic sections of the bourgeoisie were making to subordinate the newly-won independence and the anti-imperialist, anti-colonial national bourgeois state completely to their own narrow interests and to place the entire economy of the country under their private control.

Party theoretician Adhikari summarized the thrust of this paragraph: "All these measures taken together foiled the bid of the monopolists to *completely* subordinate the state to their narrow interests." Further, he rejected, by implication, the thesis of "lack of will" on the part of this section, pointing out that "Indian monopolists headed by Tata and Birla had produced another plan in 1944 in which they had argued that the state should leave the main manufacturing industries, including heavy and basic industries, to the private enterprise and concentrate on power, transport, communication and defence industries only." He also argued, quite rightly, that government efforts to obtain aid from Western countries for setting up an iron and steel industry in the public sector "met with a refusal from the imperialists. They were prepared to give aid to set up these industries in the private sector." He further underlined the Vijayawada resolution which stated, in part: "the public sector has grown despite every effort by imperialist and certain monopolist circles inside our country to thwart the growth." Adhikari reiterated: "In the form it arose the state sector was opposed by imperialists and by a section of monopolists."

Equally, Adhikari rejected the thesis of "incapability" of the Indian bourgeoisie to erect the major industries that were built in the public sector, stating: "Indian monopolists had the resources to put up additional iron and steel plant and the foreign monopoly capital was prepared to help them. This was their ambition as announced in the Tata-Birla Plan of 1944. If this were allowed to be realised they would have been in a position to get a grip over the government. This was prevented for the time being by the measures taken by the government." To be sure, the majority maintained that "the state sector also develops state capitalism in as much as this state is an organ of the class rule of the national bourgeoisie," but still it endorsed it:

> However, since this developing economy especially in the state sector, facilitates India's march towards economic independence and has an anti-imperialist and anti-colonial aspect it fulfills a national purpose and as such is progressive.

At the same time, the majority felt that the state sector facilitated the development of the private sector, even strengthened the monopoly sector, which returned with a renewed attack on the public sector. But its answer was to foster rapid expansion of the state sector so that it would "become the dominant sector of our national economy in every possible manner" and to "extend the sphere of nationalization to cover banking and general insurance, coal and other mining, oil distribution and plantations." In other words, its remedy was: more of the same.

The position of the minority was that the majority had essentially adopted a non-class approach, the end result of which would be to tail behind the national bourgeoisie, that it tended to exaggerate the threat of reactionary forces whereas the threat was lodged within the state itself, which represented and served the monopolists, and that its stress on the progressive significance of the public sector was uncalled for since the public sector was an instrument of the bourgeoisie to advance capitalism and to foster monopolies. Not only that, in its view, there was the danger of the state sector becoming the economic bulwark of a reactionary regime under the control of monopolists and equally of its not always serving anti-imperialist purposes, and therefore

> Failure to take account of such dangerous possibilities inherent in the public sector under a capitalist framework forms fertile soil for revisionism -- it amounts to ideological surrender to the national bourgeoisie.

Consequently, the correct strategy was seen as a militant one of exposing the capitalist path and advocating "people's democracy". [52]

The differences between the two groups were not resolved but papered over at Vijayawada. Following the Sino-Indian border war of 1962, the Communist movement split, with two parties -- Communist Party of India and Communist Party of India (Marxist) -- coming officially into existence in 1964, often stigmatized as revisionists and dogmatists, respectively. Still later, another and more radical group left CPI(M) in 1967 and founded the Communist Party of India (Marxist-Leninist) in 1969. Three different versions of the Marxist analysis of the state in India came to inhere in these three parties.

Three Contemporary Interpretations of the Indian State

The three-way split in the Communist movement in India in the 1960s resulted in three different Marxist interpretations of the Indian state. Where such interpretations do not drastically run counter to reality, however, they are so broad, ambiguous and flexible as to make any possible policy outcome consistent with the interpretation. Furthermore, in the case of two of these interpretations the additional attribution of "duality" to the bourgeoisie -- whereby it is both anti-imperialist and at the same time given to compromise with imperialism, that is, it is both progressive and reactionary -- means that the interpretation can never be wrong as an explanation, for all eventualities are covered. However, representing as they do the positions of parties that constitute important political forces in the country they call for serious attention. Here the position of each party is, first, discussed as embodied in its official programme or party documents; second, where possible, this discussion is supplemented by variations on the theme as provided unofficially by party officials and scholars sympathetic to the party's viewpoint; and, third, there are included critiques of the party position by other Marxist scholars.

The Communist Party of India

Official Position

The Communist Party of India in its Programme of 1964, which superseded the 1951 Programme, states:

> The state in India is the organ of the class rule of the national bourgeoisie as a whole, which upholds and develops capitalism and capitalist relations of production, distribution and exchange in the national economy of India. In the formation and exercise of governmental power, the big bourgeoisie wields considerable influence.
>
> The national bourgeoisie compromises with the landlords, admits them in the ministries and governmental composition, especially at the state levels, which allows them to hamper the adoption and implementation of laws and measures of land reform and further enables them to secure concessions at the cost of the peasantry.[53]

Several things can be noted in this definition. First of all, "the national bourgeoisie as a whole" of CPI's conception is an extraordinarily heterogeneous class. It includes not only the urban capitalist class but also the *rural* one, such as the rich peasantry.[54] Furthermore, all sections of the urban capitalist class -- whether they are big, middle or small; monopoly or non--monopoly -- are included. In this manner, if the state adopts policies against monopolies they are a result of the state power of the middle and small, or rural, bourgeoisie; if it adopts policies favouring monopolies they are a result of the influence of monopolies. In either case, they manifest the state power of the broader category of the national bourgeoisie. Secondly, the CPI definition excludes feudal landlords from the coalition holding state power. This is because of CPI's assessment that the mode of production being built by the state in India is capitalist[55] and, even though landlords are included in the administration and the state also compromises with landlords, yet feudalism had been substantially curbed, thus "a conversion of feudal landlords into capitalist landlords and a development of capitalism in the countryside. No class, certainly not the feudal landlords, would simultaneously share power in the state and allow that state to considerably diminish its economic base and social-political influence."[56]

Thirdly, the CPI interpretation definitely excludes domination of the state, or even a leading role in the state, by the big or monopoly bourgeoisie. While it admits that this bourgeoisie wields considerable influence, such influence is not decisive except on occasion, but not often. [57] On the other hand, "certain developments have taken place contrary to the desire of the monopolists," such as the public sector. Here, party theoreticians are likely to recognize, though they are unwilling to concede explicitly, that the masses in India also exercise state power; says Mohit Sen: "It cannot be precluded that, under heavy mass pressure, there will be further extension of the state sector more directly in the fields which are the exclusive preserve of the monopolists." Again, this emerges in the recognition of parliamentary democracy as "a historic advance for the people of India," for it affords them the right "to send representatives to the assemblies and Parliament, to intervene in matters of policy, to mobilise to change policies in favour of the people", and the party "considers that new possibilities exist for popular intervention in matters of state policy." [58] Still again, it is manifest in the statement that

> During these last few years, the working class has succeeded in forcing the employers and government to introduce some order and standard in the anarchy of wages prevailing in the capitalist system by means of wage boards, commissions, tribunals, tripartite conventions and collective bargaining. Sickness insurance, provident fund schemes, holidays with pay have been secured in organised industries. A well-defined national minimum wage has been accepted in principle. The organised strength of the trade unions and the striking power of the working class have increased. [59]

Since the party infers state power from policy, it is interesting to note that the power of the masses in a parliamentary system, which is otherwise acknowledged, is not incorporated explicitly in its class analysis of state power. If that were done, it would make the Marxist model a liberal-pluralist one.

The implications of CPI's particular class interpretation of state power for policy are to be seen in its analysis of the national bourgeoisie minus the monopoly bourgeoisie, because that is the class that, in the main, wields decisive influence. But a key characteristic of this class is its "dual" nature. The CPI programme talks about "the national bourgeoisie, excluding its monopoly section, which is objectively interested in the accomplishment of the principal tasks of the anti-imperialist, anti-feudal revolution, without which it knows truly independent national economy cannot be built, nor backwardness and impoverishment eradicated. But this class is also an exploiting class in the present society and as such has a dual nature. While it strives to eliminate the imperialist grip and the feudal remnants from our economy in its own interests, it vacillates and is inclined to compromise with these elements and pursues anti-people policies." [60]

The Programme accordingly discusses both aspects of the national bourgeoisie: its anti-imperialist and national actions as well as its compromises with imperialism and anti-popular actions. First of all, the national bourgeoisie foiled the attempt of imperialism to reduce the country's national independence -- which had been "a historic event" and had "opened a new epoch" -- into a formality. Then it launched on a programme of industrialization and agrarian reforms as also controls over industry and economy and of nationalization of major banking institutions which "gave the government a grip over finance and initiated the establishment of a state sector in industry." However, these measures, even though they constituted state capitalism and not socialism, were not only opposed by the imperialists and feudal classes but they "were also not to the liking of the top monopoly groups of Indian capitalists, who wanted the state sector of independent India to be restricted to defence industries, transport and public utilities, leaving the whole field of industry free for the private sector." Despite their opposition, "the state sector developed not only in these industries but also in finance and to a certain extent in trade. Thus the state sector becomes an instrument of building independent national economy and of weakening the grip of foreign monopoly capital and to a certain extent the Indian monopolies." As a result of India's ability to get aid from the socialist bloc, "the sabotage of India's plans for building heavy and basic industries at the hands of the imperialist monopolies did not meet with success.... There can be no doubt that the policy of the imperialists to keep Indian economy within semi-colonial bounds has received a rebuff. India, no longer linked and dependent solely on the world capitalist market, has been able to advance along the road of independent industrial growth." Besides, the state has pursued "in the main, a policy of peace, nonalignment and anti-colonialism," a policy that "conforms to the interests of the national bourgeoisie, meets the needs of India's economic development and reflects the sentiments of the mass of people of India."[61]

But the CPI holds that the adoption of the capitalist path has resulted in: making development a slow and halting process, imposing burdens and sacrifices on the people, concentrating economic power in a few monopolies, increasing inequalities and sharpening disparities, not developing a fully self-reliant economy, being tied in unequal trade relationships with imperialists, subverting agrarian reform, creating new reactionary vested interests in agriculture, bringing about the failure of the public sector to assume a commanding position, fostering bureaucratic and undemocratic management of the public sector and, in general, in developing the crises and contradictions associated with capitalist development. Moreover, the capitalist path has brought about the rise of right reaction based on monopoly groups which have emerged as a result of the growing differentiation within the national bourgeoisie, "harming the interests of broad sections of the national bourgeoisie and endangering India's march towards economic independence itself. In the economic sphere, they seek to annul the dominant role of the public sector, so essential for the development of national economy." [62]

In the eyes of CPI, India has accepted the goal of socialism but it has yet to complete even "the anti-imperialist, anti-feudal democratic revolution"; the national bourgeoisie which has pursued, through the Indian state, "the path of building independent national economy along the path of capitalist development, is incapable of implementing this programme." As a transitional stage, the party recommends the completion of the tasks of national democratic revolution through the non-capitalist path of development by building a National Democratic Front which would bring together "all the patriotic forces of the country, viz., the working class, the entire peasantry, including the rich peasants and agricultural labourers, the intelligentsia and the non-monopolist bourgeoisie" to jointly exercise state power. In the field of industry and commerce the national democratic state will totally eliminate foreign capital from the economy, break up the monopoly houses, nationalize banks and credit institutions, and "rapidly expand the scope of the state sector and make it the dominant sector in our national economy, by vigorously developing the key and heavy industries in the state sector and also by extending the sphere of nationalization to banks, general insurance, foreign trade, oil, coal and other mines, and plantations." At the same time, "it will give facilities to all non-monopolistic private-sector enterprises and small-scale industries by providing them with raw materials at reasonable prices, credit and marketing facilities, and allowing them reasonable profits."[63]

The pivot of this National Democratic Front will be the worker-peasant alliance, and with it the monopoly of state power of the bourgeoisie will be broken, even though the working class would not yet have hegemony, as it would in the socialist stage. It is interesting to note that, whatever its failures and weaknesses, it has been the claim of the Congress party since its foundation that it is precisely such a national democratic front. Even the CPI Programme cannot help but admit that the Congress party's struggle for independence and its consolidation "has given it a big mass base, which extends to all classes, including big sections of the working class, peasantry, artisans, intellectuals and others. The influence of the Congress, though much less than what it was in the days of the freedom struggle, is still vast and extensive." It has indeed been the intent of the CPI -- in an implicit recognition of the Congress "system", even before its articulation by Rajni Kothari -- to have as "the task of the Communist Party to make ceaseless efforts to forge unity with the progressive forces within the Congress, directly and through common mass movements to bring a leftward shift in the policies of the government, to fight for the realisation of the demands of National Democratic Front." This is parallel to its recognition of the aim of the monopolists and feudal classes, in their aim to exercise exclusive state power, "to capture the leadership of the Congress through the extreme right within the Congress" so as "to reverse the policies of the Congress in reactionary directions."[64]

Variations on CPI Themes

The CPI has often been accused, particularly after the Sino-Indian border war in 1962, of being nationalist or bourgeois nationalist. Regardless of that accusation, it is true that, either because of its own volition or because of encouragement by the Soviet Union, it has had some appreciation of the problems the state has faced in India and some sympathy for certain of the policies pursued by the state under Nehru and Mrs. Gandhi, at least until 1977. The CPI endowed part of the Congress party with having progressive tendencies; it is this belief that drove it to support Mrs. Gandhi's group in its struggle against the more conservative faction, her minority government when the Congress party broke up in 1969, her opposition to the movement launched by Jayaprakash Narayan in 1974 and 1975, and finally her declaration of Emergency in 1975. Although the experience of the Emergency apparently disillusioned the CPI, there is still appreciation of what the state under Nehru did, notwithstanding its criticism of the failings of his regime. The party's position on the public sector is also different from the other two communist parties. Here two private interviews are drawn on to illustrate the distinctive position of the party and to bring out nuances that are expressed in private; one interview is with a senior official of the trade union front of the CPI and the other is with an independent-minded and internationally renowned economist at Delhi University who, while not a party member, is sympathetic to the party.

In the interview with the party official what emerges is the emphasis on the anti-imperialist role of the public sector, the heterogeneity of the national bourgeoisie, the dual nature of the bourgeoisie, the growth of the public sector against the opposition of local and foreign monopolies, the salience of national interests beyond class interests, the importance of ideas and heritage of the nationalist movement, the implicit share in state power of classes other than the bourgeoisie and landlords, the varied interests served by the public sector beyond those of the bourgeoisie, the compulsion for the bourgeoisie to satisfy the population in a parliamentary system, and the poignant situation faced by the party in determining a political strategy in relation to the governing party: [65]

> Here the major issue is to bring about industrialization and overcome backwardness, both in agriculture and industry, and for that public sector, particularly in heavy industry, steel and basic industries, has a special role. The public sector has to grow against the opposition of developed countries who are interested in keeping countries like India underdeveloped and agricultural and will prevent, overtly or covertly, a country from coming up with modern industries, with a self reliant economy, not dependent on imperialist powers or capitalist countries. Therefore in such countries like India the public sector has a special anti-imperialist role although the state is controlled by the capitalist class and it can be termed as the state capitalist sector. The history of the development of the public sector in India shows that it has grown against the opposition of developed capitalist countries, of vested interests inside the country and above all with the help and support of the socialist countries. The public sector has been a vehicle or means of strengthening our independence from foreign imperialist interference and India's dependence on them. Foreign interests and their collaborationists in India, big business, did not defend it, they opposed it tooth and nail.
>
> (Why would the capitalists oppose the public sector if they control the state?) The capitalists are not homogeneous as a class. The capitalist class as a whole do not control the state, but as big business and monopolies, who are in league with the foreign big concerns or multinationals, they opposed it. (Who supported it then?) Those who are in the government. We call it national bourgeoisie; of course, that includes the monopolists also. They are not homogeneous and they cannot follow a consistent policy. Even though the state represents the entire class, there are conflicting interests. Sometimes they collaborate, sometimes they oppose, therefore it is a dual role of the capitalist class, particularly in a country like India. The compulsions of development and contending with the forces they confront, sometimes they vacillate and sometimes they have to adopt a zigzag course, therefore they cannot follow a very consistent policy. (Why do they collaborate with foreign multinationals?) Because they are a capitalist class, ideologically they are aligned together.
>
> (Why do they oppose the foreign multinationals?) Because of the compulsions of development. Secondly, the experience of the freedom movement, the legacy of the freedom movement -- they cannot shrug off these, because at that time they committed themselves. There were ideas like National Planning Committee, they had some blueprint, they had some ideas thrown

out before the country that these are the steps to take after independence. Nehru always harped on industry, industrialization, basic industry, self-reliance -- all these are legacies of the freedom movement. (Why does the capitalist class oppose imperialism and collaborate with it?) It is in the nature of the capitalist class to vacillate, particularly in countries economically weak like India.

(Since in your interpretation the capitalist class controls the state, then why not just give the public sector over to the private sector?) You mean denationalization. We would not support it, we will oppose it. (Why?) Because if it is in the public sector then the people and its representatives have some say in its functioning and working. If it goes to the private sector, people have no scope to say anything about its functioning. You can discuss in parliament the various proposals, its working. (But if the private sector controls the state and the state controls the public sector, then the public sector must serve the interests of the private sector?) Why entirely the private sector? Why should it be entirely the private sector? The country's interest is served, workers' interests are also served; for example, coal miners, their condition has improved, only because of nationalization. (Then it serves the interests of more than the capitalist class?) Of course. (But why should the capitalist class controlling the state serve these interests?) The capitalist class has to serve these interests because they have to be in power. People have to be satisfied to some extent with them although they are not completely satisfied. In order for them to stay in power they have to satisfy. (Have they satisfied the people?) Not yet. Because of that the ruling party was thrown out for three years in 1977. In some states they are not in power; some other parties are in power.... The crisis is increasing, that is the trouble. (Why, that should be good for you?) The crisis itself cannot bring about a revolution. It can bring reaction also. Fascism also came about through a crisis. What you require is a proper alternative. Who will come into power in Haryana if the Congress is overthrown -- the Jan Sangh. How to choose between the two. (It is a difficult proposition?) Yes. The only problem is the problem of alternative, and the left forces are weak, confined to some pockets in some states. That is the whole weakness of the situation.

The economist at Delhi University did not simply repeat the party line. He discussed in detail how the whole framework of politics and economics with which Marxists used to approach the issue of state power was in a state of flux; his own aim was, while fundamentally rooted in the Marxist tradition, to figure out what modifications must be made in that tradition to cope with new features of social reality and the liberal challenge but "without giving up my whole framework." Noteworthy in his discussion specifically of India are: skepticism about the bourgeoisie being the dominant class, priority for the removal of economic backwardness rather than class divisions, emphasis on national interests served by the public sector, and indeterminateness of the future class alignment of the public sector: [66]

What happens in a country where the bourgeoisie is too weak to become a single leading class or the dominant leading class and where the working class is also not able to lead a left class coalition? How do you deal with such a situation? (Some maintain that the bourgeoisie is in control of the state?) The aspiration of the bourgeoisie to have control is not the same thing as actual control. It is a multiclass state.

One would say perhaps, historically speaking between yesterday and today, if you see the direction of the movement, it is towards more and more bourgeois control or the emergence of a bourgeois state. Maybe that would be right, that is how it is moving. The point is that as a tendency it is towards that but at this stage the bourgeoisie is not able to, howsoever it might try, become the dominant force in state power, to exercise dominance, without the cooperation of other classes. The dead weight of the past is so great, economic and cultural, that the bourgeoisie itself has come out not in a revolutionary way but in a late reformist way, so it is not able to. On the other hand, the working class is handicapped because of so many things; therefore, it has to have an alliance between the farsighted sections of the bourgeoisie and the working class. It looks like a very reformist thesis. Crucial is the support of the peasantry as the biggest intermediate class. Who does it support?

(What is the role of the public sector in this?) It is the base of bourgeois power, it provides external economies to capitalists. It has a very great role in a backward country. It has a multifold potential but the potential that gets more realized is that it helps bourgeois growth. I am fundamentally of the opinion that there is a kind of historical phase in which that which divides the bourgeoisie and the working people is subordinate to what unites them to fight the whole heritage of backwardness. We have not yet come out of that phase in spite of the fact that we have to take more account of the differentiation. We should not only put pressure on other classes for the bourgeoisie but also against the bourgeoisie, against its exclusive privileges.

(If the public sector serves the interests of the bourgeoisie, why then have the pubic sector?) It also serves the interests of the nation, it also serves the national interest. The public sector must be preserved. It supplies steel, coal for small industries, medium industries, the things which are required for the regeneration of agriculture. It serves the national interest.

The public sector plays many roles. Public sector can give an economic base but also a political base. At the moment, the public sector is the ally of the capitalists but it is not their base, their instrument, it is not the source of their power. The battle for the class content, class alignment of the public sector is still being fought. It may be an ally but it is also the ally of someone else. It is not their exclusive instrument as yet. This battle is going. Sixty per cent is going to them, 40 per cent to someone else; 100 per cent is not going to them, at least in terms of potentialities.

Critiques of CPI Position

Two Marxist scholars take a critical stance towards the CPI position. One of them is Ajit Roy, a toughminded and orthodox Marxist in the mould of Charles Bettelheim; he has been editor of the *Marxist Review* (Calcutta) and has published over a dozen books. In developing his own distinctive position on state power in India, discussed later, Roy characterizes the CPI definition as "ambiguous". Decisively rejecting the notion that India has ever had a comprador bourgeoisie, he states that the CPI definition is tantamount to saying that the bourgeoisie as a whole holds state power. His major criticism is that, "even if broadly correct, this characterization tends to gloss over the growth of sharp socio-economic differentiation within the Indian bourgeoisie with the consequent differences in the degree of control over and share in the state power. Hence it is somewhat imprecise and unhelpful as a guide to the formulation of revolutionary strategy."[67] He is also critical that CPI does not "take any note of the fast-rising agricultural bourgeoisie and its relation with the state power." In this, Roy is mistaken, for CPI's definition of the national bourgeoisie includes the agricultural bourgeoisie. However, such an error is understandable because the term bourgeoisie itself is so flexible, rather it is used so flexibly.

The other Marxist scholar is Biplab Dasgupta, who has taught at the Institute of Development Studies at Sussex and published several major works on Indian economics and politics. He explicitly identifies himself with the position of CPI(M) and is dogmatically critical of the CPI approach to state power. He questions the CPI assumption of national bourgeoisie as a whole being the dominant class and of the monopoly bourgeoisie not being the leading element. Writing in the context of the late 1960s and early 1970s he debunks bank nationalization, public sector expansion, and other anti-monopoly measures as possible supportive evidence in favour of the CPI interpretation, holding that these "do not necessarily constitute 'anti-monopoly' actions and these are consistent with a 'pro-monopoly' policy of the government."[68] Employing the commonly-used argument about the *"overall interest"* and the "long-term interests" of the ruling class, he typically takes the position that "so long as the State remains under the control of big business, the latter has no reason to feel insecure because of the extension of public sector in certain lines of production." As for assurance on big business controlling the state, he simply offers the assertion that the Congress party "has always been the vehicle through which the interests of Indian big business and feudal elements were served." He holds that big business has continued to receive credit after bank nationalization but notes, interestingly, that "the main beneficiary of the extended amount of credit to agriculture are the rich farmers, who are the staunchest allies of big business in the rural areas." Equally, he dismisses the government's public sector policy as one capable of deceiving "only those who want to be deceived," saying that it forms a small part of the total economy. With a sleight of the pen, he transposes the public sector into the government sector and comparing the incomparable -- agricultural and industrial countries -- he then holds India's government sector as "smaller than the government sector in most countries of the western world including the United States."

As for India's policy of non-alignment, he maintains inexplicably that "the class character of the ruling class of a country under the present conditions, cannot be inferred from its foreign policy." He notes that the Indo-Soviet treaty of 1971 received the "whole-hearted endorsement" of big business; while one cannot be too certain about that, he fails to note the even more enthusiastic endorsement of the working class, indeed of every class in India, at the time. Again, the split in the Congress party in 1969 was for him simply the result of the fact that "the two wings of the Congress represent two different tactical views of the ruling class," with one emphasizing short-term interests and the other, being more farsighted, the longterm interests of big business. With this, the author's position against CPI(M)'s rival emerges unscathed, never touched by any measure of self-doubt or qualification.

The Communist Party of India (Marxist)

Official Position

The position of the Communist Party of India (Marxist) differs from that of the CPI in many essential ways, even though the two parties also take a common position in some respects, including repetition of the same language. In general, CPI(M) takes a more negative view of developments. Its interpretation of state power is as follows:

> The present Indian state is the organ of the class rule of the bourgeoisie and landlord, led by the big bourgeoisie, who are increasingly collaborating with foreign finance capital in pursuit of the capitalist path of development. This class character essentially determines the role and functions of the state in the life of the country.[69]

It is immediately obvious that here state power is held, unlike the CPI interpretation, by a coalition of the bourgeoisie and the feudal [70] landlord classes, which are taken to share power. The presence of feudal classes in the coalition would seem to suggest that feudalism has not been curbed and the state acts to preserve it. The party's rival, CPI, on the other hand, is critical of this assumption of bourgeois-landlord coalition because that would make the Indian state one of right reaction, corresponding to the state power of Chiang Kai-shek's Kuomintang China or neo-colonialist states like South Korea, whereas "the Indian state refuses to behave as the CPI (Marxist) Programme dictates. It goes on with its policy of building an independent capitalist India. It follows a foreign policy of non-alignment, of anti-colonialism and of friendship with the Soviet Union and most other socialist countries." [71]

Secondly, it is the big bourgeoisie which is the leader of the bourgeois-landlord coalition; this implies a restricted view of the potency of the medium and small bourgeoisie. In this manner, CPI(M) limits the progressive potential of the national bourgeoisie holding state power. The big bourgeoisie came to head the state following "a settlement" with the British in 1947 when India attained national independence as a result of revolutionary upsurge in the context of a weakened imperialism in the postwar period. In one of its few moments of recognizing anything positive in the big bourgeoisie, CPI(M) states that the British imperialists had hoped to render independence into a formality by using their economic dominance in India, "but the course of historical development since then has been disappointing to the imperialists and their hopes were belied." [72]

Thirdly, again, unlike the CPI case, the bourgeois-landlord coalition collaborates with foreign finance capital, even though it is not comprador -- a point by which the party distinguishes itself from more radical fractions of the Communist movement. Like the CPI, the party recognizes that the country's parliamentary system "embodies an advance for the people. It affords certain opportunities to them to defend their interests, intervene in the affairs of the state to a certain extent, and mobilise them to carry forward the struggle for peace, democracy and social progress." More specifically, it admits that in particular instances the power of the masses has been decisive: "the solution of the problems came ultimately, though haltingly, under the stress of the struggle of the democratic masses."[73] But it does not incorporate this acknowledgment into the definition of state power. In reality, however, even though this characterization would be unacceptable to the party, CPI(M) itself is a co-sharer in state power since it holds power in the states of West Bengal and Tripura and has held power in the state of Kerala. For, the Indian state consists not just of the central government but also of state governments, regardless of the degree of autonomy of the latter.

The CPI(M) also endows the Indian bourgeoisie with a "dual" character, but its discussion of the nature of this duality is heavily weighted against a substantial progressive role, rather the emphasis is on the negative aspects of its conduct. Thus:

> Despite the growth of contradictions between imperialism and feudalism on the one hand and the people, including the bourgeoisie, on the other, and despite the new opportunities presented with the emergence of the world socialist system, the big bourgeoisie heading the state does not decisively attack imperialism and feudalism and eliminate them. On the other hand, it seeks to utilise its hold over the state and the new opportunities to strengthen its position by attacking the people on the one hand and on the other, to resolve the conflicts and contradictions with imperialism and feudalism by pressure, bargain and compromise. In this process, it is forging strong links with foreign monopolists and is sharing power with the landlords.

It would seem that, in the eyes of CPI(M), notwithstanding its "duality" the bourgeoisie tends towards surrender to imperialism so that in the final analysis its position is tantamount to a junior partner of imperialism. In the view of CPI(M), the national state under the big bourgeoisie has failed to perform the "urgent tasks of the Indian revolution," including the elimination of feudalism and semi-feudalism. To begin with, it compromised with imperialism and made huge concessions to feudal princes and landlords, because it was afraid of the popular outcome. Later, it took to building capitalism in order "to further strengthen its class position in society." However, the Indian bourgeoisie has lacked both the technical base and the free capital that goes with the possession of colonies to undertake capitalist development on its own; consequently, "the bourgeoisie has employed state power" for the purpose.[74] The CPI(M) seemingly accepts the "incapability" thesis, stating

In the conditions prevailing in India, such heavy machine-building and other vital industries as have been built in the state sector, would not have otherwise come to fruition, for private capital was not in a position to find the required resources for these huge industrial projects. The building of these undertakings in the state sector has, therefore, helped to overcome, to a certain extent, economic backwardness and the abject dependence on the imperialist monopolies, and in laying the technical base for industrialisation.

Economic planning and the public sector are thus devices to advance capital accumulation in building capitalism. In relation to the public sector, the party adopts a somewhat contradictory and inconsistent position. On the one hand, it says that it serves and entrenches the interests of the bourgeoisie and is meaningless from the viewpoint of socialism, for "the actual realities show that the state sector itself in India is an instrument of building capitalism and is nothing but state capitalism." In elaboration, it states:

the influence of big business in our state sector has steadily grown, leading to increasing utilisation of it for further bolstering up big capitalists. The bulk of credit facilities from the financial institutions has gone to build them up still further. All major contracts under the plan and otherwise emanating from government go to big business. It is big business again that controls the distribution of the products of several state undertakings. Apart from the growing links between state capitalism and the monopolies, government now invites capitalists including foreign monopolists to participate in the share capital of state-owned undertakings. This further distorts the growth of the public sector.

On the other hand, without seeing anything redeeming in the public sector, CPI(M) is critical that "only an insignificant part of our economy is under the state sector and vast fields of industrial, commercial and other activities are left under private enterprise," that the government has not nationalised large-scale industries which by government policy should be in the state sector, and further that the government has relaxed its own restrictions and controls on the private capitalists.[75] In this fashion, CPI(M) is negative on the conduct of the government and is able to attack it both for having the kind of public sector it has and also for not making it comprehensive enough.

India's economic planning, capitalist industrialisation and public sector have led, according to CPI(M), to the growth of big business and to the plunder of India's economy by foreign monopolies which is being drained out of the country. They have led also to growing dependence; India is "in many respects precariously dependent on western assistance and particularly U.S. assistance" and "far from this dependence getting reduced, it is actually increasing year by year." [76] On the other hand, taking this to mean that India is more dependent than it was at independence, and that as a semi-colony the country is being completely subjugated, CPI characterizes it as "a palpably absurd formulation." [77] But CPI(M) is critical of the CPI for "the gross underestimation of imperialism and the deliberate underplaying of the role of foreign capital." [78] For CPI(M), American capital and "aid" are "creating a dangerous situation for our country also"; it continues: "they are penetrating all spheres of our national life."[79]

As for India's foreign policy of peace and non-alignment, it is seen as a reflection of the duality of the Indian bourgeoisie, opposing and collaborating with imperialism. No doubt, it is to preserve the country's independence but also "to advance its own class interests." For the latter, the bourgeoisie needs peace -- as if it does not need it in pro-American countries -- and also to exploit the contradictions between imperialism and socialism. However, while the bourgeoisie continues to employ the broad framework of non-alignment, its various policies "objectively facilitate the U.S. designs of neocolonialism and aggression and lead to India's isolation from the powerful currents of peace, democracy, freedom and socialism and as such is harmful to our interests." [80] In practice, however, CPI(M) is driven to endorse many aspects of the country's foreign policy; thus, in 1982, it concluded that "the Government of India has taken a correct stand on certain issues," such as American military aid to Pakistan, Cambodia and Diego Garcia. [81]

Overall, for CPI(M), the capitalist path under economic planning pursued by the big bourgeoisie, even though it has meant giving the economy a certain "tempo and direction", has led not only to dependence and penetration but also to imposition of suffering and misery on the population, to complete failure in agrarian reform, and to concentration of economic power at the top. In brief, India's experience with development demonstrates the futility of the capitalist path for development. The bourgeois-landlord state itself is a political and economic menace; the threat to the country is not from right reaction outside the government, for the rightist forces are lodged in the state. The way out is the "People's Democratic Front" to peacefully replace the present state by people's democracy in order to complete the "anti-feudal, anti-imperialist, anti-monopolist and democratic tasks". The Front will be a four-class coalition comprising of the working class, peasantry, middle classes and the national bourgeoisie, but what distinguishes this Front from the National Democratic Front of CPI conception is that it will be under the hegemony of the working class, that is, the Communist party. The reason that CPI(M) refuses to share power with the national bourgeoisie is that capitalism is already strong and well entrenched in India; it is therefore "unreal" to count on a non-capitalist path of development. Besides, the national bourgeoisie is for CPI(M) basically "unreliable and exhibits extreme vacillation."[82] This posture on the people's democratic front in reality springs from the fact that CPI(M) has concentrated electoral strength in certain states of India where it can come to power on its own unlike CPI whose electoral strength is dispersed across India.

Variations on CPI(M) Themes

While asserting that the state is under the control of the big bourgeoisie and feudal landlords, CPI(M) takes basically a negative position, maintaining that it serves the interests of the big bourgeoisie. But is its position really that different from CPI's? Do party officials express only its officially stated position? It would seem that its officials are, in private, closer to the CPI view in terms of the role the public sector performs and the interests it serves. In an interview, a member of CPI(M)'s Central Committee stated that the public sector was created as an instrument of capital accumulation because the Indian bourgeoisie, lacking in resources, was not capable of following the path of independent capitalist development and so it created, on the basis of resources mobilized from the public, a public sector for the purpose of serving its needs to build capitalism. However, even though he parried the question of selling it to the private sector, he saw it as fulfilling a more positive role than is incorporated in official statements, and therefore extended support to it:[83]

> The second aspect: since the public sector does enable the bourgeoisie to cut down its dependence on foreign imperialism, foreign capital, to a certain extent it plays a slightly positive role vis a vis imperialism. This will be our basic position. We do not consider it as a socialistic sector, we do not consider this brings in socialism, we do not think it has commanding heights as claimed by the government. (Do you favour its expansion?) This is a very blunt question. They do it according to their needs. They do it to the extent that it helps them not to go to foreign finance capital. Up to that extent we support it.... Nationalization of banks, the banks were owned by monopoly houses, but they were nationalized in the interests of the class as a whole. The section of the bourgeoisie which is not affected by it supports it whereas the bourgeoisie which is being hit opposes it. (Who supported it?) People of India under the influence of the bourgeoisie, the small-scale sector, they will get loans, the non-monopoly section supported the nationalization.
> (If the state is a tool of the bourgeoisie and the state controls the public sector, would you object to handing over the public sector to the bourgeoisie?) Nobody is interested in selling it, nobody is interested in buying it.... Public sector is a good thing for them. We support it to the extent that it helps fight imperialism, otherwise the dependence and surrender becomes complete.... If the public sector had not been there, there would have been abject dependence on imperialism.

A senior official of the labour front of CPI(M), however, went further than the above-mentioned committee member in his appreciation of the public sector. Even though he reiterated the CPI(M) position on state power, his overall response demonstrated the sharing in that power by the working class and its representatives, a situation that corresponds more to the pluralist model of the state:[84]

> (But if the big bourgeoisie controls the state then it must control the public sector, so what is the big difference?) It is not so, there is a little difference, the difference being that the government is directly responsible for whatever is happening in the public sector which it is not as far as the private sector is concerned, there it is a third party. The difference is that they are answerable in the Parliament. We can raise a big hue and cry when something happens in the private sector. When we raise a hue and cry we say that Government is helping the private sector, but here we say that the Government is directly involved. There is a second part; it comes in regard to development of self-

reliance and R&D as far as our country is concerned..... We are for the public sector, but with better management; it must be run with a view to attain self-sufficiency in matters of development of technology with a view to making our country self-sufficient and self-reliant. There is a third factor involved with regard to the wages and other things as far as workmen are concerned. The government is more prone to accept some legitimate demands of the workmen whereas in the private sector it is entirely related to profit. There is a difference.

(So you would favour expansion of the public sector?) Of course, any time. We have demanded nationalization of jute industry, of the sugar industry, wholesale trade of all essential commodities, so that the public distribution system would be strengthened.

A similar attitude emerges from intellectuals who espouse the CPI(M) party line. A professor of political science at Delhi University who subscribed to the CPI(M) position on state power and on the public sector as merely serving the interests of the bourgeoisie was nonetheless strongly in favour of the public sector and seemed implicitly to modify somewhat his position on the state:[85]

(Why not then let the public sector be given over to the private sector? Would you object?) I will have objection, nonetheless. I do not believe that people have no influence at all. Popular movements have some influence, and there is some accountability in terms of popular movements, parliament, and therefore if we increase our influence in parliament and create difficulties in the way of the bourgeoisie, we can make some use of the public sector. Once it is handed over to the private sector then it would be difficult to do that.

(But if the bourgeoisie is merely using the state to ransack the public sector why not just give it to them?) I do not say it is only that. It is not a means to socialism. It is not a means for achieving socialist transformation in India. It is a part of the capitalist state apparatus and capitalist political economy serving the development of capitalism. I am saying that there is now at least a possibility of influencing that process, parliamentary process can be used to influence that process, to a limited degree. I would therefore like expansion of public sector, even though it is running into losses. Because at least there is the possibility of people influencing the economy and the decision making process with the public sector.

(It does make a difference then?) Most bourgeois state institutions serve the development of capitalism, all bourgeois institutions, but they also have utility for working class because they provide channels through which working classes can influence.

The above quotation makes one curious to inquire as to why, if the bourgeoisie is in control of the state, it builds a political framework that provides popular movements the instruments to create difficulties for it. Equally, it persuades one to ask why simply harp in public on the malign interests the public sector serves when in private it is admitted that other interests are served as well and its expansion is desired. What is apparent is that in private the appreciation of the public sector is different than as the singleminded servant of the monopoly bourgeoisie, and similarly the notion of state power does not strictly follow the party's publicly expressed position, for other classes are acknowledged to partake of that state power.

A similar conclusion emerges, in a more thoughtfully elaborated form, from an interview with a senior economist at Jawaharlal Nehru University in New Delhi who can be considered to be an intellectual spokesman for CPI(M). He initially took the position, following CPI(M), that a bourgeois-landlord coalition held state power in India but said that he took it as a hypothesis to serve as a starting point to emphasize the point that capitalism was being built on an unreformed agriculture based on concentrated ownership of land. Of course, cast in this manner this descriptive hypothesis is not very enlightening; besides, it is not simply a hypothesis as conventionally understood, for people have been ready to die and kill for it. Proceeding further, while stating that the public sector was built to serve bourgeois interests, he touched upon other aspects but which are not incorporated in the public posture:[86]

> I do not think that the existence of the public sector is in any way indicative of any kind of socialism that is happening in India. By now even non-Marxists have shown adequately that it was one of convenience to the Indian bourgeoisie. What is, however, true is that unlike Pakistan the public sector was not auctioned off to the private sector. The existence of the strong petty bourgeoisie means that the public sector remains in the public sector. There are variants of capitalism, which variant of state capitalism is chosen depends on the strength of the various classes.
>
> (But is the strength of the petty bourgeoisie not the noteworthy feature of the Indian political scene as the notion of the intermediate regime suggests?) That brings me to the intermediate regime. It is undeniable that in India you have a fairly vocal, powerful urban petty bourgeoisie. All these populist measures partly cater to them; on the other hand, the intermediate regime hypothesis which actually says that the petty bourgeoisie is actually the ruling class is going too far. State policies have to take into account the existing class phenomenon as a constraint. For instance, the fact that the state controls inflation, or attempts to, it does not necessarily mean that those hit by inflation are the ruling class. The strength of the petty bourgeois class lies in the voting. The state cannot even ignore the demands of the working class, or even the agricultural laborer.
>
> That still would not mean that the state does not have a class nature. The fact that in a framework of parliamentary democracy the state has to make certain concessions, even to the workers, petty bourgeoisie and the small peasantry does not go against Lenin's theory of the state. Ultimately, the state safeguards property and ultimately it is promoting capitalism and strengthening monopolies. Precisely the crisis of this system arises that in the promotion of capitalism and strengthening monopoly capitalism and

giving concessions the state finds that it has less and less to offer and therefore it moves to more authoritarian forms. If it was possible to reconcile all class interests there would be no crisis.

(If the public sector builds monopoly capitalism, why not then just sell it to the private sector?) No. This is what I am saying, the state has to take into account the constraints upon its actions which are imposed by the strength of the lower classes. Why does the state go in for populist measures. Why were banks nationalised, for example. Various theories, but it helped to project an image of radical populist government. (So you would favour a public sector even if it is controlled by a state which is in turn controlled by the bourgeois-landlord coalition?) For two reasons I would. One, public accountability: it can be raised in parliament, these things can be raised in parliament, exposure of misdeeds if they are in the public sector may be relatively easy than it would be in the private sector. Two, it serves to fulfil an ideological role which is important. The ideology of capitalism is that business cannot be run unless they are owned by private persons. That ideology is getting subverted, public sector is a subversion of the idea. If the public sector is run badly, one would have to go further: private sector's record is worse.

Several critical issues are raised in the above interview which call for comment. First, it is maintained that the public sector enterprises have not been auctioned off in India as in Pakistan and that this is a result of the strength of the petty bourgeoisie in India. An important question arises: if the petty bourgeoisie is the cause of the continued retention of the public sector, could it not be that it is also the cause of the creation of the public sector? Why must its influence or power be limited to the retention of the public sector? Why should it then be going too far to take the petty bourgeoisie as the ruling class? Despite the recognition of the importance of the petty bourgeoisie, however, there is continued insistence in public, and formally in private, that the public sector is a creation by and for the bourgeoisie. Of course, one can extend the flexible term bourgeoisie to encompass the petty bourgeoisie as well,[87] and all intellectual problems of importance are then resolved. Second, the fact that the state, through its policies, meets certain of the demands of workers, petty bourgeoisie and small peasants needs to be -- if it is correct as is acknowledged -- incorporated in any definition of state power; it is besides the point whether or not it is inconsistent with Lenin's theory of the state, whether or not it leads to a crisis, and again whether or not it is done for the purpose of winning popular support. As long as it is admitted that the state does take into account, in respect of workers, petty bourgeoisie and small peasantry, "the existing class phenomenon as a constraint" then that fact needs to be openly incorporated in any summary definition of state power, for the state does not act much differently in respect of the bourgeoisie. The problem, of course, is that within the Marxian framework with its assumption of a zero-sum game this fact cannot be accommodated without doing violence to the entire framework and driving it into the arms of pluralist theory.

Third, it is claimed that the existence of the public sector fulfils the ideological role of showing that there is an alternative way to run an economy. Rather than this being an incidental function of an existing system supposedly brought into existence for other ends, there is evidence to show that this was precisely the purpose -- to demonstrate that there was, indeed, an alternative way to running an economy but an alternative to both capitalism and communism. In this light, regardless of the function that the public sector may perform ultimately in history, the insistent claim that the public sector was set up by, and with the purpose of aiding, the bourgeoisie -- the big bourgeoisie, at that -- is a questionable one.

Critiques of CPI(M) Position

Apart from CPI theoretician Mohit Sen's critique of the CPI(M) position from a partisan perspective,[88] the noted Marxist Ajit Roy has appraised critically the CPI(M) stand on state power from a theoretical and empirical viewpoint. Roy notes that CPI(M)'s definition of state power in India, while more elaborate, is still defective insofar as (1) it fails, in its emphasis on landlords as co-sharers in state power, to take into account the fact that the state has been largely successful in eliminating semi-feudal landlords as a class, who could therefore not be treated as a partner in the power coalition; (2) it confuses the issue of the pursuit of the capitalist path by making it seem a function of collaboration with foreign finance capital; (3) it suffers from imprecision in that big bourgeoisie, too, is a generic term and does not adequately differentiate the different sections within it; and (4) critically, it neglects "the fast-rising agricultural bourgeoisie and its relation with state power." Differing with both CPI and CPI(M) but rejecting also the CPI(M-L) formulation (considered in the next section), he carves out a distinctive position that deserves attention.

Roy distinguishes four different strata among the Indian bourgeoisie: (1) monopoly bourgeoisie, which has the entire country as its sphere of operation; it is able, through its lobbying, "to influence and direct national policies" (there is no mention of its ability to determine such policies), and therefore favours centralization of the state structure; (2) big industrial bourgeoisie, whose sphere of operation is limited to the states; it has access to administration at that level and accordingly, even though dependent on the monopoly bourgeoisie, it is against centralization, favours state autonomy and "often voices its criticism against the domination of the monopoly bourgeoisie and of government policies favouring the latter and militating against its own specific interests"; (3) small industrial bourgeoisie, which has no access to either central or state governments; "this stratum is firmly opposed to many of the basic policies of the present regime and a section of it is sympathetic to the Left and radical movements, as at least in West Bengal"; and (4) agricultural bourgeoisie; constituting "the largest section of the Indian bourgeoisie", it is the result of agrarian reforms implemented by the state and thus is a creation of the monopoly bourgeoisie; composed of former landlords who have taken to capitalist agriculture and of rich peasants, it has "gained its strength and consolidated its position in recent years."

Although the state provides aid to all four strata, Roy maintains that it has really favoured the monopoly and agricultural bourgeoisie, who have also been increasingly brought into closer association through its various institutions. But he sees some difference in the locus of their power as "the top monopolies hav. a firm grip over the Central Government and its policies, the agricultural bourgeoisie dominates the state governments and through the latter exerts a powerful influence over the Centre too." His considered statement on state power is as follows:

> The state in India is essentially the organ of dictatorship of the Indian monopoly bourgeoisie and the rural bourgeoisie, led by the former. In other words, it is not the organ of class rule of the entire Indian bourgeoisie, but of a section of it. Further, the state is not pursuing the aim of capitalist development in general or on a broad basis, but of a particular character, on a limited scale, specifically suited to the interests of particular sections of the class.[89]

Interesting here is the absence of collaboration with foreign finance capital. Roy does not regard it as a fundamental characteristic of the Indian bourgeoisie, for while the monopoly bourgeoisie no doubt uses the state "to secure the most advantageous terms for collaboration...it also uses the state power to prevent the penetration of foreign finance capital in spheres or on terms considered disadvantageous to its interests." He seemingly, by implication, thus rejects the thesis of "dual nature" of the bourgeoisie that both CPI and CPI(M) have put forward. Elsewhere, Roy has apparently endowed the Indian bourgeoisie with an urge for independence, noting that it "has embarked on a course of *relative independent* economic (capitalist) growth. As a part of this process, it has undertaken a simultaneous and balanced (again *relatively*) development on a wide front." However, at the same time, he points out that, lacking in adequate financial resources -- from either internal mobilisation or foreign aid -- the bourgeoisie has had to function within severe constraints, and he starkly underlines the alternatives for it:

> a) Either, to give up the ambition of developing an independent and balanced economic structure (and its associated foreign policy goals) and to bring down the sights to the much more modest objective of developing with the more readily available resources only light consumer goods industries, conforming to the docile colonial pattern;
> b) Or, to seek to accumulate a larger *mass* of internal resources by effecting a very sharp increase in the rate of surplus value, something generally comparable to the process of the primitive accumulation in the period of industrial revolution in the developed capitalist countries.
> The first course presupposes a shift of political power from the now dominant industrial and monopolistic bourgeoisie to sections of the bourgeoisie with latent comprador inclinations. The second course implies a much more authoritarian political system and a more naked subversion of the bourgeois democratic framework. Both the courses involve long drawn and bitter political struggles within the bourgeoisie and between the bourgeoisie and the toiling masses, headed by the working class.[90]

Roy, however, supplements his basic definition of the Indian state with reservations that make the power of the big bourgeoisie seem less awesome than what appears at first glance. He holds that the fusion between monopoly capital and state power, typical of state monopoly capitalism, has not occurred so far in India; that, while the Indian monopoly bourgeoisie has "secured a firm hold", it "has not achieved exclusive domination over state power"; that, indeed, it has "at this stage to work from peripheral positions." Still, he maintains that, "despite the inconvenience of somewhat remote control, it can successfully dictate the major policies." This is simply asserted, and no evidence is provided in support. If, however, it must work peripherally and through remote control, that is really no testimony to its strength but to its weakness.

But there is still more to its weakness, for its alliance with the agricultural bourgeoisie is no alliance but a checkmate system. Here, Roy recognizes the power of numbers in a parliamentary system. He notes that the monopoly bourgeoisie would no doubt want to transfer part of the tax burden to the agricultural bourgeoisie, "but the latter, confident of its key position in the power structure as the most numerous stratum and possessor of the largest social base among the entire bourgeoisie, has been successfully resisting it. Herein lies a crucial distinction. As long as the monopoly bourgeoisie needs mass support or is unable to overthrow the bourgeois democratic framework, it cannot but submit to blackmail by its rural counterpart."[91] In the ultimate analysis, then, it is not the monopoly bourgeoisie but the rich and even medium peasantry -- by virtue of its numbers combined with relative economic and social privilege -- that turns out to be the key power wielder and it is by courtesy of definition, which includes such peasantry under the term "bourgeoisie", that the bourgeoisie is endowed with state power in India. Roy's effort at coming to grips with Indian reality demonstrates the complexity of the problem of accurately defining state power within the Marxist framework, but it is far superior to the simplistic formulations of the communist parties -- which is understandable -- but also of intellectuals on the left. For example, Dasgupta notes the important consequence of numbers and the strategic position of the rich and medium peasantry without letting it affect his support for the CPI(M) formulation on landlords as co-sharers in state power:

> But the vast majority of the kulaks -- who are individually small but collectively enormous, who dominate the rural scene in the economic, sociological as well as political terms, who serve as 'vote-banks' for the ruling class party by mobilising the voters to its side, who are opposed to these measures (land reform, co-operative farming) -- are also indispensable allies of big business....[92]

Apropos the rich peasantry, however, it is intriguing that its interests are sought to be protected not only by conventional political parties but as well by CPI(M), the most left of major parties within the parliamentary framework. Not only did CPI(M) during 1977-79 support the Janata government, which had as one of its important constituents the "kulak" party Lok Dal, and then help its leader Charan Singh to become Prime Minister, but it has also joined movements demanding higher support prices for agricultural commodities, which are largely marketed by the privileged peasantry. A Central Committee document apologetically notes:

unfortunately, even after a thorough discussion on this issue at various levels, a criticism has come that it is not correct to support the movements since they serve the interests of the landlords and kulaks and go against the interests of the consumer. The criticism is neither based on our programmatic positions, nor on the accepted policy of the Party and the Kisan Sabha.[93]

Notwithstanding his orthodox position, Roy significantly also acknowledges the power of ideas and legitimacy needs in policy making: "the ruling classes cannot altogether repudiate their socialistic promises in a country like India, in which historical circumstances obliged the ruling party itself to inscribe all sorts of egalitarian ideals on its fundamental documents."[94]

Communist Party of India (Marxist-Leninist)

Official Position

If CPI(M) is to the left of CPI, then CPI(M-L) is to the left of CPI(M) and unabashedly swears loyalty to the path set out by Mao Zedong and the Chinese revolution. It represents a reincarnation of the Andhra thesis under Rao during 1948-50, and is testimony to the immense appeal of the Maoist revolutionary line in a situation of deep poverty. CPI(M-L) as a party was founded, with the blessings of China, in 1969 by a splinter group. It was established out of dissatisfaction with what was perceived as betrayal of revolutionary Marxism by CPI(M)'s participation in the state government of West Bengal, following the 1967 elections, and out of the revolutionary experience of this group in the small area of Naxalbari in northern Bengal. Essentially, CPI(M-L) was a party confined to West Bengal though similar groups existed in other parts of India, particularly Andhra Pradesh. As a revolutionary party, CPI(M-L) was finished by the end of 1971 but its approach to the state deserves attention because of the distinctive line adopted, because of its continuing appeal as a result of rural poverty, and because of the awe and admiration evoked by the heroism and spirit of sacrifice of its followers in the years of Naxalite revolt from 1967 to 1971.

In the first place, the Indian bourgeoisie is not seen as having a "dual" character; rather, for CPI(M-L), its surrender to imperialism has been complete; it is, and has always been, simply comprador in nature. Because of that characteristic, the bourgeoisie is alleged to have intervened "to direct the national liberation struggle from the path of revolution to the path of compromise and surrender."[95] The Congress party under Gandhi's leadership remolded the struggle "to serve the interests of the British imperialist rule and its feudal lackeys." The Communist party was no better during the course of the nationalist movement, for its leaders were traitors to the revolution and followed the Congress party; indeed, they were "agents of imperialism and feudalism." According to CPI(M-L), World War II had already weakened imperialism; then the postwar revolutionary upsurge in India shook British rule, which therefore "pressed into service its tried agents -- the leaders of the Indian National Congress" in order to crush it. With this, the British installed into state power "the Congress leadership representing the comprador bourgeoisie and big landlords" under a "sham independence" which means basically indirect rather than direct imperialist rule, simply "a replacement of the colonial and semi-feudal set-up with a semi-colonial and semi-feudal one." Accordingly, for CPI(M-L), India was not a capitalist country and could not be said to have democratic rights while the parliamentary system was considered meaningless.

The CPI(M-L) interpretation of the state in India comprises not only a bourgeois-feudal coalition of big bourgeoisie and big landlords, but also imperialists whose agents and pawns the former two are. The big comprador-bureaucrat bourgeoisie and big landlord ruling classes, in the view of CPI(M-L), are "lackeys of imperialism" but, with the decline of British imperialism, they have "hired themselves out to U.S. imperialism and Soviet Social imperialism," to whom they have mortgaged the country and turned it into a neo-colony of the new overlords. As a consequence, the Indian people are economically crushed under the exploitative weight of four huge mountains -- U.S. imperialism, Soviet Social imperialism, feudalism, and capitalism. India's foreign policy, according to CPI(M-L), is dictated by the United States and the Soviet Union and serves their strategic aim of encirclement of China and suppression of national liberation struggles in the Third World. Further, the comprador-bureaucrat bourgeoisie and big feudal landlords are by definition incapable of carrying out independent capitalist development; instead, they have allowed the domination and penetration of the national economy by U.S. imperialism and Soviet Social imperialism which have carved overlapping spheres of influence:

> U.S. imperialism and Soviet Social imperialism have brought the vital sectors of the economy of our country under their control. U.S. imperialism collaborates mainly with private capital and is now penetrating into the industries in the state sector, while Soviet Social imperialism has brought under its control mainly the industries in the state sector and is at the same time trying to enter into collaboration with private capital.

The two types of imperialism have promoted development of comprador-bureaucrat capitalism as a means to their relentless exploitation of India's masses. The promotion of the public sector, rather than being counter-imperialist is part of the same design of imperialism and social imperialism. Thus, the CPI(M-L) programme states:

> The much trumpeted "public sector" is being built by many imperialist exploiters for employing their capital and exploiting cheap labour power and raw materials of our country. The public sector is nothing but a clever device to hoodwink the Indian people and continue their plunder. It is monopoly capitalism i.e., bureaucrat capitalism.

Through their various activities the U.S. imperialism and Soviet Social imperialism are alleged to have acquired an "octopus-like grip on India's economy" and control "the political, cultural and military spheres of the life of our country."

Faced with comprador-bureaucratic capitalism, feudalism, imperialism and social imperialism, the country is said to be beset with many contradictions, but the principal contradiction is between the masses and feudalism. However, since feudalism is propped up by comprador-bureaucratic imperialism and social imperialism, the resolution of the contradiction therefore calls for a democratic revolution against all of them. The path to this democratic revolution, whose heart is the agrarian revolution, lies not in participation in the parliamentary system; indeed, CPI(M-L) thought that CPI(M) was a partner of the ruling class because of its participation in the parliamentary system and therefore ought to be unmasked. Rather, the path lies in protracted guerrilla warfare organized in rural bases under a four-class united front of working class, peasantry, petty bourgeoisie and small and middle bourgeoisie under the leadership of the working class through its political party, CPI(M-L). Armed struggle is not something for a future stage, urged CPI(M-L), but should be undertaken here and now. However, to the sorrow of CPI(M-L) the Indian state in the 1960s and 1970s proved to be too strong to be overthrown by Naxalites and their endeavour turned out to be "adventuristic" as CPI and CPI(M) had said it would. In defeat, CPI(M-L) itself became fractionated, and parts of it decided to participate in the parliamentary system, since people still held illusions about it, in the hope of overthrowing it.

Critiques of CPI(M-L) Position

The CPI(M-L) case centres around three elements: (1) the country never achieved independence; (2) its mode of production is semi-feudal like China's in the 1930s; and (3) its bourgeoisie is comprador. On the first point, regardless of their position on the country's reliance on foreign credits in its development programme, few serious scholars countenance the view that its independence is a sham. Even Ajit Roy, who quibbles over taking August 15, 1947 as marking the transfer of power, avers that "with the withdrawal of the British troops from Indian soil in early 1948, a real transfer of power was effected.... India became independent."[96] Both CPI and CPI(M) acknowledge, in retrospect, that the party's line during the 1948-1953 period on this question was not only empirically wrong but politically a mistake; the former states in regard to that line that "far from overthrowing the government of the national bourgeoisie, it actually aided the consolidation of the hegemony of this class over the Indian people as well as the strengthening of various reactionary elements," while CPI(M) says that finally the party "decisively rejected the pet theme of 'formal independence' with which we played for a time in our Party to the detriment of our working out a correct strategy and tactical line."[97]

On the second point, Dasgupta argues that, howsoever half-hearted the attempt at agrarian reforms in India, "the CPI(M-L) totally fails to appreciate the changes which have taken place in the Indian agriculture," that "there are very few 'very big landlords' left in India and their mode of exploitation is also different" than it was in pre-war China, that "the Indian villages have become more market-oriented than the villages of China during the thirties" and that "we must also note the phenomenon of 'capitalist farming' which is becoming more and more significant in certain regions of India."[98] Similarly, as already discussed above, Ajit Roy also takes the position that the semi-feudal landlords were eliminated as a class.

On the third point, Dasgupta and Roy both decisively reject the notion of Indian bourgeoisie as comprador. The former follows the standard argument to note both the competitive and complementary features of its relationship with foreign interests,[99] while the latter attributes to the monopoly bourgeoisie the intent of sponsoring relatively independent development, even as it functions under severe constraints.

3. The Autonomous Bureaucratic State

One radical interpretation takes the state to be an autonomous entity in relation to society in India. This has been put forward by Anupam Sen, who himself describes it as "a neo-Marxist perspective", even as he acknowledges that the "contention is so contrary to the 'traditional Marxist class theory of the state', in which the state is inevitably a means of class hegemony."[100]

The lineage for this position is, of course, traceable back to Marx, but not to Marx the philosopher of the universal theory of history but the propounder of the notion of the Asiatic mode of production. For the pre-colonial situation, Sen's thesis is quite understandable, for Marx's analysis was quite specifically addressed to India. The pre-colonial bureaucratic state opposed and suppressed any incipient capitalist tendencies, since the rise of the capitalist class would have threatened its autonomy and therefore its interests. The state autonomy thesis is equally understandable for the colonial period, since the essence of that period is the imposition of a political system by the metropolitan power without any organic links to local social classes over whom it dominated. While it transformed India from an exporter of manufactures to a supplier of raw materials and a market for industrial goods of the metropolitan power, and while it in the process created a comprador commercial bourgeoisie, the colonial state thwarted the development of an independent industrial bourgeoisie.

But Sen maintains that "the state in India, conditioned by the nature of its social formation was and still is autonomous." This state autonomy is attributed to the weakness of the bourgeoisie emergent from colonial rule, a weakness which facilitated the successor bureaucratic state to perpetuate its autonomy, now under the garb of socialism, "an ideology of state hegemony." As he puts it: "The 'socialistic form,' or the name 'socialism', was given only to legitimate what was already there, weak social classes dominated by a strong state which wanted to further consolidate its position by strengthening its economic power. The aim, indeed, was to make the state independent of economic subservience to the capitalist class." He adds emphatically: "In short, the Indian state was not a capitalist state" and "the state in India manifested itself over almost all social classes, the bourgeoisie, the peasants and the workers."[101]

Sen elaborates that "the attempt by post-independent state in India to maintain its autonomy" reflected itself in a wide range of activities to consolidate its own power and to weaken and control the bourgeoisie; it "resulted in extensive state control of the private corporate sector, the concentration of basic industries in the state sector, the support and encouragement of the artisan and petty industries as a counterpoise to the private corporate industries, and the failure of the bourgeoisie to transform agriculture into a capitalist undertaking." He holds the public sector in India to provide "formidable weapons for the state" to maintain its autonomy against the bourgeoisie.[102] It is this overall nature of the Indian state that, for him, has been responsible for the failure of capitalist transformation to take place and for the worsening of the condition of the masses.

There is considerable merit in Sen's analysis insofar as it underlines the enormous and overbearing strength of the Indian state in its relations with diverse social classes and equally insofar as it rejects the standard Marxist thesis of the Indian state as the state of the bourgeoisie. But it is mistaken in equating the lack of class hegemony of the bourgeoisie with the state as being autonomous from society. The analysis seems completely oblivious to the role of politics in linking society and state. In its eagerness to homogenize pre-colonial, colonial and post-colonial situations for the sake of developing a single global generalization, it simply neglects the electoral context which differentiates the post-colonial Indian state not only from earlier state formations but also from most other post-colonial states in the Third World. At the level at which the argument is cast, the analysis cannot discriminate between India and Pakistan over some four decades after independence.

Although the post-colonial state may initially have started out as autonomous from society, the subsequent compulsion of political legitimacy through elections has forced it to come to terms with social classes, even if no single class exercises hegemony. Despite its Marxist commitment, Sen's analysis is singularly innocent of dialectics and sees only a one-way relationship between state and society. It therefore misleads in concluding from what is obviously a political response to the electoral power of social forces, especially of the middle classes -- the creation of a vast sector of small-scale industries -- as being intended "for the state to maintain its hegemony over the two contending classes, the bourgeoisie and the petit-bourgeoisie". It is equally in error in neglecting the enormous social and political power of the middle peasantry, while insisting on the state's perception of itself as the protector of the vast numbers of small-holding peasants in order to maintain its independence. The same neglect characterizes Sen's treatment of the middle classes as well. At one place, he does make a minor concession to their power, acknowledging that if the state "had any relationship on the basis of inputs with any class, it was, to some extent, with the petit-bourgeoisie", but immediately adds: "However, in terms of output, it would be very difficult to locate the state in any class, as it appears, its policies were (and are) principally directed towards the augmentation of its own power and not the power of any social class."[103] For a more adequate recognition of the power of the middle peasantry and the middle classes, one would then have to examine a different genre of studies.

More critically, for a study which has the state as its centrepiece, the state remains strangely undefined, undescribed and unelaborated by Sen. What or who constitutes the state? Is it the top-level bureaucrats, or political leaders, or both? If it is the bureaucrats, why are they so powerless in their relationship with the political leadership? If it is the political leadership, what social classes does it represent politically? Sen, however, remains silent on these issues.

4. The "Intermediate Regime" Models

Kalecki's model of the "intermediate regime", with the ruling class consisting of urban middle classes and the rural rich and middle peasantry, was brought to the attention of students of Indian economy and politics by Dr. K.N.Raj in a public lecture in 1973.[104] In that lecture, Raj elaborated the model at an analytical level -- without any specific discussion of the Indian situation -- underlining the contradictions inherent in the heterogeneity of the class coalition of that regime, the resulting resource constraint on its economic instrument of state capitalism, and the tendency toward crisis in the regime. Despite the contradictions and the tendency toward crisis, such regimes had seemingly kept going and Raj related their persistence to the lack of sufficient strength on the part of the "upper middle classes" (big bourgeoisie) to overthrow and seize power and the lack of any better alternative offered by radical forces. Nonetheless, besides the gain from state capitalism in blocking economic domination by foreign MNCs, he felt that the intermediate regime -- insofar as it was based on adult franchise and thus ultimately dependent on political support of the lower classes -- provided "a framework within which further political evolution can shift the balance of power in favour of strata that are below the middle class now in position, it offers an opportunity that might not otherwise be available." However, his lecture was an implicit warning against the intermediate regime moving in the direction of private capitalism in its struggle to resolve the resource constraint problem, and simultaneously an implicit call that "if such a shift is to be prevented, and intermediate regimes made a transitional phase in the evolution towards a genuinely more broad-based political and economic system, they may need to shed altogether the alliance with the rich peasantry (and the 'upper middle class' in general) and secure stronger (and more enlightened) organisational support from among the lower classes."

At a more empirical level, political scientists Weiner, Arora and Kochanek and sociologist Beteille had earlier demonstrated the importance of the middle section of the stratification hierarchy in the politics of the country[105] and in 1970 Nayar underlined the thesis of the "middle sectors" as the ruling class thus:

> the levers of political and state power have rested in the hands of what may broadly be termed the 'middle sectors' of economic and social life in both the urban and rural areas -- the educated and professional groups, the town merchants and small businessmen in the urban areas; and the middle peasantry or kulaks in the villages. These are highly privileged groups in the context of the general economic backwardness of the country and the source of their power lies in the strategic combination of considerable population size with extensive economic resources and significant social status, as against the greater economic power but small numbers of the upper business and landowning classes and the large numbers but economic destitution of the lower classes. Socialism to the middle sectors has meant...the bringing down of the upper classes to the middle level, but no redistribution or levelling down below that level. Democracy has served these classes well in this

regard by facilitating the conversion of economic privilege and numerical strength into political power at the same time giving it an aura of genuine legitimacy.[106]

Although not based on rigorous, primary research, Nayar used this notion to explain policy in agriculture, industry, education and administration. Subsequently, he included in the middle sectors "the upper levels of the working class or proletariat, especially in the public sector," and underlined that the ruling coalition emerging out of the middle sectors has not been a static one:

> There is tension and conflict between its various constituents, which has been manifest in politics; indeed, it has provided the motive force for change in Indian politics. Changes in the power and influence of the constituent segments have been reflected in control over parties and government, and then in public policies. One dimension along which conflict among the middle sectors has become increasingly salient is that of the urban-rural division, setting the rising middle or intermediate castes in the rural areas against the middle classes of the urban areas. Since the late 1960s the intermediate castes of the rural areas, overlapping the kulak class, have become more assertive as they have prospered economically, and have forcefully demanded a greater share in political power. Outside the middle sectors, meanwhile, by virtue of its demonstrated efficiency as against the huge financial losses made by public sector enterprises, business has been demanding with greater self-confidence that government remove restrictions on its functioning for the sake of injecting more dynamism into the economy. At the same time, the working class has, as a result of enhanced social mobilization, been acting with better organization in the advancement of its interests.[107]

Nayar also saw the public sector as "the expression of the will to power of the middle sectors, particularly the middle classes of the urban areas, even though these classes have not been particularly successful in managing the institutions that they control, such as the universities. Although the justification for it comes in the name of socialism, it also represents some older values -- those of the intellectual caste of the Brahmins and of the ruling and administrative caste of the Kshatriyas as opposed to those of the lower but money-making merchant class. Nehru, who built the system, was an intellectual par excellence and a hero of the political left but also, by birth if not by belief, a Brahmin. In part, the sacred commitment to the public sector no doubt springs from idealistic motives, but in part also from the resentment of the intellectual and administrative classes against the economic success of the assumedly inferior classes. The patently sadistic satisfaction in placing controls over the private sector even at the cost of the growth of the national economy as a whole is too obvious to be missed."[108] However, he noted as well the purpose of countering economic dependence as motivating the national elite in establishing the public sector.

The importance of intermediate classes has been recognized by some Marxist scholars as well, though more narrowly conceived as the new middle class of the urban areas. This importance is situated in the context of the failure of the Indian bourgeoisie to be a revolutionary class and therefore forced to compromise with feudalism and imperialism. Drawing on the work of Gramsci, Asok Sen has theorized, without specifying how it managed to get hold of state power, that the resulting ruling bourgeois-landlord coalition lacked in social hegemony, that is, in "the capacity of the ruling class to prevail over the entire society in terms of its economic and cultural leadership."[109] In the situation of unrealized social hegemony, "the coalition therefore takes resort to bureaucratic domination," but he does not draw any conclusions about sharing of state power by the bureaucracy as an institutional group or as representative of the intermediate classes.

However, Pantham pushes this analysis further, and states in reference to the Indian bourgeoisie:

> Hence it has lacked hegemony over the society, and is able to rule over it only through the support of elite groups which perform the functions of coercion, bureaucratic control and political manipulation and legitimation. This situation gives a non-classical role and significance to the intermediate or "middle" class, and especially to such elite groups within it as the professional, bureaucratic-military, political and intellectual groups. Thus India has come to be ruled by a complex of dominant classes and elite groups. Hence neither the classical model of classes nor any simple elite-mass model of society can shed proper light on the socio-political origins of the crisis of the economic transition in India.[110]

In the first instance, for Pantham, the weakness of the bourgeoisie drove it into a compromise with feudalism and a coalition with the landlord class; this class coalition of bourgeoisie and landlords is simply assumed to hold state power, as if it were the natural order of things. However, even this "ruling class coalition has no hegemony or cultural and moral supremacy over the society. The rule or domination of this non-hegemonic class coalition is therefore crucially dependent on the support it gets from the elite groups belonging to the intermediate class."

Furthermore, the ruling class coalition is not simply dependent on the intermediate class and its elite groups, but now includes them within itself in a larger alliance based on convergence of interests of all; says Pantham: "The incorporation of the intermediate classes and elite groups into the ruling class coalition is not based merely on the bureaucratic requirements of the non-hegemonic bourgeoisie; it is also based on the narrow class interests of the heterogeneous intermediate class itself." The consolidation of the state apparatus and the overexpansion of the intermediate class under colonial rule not only made this class "unusually large in size" (exceeding the number of industrial workers), but also, combined with the lack of hegemony on the part of the bourgeoisie and top peasantry, "has given the petty bourgeoisie an ambivalent position between parasitism and quasi-autonomy." This quasi-autonomous position of the petty bourgeoisie is, according to Pantham, manifest in state capitalism becoming "a permanent feature in India," thus helping "the bureaucracy and, to some extent, the top-level political leadership to increase their power vis-a-vis the bourgeoisie."

But there is an element of mutuality in "the alliance of the professional and bureaucratic groups (which belong to the intermediate or 'middle' class) with the bourgeoisie" in that their high incomes and "the salary-hikes of the upper echelons of the professional bureaucratic class"-- which some would consider a questionable assumption -- provides the market for manufactured goods in a situation of mass poverty. Pantham holds that "formal state power at the all-India level is largely in the hands of the professional, bureaucratic and politico-intellectual groups," but it has been exercised to favour conservative class interests through conventional political brokerage rather than to push forward radical modernization.

More than Asok Sen, Pantham accords recognition to the autonomous political power of the intermediate class, especially the bureaucracy, while still staying within the Marxist formulation of bourgeois-landlord coalition as holding state power in India. Somewhat similar to Pantham's line of thinking, Bardhan in a neo-Marxist analysis posits a three-class dominant coalition of industrial capitalists, rich farmers and professionals in the public sector as holding state power in a relatively autonomous state. His argument is that India's slow economic growth is related to the heterogeneity of this class coalition, where resources are wasted in subsidies in behalf of the interests of the constituent classes rather than invested in accumulation. His position on state autonomy does not, however, fit well with his emphasis on the sensitivity of the state to inflation and its concern for anti-poverty programmes because of legitimacy needs; of course, "relative autonomy" is flexible enough to cover this situation.[111]

None of the Marxist interpretations on the intermediate regime are sensitive to differentiation within the bourgeoisie which sets the small-scale bourgeoisie in opposition to big bourgeoisie and into alliance with other intermediate strata. That issue has been ably addressed by economic columnist Jha in a booklength study, which is by far the most extensive, if not definitive, treatment of the "intermediate regime" in India. The book is centered around the major proposition that India's economic stagnation after the mid-1960s is related to the rise to dominance of the intermediate class. For Jha, this class consists of the self-employed, such as small-scale industrialists, the traders and the rich peasants. But, going beyond Kalecki and Raj, he isolates this heterogeneous group as a single class by its distinguishing characteristic of self-employment, which means, for example in the case of the urban areas, that "there is no divorce, or at best a nascent divorce between labour and capital on the one hand, and between capital and management on the other. Correspondingly, its earnings can neither be classified as a reward for labour, nor as a payment for risk-taking (i.e., profit), but are an amalgam of the two. The self-employed thus lie midway between the large-scale professionally managed capitalist enterprises of the private sector, and the working classes."

This is quite a comprehensive class, including not only traders and merchants, truckers and taxi operators, restaurant owners and small-scale industrialists, lawyers and doctors, but also other professionals and, importantly, "the owner proprietors of closely held companies who have no shareholders to answer to." In that manner this intermediate class is -- barring a few professionally managed companies and foreign corporations -- co--extensive with the definition of the national bourgeoisie by CPI, even though Jha attempts to delineate for himself a non-Marxist position albeit based on class analysis. Further, it includes the agricultural bourgeoisie as well, for Jha states "self-employment is not confined to the towns alone. The peasant farmer tilling his land, with or without hired labour, is also self-employed in precisely the same sense as the trader or the small-scale industrialist."[112] Not only does Jha conceptually isolate the intermediate class through its distinguishing characteristic of self-employment but he transforms it into a single class by a common interest in the existence of inflationary conditions stemming from a shortage economy, which provide enormous unearned tax-exempt or tax-evaded profits that, in turn, feed into a parallel black economy.

This parasitic intermediate class, according to Jha, sees as its enemies the foreign multinationals and the large business houses with professional management. It has therefore favoured a system of physical controls on their growth and expansion, even though such controls may have originated independently for other reasons, while it has encouraged the establishment of the public sector as an additional curb on them. In this it is alleged to have been joined by the Left, including the communists, which has, despite its professed radicalism, served as a Trojan horse for the intermediate class. The intermediate class further made the working class of the organized sector, which was protected against inflation by dearness allowances, into an ally against its enemies. The system of controls that the intermediate class has favoured has perpetuated a shortage economy which increases its profits while the financial burden is shifted to fixed-income and unorganized groups. Cornering an increasing share of the national product, the intermediate class thus expanded its economic power. Enhancing its total power on the social scene was its enormous size:

> ...the total number of wage earners in industry does not exceed 13 to 14 million. Against this, the number of farmers cultivating 10 acres or more amounts to nearly 11 million. Add to this five million shopkeepers, a million or so bus, truck, taxi and scooter rickshaw operators, around four million self-employed and their relatives in the unorganized sector, a quarter of a million or more professionals, and a few million corrupt civil servants, and it becomes clear that with at least 20 million income earners and eight to ten times as many dependents and relatives in joint families who have their interests, the intermediate class is easily the largest single class in the country.[113]

On this potent mixture of economic power and numerical size, the intermediate class -- a creation of post-independence policies -- rose to "dominant status in society"[114] in the mid--1960s. It is Jha's thesis that the subsequent economic stagnation and increased curbs on big business, as also increased subsidies and protection afforded to surplus farmers and small industry, following the 1967 elections -- in the aftermath of which the intermediate class is said to have captured the Congress via the route of first occupying the opposition parties -- are related precisely to the rise to power of the intermediate class; the intermediate class was strengthened by economic stagnation and, in turn, has used its strength to perpetuate stagnation, but has triggered a political revolt on the part of other groups that have suffered from stagnation.

While Jha has made a strong case for the recognition of the economic and political power of the intermediate class, perhaps there is an element of overreach in his explanation. Although his work is quite persuasive in terms of treating intermediate strata as a class, it is less satisfactory in relating this class to state power. For one thing, he does not situate his discussion of the intermediate class in any systematic theory or overall proposition on class and state power. For another, he himself trivializes his work by saying that his conclusions on the relationship of the intermediate class to state policies "are highly tentative, being more in the nature of promising hypotheses than firm conclusions" and by claiming only that the influence of the intermediate class extended to *reinforcement* rather than imposition of controls.[115] If that is the case, then the more important question relates to the class basis of state power when controls were imposed, when the heavy industry strategy was initiated, when the public sector was launched, when big business was excluded from key industrial sectors, when the socialist pattern of society was proclaimed as India's ultimate goal, when modest agrarian reforms were introduced which created the rural component of the intermediate class -- all a decade to a decade and a half before the rise of the intermediate class to dominance in society. In this regard it would seem that Jha slights the independent influence of the upper-caste, English-educated, urban middle class -- with its modernizing, if not socialist, ideology -- that was pre-eminent in the Congress party and government in the 1950s. He does not ask why this class undertook the policies it did that were eventually to enhance the power of small-scale industrialists and rich peasantry but diminish its own. Besides, even the economic stagnation which is so central to Jha's argument began with the drought years of 1965 and 1966, much prior to the 1967 elections which, he claims, brought the intermediate class to political prominence even though initially in the opposition parties.

Again, Jha overreads the significance of the intermediate class in the Emergency, holding Mrs. Gandhi to be the representative of this class and therefore interested in protecting its position while the Janata forces were rising against it; an opposite but equally plausible case could be made, on the basis of evidence in his book, that Mrs. Gandhi was moving toward a growth-oriented policy which alienated the intermediate class and drove it into the arms of the Janata forces; for, after all, who did the Jan Sangh and Lok Dal represent? Such an interpretation would be more consistent with the attempts by Mrs. Gandhi, after her return to power in 1980, at greater liberalization of the economy and a more hardheaded attitude toward the public sector. Regardless of these deficiencies, Jha's work is significant for underlining the importance of the intermediate class though, in the final analysis, that class is largely equivalent to the national bourgeoisie of CPI conception.

Another independent-minded economist, Ashok Desai, also takes the economic stagnation of India, more particularly the "persistently low rate of industrial growth for almost 15 years," as a starting point for analysis but comes out with somewhat different results in terms of the class basis of the state.[116] He dismisses "kulak power" as an explanation, advanced by Ashok Mitra, even as he maintains that there has been a persistently high rate of industrial investment, thus holding high capital-output ratios as the critical variable in industrial stagnation. Here, the villain of the piece is the public sector, not simply because it involves high capital intensity industry but also because "capital-output ratios are higher in the public than in the private sector in the same industries." What then is the explanation? He sidesteps the argument about high capital-output ratios being a function of (1) corruption among politicians and bureaucrats which increases the capital costs of public sector projects and (2) the additional "social responsibilities" assumed by the public sector. Instead, he provides a class explanation; he does not elaborate precisely the constituents of his class but it obviously refers to the urban educated middle class and the salariat as also, distinctively, the labour aristocracy or labour in highly skilled occupations and in the organized sector.

Desai states: "the strongest political force in this country, even more powerful than the kulak, the bourgeois and the capitalist combined, is the petty-bourgeois haute-proletarian with a little education and less property." Inheriting power at independence, he continues, this class "set about multiplying itself by setting up schools and colleges," as a result of which within two decades a vast army of job aspirants belonging to this class was created, "all clamouring for jobs." Under the pressure of this class, politicians -- most of whom belonged to the same class -- sought job creation for their friends and relatives through influence with private industrialists but, having exhausted this channel, "finally, they turned to two institutional means -- small enterprises and public enterprises." For Desai, the small-scale industrial sector is "another equally scandalous story" but, concentrating on the public sector, he argues that "the same forces that led to the proliferation of public enterprises also led to their overmanning, which became irremediable because, once appointed, no employee of a public enterprise could be dismissed." This overmanning, the result of the effort of the petty-bourgeoisie and haute-proletariat "to squeeze in evergrowing numbers into an industrial base that is not growing fast enough", was corrosive for productivity and technology development: "since no one wanted to be found overtly doing nothing, a work ethic grew up among employees by which everyone worked less than he could." The minimum-work ethic is what has brought about the high capital-output ratios and has made the public sector unprofitable, drawing then on general revenues for investment. But Desai predicted that a situation would develop where government revenues would be sufficient only "to maintain a vast and inefficient industrial base and no expansion of industrial employment is possible." At that point, he said, the petty-bourgeois haute-proletarian class will "rise in revolt and destroy the system that employs them, but fails to exploit them."

Summary and Conclusions

It is a tribute to Marxism in terms of its concern with vital issues of social relevance that it is its followers who have grappled the most with the question of who controls state power in India. No other ideology or paradigm matches it in its attention to this question. It should occasion little surprise then that most of the interpretations of state power considered here happen to be Marxist. Indeed, any general inquiry into interpretations of state power ends up being essentially an inquiry into Marxist interpretations of state power. Consequently, one must approach Marxist contributions on the subject with a degree of humility.

On the other hand, one need not be awed into accepting the Marxist model; indeed, Marxism provides no single model of state power. Marxism today is like Hinduism, encompassing the most varied interpretations of state power, often contradictory; it even has its own secular version of *maya* with its position that -- actual powerholders can be ignored because they camouflage the real wielders of power. Nor does Marxism possess any special or unique tools of analysis that could enable it to provide a single thesis or theory on state power. The very diversity of interpretations among self-professed Marxists and the rapidity with which their positions can shift are testimony to the absence of such analytical skills and tools.

Moreover, there is in Marxism even lack of clarity in relation to the most fundamental concepts employed. There can be no more central concepts to Marxism than class, state and mode of production; indeed, the entire edifice of Marxism rests on them. Yet Chilcote, a scholar sympathetic to Marxism, repeatedly comments that "Marx did not fully elaborate a conception of class," that "Marx did not fully develop a theory of class," that "Marx never fully elaborated a theory of state and class," and that "Marx offered no explicit definition of class." Even the term "middle class" is used by Marx flexibly, at times to refer to capitalists, at others to strata between capitalists and workers, and Raj rightly notes that there is "some ambiguity about the concept in Marxist literature itself." Again, Alavi bewails the fact, that "Much of the debate in this area has been vitiated by misunderstandings and disagreements over some basic concepts. The difficulty stems, in no small measure, from the fact that Marx does not provide us with a concise and precise definition of the basic concept of 'mode of production,' as a concept of social structure, which he alludes to....".[117] Beyond this, concepts are so flexibly interpreted as to make them mean anything.

More specifically in relation to state power in India, there is among Marxists a fundamental division between (1) those who see it as a function of classes in society, and (2) those who see the state as a bureaucratic phenomenon, autonomous and independent, standing above and over all social classes. Most Marxist interpretations fall in the first category, but the second category is represented here by Anupam Sen and Bardhan and, to some extent, by Asok Sen and Pantham; more importantly, the second category traces its descent back to Marx himself and that too in respect of his work specifically on India, albeit pre-colonial India, relating to the Asiatic mode of production. One thing is clear, however, that, unlike many Third World countries, India has not seen the military as a bureaucratic institution become a sharer in state power for nearly four decades after independence.

In considering the standard Marxist approach which posits state power as a function of social classes, one is again struck by the diversity of interpretations. At the simplest level, one can with Marxism conceive of society as being divided dichotomously between the ruling or dominant class and the ruled or dominated class. Most Marxist interpretations that have been considered here would no doubt be consistent with the notion that the state is an administrative organ of the ruling class but, at that level of generality, the instrumental state is of little relevance for separate discussion.

One can, however, further look at society as being divided into (1) upper classes, (2) intermediate classes, and (3) lower classes. One could refine this description even more by distinguishing between rural and urban classes, and again on the basis of whether they are set in modern (industrial) or traditional (feudal) sectors, but for purposes of discussion here the simpler three-fold classification will be used in order not to overcomplicate or clutter analysis.

Most orthodox Marxist interpretations of the state pertain to the upper classes (bourgeoisie or landed aristocracy) holding state power. Such is the case with analyses of state power in India where either a single class like the bourgeoisie (as in the instance of CPI) or a class coalition of bourgeoisie and landlords -- as in the examples of CPI(M) and CPI(M-L) -- is said to hold state power. At first blush, such a position seems precise enough in regard to the upper classes, but in actual fact the terminology is vague and flexible to encompass the middle strata as well. The bourgeoisie is in the case of CPI a rather comprehensive class, incorporating both urban and rural elements and also fractions across the entire spectrum in terms of size -- big, medium or small. Thus this definition incorporates elements of what may be considered as belonging to the middle classes. Furthermore, even the lower classes are recognized as having an input into state policy in the form of "heavy mass pressure" and "popular intervention." In the case of CPI(M), too, the bourgeoisie is similarly a broad and comprehensive group. Furthermore, the CPI(M) also does not ignore the role of the masses in state power. Where the two parties differ is in relation to the role of big or monopoly bourgeoisie; whereas for CPI "the big bourgeoisie wields considerable influence," for CPI(M) the class coalition is led by "the big bourgeoisie."

More importantly, the two parties differ on the nature of the bourgeoisie. For CPI the national bourgeoisie is opposed to imperialism and wants to build an independent national economy and polity even though it is driven to compromise with imperialism. What is even more interesting is the fact that the big or monopoly bourgeoisie of CPI's conception -- the very top of the economic hierarchy -- is the target of measures to control and check it, these measures issuing out of a state under the control of the national bourgeoisie. For CPI(M), on the other hand, the bourgeoisie forges "strong links with foreign monopolists" in order to strengthen its hold on state power. For CPI(M-L) the class coalition of bourgeoisie and landlords is further a comprador coalition, subordinate to imperialism of both the American and Soviet varieties.

Although there is some furtive recognition of the share of intermediate classes (rural and urban) in state power in India by CPI and CPI(M) through the flexible usage of the vague term bourgeoisie but without an open and explicit acknowledgement of it, the power of intermediate classes is more openly posited by Marxist scholars, rather than Marxist parties. Among such scholars are Kalecki, Asok Sen, Pantham, and Bardhan. Even more orthodox Marxist scholars, such as Ajit Roy and Dasgupta, are also apt to implicitly acknowledge intermediate classes as co-sharers in state power in India. Non-Marxist scholars using Marxist concepts to underline the power of intermediate classes include Raj and Jha. The policy manifestation of intermediate classes controlling state power is considered to be state capitalism.

When it comes to the lower classes, Marxism does not countenance any notion of them as sharers in state power. But, as has been seen in the position of CPI and CPI(M), they are implicitly acknowledged to have a share in state power, at times against the interests of the big bourgeoisie. Especially critical in this regard is the share of these two parties themselves in state power as parties of the proletariat, which then are able to use state power to enact policy measures to advance or protect the interests of the lower classes, as in Kerala and West Bengal. It is noteworthy, too, that those who articulate the notion of "intermediate regime", or of the state under the control of intermediate strata, include in that strata also those sections of the proletariat that are in the organized corporate sector. Interestingly, in private, partisans of CPI and CPI(M) are ready to acknowledge that many state policies have emerged from a state that is, in effect, a multi-class coalition which is responsive to "national" interests or the interests of classes nearly all along the spectrum of the social hierachy. Equally, they concede the importance of ideas and ideological heritage in the formulation of state policies.

The wide variety of interpretations of state power in India point to the fact that Marxism, much like the social sciences in general, is more important for the questions it asks than for the answers it provides. Perhaps Indian social reality is too complex to be encompassed within the orthodox framework of Marxism, especially when the Indian state is a state-in-formation, not a consolidated stable phenomenon. As Hanson once concluded: "In sum, India has an exceptionally complicated and highly fluid class structure, in which it would be futile to search for any unique social source of political power."[118] Meanwhile, concessions to social reality make Marxist interpretations come close to the liberal-pluralist model. However, that model, while in some sense a more appropriate description of the Indian state, is not very sensitive to the degree of power held by different classes. On the other hand, Marxist interpretations often come more as assertions, if not dogma, rather than theses founded on rigorous and empirical study. And it is in the direction of a more empirical focus in the investigation of the relationship between class and state that the work of future researchers ought to proceed. Empirical research could concentrate on both the social background of decision makers and policy analysis of a sample of key decisions. However, it would seem that such an endeavour would be more fruitful if it differentiated India's social reality in terms of political space and time rather than attempting to compress it into a single mould for purposes of a grand generalization.

In terms of political space, Indian social reality could be differentiated among local, state and national levels. Tentatively, it would seem that at the local level (village, district) economic power, especially of the landed classes, is closely associated with social and political power. Accordingly, at this local level political power would appear to correspond closely to the Marxist model. At the state level, the power of economically dominant classes in the rural areas is obviously manifest, but universal franchise facilitates the aggregation of power by other groups so as not only to be taken into account in political decision-making but to have a share in it. It is because of the operation of this principle of power aggregation that it has been possible for Communist parties to come to power in the states of Kerala,

Tripura and West Bengal or for them to be part of ruling coalitions. Such power achievement can be of tremendous importance; as Kohli has shown, state governments can, as for instance in West Bengal and Kerala, enact measures of reform on behalf of the lower classes and furthermore bring about a disassociation of economic and social-political power at the local level.[119] In this fashion, the model of governance at the state level moves considerably in the direction of the liberal-pluralist model. One needs to keep in mind as well that the lack of monopoly of political power by economically dominant classes that has become manifest in West Bengal operated also earlier but within the Congress party -- a pattern typical of most states -- although perhaps in not as rigorous or adversarial a fashion as under Communist auspices.

At the national level, the vast diversity of India diffuses political power even further among a large number of groups so that they share in power and are taken into account in political decision making, and consequently the liberal-pluralist model is operative to a greater extent than in the states. The economically dominant rural classes, while obviously influential, do not carry the same weight here as at the state level, which has become a point of grievance with them. Big business has a presence, but it is one among many other groups, including the middle or intermediate classes.

In terms of political time, a major divide can be set at the mid-1960s, distinguishing between the Nehru era and the post-Nehru period. Again, one could tentatively suggest that, during the Nehru era, the upper landowning classes (the rich and middle peasantry, and large landowners but not feudal landlords who had allied earlier with the British) and middle classes were in power in the states. But gradually power shifted to the new kulak classes (rich and middle peasantry originating in the intermediate castes but increasingly taking to capitalist farming) as the dominant partner, with a receding role for the middle classes -- a pattern that became more fully manifest in the post-Nehru period. At the national level, the Nehru era was characterized by political power in the hands of a largely autonomous, modernizing political elite, headed by a charismatic leader, but power rested on the base of an alliance between the urban middle classes and upper landowning classes in the rural areas. With the passage of time, the political elite's autonomy from society eroded as a result of periodic elections, while the rural kulak classes became more powerful at the cost of the urban middle classes; still, the middle sectors together constituted the major sharers of political power. However, the base of support of the ruling Congress party in the 1970s shifted to the minorities, Harijans and the poor, as a result of the party's turn to a more populist stance. This populist stance brought about programmes to help the poor, whose implementation got subverted at the local levels because of the association of economic and political power in the hands of landowning classes. Although big business hovered around with demands, it had no direct share in political power. It may be suggested that the political leadership is simply a front for big business, but political leaders are interested in election and, while money is important, election is not simply a function of money and the electorate also demands its price for political support. There is no greater testimony to the failure of money to buy votes than the election of communist governments to power.

If the question of what social classes hold state power in India is hostage to a bewildering number of answers, no less so is the question of the class forces behind the creation and maintenance of the public sector. Some of the different interpretations on the public sector can be summarily noted here:

1. *Public Sector as a National Project*: The public sector was developed by an autonomous and benign modernizing national elite as part and parcel of the overall necessary effort for nation-building. (Kothari, 1970)
2. *Public Sector as an Instrument to Counter Big Bourgeoisie and Imperialism*: The public sector, even though built by a state under the control of the bourgeoisie, was an instrument of the national bourgeoisie -- directed at undercutting the attempt of the monopolist section of the bourgeoisie to acquire complete control of the economy and the state -- as also an instrument aimed against imperialism. (CPI)
3. *Public Sector as an Instrument to Advance the Interests of the Big Bourgeoisie:* The bourgeois-landlord state under the leadership of the big bourgeoisie has established the public sector "as an instrument of building capitalism and is nothing but state capitalism." (CPI-M)
4. *Public Sector as a Tool for Plunder by Imperialism :* Rather than being counter-imperialist, the public sector had been promoted by imperialism (American and Soviet) to develop comprador-bureaucrat capitalism in India. (CPI-ML)
5. *Public Sector as an Economic Power Base of an Autonomous Bureaucratic State* : A strong state in the context of weak social classes has through the public sector, under the mask of socialism, attempted "to further consolidate its position by strengthening its economic power." (Anupam Sen)
6. *Public Sector as a Project of the "Intermediate Classes"*: The public sector as part and parcel of 'state capitalism' is the economic expression of state power in the hands of a class coalition of the "intermediate" or middle sectors (middle classes, rich and middle peasantry, organized labour) and is directed against the bourgeoisie above and the less privileged below. (Kalecki, Raj, Nayar, Jha, Desai)
7. *Public Sector as a Function of the Alliance of the Middle Class with Bourgeois-Landlord Coalition* : State capitalism is either a concession to the middle class on the part of a bourgeois-landlord coalition that lacks hegemony, or it is created to serve the interests of the alliance of bourgeoisie, landlords and petty bourgeoisie. (Asok Sen, Pantham, Bardhan)
8. *Public Sector as a Function of the Ideology of Prime Minister Nehru*: The public sector was founded less because of historical necessity for nation-building or because of its necessary instrumentality for the class interests of a particular class or class coalition but because of the ideology of one man, who emerged as a powerful leader at the helm of affairs at a critical juncture in the history of the nation. (composite position)

As with the question of state power, the resolution of the conflict among these varying views on the establishment of the public sector in India requires empirical treatment.

NOTES

1. Rajni Kothari, "Form and Substance in Indian Politics," *Economic Weekly*, (1961), "The Congress System," *Asian Survey*, IV, no. 12 (1964), pp. 1161-73, and *Politics in India* (Boston: Little, Brown and Company, 1970), 461 pp. The foreword to the book is by Gabriel A. Almond, James S. Coleman and Lucian W. Pye.
2. *Ibid.*, pp. 6, 10.
3. *Ibid.*, pp. 112, 222, 417-18, 434, 447.
4. Rajni Kothari, "A Fragmented Nation," *Seminar*, No. 281 (January 1983).
5. Unlike the major work *Politics in India*, the critical writings on the Indian state have appeared in the form of articles and contributions. Here I have drawn on: (1) "Crisis of the Moderate State and the Decline of Democracy" (38 pp; typescript, 1982), which was written for a festschrift in honour of Professor W.H. Morries-Jones; (2) "Rebuilding the State," *Seminar*, No. 257 (January 1981), pp. 1-8; and (3) "Democracy and Fascism in India," (28 pp; typescript, 1981), which is a larger version of the article "Where Are We Heading?" that appeared in *Indian Express*, November 29, 1981.
6. In a private conversation in 1982, Kothari explained that the Emergency was instrumental in bringing liberals and Marxists together; the Marxists recognized the genuine commitment of liberals to democracy and liberty, while the liberals came to understand, through their personal experience in jails, the earlier agony of the Naxalites.
7. "Rebuilding," p. 2.
8. "Democracy," *op. cit.*
9. *Ibid.*
10. "Crisis," pp. 25-26.
11. *Ibid*
12. Marcus Franda, *Radical Politics in West Bengal* (Cambridge, Mass.: MIT Press, 1971).
13. P.C. Joshi, "Reflections on Marxism and Social Revolution in India," in K.N. Panikkar, *National and Left Movements in India* (New Delhi: Vikas, 1980), pp. 183-84.
14. Bipan Chandra, "Karl Marx, His Theories of Asian Societies and Colonial Rule," *Review*, V, no. 1 (1981), 13-91.
15. Jairus Banaji, "The Comintern and Indian Nationalism," in Panikkar (ed.), pp. 213-265.
16. "Draft Platform of Action," in Democratic Research Service, *Indian Communist Party Documents 1930-1956* (Bombay: Democratic Research Service, 1957), pp. 1-21.
17. Banaji, pp. 241-42.
18. Gene D. Overstreet and Marshall Windmiller, *Communism in India* (Berkeley: University of California Press, 1960), p. 158.
19. Banaji, p. 261.
20. Overstreet and Windmiller, p. 260.
21. G. Adhikari, *Communist Party and India's Path to National Regeneration and Socialism* (New Delhi: Communist Party of India, 1964), pp. 88-89, 93.
22. Overstreet and Windmiller, p. 260.
23. M.B. Rao (ed.), *Documents of the History of the Communist Party of India* (New Delhi: People's Publishing House, 1976), vol. VII (1948-1950), pp. 1-118.
24. *Ibid.*, pp. 15-16, 31, 40.
25. *Ibid.*, pp. 31, 55.
26. *Ibid.*, pp. 30, 69.
27. *Ibid.*, pp. 115, 46, 48.
28. *Ibid.*, pp. x, 76, 806.
29. *Ibid.*, pp. 669-774, 806-808.
30. *Ibid.*, pp. 747-51, 790-91, 1042-43.
31. *Ibid.*, pp. 640-44, 746-57, 812, 837, 1042-47, 1066, 1086.
32. *Ibid.*, pp. 968-69, 975, 979, 987, 989-90, 994, 1007. The authors were: Ajoy Ghosh, S.A. Dange, and S.V. Ghate.
33. Mohit Sen (ed.), *Documents of the History of the Communist Party of India* (New Delhi: People's Publishing House, 1977), vol. VIII (1951-1956), pp. 1-18, 42-54.

34. Adhikari, p. 120.
35. Victor M. Fic, *Peaceful Transition to Communism in India* (Bombay: Nachiketa Publications, 1969), p. 137.
36. Mohit Sen (ed.), *Documents*, pp. 1-18.
37. *Ibid.*, pp. 42-54.
38. See Fic, pp. 96-104, 116-63; also Overstreet and Windmiller, pp. 314-17.
39. Adhikari, p. 126.
40. Democratic Research Service, *Communist Party Documents*, p. 262.
41. Adhikari, pp. 127, 128, 130.
42. *Ibid.*, pp. 131, 133.
43. *Ibid.*, pp. 133; emphasis in the original.
44. "Report to the Party Congress," in Democratic Research Service, *Indian Communist Party Documents*, pp. 232-33.
45. Adhikari, p. 134.
46. Adhikari, pp. 133-34; see also Fic, pp. 153-54.
47. Adhikari, p. 139.
48. Mohit Sen (ed.), *Documents*, p. 426.
49. Democratic Research Service, pp. 282, 292.
50. Adhikari, pp. 140, 144, 152-53.
51. *Ibid.*, pp. 156, 157, 160.
52. Adhikari, pp. 162-69.
53. Communist Party of India, *The Programme of the Communist Party of India* (As adopted... December 1964) (New Delhi: 1965), pp. 25-26.
54. Mohit Sen, *Aspects of CPI Programme* (New Delhi: Communist Party of India, 1966), p. 18.
55. Communist Party of India, *Proceedings of the Seventh Congress of the Communist Party of India: Bombay, 13-23 December 1964: Volume 3, Discussions* (New Delhi: 1965), p. 44.
56. Mohit Sen, *Aspects*, p. 18.
57. CPI *Proceedings of the Seventh Congress*, p. 44
58. Mohit Sen, *Aspects*, pp. 19-20.
59. CPI *Programme*, p. 15.
60. CPI *Programme*, p. 38.
61. *Ibid.*, pp. 7-9, 32.
62. *Ibid.*, p. 14.
63. *Ibid.*, pp. 36-39, 50-51.
64. CPI *Programme*, pp. 43-45; see also Mohit Sen, *Aspects*, p. 33.
65. Interview No. 111 in New Delhi on November 4, 1982.
66. Interview No. 73 in New Delhi on October 4, 1982.
67. Ajit Roy, "Sharers in Indian State Power," in K. Mathew Kurian (ed.), *India—State and Society: A Marxian Approach* (Bombay: Orient Longmans, 1975), pp. 129-43.
68. Biplab Dasgupta, "Class Character of the Ruling Class in India," in Kurian (ed.), 115-28.
69. Communist Party of India (Marxist), *Programme: Adopted... 1964* (New Delhi: 1979), pp. 22-23.
70. A party publication refers to the big bourgeoisie's "alliance with feudal and semi-feudal landlordism." See Communist Party of India (Marxist), Polit Bureau, *Ideological Debate Summed Up* (Calcutta: 1968), p. 105
71. Mohit Sen, *Aspects*, p. 47.
72. CPI(M) *Programme*, p. 27.
73. *Ibid.*, p. 23.
74. *Ibid.*, pp. 4-6.
75. *Ibid.*, pp. 6-9.
76. *Ibid.*, p. 10.
77. Mohit Sen, *Aspects*, pp. 7, 44.
78. CPI(M) Polit Bureau, *Ideological Debate*, p. 104.
79. CPI(M) *Programme*, p. 11.
80. *Ibid.*, pp. 18, 22.

81. Communist Party of India (Marxist), *Documents of the Eleventh Congress of the Communist Party of India (Marxist): Vijayawada, January 26-31, 1982* (New Delhi: 1982), p. 293.
82. *Ibid.*, pp. 12, 32, 45.
83. Interview No. 108 in New Delhi on November 2, 1982.
84. Interview No. 105 in New Delhi on October 30, 1982.
85. Interview No. 75 in Delhi on October 5, 1982.
86. Interview No. 169 in New Delhi on December 4, 1982.
87. Note the description by an economist at Jawaharlal Nehru University: "State power continues to be based on a coalition between the bourgeoisie and large landowners. The coalition has three specific elements: the monopoly bourgeoisie; the small urban bourgeoisie consisting of businessmen confined to single industries or states and professional groups who are not direct exploiters but integrated into the system of exploitation like lawyers, managers and upper bureaucracy; and finally the class of landlords and rich peasants..." Prabhat Patnaik, "Imperialism and Growth of Indian Capitalism," in Kurian (ed.), p. 148; emphasis added.
88. Mohit Sen, *Aspects*, especially pp. 43-53.
89. Roy, pp. 133, 136.
90. Ajit Roy, *Political Power in India: Nature and Trends* (Calcutta; Naya Prakash, 1975), pp. 140-44.
91. Roy, "Sharers in Indian State Power," pp. 136-37. He adds: "But, if and when it considers this support dispensable or loses it for some reason or other, it may seek to alter the form of state power......its culmination into the naked dictatorship of Indian monopoly capital cannot take place without a showdown between the two partners." Another scholar also sees the ruling class as split between the industrial bourgeoisie and the rich peasantry along with the latter's middle peasant and landlord allies; see T.V. Sathyamurthy, "State Power and Class Conflicts in India," *Mainstream*, (June 4, 1983), pp. 11-17.
92. Dasgupta, p. 128.
93. Communist Party of India (Marxist), *Documents of the Eleventh Congress*, p. 125.
94. Ajit Roy, *Economics and Politics of Garibi Hatao* (Calcutta: Naya Prakash, 1973), p. 137.
95. See "Programme of the Communist Party of India (Marxist-Leninist)," *Liberation*, IV, no. 4 (April-June 1971); reprinted in *Chingari* (Toronto), IV, no. 3 (July-October 1971), pp. 60-63, in Samar Sen et al., *Naxalbari and After: A Frontier Anthology* (Calcutta: Kathashilpa, 1978), pp. 275-84, and also in Sankar Ghosh, *The Naxalite Movement: A Maoist Experiment* (Calcutta: Firma K.L. Mukhopadhyaya, 1974), pp. 190-99. For a more elaborate version of the Maoist line in India, see T. Nagi Reddy, *India Mortgaged: A Marxist-Leninist Appraisal* (Anantapuram, Andhra Pradesh: Tarimela Nagi Reddy Memorial Trust, 1978). See also Suniti Kumar Ghosh, "The Indian Bourgeoisie and Imperialism," *Bulletin of Concerned Asian Scholars*, XV, no. 3 (1983), pp. 2-16.
96. Roy, "Sharers in Indian State Power," p. 129.
97. Mohit Sen, *Aspects*, p. 1, and Polit Bureau, *Ideological Debate*, p. 112.
98. Dasgupta, p. 127.
99. *Ibid.*, pp. 118, 126.
100. Anupam Sen, *The State, Industrialization and Class Formations in India: A Neo-Marxist Perspective on Colonialism, Underdevelopment and Development* (London: Routledge & Kegan Paul, 1982), p.209.
101. *Ibid.*, pp. 6, 87, 104.
102. *Ibid.*, pp. 7, 221.
103. *Ibid*, pp. 124, 105.
104. K.N. Raj, "The Politics and Economics of 'Intermediate Regimes'," *Economic and Political Weekly*, VIII, no. 27 (July 7, 1973), pp. 1189-98.
105. Myron Weiner, *Party Building in a New Nation: The Indian National Congress* (Chicago: University of Chicago Press,1967); Satish K. Arora, "Social Background of the Indian Cabinet," *Economic and Political Weekly*, VII (August 1972), pp. 1523-32; Stanley Kochanek, *The Congress Party of India* (Princeton: Princeton University Press, 1968), pp. 319-404; Andre Beteille, *Caste, Class, and Power* (Berkeley: University of California Press, 1965), pp. 185-225.

106. Baldev Raj Nayar, "Business Attitudes Toward Economic Planning in India," *Asian Survey*, XI, no. 9 (1971), pp. 850-65 (originally presented as a paper at AAS meeting at San Francisco, April 1970), and "Political Mobilization in a Market Polity: Goals, Capabilities and Performance in India," in Robert I. Crane (ed.) *Aspects of Political Mobilization in South Asia* (Syracuse, N.Y.: Syracuse University, Maxwell School, 1976), pp. 135-59.
107. Baldev Raj Nayar, *India's Quest for Technological Independence: Vol. 1 Policy Foundation and Policy Change* (New Delhi: Lancers Publishers, 1983), pp. 116-23.
108. Baldev Raj Nayar, *India's Quest for Technological Independence: Vol. 2 The Results of Policy* (New Delhi: Lancers Publishers, 1983), p. 21.
109. Asok Sen, "Bureaucracy and Social Hegemony," in *Essays in Honour of Prof. S.C. Sarkar* (New Delhi: People's Publishing House, 1976), pp. 667-85.
110. Thomas Pantham, "Elites, Classes and the Distortions of Economic Transition in India," in Sachchidananda and A.K. Lal (eds.) *Elite and Development* (New Delhi: Concept Publishing Company, 1980), pp. 71-96.
111. Pranab Bardhan, *The Political Economy of Development in India* (Oxford, UK: Basil Blackwell, 1984). It needs to be added that the position these scholars represent is of longer standing than is assumed. Note the description of the "structuralist" position in 1974: "the balance of political and economic power has been tilted in favour of a triple alliance comprising of (a) the rich sections of the farming community, (b) the affluent segments or urban industry and trade, and (c) the salariat and the relatively well-paid sections of the organized labourforce; this alliance has taken control of the levers of power...." C.N. Vakil and others, *A Policy to Contain Inflation with Semibombla* (Bombay: Commerce, 1974), p. 31.
112. Prem Shankar Jha, *India: A Political Economy of Stagnation* (Bombay: Oxford University Press, 1980), p. 95.
113. *Ibid.*, p. 103.
114. *Ibid.*, p. 132.
115. *Ibid.*, pp. vii, 120.
116. Ashok V. Desai, "Factors Underlying the Slow Growth of Indian Industry," *Economic and Political Weekly*, XVI, nos. 10-12 (March 1981), pp. 381-392.
117. Ronald H. Chilcote, *Theories of Comparative Politics: The Search for a Paradigm* (Boulder, Colorado: Westview Press, 1981). pp. 122. 348, 381; K.N. Raj, "The Politics and Economics of 'Intermediate Regimes'," pp. 1190, 1195; and Hamza Alavi and Teodor Shanin (eds.), *Introduction to the Sociology of 'Developing Societies'*, (New York: Monthly Review Press, 1982), pp. 176-77.
118. A.H. Hanson, *The Process of Planning* (London: Oxford University Press, 1966), p. 243.
119. Atul Kohli, "Parliamentary Communism and Agrarian Reform: The Evidence from India's Bengal," *Asian Survey*, XXIII, no. 7 (July 1983), pp. 783-807.

Chapter III

Class and Ideology in the Age of Nationalism

The two distinctly archetypal forms for the organization of a nation's economic activity -- capitalism and socialism -- have undergone considerable modification in recent history in the direction of "mixed economy". This has been, in good measure, the consequence of a combination of the requirements of legitimacy and effectiveness. What is striking is the assumption of an active role by the state in capitalism in response to "market failures" as well as the allocation of an important role to the market in socialism for reasons of "government failure." The increased role of the state in capitalism is manifest not simply in regulation or management of the economy, but also in entrepreneurship.

If the state has become increasingly involved in entrepreneurial activity more generally, this is especially so in the case of the Third World; as Evans points out, "the centrality of the state to accumulation on the periphery is incontrovertible," and "unless the state can enforce a priority on local accumulation and push local industrialization effectively, there is no effective sponsor for peripheral industrialization." Indeed, earlier, Gerschenkron had held the state's entrepreneurial role to be critical for economically backward societies; in the absence of it, opportunities for industrialization would be missed. He further hypothesized that the extent of the state's entrepreneurial role was, as part of the compulsions of economic backwardness, directly proportional to the extent of the economic gap between the backward society and advanced industrial societies. Similarly, Jones and Sakong contend that "the need for government intervention will tend to be greater the lower the absolute level of development, since market failures will be more widespread."[1]

However, there is considerable variation in the entrepreneurial role assumed by the state in Third World countries that are similarly placed economically. One cannot, therefore, simply assume the degree of economic backwardness to be directly related to the state's entrepreneurial role in some pragmatic, but simply automatic, adjustment to the fact of economic backwardness. One needs additionally to take account of the factor of ideology. To be sure, Gillis remarks that "the rationale for establishing public enterprises has usually had very little to do with socialist ideology or, for that matter, any coherent ideological framework of any kind." However, this represents too sweeping a generalization across the entire Third World, and even Gillis is forced to acknowledge that there are exceptions. On this issue, of course, there is considerable difference of opinion; indeed, another economist holds that "the reasons for which public enterprise has been adopted in developing countries are probably more often social or abstractly ideological than economic." The role of ideology in relation to the public sector becomes manifest particularly strongly in respect of "denationalization," for, as Lamont pithily points out, "denationalization is a problem of ideology."[2] The relationship of ideology to the public sector therefore demands adequate attention in the study of specific countries. Here, three different models are worthy of note.

The "mixed economy" may be, for a variety of reasons, an ideological or consummatory value in itself. Many years ago, the Marxist scholar Sachs isolated a model of mixed economy where "the public sector is given a permanent place in the economy and its rate of growth is to be higher than that of the private sector, so as to achieve progressively a situation of predominance of the public sector in the whole economy." Along with that as well as comprehensive economic planning, certain strategic sectors are reserved solely for the state and here "the predominance of the public sector is to be achieved as soon as possible." As a result, "the development of private monopoly capital will be restricted or at least slowed down." Sachs believed at the time that this model "would assure a quicker rate of growth of the economy."[3] Sachs' position is somewhat ambiguous and his model -- which he characterized in a culturally specific manner as the "Indian pattern" -- may well have been intended to describe a stage of gradual transition to socialism. But it can be treated here as a more or less permanent pattern, even if with the public sector in a predominant role, and characterized more generally as the *consummatory "mixed economy"* model.

However, the public sector in a mixed economy has been viewed in instrumental terms as well. On the one hand, it has been treated as part of, and a route to, the building of socialism. The lack of an immediate full-scale installation of a socialist system, to which the political leadership is ideologically committed, is then attributed to the presence of constraints, economic and non-economic. One of the two patterns that Rao distinguished in the case of "mixed economies" was where "the State has the control of strategic points of the economy which are used as commanding heights for determining the main direction of development of the economy. The private sector in this pattern is subordinated to the overall demands of national development on socialist lines."[4] This pattern can be characterized as the *instrumental-socialist "mixed economy"* model. In this case, as in the case of the earlier consummatory "mixed economy" model, the state may resort to nationalization of private enterprises besides undertaking new industrial enterprises; in addition, the state may subject the private sector to a whole spectrum of controls, with nationalization representing "often the most extreme in such a spectrum."[5]

On the other hand, the ideological commitment to capitalism is no bar to a vigorous public sector, which may be fostered in order precisely to stimulate the private sector in the context of economic backwardness. Sachs described as "Japanese" pattern the situation where the public sector subserves the purpose of aiding capitalism. It can be characterized more generally as the *instrumental-capitalist "mixed economy"* model. Here, the public sector is meant to be only a transitional feature in order to assist the advance of capitalism. It is, in Sachs' terms, permanent only in respect of "social overheads" by way of public utilities and public finance for private enterprise. Any new industrial enterprises started by the state are intended for subsequent "privatization," with the state pursuing "a policy of the conscious formation and strengthening of the capitalist class" and thus facilitating "the formation of monopoly groups." Similarly, the second of Rao's two patterns was where "the State agencies and the public sector provide external economies to the fast growing private sector or to private concentrators of economic power. The basic decisions of production, distribution, saving and investment in this case are taken either by the private sector directly or in response to the pressure of this sector by State agencies."

Interestingly, evidently associating "mixed economy" politically with Kalecki's "intermediate regime", Rao averred that this particular pattern was a function of an alliance of the intermediate classes with the upper classes, with the coalition "resorting to socialist ideology only to win mass support but using all levers of power to facilitate a type of capitalist development in the interest of a narrow section" of society. On the other hand, he held the "instrumental-socialist" pattern to be a function of an alliance of the intermediate classes with the have-nots, with this coalition "using socialist ideology to release mass energy and initiative on a vast scale and using the levers of power to promote the process of socialist transformation of the economy in the interest of the widest sections" of society, more particularly of those at the bottom. In contrast, some hold the public sector to be simply a "form of public works for the middle class" alone.[6]

It is evident again that both (1) ideology and (2) interests are critical in the understanding of the creation and expansion of the public sector. Any serious examination of the issue for specific countries must assess then the relative role played by each factor. In regard to ideology, one must examine its nature and the particular bearers of it. In regard to interests, it is important to evaluate the nature of the class or class coalition holding state power. But equally important is (3) the structural context in which ideology and interests operate and interact; the context may facilitate or inhibit particular lines of policy. Of particular relevance in this connection are the nature of historical legacy and the kind of social and political institutions that exist.

In order to understand the forces that would subsequently have an impact on shaping state policy on the public sector in post-independent India, this chapter looks at the pre-independence situation. More specifically, it examines: (1) the colonial legacy; (2) the ideological and social bases of the nationalist movement; (3) the development of Nehru's socialist vision; and (4) the posture of the capitalist class on Nehru and economic planning before independence.

1. The Colonial Legacy

Colonial rule over nearly two centuries had a momentous effect on the political economy of India. Most fundamentally, it integrated India's economy with the world capitalist system, which had Great Britain as the hegemonic power over most of this period, with India playing a significant role as a resource in Britain's rise to that position. In the process, India became a source of raw materials and a captive market for British manufactures. It was further turned into a strategic lynchpin in Britain's worldwide empire and a military tool in the maintenance of that empire in

Asia and Africa. Once a great manufacturing country itself, India also became de-industrialized and ruralized.[7]

In order to protect its own industrial interests, Great Britain was hostile to the rise of competing industries elsewhere, for which purpose the doctrine of free trade and laissez-faire became eminently useful. While many other nations were able to foster industry through protective tariffs and interventionist policies, Great Britain as the colonial power was able to deny India this option. Even mild duties on the import of British cotton goods for the sake of raising revenue, not for protecting local industry, led the British parliament in 1877 to pass a unanimous resolution for their repeal because of the protective element involved. With the growth of the nationalist movement in India, the British administration was compelled to pay some heed to Indian economic interests through a moderate tariff protection. But Great Britain remained negative towards public support for local industry in India. When around the turn of the century some provincial governments acted to foster industry with state resources, London overruled such policy.[8] Typical of the British attitude right to the end was the following comment by a British general in 1946, when he was requested that war-trained Indian technicians be continued to be employed in making sophisticated components and machinery but now for civilian use: "Surely India can import whatever items of manufacture she requires from the U.K." The British response to India's request in the early 1950s for technical assistance in building a steel plant was simply "not interested."[9]

The break in India's integration with the British economy during World War I provided for some industrial advance, but the low level of industrial development is demonstrated by the fact that by 1919 all factory employment amounted to 1.17 million, constituting less than 1 per cent of the total labour force.[10] India's pathetic lack of industry was the subject of a ringing indictment at the end of the war by the Indian Industrial Commission, which advocated active government efforts to promote industrial growth. However, not much occurred by way of a positive policy or programme for industrial development. There was nonetheless a breakthrough of sorts in relation to tariff protection when, following a recommendation by the first Indian Fiscal Commission, the government adopted in 1923 a policy of discriminating protection, aimed less against British industry than at the rising industrial powers of Germany and Japan.[11] Even though the protection provided was a limited, moderate and fitful one, and the actual implementation of policy rendered it "even more ineffective in operation,"[12] it had some beneficial effect on the development of industry in India. The real help to industry in India, however, came less from government policy than from the two wars, which "disengaged" the economy from Great Britain, and the interwar depression which weakened the British economic position globally while the gathering nationalist movement strengthened the hands of local industry.

The change on the industrial scene as a result of the wars and depression took place in two respects: one, there was expansion and diversification of India's industrial base on a considerable scale; and, two, the Indian business class was able, through purchase of foreign enterprises and through new plant, to move abreast and achieve a position of equality, if not of superiority, in relation to foreign-controlled, mainly British, industry in India. The role of Indian business was especially noticeable in new lines of production, such as cement, paper and bicycles, where it assumed a leadership role. Indian industrialists thus came to dominate the more modern sectors of consumption goods geared to the internal market, while British business continued to dominate the traditional lines, such as tea and jute, addressed to export. Indian steel was able to consolidate its market in India, and sugar refining expanded vastly to provide for local consumption. Indian business also moved into jute manufacturing though the British retained dominance. The country was now able to supply, though admittedly for a narrow market, its needs of simple consumer goods like cloth and sugar from local manufacture. As Bagchi points out, "between 1900 and 1939, India more or less completed the 'textile revolution'; she became nearly self-sufficient in the production of cotton textiles, and emerged as a major exporter of cotton piecegoods during the Second World War."[13] India additionally acquired the capabilities to produce some consumer durables, such as bicycles and sewing machines, but most of the needs here were still met through imports.

The impact of the wars on industrial development, however, was limited by the single most important weakness of its industry -- the lack of a capital goods base. The increase in local manufacturing during the war years was more a function of utilizing idle capacity than large-scale expansion of industrial plant. Indeed, the effect of World War II was the material exhaustion of existing plant rather than its strengthening. After the war India was still dependent for industrial development on import of machinery and equipment, machine tools and chemicals. Some development in capital goods industry, such as textile machinery, was initiated during the war, but it was only after independence that significant movement would take place in this area. Indian industry thus remained, in the main, confined to the production of simple consumer goods.

Furthermore, notwithstanding the expansion and diversification of Indian industry, the total effort was small, perhaps insignificant. Although the country's size made India the tenth largest producer of manufactured goods in the world, its per capita output of such goods at independence was only a quarter that of Egypt and one-tenth that of Mexico, not particularly great exemplars themselves of industrialization at the time.[14] As a share of total national income, factory production amounted to hardly 6 per cent.[15] Factory employment of 2.6 million in 1947 constituted less than 2 per cent of the labour force,[16] and was concentrated in jute and cotton manufacture and other agriculture-related processing. For a country of some 350 million in 1947, electricity consumption was only 3,356 million kwh, that is, about 9.5 kwh per capita.[17] Its technical manpower base was insignificant. As for living standards, the dispute among economists has been whether over the preceding three decades, there was deterioration of 5 to 15 per cent in per capita income (the received view) or no change at all (the revisionist view).[18] To the end, then, "the economy of India remained poor, basically agricultural and colonial";[19] to repeat, "at the time of Independence, India was still largely non-industrial and one of the world's poorest states."[20]

The industrial structure that the independent government of India inherited in 1947 thus had two major features. First, even though it was in the aggregate substantial and undoubtedly important, industry had a minor place in the total economy, which was overwhelmingly agricultural. Second, while Indian industry, had made considerable advance in simple consumer goods and some progress in the direction of consumer durables, it was utterly deficient in capital goods. The industrial structure was dominated by cotton and jute textiles, the two together constituting nearly 65 per cent of total value added by manufacture through factory production and employed about the same proportion of the factory labour force.[21] This situation was one that would be unacceptable to a successor nationalist regime and would provide a compelling incentive for an ambitious programme of rapid industrialization, especially in relation to capital goods industry. By the same token, it would raise the important question of the mechanism through which industrialization should take place -- public sector or private sector. The fact that the lack of adequate industrialization was identified in the nationalist mind with not simply colonial rule, but also with the policy of laissez-faire, reliance on private sector and capitalism, created a certain prejudice in favour of economic planning and public sector among certain critical groups, such as segments of the new middle class.

This prejudice in favour of the public sector was aided by the experience of World War II, when a whole complex of physical controls over the economy was adopted as part of the war effort. Indeed, the colonial state shifted during the war from a laissez-faire model to an interventionist model. Besides rationing of food and cloth, controls were imposed on production, distribution, and pricing of other essential commodities. The bureaucracy acquired, as a result of these changes, considerable skills to control the activities of the private sector. Some would maintain that perhaps it also acquired a vested interest in controls because of the power they gave it over the private sector, and may thus have become less resistant to the idea of planned economy with a public sector. The colonial state was a bureaucratic state presiding over a vast far-flung administrative apparatus whose top officialdom was used to the exercise of extensive powers and may not have been that resistant to state control over the economy to begin with. In any case, the system of wartime controls had potential value for any political leadership interested in a planned economy and a vigorous public sector.

World War II had another impact of considerable significance. In India, as in most traditional societies, economic activity, especially pertaining to trading, commanded low prestige compared to aristocratic, soldierly or intellectual activity. The top two castes in the ideal Indian social hierarchy were the intellectual caste of the Brahmins and the warrior and administrative caste of the Kshatriyas. The merchant caste of the Vaishyas and the set of menial castes known as the Sudras, incorporating the peasantry, lay below them socially and were also subordinate politically. As Brahmins and Kshatriyas took to Western education under colonial rule, they emerged as the new middle class which not only came to fill the ranks of the bureaucracy but also to generate and dominate the nationalist movement. In their new roles, the upper castes continued to carry the traditional low opinion of the business class which, in its new garb, emerged out of the earlier trading communities. World War II made available to an expanded business class immense opportunities for economic exploitation in a situation of great scarcity; these opportunities were utilized to the hilt and enormous fortunes were made by many. This development did not serve to enhance the reputation of the business class, and made it vulnerable to strong regulation and control by the state without many likely to rise to its defence.

Certain segments of nationalist opinion had even before World War II advocated economic planning and an expanding public sector, indeed socialism, and this opinion became increasingly more widespread during the war years. Partly perhaps to pre-empt this nationalist opinion, even the colonial state, now in its last days, felt obliged to accept in principle a role for the state in economic planning and in fostering a public sector. Acknowledging in a tone of repentance that "the attitude of Government towards industry in the past was for many years one of laissez faire," the government belatedly declared in 1945 that "the continuance of their existing policy, in the conditions in which India will find herself after this war, will not meet the objectives of sound post-war development" and that now "a vigorous and sustained effort is necessary in which the State no less than private industry must take a part".[22] The government came to believe that "the development of industry must be planned by Government in cooperation with industry and every effort made to make the plan effective." Even though it felt that controls should be kept to the minimum, it averred that "in a planned economy it is impossible to do without controls."

The government now perceived a special role for itself in relation to basic industries -- such as iron and steel, heavy engineering, machine tools and heavy chemicals -- holding that such "basic industries of national importance may be nationalised." However, the government's approach was essentially non-ideological, for such nationalization was envisaged only "when adequate private capital is not forthcoming and it is regarded as essential in the national interests, to promote such industries." At the same time, "all other industries will be left to private enterprise under varying degrees of control," including licensing, necessary to assure balanced investment as among agriculture, industry and social services, fair conditions for labour, avoidance of excessive profits and unhealthy concentration of economic power. Basically, the government's approach was that of an *instrumental-capitalist* mixed economy. There was no pretense at ushering in socialism, or even making the public sector dominant in the economy as a desirable goal. The declaration, moreover, was a rather hurried public relations effort: it represented less a definite policy and more a statement of provisional intentions preliminary to entering into consultations with provinces and princely states. It came, furthermore, at a time of great political and constitutional uncertainty. Still, it underlined the emerging importance, albeit non-ideological, of economic planning and public sector, even for a colonial administration though one in its dying days.

Until this late awakening to the importance of industry, the British attitude to industrial development in India had been negative. This was largely the result of the interests of the metropolitan bourgeoisie, which wanted India retained as a captive market for its industrial goods. Yet the interests of the same bourgeoisie demanded that some modernization take place in limited sectors in order precisely to hold the conquered country strategically as well as to exploit it economically in its role as a supplier of raw materials and as a market for British manufactures. On the other hand, investments that were necessary in the economic infrastructure to fulfill these purposes were not available in private hands in India, while the risks attendant on such investments were not acceptable even to British investors. Consequently, in the building of economic infrastructure the state was compelled to undertake such activity directly itself through the instrument of government departments or indirectly through state-backed incentives to the private sector. Two important areas for such effort were the building of several canal irrigation systems under government departments, to ensure increased agricultural supplies, and the extensive railway network -- built under state-guaranteed returns on investment but over the years operated under every conceivable type of management until eventually most of it was nationalized -- in order to ensure the security of the empire and access to the local market. In addition, government retained ownership and management of ordnance factories producing ammunition and small arms. During World War II, the government also initially commandeered and subsequently took over shipbuilding and aircraft repair facilities. Furthermore, there was government ownership and control of communications, such as posts and telegraphs and roads, and a considerable share (over 25 per cent) in power generation. However, "the importance of the public sector in relation to total economy was marginal: government expenditure accounted for barely 8 per cent of aggregate national expenditure" by 1950-51.[23] Although the aggregate economic activity under the state was not much, there was nonetheless a tradition of a public sector under colonial rule, though limited to economic infrastructure and defence production. Here, it is noteworthy that, though the colonial state in general stayed aloof from industry, some of the more progressive princely states -- such as Mysore and Travancore -- fostered local industry, even under state ownership and control.

The end result of some two centuries of colonial rule then was that, despite the growth of an industrial complex of considerable size in the aggregate, India remained basically an agrarian economy, with over 80 per cent of the population living in rural areas. What is more, during the first half of the 20th century, the economy remained stagnant. Consequently, the successor government was left the task of industrialisation, even though it remained an open question as to whether that should be accomplished through the public sector or private sector, or both.

2. Ideological and Social Bases of Nationalism

Economic planning and the public sector in India after independence are intimately linked with the development of the nationalist movement whose organizational expression was the Indian National Congress. They are most especially associated with Jawaharlal Nehru, the political hero of that movement second only to Mahatma Gandhi, its intellectual spokesman and its ambassador to the world during the mass phase that followed World War I. The movement under Gandhi's leadership drew him in and, rising quickly to its top inner leadership group, Nehru in turn attempted to guide the movement in directions that he perceived ideologically desirable.

Apart from the indefatigable campaigning across the length and breadth of the country to mobilize the masses behind the nationalist movement, two of Nehru's most noteworthy efforts in attempting to mould that movement pertained to giving it an international outlook and to focusing concern on a future post-independence economic programme. On the first, he created an awareness of the world beyond India and of historical forces interacting on the international stage, situating India's nationalist movement in the larger flow of progressive social forces in the world, especially in Asia. On the second, he convinced his fellow nationalists that it was not sufficient to work for independence through the overthrow of colonialism but also for a more purposeful vision for reorganization of the social and economic structure. Here his own stress was on a socialist society with an active and expanding public sector. As a consequence, the question of the preferred post-independence economic system became a matter of some debate within the nationalist movement. However, that movement was no simple affair but one of considerable social and ideological complexity, and the positions taken in the debate and the consensus formulas tenuously arrived at reflected the interaction of multiple social forces and other contingent factors, such as factional conflicts and personal value commitments.

The organizational beginnings of the nationalist movement can be conveniently dated to 1885 when the Indian National Congress was founded. The freedom struggle that followed was not a short-term affair but a protracted one, lasting over some six decades during which it went through several phases. In the process, it became transformed from (1) a narrow interest group affair of the highly educated demanding widening of opportunities, during the moderate phase of liberal nationalism from 1885 to about 1900, to (2) a wider militant protest movement opposing colonial rule, drawing more on the appeal to local traditional values, during the extremist phase from around 1900 to 1920, and to finally (3) a mass movement during the Gandhian phase from 1920 to 1947. Characteristically, the emergence of one phase did not mean the elimination of the dominant political trend of the earlier phase. Rather, the Congress party displayed a cumulative tendency whereby a new ascendant political trend was simply added on to earlier ones. This was true equally of the leadership of the movement "which was itself subject to modification and adaptation over the years as new and more diverse social elements were politically activated and recruited into the movement."[24]

During the first phase of moderate nationalism (1885-1900), dominated by the values of secular liberalism close to those of the colonial rulers, the nationalist movement was initiated and led by the new middle class, largely of Hindu origin. Itself a creation of colonial rule, the new middle class was eventually to become its nemesis. This new middle class had arisen not because of any drastic economic change within India but because of a state edict in 1835 making English the language of administration and education, thus propelling thousands of young men to pursue university education in England and at home. There were some 55,000 people with university education by 1885, representing a narrow social base of "upper-caste, middle and low income urban groups rather than the old native aristocracy."[25] This educated elite was mostly concentrated in government services or the new modern professions, with law the predominant choice. At the first Congress session in 1885, the 72 delegates from the various provinces had occupations that manifested a "remarkable correlation with those pursued by the educated elite of India."[26] The Congress elite displayed considerable continuity over the first two decades of the organization's history. Of the 13,837 delegates to Congress sessions over the period 1892-1909, nearly two-thirds came from the professions and of them some 60 per cent were from law, thus demonstrating the dominance of the new middle class and within it of lawyers. The landed gentry and the commercial classes formed 19 and 15 per cent respectively of the delegates. There was no local bourgeoisie to speak of at the time. Even though some Indians had taken to industrial activity in the second half of the 19th century, most modern economic activity was controlled by British investors, and the rise of a local bourgeoisie would have to wait until after World War I. In the meantime, in the first phase the Congress delegates were "clearly a homogeneous group drawn from the new English-educated elite which was predominantly Hindu, upper caste, and professional."[27] Noteworthy is the predominance of Brahmins, constituting as they did about half of the Hindu delegates, who in turn were nearly 90 per cent of the total.

Within the Congress party there was uniform concern at the poverty of India and the siphoning of the country's wealth by Great Britain. The period is noteworthy for the articulation of the goal of modernization and industrialization of India, a cause consistent with the middle class nature of the nationalist movement. Some like Dadabhai Naoroji of the "drain of wealth" theory fame appealed for a recourse to the principles of laissez faire without discriminatory treatment aimed at India, while others like M.G. Ranade and R.C. Dutt advocated a Listian prescription.[28] But there was at the time little impact of Marxist ideas, which had to await the Russian revolution. The preferred economic model then was that of capitalism under state sponsorship or encouragement while the political ideology was constitutional liberalism, holding parliamentary institutions of the British model as the appropriate ones for India. The legacy of this phase was the ideal of modernization of India and the faith in democratic institutions.

The phase of extremism over the period 1900-1920 represented a widening of the nationalist movement from the provincial capitals to the district towns, from the more educated to the less educated (the product not of university but of the expanded secondary education system), from the higher income to the lower income segments of the new middle class. This populist phase was closer to indigenous values and was accompanied by and drew on religious revivalism on the part of the Hindu community; it rejected colonial rule and colonial institutions, stressing instead moral renewal on the basis of traditional values and institutions rather than their reform and modernization.

The last phase of nationalism as a mass movement following World War I reflected the impact of the greater spread of education, the consolidation of the bourgeoisie, and wider urbanization but, most importantly, the political and organizational penetration of the rural areas by the Congress party under Gandhi's leadership. It was the genius of Gandhi, as it was that of Mao Zedong, to conceptualize the mobilization of the peasantry in a largely rural society as an essential prerequisite for the overthrow of imperialism, to evolve a political strategy for such mobilization, and then to effectively implement that strategy, bringing this powerful social force into the Congress party. Gandhi's contribution to the Congress lay not only in transforming it into a mass movement but doing it through a system of ethics that centered on truth, non-violence and the principle that the end does not justify the means. Gandhi's involvement of the masses made it imperative for Congress to come to terms with the question of their economic uplift, even though his own programme may not have been acceptable to many. Incorporating indigenous ideological and institutional elements, Gandhi's economic model focused primarily on the rural areas, envisaging basically a decentralized, agricultural society consisting of self-sufficient villages, under a philosophical approach that emphasized the limitation of wants.

The series of agitations that Gandhi led expanded the freedom struggle in ever-widening concentric circles, transforming it into a mass movement. As Kochanek says, "the decade from 1932 to 1942, more than any other period in Congress history, was responsible for broadening the base of the movement by making possible the recruitment of non-urban elements and lower, non-Brahmin, castes." As a result, "by the time of independence it was no longer the strictly middle-class urban movement it had been before 1920." For the entire stretch of the nationalist movement, Kochanek observes: "From decade to decade there was a gradual broadening of the recruitment base for the Congress elite. The trend was from urban to rural, from high caste through the middle castes to the lower castes, from the modernized professions to the more traditional pursuits."[29]

But the Gandhian phase also saw the adherence of the bourgeoisie, as well as of labour, to the nationalist movement. No doubt, Gandhi's pro-peasant Narodnik-type political philosophy incorporated certain anti-capitalist and anti-industry values, but several other factors led the business class to support the movement: his theory of trusteeship -- under which those who happen to be wealthy would look upon their assets in society as a trust to be used productively on behalf of the public and not in their own private interest -- rather than nationalization and public ownership; his opposition to a strong overpowering state since he perceived the state, as did Marx and Engels, to be the embodiment of violence; his creed of non-violence which barred both class conflict and confiscation of property; the assurance of a protected domestic market through the social boycott of foreign goods; Gandhi's charisma issuing out of his personal asceticism and sacrifice; perhaps his caste origins (Vaishya subcaste of the Banias); and the conviction on the part of the bourgeoisie that its own growth would be blocked or stunted by imperialism. Despite its political and financial support to the nationalist movement, however, business could not be aggressive about it; rather, it had to be cautious in its support so as not to arouse the hostility of the colonial administration.

The transformation of the Congress into a mass movement, incorporating the peasantry and labour, saw not only the rise of Gandhism, but also that of a hesitant social revolutionary trend inspired and led by Nehru. This trend became organizationally manifest within the movement through the founding of the Congress Socialist Party in 1934. A year later the Communist party, as part of its somersault change to a new United Front strategy, entered this group in large numbers and came to dominate several of its branches. Nehru and the left in general valiantly attempted to move the nationalist movement in a socialist direction, but it was an arduous undertaking because of the complex social composition of the Congress movement. Socialism appealed largely to segments of the new middle class, and it supplemented other ideological trends within the Congress rather than supplant them.

Through a process of ideological accretion over the course of some six decades of the nationalist movement there came into existence three dominant ideological trends -- liberalism, Gandhism, and socialism. While in some sense these trends stood in opposition to each other, they also served as integral components in varying proportions of a vague overall value consensus in the Congress; furthermore, each trend itself incorporated to some degree elements from its competitors. The overwhelming ideological orientation within the Congress, however, was non-revolutionary and non-socialist. This really meant opting for the capitalist model because of the perception that Gandhism was utopian in its rejection of industrialization and in its reliance on moral conversion rather than institutional mechanisms in regard to trusteeship. On the other hand, the socialist trend was "extremely weak", but "the real weakness of the social revolutionary element in India was disguised by the political and intellectual eminence of Nehru," which was rather a "testimony to the traditional Indian tolerance for ambiguity and amorphous organizations."[30]

In terms of its social basis, the Indian National Congress was what its name suggested, a congress of the Indian nation -- even if a nation in the making -- a multiclass coalition, encompassing the key social classes. It was a broad coalition whose core was the alliance between the new middle class and the peasantry, but under the hegemony of the new middle class, with the bourgeoisie and labour as subsidiary allies. Although the influx of the peasantry with its massive numerical base created a dynamic situation in respect of the balance of social forces within the Congress party, the hegemony of the new middle class can be taken to extend throughout the course of the nationalist movement and, indeed, through the 1950s. In terms of the caste background of this coalition, one can tentatively suggest that the new middle class at the time coincided with the upper castes of Brahmins and Kshatriyas, while the bourgeoisie and peasantry consisted, respectively, of the middle castes of Vaishyas and Sudras, and labour that of Sudras and untouchables.

The social coalition under the Congress is significant as well for what lay outside it. First and foremost, excluded from it were those classes and groups that were allied with the colonial bureaucracy, chief among them the landed aristocracy and princes (except for some patriotic members) plus substantial sections of religious minorities, primarily Muslims. The Muslim community largely stayed away from the Congress, not simply because the latter espoused secular nationalism or because of any tactical errors in accommodating Muslim aspirations, but because Muslim elites led by Sir Saiyyid Ahmed Khan in the mid-19th century saw their destiny linked to collaboration with the British and asked their followers to stay away from the nationalist movement. Furthermore, despite the Congress party's claim to being an organization of the peasantry in general, also excluded from it were the lower peasantry and agricultural labour (the latter coinciding to a considerable extent with the untouchables) because these were not yet enfranchised and remained largely unmobilized by the nationalist movement. Consequently, the peasantry allied with the Congress meant basically the rich and middle peasantry in Hindu-majority areas. Nonetheless, partly as a result of its ideological commitment and partly in reaction to sectional movements encouraged by the colonial bureaucracy, the Congress assumed the role of protector of the untouchables or scheduled castes. As a result of this posture and its notion of a unified single peasantry, along with the conception of India's villages as harmonious consensual social units, the Congress was able to present itself as the embodiment of the nation and national unity and, by the same token, to prevent "the fuller emergence of class struggle in the countryside."[31]

The foregoing analysis of the social basis of the nationalist movement in India as a broad multi-class national coalition has been the staple of mainstream analysts.[32] It has, however, in the past been rejected by Marxist analysts in favour of a simple declaration that the Congress was in the era of nationalism a bourgeois organization under the hegemony of the bourgeoisie. This assertion was managed simply by the heroic equation of the role of the Congress leadership with the role of the bourgeoisie. No doubt, as Chattopadhyaya observes, such an equation is "oversimplified, mechanical, unverifiable and of limited predictive value in analysing the ambivalences of the Indian national movement", especially as expressed through Gandhi and Nehru; besides, it ignores the peasantry as the main force of the nationalist movement.[33] The Marxist line seems to have rested on a mechanical analogy with Europe where nationalism is considered to have been a function of the rise of the bourgeoisie to economic and political power. Undoubtedly inspired by the Comintern, the characterization of the Congress party as a bourgeois organization by Marxist analysts is of long standing though the consequences ascribed to its domination have shifted over time. At first, in the 1920s, the Communists in India maintained that the Congress could not be anti-imperialist because the bourgeoisie had already gone over, or was in the process of going over, to imperialism. Over the period 1929-1935, not only were the Communists anti-Gandhian, they additionally attacked Nehru as a left demagogue who attempted to "bamboozle the mass of workers" and "keep the masses within the framework of British imperialist constitution and legislation," "faithfully carrying out the instructions of landlords and capitalists." They likened Nehru to Kerensky and Chiang Kai-shek, declared him to be an enemy of independence and socialism, and branded him a "social-fascist." The task of the Communist party was then conceived of as exposing Nehru and other left nationalists as "assistants of British imperialism", and of launching a "ruthless war" against them.[34] Nehru himself had spoken with some sarcasm about this line:

> According to the communists, the objective of Congress leaders has been to bring mass pressure on the Government in order to obtain industrial and commercial concessions in the interests of Indian capitalists and zamindars....The Indian capitalists are supposed to sit behind the scenes and issue orders to the Congress Working Committee first to organize a mass movement and, when it becomes to vast and dangerous, to suspend it or sidetrack it. Further that the Congress leaders really do not want the British to go away, as they are required to control and exploit a starving population, and the Indian middle class do not feel themselves equal to this.
>
> It is surprising that able communists should believe this fantastic analysis, but, believing this as they apparently do, it is not surprising that they should fail so remarkably in India. Their basic error seems to be that they judge the Indian national movement from European labour standards; and, used as they are to the repeated betrayals of the labour movement by the labour leaders, they apply the analogy to India.[35]

Subsequently, over the period 1935-1941, the Communist party joined Congress because it now perceived the latter to have become anti-imperialist. Still later, in 1941, it defected from Congress to collaborate with the colonial state in aid of the war effort because World War II had, for it, now become transformed into a people's war. All through these shifts, the Congress party was taken to be under the hegemony of the bourgeoisie.

More recent revisionist Marxist writing on Indian nationalism has come to depart from the traditional Marxist interpretation -- often later characterized as vulgar or mechanical Marxism -- and to move essentially close to the mainstream analysis of the social basis of nationalism. In the 1960s, in retrospect and as a minority dissident viewpoint in the party, one Communist leader, Mohan Kumaramangalam, chided his comrades for failing to comprehend the real nature of the nationalist movement, or the depth of support for it, with their glib characterization of it as a Hindu bourgeois movement; on the contrary, he described it as a genuine movement of national liberation.[36]

Another Marxist writer, Baren Ray, has criticized Indian Marxists for assuming earlier that the national bourgeoisie could not be anti-imperialist and, equally, for believing "the national movement essentially as the *doing* of the bourgeoisie." However, even as they later saw the contradiction between the bourgeoisie and imperialism, they "continued to view the growth of the Indian national forces -- a complex of developments in a multi-structured society -- in an oversimplified bipolar way," viewing "the national bourgeoisie and the working class as the only significant contending forces in Indian society." For his part, Ray holds the origins of the nationalist movement to be rooted in changes in the superstructure in India in interaction with Europe: "The developments in the Indian superstructure -- the growth of bourgeois-democratic values and strivings -- were the result of the impact of the European superstructure rather than the reflection of changes in the base brought about by an indigenous bourgeoisie. Historically speaking the beginning of modern national awakening in India very much predates the Indian bourgeoisie becoming national."[37] Ray takes the social vehicle of the nationalist movement to have been "the intelligentsia-cum-middle class" representing "the earliest forces," later joined by "the peasant upsurge, the national bourgeoisie and lastly the working class." He castigates the earlier "left-sectarian position" among Marxist writers, which took the "national movement as the *product* of the bourgeoisie and not as a movement of the new intelligentsia and other patriotic sections representing the growing national awareness that had grown *earlier* than and independent of any move on the part of the bourgeoise." He goes on to acknowledge: "By not distinguishing between the first three components of the national movement we viewed the entire movement as of the influence of the national bourgeoisie and thus failed to distinguish between the revolutionary-democratic elements who constituted the active core of the national movement and the national bourgeoisie as an economic class." As for the bourgeoisie, "it wielded comparatively greater influence on the national independence movement and later on the pattern of economic development," being a relatively developed class, "but it was not the main, much less the only determinant."

Similarly, Bipan Chandra, the dean of independent or revisionist Marxist historians in India, holds that the powerful struggle against imperialism in India reflected "the contradiction between imperialism and all the Indian people, of whom the bourgeoisie constituted only one important segment," adding emphatically: "At no stage in its origin and development was the capitalist class the driving force behind this struggle or its militancy."[38] For him, basically, while the Indian capitalist class stood in a long-term contradiction with imperialism, it also suffered from a short-term dependence on imperialism and therefore worked out a relationship of short-term accommodation with it within the context of a long-term opposition. Starting belatedly, developing slowly from a very poor base -- even if in a non-comprador manner -- and spread thinly, it did not have the confidence to challenge imperialism, which is what explains "the late entry of the class into active politics and its mild political stance." By the time it acquired reasonable strength after World War II, "precisely at this time political freedom came as a result of popular political pressure and changes in the world balance of forces." Elsewhere, Chandra and his associates characterize the Congress as "the historic bloc" of the people and "not a class party" of the bourgeoisie, rejecting the notion of characterizing

> such a party, which permitted Communists and socialists to function within it in this manner, as a class party of the bourgeoisie. The fact is that all ideological tendencies and political programmes could freely contend within the ranks of the Congress for general acceptance by the mass of Congressmen and it is this which made it a political expression of the historic bloc of the Indian people in their struggle against imperialism.

Further, while still continuing to subscribe to the notion of bourgeois hegemony, they maintain:

> The Congress in the *Colonial period* was not a class party of the bourgeoisie, much less of the "feudal castes and classes," but was a party of the Indian people as a whole including the bourgeoisie, petty bourgeoisie, sections of landlords, peasants and workers. We do not wish...to hand over the entire heritage of the Congress-led anti-imperialist movement, and that of some of its towering leaders such as Dadabhai Naoroji, Gandhi and Nehru, to the bourgeoisie and the feudal classes whose role in the movement was minimal. At no stage from its inception to its later development was the capitalist class the driving force behind the struggle for freedom or its militancy.[39]

In the final analysis, what we have here is a definite narrowing of the distance between mainstream and more recent Marxist analysis and a tacit consensus around the "intermediate regime" of Kalecki's conception. But, in the Indian case, the core coalition of the new middle class and the rich and middle peasantry under the hegemony of the new middle class [40] was, during the Gandhian phase of the nationalist movement, in alliance with classes both above and below -- above with the bourgeoisie (but not with the landed aristocracy) and below with labour -- rather than being antagonistic to them. At the time, the lower peasantry and landless labour stood outside the alliance.

It is noteworthy that, by virtue of the fact that it was a national movement, the Congress party was in its class composition largely similar to the Communist Party of China, where the formula was a four-class coalition of the proletariat, the peasantry, the national bourgeoisie and the petty bourgeoisie. Interestingly, Harris underlines that "Mao warded off the demands of poor peasants in order to keep the rich peasants and small landlords in his coalition, stressing always that domestic issues of class conflict must be subordinated to the central task of evicting foreign imperialism." What apparently distinguished the Communist Party of China was the hegemony of the proletariat, by which was really meant the hegemony of the party as a surrogate for the proletariat but whose leadership was essentially from the new middle class as well. The critical difference between the two parties, apart from the use of violence, lay in the fact that, growing at the periphery of society, the Chinese communist party could develop as an organizational weapon and evolve a rigorous economic programme in relative autonomy from society.[41]

The Congress party, on the other hand, grew at the centre of society with little autonomy from it. Besides, the nationalist movement in India did not develop into a social revolution, rather it represented a largely non-violent war of independence. Accordingly, in working out an economic programme, the Indian National Congress had to accommodate the interests of various social classes and competing ideologies among them.[42] But equally important in evolving this programme was the role of particular individuals, none more important than that of Jawaharlal Nehru. His ideology or set of ideas could to some extent be related to the aspirations of the new middle class, but other competing ideas appealed to that class as well. Other leaders like Patel, Prasad and Rajagopalachari, or indeed Gandhi, who espoused ideologies different from that of Nehru, were no less middle class than Nehru either in social background or appeal. Importantly, Nehru undertook a strenuous campaign to socialize the new middle class into the particular ideas that he himself held. Therefore the relationship of the new middle class to the public sector, state capitalism or state socialism has to be understood not in a static sense of automatic reflection of class but in a dynamic sense of the change wrought in its orientation by particular individuals like Nehru.

3. The Development of Nehru's Socialist Vision

In relation to his ideological position, Nehru has been criticized for lacking a coherent and consistent ideology as well as for being a disciple of scientific socialism, for his betrayal of Marxism as well as for following the Communist path, for his apparent hostility to business as well as for being a captive and protector of business, for launching India on the path of capitalist development as well as for initiating socialist bureaucratism in India. Sufficient evidence can be advanced to convict him on any of these points. It is not difficult to find such evidence, because Nehru's life spanned three-quarters of a century, some forty-five years of which were spent in active politics. During those years, he was not simply involved actively in politics but was in the front rank of nationalist leaders, second only to Gandhi in the 1930s and 1940s, and was subsequently the chief nation-builder as head of government in independent India. Throughout he remained an indefatigable speech giver, while before independence he wrote prolifically, producing many books and innumerable articles and statements. Given to thinking aloud, proud of his individuality and therefore refusing to merge his personality in a group, interested in the solution of problems in an existential world rather than in upholding the purity of doctrine, Nehru provides the critic enough evidence to support any particular interpretation of his ideological position. Often such analysis, however, tells us more about the ideology of the critic than that of Nehru.

The question of the precise nature of Nehru's ideology cannot be resolved simply by reference to socialism. For, although there can be no doubt about Nehru's commitment to it, socialism did not exhaust all of his ideological convictions. It was only one among a complex set of ideological beliefs which were not reducible to socialism or to the sphere of economics. While according to some Nehru had worked this set of beliefs into a synthesis,[43] others believe it remained unintegrated[44] or muddled.[45] At the least, there was tension among its constituent elements, and to single out any one of them as the test of Nehru's ideological integrity is to fail to do justice to the complexity of his thought.

Nehru's ideology rested on four different and independent pedestals or foundations, none of which was reducible to the other but remained in dynamic interaction with the others, indeed countered and limited by them, so that none of them could lay complete claim to his loyalty. As his official biographer observes: "Denying himself the easy, because total, answer of Marxism, he had to work out a more untidy and complex analysis."[46] First and foremost, there was secular nationalism; while Nehru was a genuine internationalist and a key figure in broadening the horizons of the nationalist movement, India's interests came first with him rather than the demands of a universal ideology. His vision of nationalism was centered on a nation-state for the Indian people as a whole, beyond the particularistic ties of caste, language and, especially, religion. Second, there was liberalism, which involved an abiding commitment to political democracy, individualism and some limits to an interventionist state. The commitment to political democracy could not thus but constrain economic policy in terms of what was acceptable to key sectors in society. Radical critiques of the somewhat feeble thrust for social and economic reform in Nehru's ideology rest, of course, on the premise of exclusion of democracy, which though is not openly declared because of the hostility such admission may provoke in the context of a political culture that places considerable value on democracy. Third, there was Gandhism, especially in the sphere of ethics and morality. The Gandhian emphasis on the right means and non-violence strengthened the impact of liberalism in relation to the importance of the individual and limitations on the state. Fourth, there was socialism, at times scientific socialism, especially in terms of its central element of the public ownership of the means of production. Gandhism strengthened his commitment to socialism through its concern for the poor masses, while weakening it through its opposition to violence, confiscation of property, and excessive centralization of power in the state. Except for Gandhism, all the other elements represented ideologies transplanted from Europe to India. Nehru's distinctive contribution to the nationalist movement lay in introducing socialist ideas to the Congress party.

Among the various constituent elements of Nehru's ideological ensemble, some hold liberalism to be the dominant element [47] while others take socialism to be the encompassing framework.[48] Still others believe that "he acquired and maintained throughout his life, a half-liberal, half-Marxist position. He was a libertarian Marxist, whose idea of socialism encompassed at every stage a large and irreducible measure of civil liberty."[49] Or, again, "it was a little difficult to say whether he was a liberal contaminated by Marxism, or a Marxist contaminated by liberalism."[50] As one biographer observes regarding his ideology:

> It was not an intellectual framework of logical consistency but a sense of what he regarded as essential values which held together all these elements in Nehru's mental make-up. It made him a Marxist who rejected regimentation, a socialist who was wholly committed to civil liberties, a radical who accepted non-violence, an international statesman with a total involvement in India and, above all, a leader who believed in carrying his people with him even if he slowed down the pace of progress.[51]

Whether he actually succeeded in achieving a genuine synthesis of these various elements is a debatable question, but what is true is that working out a synthesis was not a once-for-all operation but a ceaseless life-long quest. Perhaps a more accurate statement would be that different balances among the elements were worked out, depending on time and circumstance. Furthermore, it was not simply a question of shifting balances but also of growth and maturation as he had increasingly to take practical account of social realities in India beyond intellectual adherence to ideology. Consequently, for a fuller grasp of his ideological position, it is necessary to examine its development over the period of time following his entry into politics in 1920.

Although intellectual and ideological development of an individual, like history, constitutes a seamless web, it would seem fruitful to divide the years in Nehru's life from 1920 to the assumption of office as head of government in 1947 into three periods: (1) 1920-1929, which marks the road to conversion and the proclamation of faith in the socialist vision; (2) 1930-1937, which is characterised -- apart from the repeated affirmation of faith -- by a continual attempt, not always successful, to reorient Congress and the masses toward the socialist vision; and (3) 1938-1946, which marks a shift from mere proclamation and affirmation of faith to conceptualizing the adaptation of the vision of a socialist society to Indian realities and of specifying the mode of transition to the socialist society.[52] A fourth period, considered in the next chapter, concerns the years in office from 1947 to 1964, which carry forward the process initiated in the third period -- culminating in the notion of "mixed economy" for a fairly long term mode of transition -- but now through the actual grappling with policy choices in the presumed transitional stage.

3a. *1920-1929 -- The Road to Conversion:* For an individual who was later to occupy the centre stage in politics for nearly four decades, Nehru had remarkably little active involvement in politics while at Harrow, Cambridge or Inner Temple over the period 1905-1912. Yet the first stirrings of nationalism were apparent even earlier in his elation at the Japanese defeat of the Russians in 1905 and his identification with the cause of the "extremists" led by Tilak, even as he was disturbed by the partition of Bengal. He was remarkably aware of the course of politics in India, and made no effort to hide his differences with his more moderate father. At Cambridge he was exposed to Fabian socialism but it was the object of little serious interest. But we can attribute his lifelong attachment to liberalism, with its values of individualism and democracy, to his years of study in England as also, paradoxically, to the nearly four decades of struggle against England for the lack of freedom and democracy in India.

After returning to India and entering the bar, he attended sessions of the Congress but the moderate tone of the nationalist movement left him unmoved. The arrest of a militant leader in 1917 spurred him into active agitational politics. However, his intervention was that of "a pure nationalist," with the "vague socialist ideas of college days having sunk into the background," but "fresh reading was again stirring the embers of socialistic ideas in my head. They were vague ideas, more humanitarian and utopian than scientific."[53] Even by the time of the Russian revolution, he "had not read anything about Marxism."[54]

Nehru's sustained involvement in the nationalist movement began with the dramatic emergence of Mahatma Gandhi to leadership in India politics, as the latter threw a direct challenge to the British empire soon after World War I. The charismatic leader, with his activist approach to the task of overthrowing colonial rule through transformation of the nationalist struggle into a mass movement, bound Nehru as if in a magic spell. It was the transformation of the nationalist movement with the induction of the masses into what had been before essentially an affair of the middle classes -- with their common style of life and common interests -- that provided the impetus for the formulation of an ideology for the movement. With the resulting heterogeneity of the movement, "it became Nehru's burden to find for nationalism an ideology which would hold the various classes together."[55] The ideology did not emerge full-blown but developed over a period of time.

In 1920 Nehru, having reached the age of 30 the previous year, had his first contact with the peasantry and its poverty; his visceral reaction to what he saw was to launch him on an ideological quest that led to socialism. But for now, with his customary concern for the underdog, it was more his conscience that had been pricked by the country's poverty: "I was filled with shame and sorrow -- shame at my own easygoing and comfortable life and our petty politics of the city which ignored this vast multitude of semi-naked sons and daughters of India, sorrow at the degradation and overwhelming poverty of India. A new picture of India seemed to rise before me, naked, starving, crushed and utterly miserable."[56] Nehru's shock at the poverty witnessed may well have been the result of his protected upbringing in a well-endowed family. Even as a grown man, Nehru did not have to work for a living, since he was amply provided for by his father. Perhaps his intellectual disdain for private profit stemmed from this background. His associates and rivals in the nationalist movement, mostly conservative, were in this respect closer to the realities of life, indeed to the earth of India's villages.

Soon after his first encounter with the peasantry, Nehru started a periodic trek to prison, where he was to spend a total of nine years during his lifetime. The sojourn in prison gave him time for reading, reflection and writing. However, in the early 1920s, despite occasional disappointments, Gandhi was not only his political hero but also almost his sole ideological guide. In 1923 Nehru equated fascism and Bolshevism, declaring "they are really alike and represent different phases of insensate violence and intolerance." He counterpoised them against his favoured choice of Gandhian non-violence and non-cooperation, wherein lay "the salvation of India and, indeed, of the world."[57] Nehru at the time was only concerned with the attainment of political freedom against the might of imperialism, not with its implications for social and economic change. Independence would simply mean Indians taking charge of the army, police and finance, and villages running their affairs under a decentralized political order.

By the mid-1920s, however, through reading and reflection, Nehru had begun to work out an independent ideological path for himself. He increasingly concerned himself with the social organization of the future, rather than with a resurrection of the past as in Gandhi. His vision of the future envisaged "a rational, educated and forward-looking society based on modernization, industrialization and a scientific temper" along with "civil liberty as an absolute value, to be safeguarded at any cost."[58] Soon to be added to this vision was Marxian socialism. This occurred with his trip to Europe in 1926-27, which marked a watershed in his ideological development.

While on this trip for reasons of his wife's health, he read a lot but, most importantly, attended in 1927 the Congress of Oppressed Nationalities at Brussels as the representative of India's nationalist organization. Brecher correctly notes that the Brussels conference "proved to be a milestone in the development of Nehru's political thought, notably his espousal of socialism and a broad international outlook. It was there that he first came into contact with orthodox communists, left-wing socialists and radical nationalists from Asia and Africa. It was there that the goals of national independence and social reform became linked inextricably in his conception of future political strategy."[59] Nehru himself in strong language condemned imperialist exploitation of India.[60] In a resolution, which he drafted and proposed, the conference hoped that "the Indian national movement will base its programme on the full emancipation of the peasants and workers of India, without which there can be no real freedom."[61] With the conference and self-education, there thus "came a conversion to the Marxist interpretation of history. In this respect he regarded himself as a full-blooded Marxist."[62]

The impact of Marxist ideology was consolidated as he visited the Soviet Union later that same year as an official guest; he wrote a series of laudatory articles on the Soviet experiment which were carried by Indian newspapers and then published in book form. As a result of the visit, "the grounding in Marxism, which he had received at the Brussels conference and after, was followed by a near-conversion to communism by practical testimony."[63] But still, "while he accepted the Marxist analysis of the past, he was not convinced by the Marxist diagnosis of the future. Coercive methods and a revolutionary dictatorship were to him neither inevitable nor necessary nor worthwhile."[64]

After his return to India, there developed a conflict with Gandhi, with Nehru maintaining that India's goal should be complete independence rather than dominion status. In the process, he emerged as a leader of consequence in the Congress party, now rivaling the old guard, with a constituency of his own given the appeal he held with the younger generation. Already a general secretary of the party, he now acquired an eminent role in the affairs of the party. Within the party the Left, in part under his example and leadership, also assumed increasing importance. In 1928, recognizing his growing stature as a leader of the Left, the All India Trade Union Congress elected him president. Everywhere he was in great demand as a public speaker. And now it was not only the message of independence that he emphasized but also that freedom was a first step to socialism.[65]

In 1929, though only 39 years of age, he was bestowed, at the insistence of Gandhi, with the greatest honour within the power of the Congress party -- it presidency. Gandhi favoured Nehru over Patel, apparently in order "to divert the radical youth away from Communism and to secure their allegiance to the Congress; and to wean Nehru himself from the drift to the far left. Nehru was then in his extremist phase, a visionary, critical of the moderate policies of the Congress and strongly attracted to the Marxist ideal."[66] Gandhi gave him high but prescient praise: "And if he has the dash and the rashness of a warrior, he has also the prudence of a statesman. He is undoubtedly an extremist, thinking far ahead of his surroundings. But he is humble enough and practical enough not to force the pace to the breaking point. He is pure as crystal, he is truthful beyond suspicion. He is a knight *sans peur et sans reproche*. The nation is safe in his hands."[67] Gandhi's role in the rise of Nehru was critical; despite his ideological and tactical differences, he built Nehru's self-confidence, provided him opportunities of office and eminence ahead of his time and over the claims of his rivals, and finally he explicitly chose him as his political successor. Not only in that limited fashion as patron and benefactor was Gandhi crucial, but also historically he constructed the essential stage upon which Nehru could act as a radical mass leader by earlier transforming the nationalist movement.

With the Congress presidency, Nehru had arrived politically, he was now in the first rank among Congress leaders. However, although a protege of Gandhi, he struck an independent ideological path which found expression in his presidential address in 1929. In a speech that was a full-throated proclamation of his conversion to the socialist vision, Nehru declared:

> I must frankly confess that I am a socialist and a republican and am no believer in kings and princes or in the order which produces the modern kings of industry, who have greater power over the lives and fortunes of men than even the kings of old, and whose methods are as predatory as those of old feudal aristocracy.

He well realized that the Congress was not ready for acceptance of a socialist programme, but he found socialism on the march around the world and had no doubt that "India will have to go that way too if she seeks to end her poverty and inequality though she may evolve her own methods and may adopt the ideal to the genius of her race." He made his ideological differences with Gandhi explicit, calling the trusteeship theory "barren" and proclaiming that "the sole trusteeship that can be fair is the trusteeship of the nation and not of one individual or a group." Nor was he any longer squeamish about violence, holding that if the "Congress or the nation at any future time comes to the conclusion that methods of violence will rid us of slavery then I have no doubt that it will adopt them. Violence is bad but slavery is far worse." His acceptance of non-violence was on practical, not ethical, grounds. Despite the conversion to socialism and its open proclamation, however, independence remained the first priority, for he acknowledged: "All these are pious hopes till we gain power and the real problem therefore before us is the conquest of power."[68]

The year 1929 marked the end of a decade during which Nehru had travelled the road from a pure nationalist to a disciple of Gandhi and then, through a process of self-education, to a convert to socialism. The same decade had marked the transition from a relatively unknown recruit to the nationalist movement, though undoubtedly the son of an eminent father, to an activist leader with a record of courage and sacrifice, and then to the holder of power at the pinnacle of the Congress party. Having reached that position at the top of a party which took its sole goal to be the achievement of national independence, it became Nehru's mission to take that party in the direction of socialism.

3b. *1930-1937 -- Reorienting the Nationalist Movement Towards Socialism:* The task of reorienting the Congress at the national level had, of course, been presaged by him earlier at the provincial and pressure group level. In 1928, Nehru had organized the Independence for India League as a pressure group of the left wing within the Congress to push for full independence. But, since that goal no longer sufficed for him, the programme of the U.P. Branch of the League specified that "the league aims at the socialistic democratic state in which every person has the fullest opportunities of development and the state control of the means of production and distribution." As part of the effort to build "our future socialist state," the League sought "to change the present capitalist and feudal basis of society" through organizing the exploited.[69] The ideological elan was not sufficient to sustain the League, however, and it withered away in a year, reflecting poorly on Nehru's organizational talents. But in 1929, under his prodding, first the U.P. Provincial Congress and then even the national Congress accepted the notion that India's economic structure was just as much responsible for the poverty of the people as imperialism, and therefore in need of "revolutionary changes."[70] But what these changes covered was not spelled out.

To convert the Congress party to socialism at the national level was, however, a formidable task, for not only was the Congress leadership resistant to ideological debate that would split the movement in its struggle against imperialism but also the organisation was basically a conservative one. The task required, "from the start, careful maneuvring. There was the prime obstacle of Gandhi, who would never accept any laws of history and change."[71] No doubt, the strength of the left wing had increased under the inspiration and leadership of Nehru, but it could not shake the Congress leadership to commit itself to socialist ideology. Thus during the entire course of the freedom struggle, the Congress party did not adopt the ideology of socialism of any variety. However, Nehru did succeed, to some extent, in having the party adopt some policy positions which could be considered part of, or at least in the direction of, a socialist programme. At the same time, he continued to affirm his own faith in socialism and to popularize it among cadres and masses through innumerable speeches; as he expressed it, "I wanted to spread the ideology of socialism especially among Congress workers and the intelligentsia."[72] It was largely a one-man mission; as Namboodiripad observed: "the impact of socialism was felt inside the Congress mainly through Nehru."[73] Nehru was "in deadly earnest" about this mission to spread the message of socialism.[74] But his mission was rudely interrupted as the decade of the 1930s opened, as Nehru made his way to jail as part of the civil disobedience movement.

In 1931, the idea of a nationalist commitment to fundamental rights evolved in conversations between Nehru and Gandhi; at Gandhi's request Nehru drafted a resolution on the subject, later incorporated changes suggested by Gandhi, and finally moved it at the annual session of the Congress, which accepted it.[75] The resolution proclaimed in the preamble that "in order to end the exploitation of the masses, political freedom must include real economic freedom of the starving millions." Apart from a long series of provisions pledging the Congress to freedom of expression and religion, to equality and labour welfare, and the secular state, the Resolution on Fundamental Rights and Economic Policy declared: "The State shall own or control key industries and services, mineral resources, railways, waterways, shipping and other means of public transport."

Nehru himself felt that "this very mild and prosaic resolution" did not spell socialism at all, and a capitalist state could easily accept almost everything contained in that resolution," but nonetheless "it took a step, a very short step, in a socialist direction by advocating nationalization of key industries and services."[76] A sympathetic historian observes that, while they might seem only "mildly socialist" in retrospect, the resolution's provisions "sounded revolutionary" at the time.[77] Certainly, the government was disturbed by it, while the business class seems to have been "shocked" at the declaration on the economic role of the state.[78] Ironically, the person who presided at that session of the Congress was Sardar Patel, the leader of the conservative, business-oriented wing. It is important, however, to recall the general attitude at the time toward Nehru's socialist planks as toward his pronouncements on international relations. As one biographer highlights it, Congressmen had developed a nonchalant posture toward Nehru's ideological predilections, agreeing to one or another resolution "without either fully examining or understanding its implications. 'Another of Jawaharlal's whims,' was the general verdict. 'Let's humour him and pass it'."[79] Nonetheless, the Karachi resolution became the fountainhead and ultimate reference point for legitimizing formulas to steer Congress subsequently in a socialist direction; it became a first principle, blocking further discussion on the ground that Congress had accepted it a long time ago. In 1933, the Congress reconfirmed its support of the resolution.[80] However, in 1934, perhaps finally aroused as to what the resolution could be interpreted to mean, the party left no doubt as to where it stood ideologically. The Congress Working Committee now provided a clarification that the resolution on fundamental rights did not imply either class war or confiscation of private property, both of which were held to be contrary to non-violence;[81] this clarification, of course, aroused Nehru's ire.

Even as Nehru was involved in the struggle against imperialism, going to jail frequently, and grappled with the challenge of communalism, he pressed on with expounding his faith in socialism, propagating the creed, and attempting to convert Congress. The task was a daunting one in the early 1930s, for those years witnessed what was for the Congress leadership a nightmare in the face of British repression against the civil disobedience movement. As he came out of jail in 1933, he picked up where he had left off in 1931. The period from 1933 to 1936 is considered to be one when Nehru's fervour for socialism was at its highest. With world capitalism in crisis in the early 1930s, Nehru found the Marxist analysis vindicated: "Marxism alone explained it more or less satisfactorily and offered a real solution. As this conviction grew upon me, I was filled with a new excitement....."[82] However, given his distaste for dogmatism, he could not give complete loyalty to Marxism. His position was characterized by ambiguity and qualification. On the other hand, he dared to see things whole against a broad canvas and make larger judgments. Thus, in relation to the Soviet Union he was distressed by the violence and terror, but perceived them to be problems characteristic of transition, declaring: "I do not approve of many things that have taken place in Russia, nor am I a Communist in the accepted sense of the word. But taking everything together, I have been greatly impressed by the Russian experiment."

Nehru was terribly influenced by the acute crises of the early 1930s, which seemed to tear the world apart in a manichean fashion between fascism and communism. Jettisoning the use of his usually preferred term 'socialism' and forgetting about Gandhi for the moment, Nehru saw no possibility of a middle road. He made his preference obvious in 1933, even though he did not favour a mechanical adoption of the communist model. Rather he urged its adaptation to Indian conditions although he did not address himself specifically to the requirements of such adaptation:

> I do believe that fundamentally the choice before the world today is between some form of Communism and some form of Fascism, and I am all for the former, i.e. Communism..... There is no middle road....and I choose the Communist ideal. In regard to the methods and approach to this model I may not agree with everything that orthodox communists have done. I think that these methods will have to adapt themselves to changing conditions and may vary in different countries. But I do think that the basic ideology of Communism and its scientific interpretation of history is sound.[83]

Nehru's assertion of a drastically different ideological path, along with differences over tactics in the conduct of the freedom struggle, disturbed many of his colleagues, frightened "to some extent the vested interests,"[84] and brought him into conflict with Mahatma Gandhi. In 1934, the Congress Working Committee declared that the principle of non-violence barred confiscation of property and class war, which Nehru took to be a disavowal of the Karachi resolution and an attack on his ideology.[85] On the other hand, inspired by Nehru's advocacy of socialism, the leftists organized the Congress Socialist Party as a political group within the Congress to take it on a leftward course. As the Congress party seemed intent on tearing itself apart through conflict between radicals and conservatives, Gandhi called on Nehru in 1936 to assume the presidency once again as a possible bridge-builder.

Perceiving a popular sentiment for change, Nehru as president attempted to press forward with several resolutions of a radical nature but was unsuccessful against conservative opposition. One of the few accepted was a moderate resolution on agrarian reform. He then took recourse to the one path that seemed open to him at the time -- socializing cadres and workers. In his long presidential address at the Lucknow session of the Congress in 1936, he linked India's freedom struggle to the global conflict between imperialism and fascism on the one hand and nationalism and socialism on the other, declaring: "inevitably we take our stand with the progressive forces of the world which are ranged against fascism and imperialism." Distressed at the disunity of the Congress, which he blamed on the middle class nature of its leadership that was cut off physically and mentally from the masses, he implored the Congress to renew the organic link with the masses and become an organization "not only *for* the masses" but also "*of* the masses." Nehru recommended the affiliation of peasant associations and trade unions with the Congress as one way of reorienting it. At the same time, he was emphatic about his own ideological position even if it did not find endorsement from Congress:

> I am convinced that the only key to the solution of the world's problems lies in socialism, and, when I use this word, I do so not in a vague humanitarian way but in the scientific, economic sense. Socialism is, however, something even more than an economic doctrine; it is a philosophy of life, and as such also it appeals to me. I see no way of ending the poverty, the vast unemployment, the degradation and the subjection of the Indian people except through socialism. That involves vast and revolutionary changes in our political and social structure, the ending of vested interests in land and industry, as well as the feudal and autocratic Indian states system. This means the ending of private property, except in a restricted sense, and the replacement of the present profit system by a higher ideal of cooperative service. It means ultimately a change in our instincts and habits and desires. In short, it means a new civilization, radically different from the present capitalist order.

He confessed that "I do not know how or when this new order will come to India," but believed that "every country will fashion it after its own way and fit it in with its national genius." He was aware that socialism did not fit in with the ideology of Congress, but wished that Congress would become "a socialist organization" even though he placed no great hope on it since "the majority in the Congress, as it is constituted today, may not be prepared to go thus far." Still he emphasized: "Socialism is thus for me not merely an economic doctrine which I favour; it is a vital creed which I hold with all my head and heart." The advocacy of socialism was tempered only by the priority of independence and, as always, by the need for national unity: "Much as I wish for the advancement of socialism in this country, I have no desire to force the issue in the Congress and thereby create difficulties in the way of our struggle for independence."[86]

The Lucknow presidential address is considered by many to mark the high tide of Nehru's socialism, after which his socialist fervour seems to ebb. It was, no doubt, a daring speech for a socialist president at the head of a conservative organization even if it provided no clue to a programme of action. It is possible, in retrospect, to mock the resort to rhetoric and the reluctance to make hard choices, but even the rhetoric was then unacceptable to the conservative leadership. Two months later, as he continued his missionary activity, six of his colleagues led by Rajendra Prasad and Sardar Patel resigned from his working committee in protest. They were undoubtedly concerned at what seemed like an organized political offensive on the part of the socialist wing and, though unstated, by the influx of communists into Congress through the Congress Socialist party. In their words:

> We feel that the preaching and emphasising of socialism particularly at this stage by the President and other socialist members of the Working Committee while the Congress has not adopted it is prejudicial to the best interests of the country and to the success of the national struggle for freedom which we all hold to be the first and paramount concern of the country.... The Congress organisation has been weakened throughout the country without any compensating gain. The effect of your propaganda on the political work immediately before the nation, particularly the programme of election, has been very harmful....[87]

It was not only the conservative wing in Congress that was upset at the spectre of socialist forces taking over Congress, but also the Indian business class.[88] In reaction to the step taken by his colleagues, Nehru decided to resign but Gandhi was able to smooth over the incident; indeed, Nehru was asked to stay on for another year because of the forthcoming election campaign, given his almost mystical appeal to the peasantry and the middle classes. The event, though, taught Nehru the limits to the pursuit of leftist goals inside the Congress. On the other hand, he could not be lured into deserting Congress, for that would have put an end to the dream of independence. What he learned from the event was the distinction between his personal ideological goal and the organization's political goal, the former to be pursued only as opportunities open. As he was to express it in 1939: "But India has not accepted this goal, and our immediate objective is political independence. We must remember this and not confuse the issue or else we will have neither Socialism nor independence." He said this even as he thought "India and the world will have to march in this direction of Socialism."[89] The Congress under Gandhi's leadership was further essential to his goal, for "as he saw it, rightly or not this provided the best, even if paradoxical, chance of making his own socialist ideas practicable."[90]

Chastened perhaps by the earlier conflict with colleagues over socialism and no doubt preoccupied with the demands of the looming election campaign, Nehru in his presidential address at Faizpur at the end of 1936 evidenced less concern for questions of ideology and more for concrete issues of abolition of princely states, opposition to possible acceptance of office by Congressmen after the elections, and bringing political change by blocking the legislatures. Reference to ideology was not altogether absent but was in low key. Ranging over the worldwide problem of imperialism, especially in relation to Spain, he averred: "Modern imperialism is an outgrowth of capitalism and cannot be separated from it." "The disease is deep-seated," he warned, "and requires a radical and revolutionary remedy, and that remedy is the socialist structure of society." However, the remedy could only be for the distant future: "We do not fight for socialism in India today, for we have to go far before we can act in terms of socialism." The immediate goal was not socialism, even though that was important and desirable, but independence:

> The Congress stands today for full democracy in India and fights for a democratic state, not for socialism. It is anti-imperialist and strives for great changes in our political and economic structure. I hope that the logic of events will lead it to socialism, for that seems to me the only remedy for India's ills. But the urgent and vital problem for us today is political independence and the establishment of a democratic state.

More concretely, Nehru underlined economic issues such as poverty and unemployment and land reform; here, indicative of a new emerging interest, he felt that "only a great planned system....with vision and courage to back it, can find a solution." However, this had to remain only a distant goal, since "the immediate goal -- independence -- is nearer and more definite."[91]

The party's progressive election manifesto, adopted earlier at Nehru's behest, had assured a new deal for the peasantry through agrarian reforms. Nehru travelled like a hurricane through India campaigning for Congress, and his mass appeal was vindicated by the tremendous victory scored by the party. The Congress assumed office in eight of eleven provinces in British India, even though Nehru was against it for fear of office dimming the spirit for independence. He could not remain unaware, though, that if Congress could come to office at the provincial level it could not be far behind from power at the central level. The installation of Congress ministries had one result which was of some consequence in Nehru's own ideological development, for it directed attention toward the solution of actual problems, not simply passing resolutions and giving speeches, as had been his custom hitherto. Since the mid-1920s, Nehru had underlined the necessity of economic planning along with socialism. Now, at his initiative, the Congress in 1938 established the National Planning Committee. This started a period of re-education for Nehru, persuading him to link his socialist vision to Indian social realities.

3c. *1938-1946 -- Towards Adaptation of the Socialist Vision:* It is important for an adequate evaluation of this period to realize that other more critical and urgent issues could not but crowd out concern for the future social and economic organization of the country -- the rise of communalism and separatism that culminated in partition, World War II and India's posture toward involvement in it, the continued struggle for independence and British repression, long terms of imprisonment during most of the war, then virtual civil war, the threat of balkanization, and negotiations toward the end of empire.[92]

The earlier period from 1929-1937 had seen Nehru's Marxism at its intensest. From the role of a revolutionary radical in that period Nehru seemed, in the eyes of some, to revert now to "the role of a radical nationalist," with his ideology "not Marxism but a mild form of Fabianism.....though once in a while there came flashes of his old Marxism." It has even been suggested that Nehru had been domesticated and, in line with the Marxist emphasis on the bourgeoisie as explanation, that a counter-strategy by the capitalist class "played an important role in first containing him and then moulding him."[93] Rather than endowing the capitalist class with such extraordinary capacity to bring about change in Nehru's ideology, it is more fruitful to examine that change as a stage in intellectual growth as he brought his earlier socialist exuberance into contact with problems of development. An enormous aid in that examination is provided by the analytical distinction that the Marxist scholar P.C. Joshi makes among socialist vision, socialist model, and socialist mobilization.[94] The vision refers to the general features of the preferred society which would prevail universally across nations, the model concerns the form of that vision specific to each country, and the mobilization relates to the instrument for bringing about the model into existence.

In regard to Nehru's vision of the preferred society, it is important first to recognize in him, negatively, the unalterable rejection of capitalism along with its underlying principles of private profit, private property and inequality. Not for him the conception of social gain through pursuit of private interest as in Adam Smith. More positively, there is in him the personal value preference for an egalitarian society that is organized directly on the basis of social gain and social ownership of property. This society, however, was to be founded on political democracy, not dictatorship. Whereas he held capitalism to be incompatible with democracy, he considered -- like Schumpeter and later Galbraith -- democracy and socialism not only to be compatible but necessary for each other. His commitment to these central features of his vision of the ultimate society remained unchanged throughout his life. However, this vision of a socialist society did not conform "to any single socialist theory or doctrine. His conception of socialism was synthetic to the point of being eclectic."[95] His official biographer observed that with Nehru "socialism was a broad tendency and not a precise body of rigid belief."[96] Of course, Nehru also believed that what was a personal value preference for him was also an "objective necessity"[97] built into the historical process.

Even during the years when his Marxist phase was at its peak, Nehru urged that socialism would have to be adapted to Indian conditions. But he provided little insight into the ways in which it was to be adapted. While his preference for the socialist vision was obvious, there was no specification of the socialist model for India. Gandhi, ever alert to the chinks in the political armour of others, had shrewdly warned him in 1934: "You have no uncertainty about the science of socialism but you do not know in full how you will apply it when you have the power."[98] Or, again Gandhi commented: "He is a firm believer in socialism, but his ideas on how best the socialist principles can be applied to Indian conditions are still in the melting pot."[99] Nehru's earlier espousal of the socialist vision, even then significantly modified by its combination of democracy with socialism, may have enthused the conventional Marxist; but it could have had little meaning to Indian society unless it could be related to that society not only in terms of ends but also means. Nehru himself advised his socialist colleagues in 1936:

> Two aspects of this question [socialism] fill my mind. One is how to apply this approach [socialism] to Indian conditions. The other is how to speak of Socialism in the language of India....the language which grows from a complex of associations of past history and culture and present environment. Merely to use words and phrases, which have meaning for us but which are not current coin among the masses of India, is often wasted effort..... This

is a question which I should like a socialist to consider well.[100]

Similarly, he advised Krishna Menon the same year: "It is easy enough to take up a theoretically correct attitude, which has little effect on anybody."[101]

It is this requirement of specific adaptation of the socialist vision to Indian society that makes Nehru's involvement with the National Planning Committee especially significant, for that is what provided the impetus to move him toward formulating a socialist model specifically for India. Nehru had found socialism attractive in the first place for the qualities of science, rationality and planning that he associated with it. But until then "Nehru's interest in economic development, planning and socialism was primarily theoretical and intellectual."[102] Of course, in planning for the future Nehru had to confront the initial starting point not only in relation to the economy but also society and polity. Furthermore, Nehru could not plan as if he were an economic czar; rather, with absolutely no power, he had to carry along a large group of other planners, drawn not only from Congress and non-Congress provincial governments and some princely states but, importantly, from business, finance, labour, education and science. The committee also covered a wide ideological spectrum; in Nehru's words, "hardheaded big business was there as well as people who are called idealists and doctrinaires and socialists and neo-Communists."[103] Nehru soon realized that "constituted as we were, not only in our committee but in the larger field of India, we could not then plan for socialism as such." Conscious of the potential divisiveness of ideology, his approach was to look at concrete issues:

> Constituted as we were, it was not easy for all of us to agree to any basic social policy or principles underlying social organization. Any attempt to discuss these principles in the abstract was bound to lead to fundamental differences of approach at the outset and possibly to a splitting up of the committee. Not to have such a guiding policy was a serious drawback, yet there was no help for it. We decided to consider the general problem of planning as well as each individual problem concretely and not in the abstract, and allow principles to develop out of such considerations.[104]

Two distinct approaches were apparent in the committee: "The socialist one aiming at the elimination of the profit motive and emphasizing the importance of equitable distribution, and the big-business one striving to retain free enterprise and the profit motive as far as possible and laying greater stress on production." But rather than resolve this conflict, the committee took as its frame of reference the achievement of an adequate standard of living for India's masses and of national self-sufficiency.

The work of the committee was never completed and no overall final report was arrived at, leave aside an economic plan in the real sense of the word. The committee had not been daunted by the lack of cooperation from the central government, but the resignation of Congress ministries in 1939 created serious difficulties. Finally, the committee effectively went into limbo with Nehru's imprisonment the next year, which lasted the duration of World War II except for an interruption of eight months.

Although agreement could not be reached on many issues, there is manifest in the committee's work, according to Joshi, "in an embryonic form the conception of a mixed economy as the Indian model of transition to socialism."[105] In other words, in this interpretation the model of the mixed economy was instrumental-socialist, not instrumental-capitalist, though some ambiguity remained over the pace of movement toward socialism. Nehru himself felt that, while it was unreasonable to expect the committee to endorse socialism,

> Our plan, as it developed, was inevitably leading us toward establishing some of the fundamentals of the socialist structure. It was limiting the acquisitive factor in society, removing many of the barriers to growth, and thus leading to a rapidly expanding social structure.[106]

In the event, the committee did not eliminate free enterprise but placed severe restrictions on its scope. There was agreement on state ownership and control of defence industries, while "regarding other key industries, the majority were of opinion that they should be state-owned, but a substantial majority of the committee considered that state control would be sufficient. Such control of these industries, however, had to be rigid. Public utilities, it was also decided, should be owned by some organ of the state."[107] There were differences over whether the emphasis should be on heavy industry or on cottage industries preferred by Gandhi. Nehru favoured heavy industry but saw merit in cottage industries from the perspective of employment over a transitional period. In regard to agriculture, the committee accepted the cooperative principle but allowed peasant farming on small plots to be continued; intermediaries, however, were definitely to be abolished. Some, including Nehru, favoured socialization of credit and state control of foreign trade.

More important than the substantive agreement over organization of the economy was the general approach that Nehru brought to bear on the work of the committee and the underlying rationale for it. For, that demonstrates the educative effect of actually planning for practical affairs, even within limits, rather than merely proclaiming universal ideologies:

> And all this was to be attempted in the context of democratic freedom and with a large measure of cooperation of some at least of the groups who were normally opposed to socialistic doctrine. That cooperation seemed to me worth while even if it involved toning down or weakening the plan in some respects. Probably I was too optimistic. But as long as a big step in the right direction was taken, I felt that the very dynamics involved in the process of change would facilitate further adaptation and progress. If conflict was inevitable, it had to be faced. But if it could be avoided or minimized, that was an obvious gain. Especially as in the political sphere there was conflict enough for us, and in the future there might well be unstable conditions. A general consent for a plan was thus of great value. It was easy enough to draw up blueprints based on some idealist conception. It was much more difficult to get behind them that measure of general consent and approval which was essential for the satisfactory working of any plan.[108]

Earlier, he had tactically separated planning from socialism, but for a purpose; as he reminded the committee's secretary, K.T. Shah, in 1939: "If we start with the dictum that only under socialism there can be planning, we frighten people and irritate the ignorant. If, on the other hand, we think in terms of planning apart from socialism and thus inevitably arrive at some form of socialism, that is a logical process which will convert many who are weary of words and slogans." Nehru at the time did not want a direct challenge to the existing structure but neither did he want vested interests in that structure to be strengthened, because that would place obstacles in the path of change later. He felt that "in India today any attempt to push out the middle class is likely to end in failure. The middle class is too strong to be pushed out and there is a tremendous lack of human material in any other class to take its place effectively, or to run a planned society." He was further afraid that: "Here, in India, a premature conflict on class lines would lead to a break-up and possibly to prolonged inability to build anything." This fear of "disruptive forces" provides an important clue to Nehru's essential moderation. He wanted to work toward his ideological goal, but not at the cost of breaking up the country.[109]

The ideologically committed socialists may have felt that Nehru should have remained a doctrinaire, if ineffectual, Marxist. But as a national leader, he perceived his purpose to bring about change within the limits placed by social realities even while remaining loyal to his larger vision. In this particular instance, he had little power; he was simply chairing a non-official committee of essentially non-governmental people, representing a wide ideological spectrum, but with a driving compulsion to demonstrate to colonial rulers -- too anxious to highlight the divisions among Indians and their lack of capacity at future governance -- that they could work together. He was not heading a party government with a popular mandate for a political platform.

While the committee's work had educative value, it did not mean that Nehru had changed his goal of socialist society. It certainly did not mean that the committee's recommendations represented Nehru's preferred choice, simply the consensus in those particular circumstances. Even taking into account social and political realities, Nehru's own position was more radical and was more truly reflected in the proposals made by K.T. Shah, his catalyst and lightning rod on the committee:

> Private monopolies in key industries or services are dangerous, and must be rigorously forbidden.... Public Utilities...should be conducted as public monopolies. New industries, suitable only for large-scale work by power-driven machinery and for standardized mass production, where no private vested interests now exist, must be established and conducted as public enterprise. All Key Industries must be state-owned and state-managed. The profit *motif* must be summarily and completely excluded from the conduct and organisation of every such industry. Where in any such Key Industry, private enterprise exists, and has created vested interests, these must be bought out, and the industry nationalized, subject to the payment of a just compensation..... The foreign trade of the country must be a public monopoly, conducted as an integral part of the National Plan.[110]

Even when watered down by the requirements of consensus, the path-breaking endeavours of Nehru in the National Planning Committee proved consequential; they were to make India "plan-conscious" and send government, business and other groups in a flurry to evolve plans "in rivalry" to his work.[111]

If the economic mode of transition to socialism was, for Nehru, by way of "mixed economy"-- though yet not termed as such -- under a regime of planning, then democracy was the political route to that socialist society, not just a feature of the ultimate society; "his view therefore stands in sharp contrast to the classical view which stresses the necessity of 'dictatorship of the proletariat' as the political form inevitable for transition to socialism."[112] Furthermore, while Nehru recognized that class conflict was inherent in society, he did not follow classical Marxism in advocating its intensification. Again, he rejected violence and, inclining towards Fabian socialism, wanted the achievement of his vision through non-violence and democracy, difficult as that may prove to be. Nehru believed that socialism could not be imposed, people had to be ready for it and to want it; thus it was that Nehru took on such an active role in the education of cadres and masses in socialism. Basically, then, Nehru believed in socialism by consent; looking at this stance more critically, the Comintern had as early as 1929 ridiculed him as a leader "who promises all the blessings of socialism without a revolutionary struggle."[113]

Nehru departed from classical Marxism also in regard to the organizational instrument for socialist mobilization. For him, the appropriate instrument for such mobilization was the Congress party as a national multi-class organization rather than any distinctive party of the proletariat. This position stemmed from his feeling that, despite the claims of other organizations and "bourgeois as the outlook of the National Congress was, it did represent the only effective revolutionary force in the country" and that Gandhi was a truly revolutionary leader. He was not against organizational independence of labour and peasant organizations, but wanted them to cooperate with Congress. Indeed, he saw himself as the link between nationalism and mass organizations, hoping to bring them "closer to each other -- the National Congress to become more socialistic, more proletarian, and organized labour to join the national struggle."

In the late 1920s Nehru felt that "the course of events and the participation in direct action would inevitably drive the Congress to a more radical ideology and to face social and economic issues."[114] Subsequently, as Congress widened its support among the masses, he believed that the party itself could be reoriented toward socialism; he felt that it was, as a result, by far "the most effective radical organization in the country and it is easier to work great changes in the mass mentality through it rather than through any other means."[115] He felt strongly, as one Congress leader relates it, that "however radical one might be... nothing could be achieved except through the Congress. In a vast country like India it was not easy to build up an organisation with branches in all the towns and villages. Therefore, making the Congress itself the instrument of socialism was the best way for those who espoused the creed."[116] He realized that there were forces hostile to socialism within Congress but felt these could be overcome by the pressure of the masses now mobilized into the party. He was overly optimistic in this regard, but Joshi notes that "it was the unique contribution of Nehru to provide a bridge between these two processes of nationalism and socialism."[117] Similarly, Chandra and his associates appreciatively refer to Nehru's paradigm with its "conceptual nexus between nationalism and socialism that he advocated and his understanding that the left had gradually to transform the Congress in a socialist direction and that to break away from the National Congress and Gandhi's leadership would be counter-productive for socialist and radical goals."[118]

Nehru was not particularly successful in uniting leftist forces behind him, largely because most leftists refused to rally around his ideas as these seemed to depart from classical Marxism. While theoretically positing that socialism had to be creatively adapted to the circumstance of different countries, they remained bound to the classical model, refusing to see any merit in his formulations in relation to India and condemning him to be a bourgeois reformist. Nehru was thus left to fight his battles alone while socialists and communists became peripheralized in Indian politics. In the process, Nehru's model was bereft of organizational and institutional support and its influence did not extend beyond sections of the middle class.[119]

If Nehru made important departures from classical Marxism in his socialism, no less significant was his break with Gandhi (though that is not fully treated here). As against Gandhi, he opted for modern industry and technology over self-sufficient villages, large-scale industry over cottage industry, and public ownership of the means of production over private trusteeship.[120]

4. Nehru, Planning and the Capitalist Class

The establishment of the public sector in Third World countries after independence has been held by some to be a function of the successor new middle class being in power, partly in reaction to the local capitalist class having allegedly been in alliance with the colonial power; in this interpretation, the public sector is understood to run counter to the interests of the capitalist class.[121] Contrariwise, it is also held by others that the setting up of the public sector could not but be in the interests of the capitalist class, because the nationalist party that succeeded the colonial power was a bourgeois organization dominated by the local bourgeoisie.[122] Alternatively, it is suggested that the public sector must be in the interests of the capitalist class, because there is little to differentiate between nationalist and bourgeois positions on the economic role of the state since both focused on an active role for the state during the course of the nationalist movement.[123]

These variant interpretations of the relationship of the public sector to the interests of the capitalist class raise questions both of logic and fact. Strictly speaking, the capitalist is an economic man whose basic aim is the maximization of private profit. In the pursuit of such profit, capitalists compete with each other; since this can lead to cutthroat competition, they have no objection, indeed deem it essential, to accord the state a regulative role in order to provide order in the economic environment through acceptable rules of the competitive game. Equally, in the maximization of private profit, capitalists have no objection -- indeed, they constantly and persistently urge for appropriate incentives and initiatives -- to the state taking on a promotive role in ensuring an economic environment in which capitalism and capitalists can flourish. When the state takes on this promotive role, no doubt not all capitalists benefit equally, but nearly all benefit if unevenly. What capitalists object strenuously to in the performance of this promotive role, however, is any attempt on the part of the state to exclude, directly or indirectly, capitalists from making profits in any area of economic activity. As O'Connor perceptively observes: "capital normally opposes the establishment of state industry that competes with monopoly capital -- that is, it opposes any program that socializes profits."[124]

If this is correct, it follows logically then that, where capitalists reconcile themselves to giving up the making of private profits in favour of socialization, it is a sign of their weakness in relation to the political environment in assuring an economic arrangement that ideally serves their interests. There may be no alternative to such socialization in a particular set of circumstances but it is not the preferred ideal arrangement, given the logic of the motive of private profit. To argue otherwise -- that this, too, in effect, is in the interests of capitalists -- is really to argue that, no matter what happens, except perhaps for the complete extinction of the class itself, it cannot but be in the interests of the capitalist class.

It may be that other motives are operative to modify the social behaviour of capitalists but then it cannot be argued that the behaviour is a function of the motive specific to capitalists qua capitalists. For example, in a situation of war, capitalists may put up with state restrictions on their economic activity but they do so not necessarily out of the profit motive but out of interest in physical survival. Similarly, in a colonial context, capitalists may take certain positions which perhaps spring out of resentment at enslavement rather than the profit motive.[125] Even if stemming from interest, those positions may correspond, not to what is preferred ideally by the capitalist with his profit motive, but what is feasible in the particular set of circumstances in which capitalists are situated. But in this case it is the social context which needs to be given adequate weight, not simply some arbitrary attribution of interests, whether by observers or capitalists themselves. Furthermore, capitalists may also theoretically promise to submit themselves to certain restrictions in the future, not so much because it is ideally in their interest to do so, but as a damage-limitation effort to undercut radical proposals for even more restrictive regimes. This, too, reflects the impact of larger social forces rather than capitalist interest. As the societal context changes, the promise of what was once considered acceptable may also change. A proper understanding of the position of the capitalist class on the role of the state thus requires not simply the assumption that anything the capitalist does is necessarily in his profit-motivated interest, but an adequate appreciation of the constraints the societal context places on the range within which that interest can be pursued.

As for the empirical aspect of the question, the enormously impressive fact is that the capitalist class as a whole did not become an ally of imperialism during the colonial period in India. It is remarkable that as widely divergent Marxists as Ajit Roy and Bipan Chandra are agreed upon the fundamental fact that the Indian capitalist class has not been a comprador bourgeoisie. Roy dismissed, as early as 1953, the application to India of the notion of bourgeoisie as comprador on the analogy of China, arguing that "the topmost stratum of the Indian bourgeoisie, the Indian monopoly capitalists, are not comprador in character. Indeed, comprador elements never played any important role in economic and political development of the country, at least since the turn of century."[126] Similarly, Chandra has argued that the Indian capitalist class, not being comprador, stood in a long-term contradiction to imperialism even though the colonial context persuaded it to accept a short-term accommodation with imperialism.[127] As a consequence, while the capitalist class provided support to the nationalist movement, it did so in a passive manner. However, the capitalist class did not provide either the leadership or the main force of the movement; even as ardent a follower of Gandhi as G.D. Birla stoutly denied that he was ever a Congressman.[128] Of course, some members of the class provided "financial assistance to the Congress, but perhaps the extent of such support is exaggerated."[129] The general support of the capitalist class to the nationalist movement, however, tended to be directed more to the right wing in the Congress.

However, even the right wing in Congress was not a creation of the capitalist class, only the recipient of the support of that class. The right wing arose, much as the left wing did, as the nationalist movement became transformed into a mass movement with the induction of the peasantry. The leadership of the right wing, belonging to the new middle class just as much as that of the left wing, rose to eminence on the basis of its record of sacrifice and organizational success in peasant struggles, not because it had the backing of the bourgeoisie. But it came to its political views no more at the behest of the business class than did the left wing leadership to its views at the instigation of labour. The fact that the capitalist class did not provide leadership of the right wing reflects a weakness of that class, arising out of the historical situation of functioning within a colonial context, where the new middle class had preempted leadership and where it had gone on to mobilize the peasant masses as a pillar of support against colonialism. The national bourgeoisie came to support (but really did not join) the nationalist movement at a later stage as it consolidated itself as a class after World War I.

The passive role of the capitalist class in the nationalist movement has to be seen not only in the light of its short-term accommodation with imperialism, for fear of reprisals against its property, but also in terms of the image held of it by the Indian populace, especially substantial sections of the middle class. That image was not one of pioneers and captains of industry, but of ruthless exploiters, blackmarketeers and adulterators, interested more in the quick rupee than in long-term development. By the same token, radical anti-capitalist ideas had considerable appeal among substantial segments of the population.

It is really the weakness of the bourgeoisie in the cultural and political environment of India that enables observers to qualify as "mature" the views of the eminent business leader G.D. Birla insofar as he recognized the limits that the particular historical setting imposed on the ambitions and interests of business. As against his more restrained posture in response to the radical views expressed by Nehru in the 1930s, many businessmen -- some jointly in a manifesto and others individually-- attacked Nehru viciously; some of them called him "a wholehearted communist" who masked his real intent behind the smokescreen of the term socialism. Homi Mody declared: "His meaning is clear and the programme is fairly definite. First, political independence, and then a Socialist State, in which vested interests, property rights and the motives of profit will have no place at all. Let those whose minds are running in the direction of intermediate stages and pleasant halting places not forget that they are really buying a through ticket to Moscow." Nehru was attacked for pushing a "destructive and subversive programme", which advanced class hatred and class war and thus split the nationalist movement, delaying national independence.[130]

In their wholesale attack on Nehru, these businessmen failed to recognize their weakness and vulnerability in the particular historical context, seemingly operating in an imaginary world where the ideal interests of capitalists could be pursued. It was left to Birla to underline for his fellow businessmen their weak position: "Evidently, you did not consider its contents carefully. The manifesto has given impetus to the forces working against capitalism -- another result which you did not intend." He acknowledged that "we are all against socialism", but pointedly reminded them of their lack of appropriate credentials for such opposition in the context of mass poverty: "It looks very crude for a man with property to say that he is opposed to expropriation in the wider interests of the country.....but the question is, 'Are you or myself a fit person to talk?'"[131] These are the words of the leader, not of a hegemonic class, but of one that is highly vulnerable.[132] Earlier, Birla had confessed that businessmen would suffer not only at the hands of the colonial government but also of "their own countrymen", with Nehru apt to be the worst thorn in their side.[133]

Notwithstanding the evidence, it is amazing that even otherwise careful and independent minded Marxist scholars, such as Bipan Chandra, can still relapse into the standard Marxian line on the "bourgeois hegemony over the nationalist movement and social development".[134] In the light of his general argument on the position of the Indian capitalist class in relation to the nationalist movement, such a stance is, in the absence of any other evidence, a complete *non sequitur* except if bourgeois refers to the petty bourgeoisie which would indicate a rather flexible use of the term.[135]

Chandra also takes the position that, while Nehru's Marxist phase, especially during 1933-1936, had caused consternation in the capitalist class, that class adopted a strategy which succeeded "in first containing him and then moulding him so that, by 1947, the capitalist class was ready to accept him as the Prime Minister of independent India and to cooperate with him in the task of building up its economy along the capitalist path."[136] And what did this strategy consist of? It consisted primarily of reliance on the leadership of the right wing of Congress, since opposition by the capitalist class itself to Nehru would have been counter-productive. However, to reiterate, this is an index, not of the strength of the capitalist class, but its weakness. If it must rely on the talented right-wing leadership of Patel, Prasad and Rajagopalachari, and perhaps even Gandhi -- all of whom had their own autonomous bases of political support -- what could be the fate of the capitalist class if, let us assume, these leaders were not around any more, or had become politically irrelevant. Should one assume that Nehru had been moulded for ever? Chattopadhyaya's criticism in another context would seem to apply here insofar as there is the assumption of "a once-over, static, irreversible phenomenon and therefore incapable of explaining the zig-zag course of the post-independence period."[137] It seems that the capitalist class was more cognisant of its weakness; as Birla confided to the Viceroy in 1936: "But the possibility of the present left-wing minority in the Congress becoming a majority at some date cannot be ignored. Most of these left-wingers are young men whereas the right wing leaders are mostly old."[138] If reliance on right-wing leadership was one index of the weakness of the bourgeoisie, so was the extreme and abiding reluctance on its part to engage in an "open confrontation with Nehru", for it "had no choice but to accept Nehru, because of his large popular base."[139] Needless to add, this element of Nehru's mass popularity inhibited the right wing of Congress no less than the bourgeoisie, especially after independence, because of fear of electoral reprisal on the part of the masses and therefore of loss of political power and stability.

Furthermore, Chandra provides no evidence that the right-wing leadership acted simply at the behest of the capitalist class and that it would not have done so if that class had not desired it. Not only that, he also assumes that, simply because the capitalist class so desired, it was successful in moulding Nehru. At the same time, a whole series of contextual factors are given the nod of recognition as having a bearing on Nehru's ideological stance -- "his failure to build a political base of his own and lack of active work among or even contact with workers and peasants after 1936; his attachment and subservience to Gandhi which was strengthened by his fear of being 'lonely' or isolated politically; his refusal to form a socialist group or join hands with existing ones or organise in any form radical activity outside the Congress framework; the weakness of the Left outside the Congress; his utter neglect of organisation, even within the Congress". Indeed, in the final analysis, the capitalist strategy boiled down to no more than merely recognizing the allegedly true character of Nehru -- his essential sobriety and realism notwithstanding his strong words. If there be containment and moulding here, then it is really self-containment and self-moulding. However, it is the capitalist strategy and its supposed success that are apotheosized, and the conclusion on the nationalist movement being a bourgeois movement under the hegemony of the bourgeoisie consequently remains unshaken.

Equally as questionable seems to be the proposition that there is little to differentiate between the pre-independence position of the capitalist class and that of the radical wing of the nationalist leadership in India on matters of economic planning and socialism.[140] The central basis for the conclusion on lack of difference between their positions is a comparison of the work of the National Planning Committee (NPC), incomplete as it was, and the economic plan worked out in 1944 by what constituted India's business and managerial elite, popularly known as the Bombay Plan or the Tata-Birla Plan. The NPC was not strictly reflective of the radical nationalist position since it incorporated a strong component from industry and finance. Even so, it has already been seen that the NPC's work focused on a mixed economy of the instrumental-socialist type. Lest this be considered a prejudiced view, there is the judgment of the Marxist scholar Sachs that: the perspective of the National Planning Committee, especially at least that of K.T. Shah and Nehru, "is being frequently underestimated in the discussions on the 'socialistic pattern of society'. It is easy to show the impracticability of some of K.T. Shah's rather doctrinaire proposals and even to argue against the theoretical soundness of some of his positions. But, it would be equally improper to minimize its contribution to an ideological formulation of a planned polity, which deserves to be considered in the connotation of a 'socialistic pattern of society'. The Report embodied a very bold and radical programme of socio-economic changes to which much thought had been given during the crucial years of national struggle for Independence."[141] Kurien would seem to agree when he says that "what the Committee envisaged for India was very much a *socialist* plan."[142]

As compared to this assessment, the position of the Bombay planners stands on an altogether different footing. The Tata-Birla Plan was an extraordinarily ambitious plan, with its chief objective being the tripling of national income and the doubling of per capita income within 15 years. The major thrust of the Rs. 100 billion 15-year plan was toward industrialization, which was to receive 45% of total investment. Here, the importance of basic industries was underscored immediately by the planners; it was their perception that the absence of basic industries had impeded India's industrial development and it was their aim "to give priority to basic industries over other industries and thus to speed up development" and to reduce foreign dependence. If the planners conceived of their planning objectives ambitiously, they were equally quite forthright about resource mobilization. Not finding visible resources for one-third of the total investment, they argued for heavy deficit-financing, but immediately reconciled themselves to its political implications. Never so clearly have a group of eminent businessmen admitted the drastic nature of measures required for a rapid programme of economic development, and concomitantly the political implications of economic planning:

> During this period, in order to prevent the inequitable distribution of the burden between different classes which this method of financing will involve, practically every aspect of economic life will have to be so rigorously controlled by government that individual liberty and freedom of enterprise will suffer a temporary eclipse.[143]

At another place, the planners recommended as a redistributive measure that "profits should be kept within limits through fixation of prices, restriction of dividends, taxation, etc." For the reduction of inequalities, they stated that "a steeply graduated income tax which would keep personal incomes within limits would obviously be the most important weapon for this purpose in the fiscal armoury of the country." State intervention was accepted as "an indispensable feature" of planning, and the point was reiterated that

> No economic development of the kind proposed by us would be feasible except on the basis of a central directing authority, and further that in the initial stages of the plan rigorous measures of State control would be required to prevent an inequitable distribution of the financial burdens involved in it. An enlargement of the positive as well as preventive functions of the State is essential to any large-scale economic planning. This is inherent in the idea of planning and its implications must be fully admitted.[144]

In such radical terms did the business and managerial elite plan for an India soon to become independent politically. There is up to this point little to distinguish the

Bombay planners from the nationalist planners. Perhaps if they had waited to formulate such a plan after independence, when nationalist sentiment usually becomes muted or eroded, they would have thought differently.

But did the Bombay planners really have a conception of the ends of planning the same as or similar to that of Nehru and his radical associates? Quite the contrary, their conception of the role of the state in economic planning was strictly one that would be promotive of capitalism. Their objective was a consummatory capitalist economy; even with the benefit of the doubt, it was perhaps only a mixed economy of the instrumental-capitalist type. Critical to this conclusion is the sharp distinction which the Bombay planners themselves made between the limited interim period -- during which development of the economy is fostered under a regime of planning -- and the normal operation of the economy subsequent to the achievement of such development. The Bombay planners left no doubt about their choice of the preferred economic system. While mindful of the shortcomings of the capitalist system, they underlined that "it possesses certain features which have stood the test of time and have enduring achievements to their credit," and warned against uprooting "an organization which has worked with a fair measure of success in several directions", insisting that capitalism has "a very important contribution to make to the economic development of India." They pointedly warned that, "if the future economic structure of the country is to function effectively...it must provide for free enterprise" apart from assuring a fair income distribution.

The Bombay planners were willing to acknowledge that the state may own certain industries permanently, such as those pertaining to defence and vital communications. They were even pragmatic enough to concede that completely new types of industries, or industries requiring location in new areas, or industries in need of large financial assistance from the state, may be owned by the state, but immediately qualified this concession: "if later on private finance is prepared to take over these industries, State ownership may be replaced by private ownership." Interestingly, one of the Bombay planners, A.D. Shroff, had chaired the NPC's subcommittee on industrial finance, which made its difference with the larger committee explicit: "This Sub-Committee are of the opinion that they cannot endorse the recommendation that the State should own and control all Defence Industries and Public Utilities, and that the Key Industries may be State-owned or State-controlled."[145] The Bombay planners thus wanted only a limited role for state ownership of business enterprises, preferring to restrict the state only to economic controls:

> From the point of view of maximum social welfare State control appears to be more important than ownership or management. Mobilization of all the available means of production and their direction towards socially desirable ends is essential for achieving the maximum amount of social welfare. Over a wide field it is not necessary for the State to secure ownership or management of economic activity for this purpose. Well-directed and effective State control should be fully adequate.

Even in terms of controls, the planners envisaged them strictly in the context of political democracy, not of authoritarian or totalitarian government, and on the pattern of war-time restrictions, that is, "of limited duration and confined to specific purposes."[146]

It is a gross misinterpretation therefore to take the adoption of controls over private enterprise for a limited interim period of time, in order to bring about a more flourishing industrial capitalism, to mean the same thing as the attempt of Nehru and his radical colleagues to impose restrictive controls permanently on capitalism, along with a rapidly expanding public sector, in order to push society toward a different type of economic organization. It was not without reason therefore that Marxist leaders and scholars had earlier concluded that the Bombay planners intended to build capitalism, not to control or curb it.[147] Furthermore, one cannot exclude a considerable measure of political calculation on the part of the capitalist class in the heyday of nationalism to forestall, by its concessions on imposition of state controls, acceptance of more radical proposals on state ownership. This suspicion is strengthened by the fact that six years later one of the signatories of the Bombay Plan, John Matthai, resigned from the government in 1950 in protest against even the establishment of a planning commission. What the capitalist class essentially envisaged, in the final analysis, was a model of development after the pattern of Meiji Japan, and in this it may well have reflected the real sentiments of the bulk of the top leadership of the Congress party outside of Nehru.

To draw attention to the difference between the nationalists, more particularly Nehru, and the capitalists in terms of the aims and modalities of economic planning, however, is to say nothing about the end results of putting their recommendations into practice; in the actual process of history such results may turn out to be consistent or quite contrary to the intentions in either case.

Summary and Conclusions

Three visions concerning the future mode of economic organization developed within the Congress party during the course of the nationalist movement, all presupposing acceptance of political democracy -- capitalism, Gandhism and socialism. It was Nehru's historic role to carry the socialist message to the nationalist movement and to lead the socialist forces in the Congress party, even as he attempted to adapt the socialist vision to the particular circumstances of India. While allowing the three visions to compete, the Congress party did not strictly endorse any of them, at least not fully. However, the real contest boiled down to between capitalism and socialism. For while there was appreciation of Gandhism's thrust for village self-sufficiency and cottage industries, this was not considered to be the appropriate prescription for the modern age. Furthermore, insofar as the principle of trusteeship was concerned, Gandhism had little to suggest by way of institutional mechanisms, relying as it did fundamentally on moral conversion.

Between capitalism and socialism, the Congress did not choose, but under constant pressure from Nehru it conceded the principle of state control or ownership of public utilities and key industries. While the Congress denied this had anything to do with nationalization or socialism, this concession provided legitimacy for subsequent efforts by radical nationalists in planning for what was believed to be a socialist trajectory through the transitional stage of mixed economy. The colonial state's own modest role in the setting up of economic infrastructure and public utilities, subsequently fortified by the imposition of a network of economic controls during World War II, served as a starting point in plans for such a trajectory. More negatively, the identification of the country's economic backwardness under colonialism with capitalism and laissez faire acted to create an orientation in favour of an activist entrepreneurial role for the state, if not socialism, during the nationalist movement.

On the other hand, this orientation was to some extent countered by the existence in India, even under colonialism, of the most developed bourgeoisie in any country of the Third World, which could not thus be altogether ignored. At the same time, the lack of a full-scale endorsement of capitalism or an instrumental-capitalist mixed economy was related to the lack of hegemony by the capitalist class in the nationalist movement. The core of the social coalition that led the movement was an alliance between the new middle class and the rich and middle peasantry under the overall hegemony of the new middle class. The bourgeoisie entered this coalition in the 1920s along with labour, but both occupied a subsidiary position in relation to the core. The domination of the social coalition by the intermediate strata or middle sectors but without excluding the capitalist class above and labour below did not, however, mean any endorsement for the socialist vision of Nehru either. Right to the end of the nationalist struggle, the Congress party not only would not accept Nehru's socialist vision but it reined him in when he attempted to use his high office in the party to advance the cause of socialism. From the behaviour of the Congress leadership it is obvious that there is no automatic preference on the part of intermediate strata for socialism or an instrumental-socialist mixed economy. The intermediate strata as a group was divided in its orientation, but the bulk of the top leadership was oriented toward an instrumental-capitalist mixed economy, despite the appearance -- flowing out of Nehru's charismatic mass appeal and public eminence -- that socialism was the favoured prescription with the nationalist movement. Among this leadership were political giants such as Sardar Patel, Rajendra Prasad and C. Rajagopalachari, besides the representatives of the capitalist class.

NOTES

1. Peter Evans, *Dependent Development: The Alliance of Multinational, State, and Local Capital in Brazil* (Princeton: Princeton University Press, 1979), p. 43; Alexander Gerschenkron, *Economic Backwardness in Historical Perspective* (Cambridge, Mass.: Harvard University Press, 1962); and Leroy P. Jones and Il Sakong, *Government, Business, and Entrepreneurship: The Korean Case* (Cambridge, Mass.: Council on East Asian Studies, Harvard University, 1981), p. 11.
2. Malcolm Gillis, "The Role of State Enterprises in Economic Development," *Social Research*, vol. 47 (1980), 248-89; John B. Sheahan, "Public Enterprise in Developing Countries," in William G. Shepherd, *Public Enterprise: Economic Analysis of Theory and Practice* (Lexington, Mass.: Lexington Books, 1976), ch. 9; and Douglas F. Lamont, *Foreign State Enterprises* (New York: Basic Books, 1979), p. 90.
3. Ignacy Sachs, *Patterns of Public Sector in Underdeveloped Economies* (Bombay: Asia Publishing House, 1964), pp. 79-80.
4. V.K.R.V. Rao, "Some Fundamental Aspects of Socialist Change in India," in Ashok V. Bhaskar (ed.), *Growth of Indian Economy in Socialism* (Bombay: Oxford and IBH Publishing Co., 1975), pp. 485-501.

5. International Legal Center (ed.), *Law and Public Enterprise in Asia* (New York: Praeger, 1976), p. 12.
6. Albert Breton, cited in Alan Tupper, "The State in Business," *Canadian Public Administration*, XX, no. 1 (1981), pp. 124-50.
7. See R. Palme Dutt, *The Problem of India* (New York: International Publishers, 1943), pp. 45-52, and Amiya Kumar Bagchi, "De-industrialization in India in the Nineteenth Century: Some Theoretical Implications," *Journal of Development Studies*, XII, no. 2 (1976), pp. 135-64. For an overview, see also Baldev Raj Nayar, *India's Quest for Technological Independence: Vol.I -Policy Foundation and Policy Change* (New Delhi: Lancers Publishers, 1983), pp. 105-113, 135-147.
8. Sachs, p. 109.
9. H.V.R. Iengar, "Role of the Private Sector," in C.N. Vakil (ed.) *Industrial Development of India: Policy and Problems* (New Delhi: Orient Longmans, 1973), p. 30; and S.S. Khera, *Government in Business* (New Delhi: National, 1977), p. 50.
10. Jagdish N. Bhagwati and Padma Desai, *Indian Planning for Industrialization* (London: Oxford University Press, 1970), p. 31.
11. G.K. Shirokov, *Industrialisation of India* (Moscow: Progress Publishers, 1973), p. 24.
12. Amiya Kumar Bagchi, *Private Investment in India 1900-1939* (Cambridge: Cambridge University Press, 1972), p. 45.
13. *Ibid.*, p. 433.
14. B.R. Tomlinson, *The Political Economy of the Raj 1914-1947: The Economics of Decolonization in India* (London: Macmillan, 1979), pp. 31-33.
15. Bagchi, p. 88.
16. Dharma Kumar (ed.), *The Cambridge Economic History of India* (Cambridge, UK: Cambridge University Press, 1983), II, p. 642.
17. India, *Statistical Abstract, India*, 1949, vol. II, p. 1756.
18. Kumar (ed.), p. 379.
19. Bagchi, p. 442.
20. Kumar (ed.), p. 553.
21. Bhagwati and Desai, pp. 39-42.
22. India, Planning and Development Department, *Statement of Government's Industrial Policy* (New Delhi: 1945), pp. 1-8.
23. Kumar (ed.), p. 963.
24. Stanley A. Kochanek, *The Congress Party of India: The Dynamics of One-Party Democracy* (Princeton University Press, 1968), p. 319.
25. Kochanek, p. 322.
26. Bruce McCully, *English Education and the Origins of Indian Nationalism* (New York: Columbia University Press, 1940), pp. 384-85, cited in Kochanek, p. 322.
27. Kochanek, p. 324.
28. Shirokov, p. 54; Sachs, pp. 107-108.
29. Kochanek, pp. 332, 335, 341.
30. Angus Maddison, *Class Structure and Economic Growth: India and Pakistan Since the Moghuls* (London: George Allen & Unwin, 1971), p. 72.
31. Bipan Chandra, "Peasantry and National Integration in Contemporary India," in K.N. Panikkar (ed.) *National and Left Movements in India* (New Delhi: Vikas, 1980), pp. 106-141.
32. See Kochanek, ch. 13; Robert L. Hardgrave, Jr., *India: Government and Politics in a Developing Nation* (New York: Harcourt, Brace & World, 1970), ch. 2; Rajni Kothari, *Politics in India* (Boston: Little, Brown, 1970), ch. 2; V.R. Mehta, *Ideology, Modernization and Politics in India* (New Delhi: Manohar, 1983), pp. 164-65.
33. Boudhayan Chattopadhyaya, "The Ambivalence of Nehru: The Narodnik Utopia vs. The Liberal Utopia," in Centre for Social Research, *The Seminar on Nehru*, (Madras: Centre for Social Research, 1974), pp. 244-59.
34. Cited in Bipan Chandra et al, "The Communists, the Congress and the Anti-Colonial Movement," *Economic and Political Weekly*, XIX no. 17 (April 28, 1984), pp. 730-736.
35. Jawaharlal Nehru, *Toward Freedom* (New York: John Day Company, 1942), pp. 233-34.
36. Satindra Singh, *Communists in Congress: Kumaramangalam's Thesis* (Delhi: D.K. Publishing

House, 1973), p.5.
37. Baren Ray, *India: Nature of Society and Present Crisis* (New Delhi: Intellectual Book Corner, 1983), pp. 52-57, 61, 107.
38. Bipan Chandra, "The Indian Capitalist Class and British Imperialism," in R.S. Sharma (ed.) *Indian Society: Historical Probings in Memory of D.D. Kosambi* (New Delhi: People's Publishing House, 1974), pp. 390-413.
39. Bipan Chandra et al., "The Communists...."
40. Even Chandra recognizes this position of the new middle class when acknowledging that the Indian capitalist class "consistently backed the petty bourgeois leadership of the national movement" while at the same time admitting that "the Indian capitalists associated themselves with the nationalist movement both as a segment of Indian society and as a separate and distinct political force; but for the most part they did not do so through direct participation." See Chandra, in Sharma (ed.), pp. 408-410. It is noteworthy that Nehru himself subscribed to the notion of the middle class thrust of the Congress: "Our direct-action struggles in the past were based on the masses, and especially the peasantry, but the backbone and leadership were always supplied by the middle classes and this, under the circumstances, was inevitable." At the same time, he underlined the "two-faced" nature of the middle class, with sections at the bottom providing the revolutionary cadres and mass, and the upper sections aligning with imperialism, while the centre shifted its sentiments and alliances between the former and the latter. See Jawaharlal Nehru, *Toward Freedom* (New York: John Day Company, 1942), p. 397; also Dorothy Norman (ed.), *Nehru: The First Sixty Years* (New York: John Day Company, 1965), vol. I, p. 430.
41. Nigel Harris, *India-China: Underdevelopment and Revolution* (Delhi: Vikas, 1974), pp. 304-305.
42. One scholar states that the "triple alliance" among the new middle class, capitalists and peasantry "was epitomized by the trinity that led the Congress in 1947 -- Gandhi with his insight into the peasant mind, Nehru the leader of the middle class, and Patel with his loyalties to the indigenous bourgeoisie." Ashok S. Guha, *An Evolutionary View of Growth* (Oxford: Clarendon Press, 1981), p. 113. Interestingly, all three leaders were members of the new middle class, and all were lawyers by profession. More to the point is the observation: "Throughout South Asia, the political leaders who took over the new regimes were moderate constitutionalists; men whose outlook was largely westernised and secular and whose social origins were upper middle class, with a heavy emphasis upon professional and intellectual values." Hugh Tinker, in A. Jeyaratnam Wilson and Dennis Dalton (ed.) *The States of South Asia* (London: C. Hurst & Company, 1982), p. 6.
43. V.K.R.V. Rao and P.C. Joshi, "Some Fundamental Aspects of Socialist Transformation in India," *Man & Development*, IV, no. 2 (1982), pp. 9-22; P.C. Joshi, "Nehru and Socialism in India, 1919-1939," in B.R. Nanda (ed.), *Socialism in India* (Delhi: Vikas, 1972), pp. 122-41.
44. Michael Brecher, *Nehru: A Political Biography* (London: Oxford University Press, 1959), pp. 598-599.
45. Walter Crocker, *Nehru: A Contemporary's Estimate* (New York: Oxford University Press, 1966), p. 81.
46. S. Gopal, "The Formative Ideology of Jawaharlal Nehru," in K.N. Panikkar (ed.) *National and Left Movements in India* (New Delhi: Vikas, 1980), pp. 1-13.
47. Brecher, pp. 601-602; Crocker, p. 71.
48. Joshi, pp. 122-141.
49. Gopal, "The Formative Ideology...," pp. 8, 13.
50. A.H. Hanson, *The Process of Planning* (London: Oxford University Press, 1966), p. 258.
51. Gopal, "The Formative Ideology...," pp. 8, 13.
52. No one can write about Nehru without acknowledging his debt to Nehru's own works and of his biographers, most particularly Michael Brecher and S. Gopal. Two other articles I found especially useful in understanding Nehru's ideology analytically are those by P.C. Joshi and Ranjit Das Gupta; both are cited in footnotes.
53. Nehru, *Toward Freedom*, p. 44; Dorothy Norman (ed.), *Nehru: The First Sixty Years* (New York: John Day Company, 1965), vol. I, p. 41.
54. Nehru, in Norman (ed.), I, p. 57.
55. Gopal, "The Formative Ideology....", p. 2.
56. Nehru, *Toward Freedom*, pp. 56-57; Norman (ed.), I, p. 54.

57. Nehru, in Norman (ed.), I, pp. 108-112.
58. Gopal, "The Formative Ideology....," p. 6.
59. Brecher, p. 109.
60. Jawaharlal Nehru, *Selected Works*, Vol. II, pp. 270-98.
61. Brecher, pp. 111-112.
62. Gopal, "The Formative Ideology...," p. 6.
63. Sarvepalli Gopal, *Jawaharlal Nehru: A Biography: Volume One 1889-1947* (Cambridge, Mass.: Harvard University Press, 1976), p.108.
64. Gopal, "The Formative Ideology...," p. 7.
65. See Norman (ed.)., I, 153-167.
66. Brecher, p. 137.
67. Cited in Brecher, p. 138.
68. Nehru, in Norman (ed.), I, 195-210.
69. Jawaharlal Nehru, *Selected Works*, Vol. III, pp. 67-78, 281-92.
70. Indian National Congress, *Resolutions on Economic Policy, Programme and Allied Matters (1924-1969)* (New Delhi: All India Congress Committee, 1969), p. 3.
71. Gopal, "The Formative Ideology...," p. 9.
72. Nehru, *Toward Freedom*, p. 138.
73. Cited in Chandra et al., "The Communists...," p. 733.
74. Frank Moraes, *Jawaharlal Nehru: A Biography* (New York: Macmillan, 1956), pp. 176-77.
75. Jawaharlal Nehru, *An Autobiography* (London: John Lane The Bodley Head, 1936), pp. 267-68.
76. Nehru, *Toward Freedom*, p. 196; Norman (ed.), I, 248.
77. B.R. Nanda (ed.), *Socialism in India* (Delhi: Vikas, 1972), p. 10.
78. Ranjit Das Gupta, "Nehru's Economic Thinking and India's Struggle for Economic Independence," in Centre for Social Research, *The Seminar on Nehru*, p. 196.
79. Moraes, p. 185.
80. Norman (ed.), I, 293.
81. Congress, *Resolutions*, pp. 9-10.
82. Nehru, *Toward Freedom*, p. 230; Norman (ed.), I, 278-279.
83. Cited in Brecher, p. 188.
84. Nehru, *Toward Freedom*, 195; Norman (ed.), I, p. 327.
85. Brecher, p. 202.
86. Nehru, *Toward Freedom*, pp. 389-416; Norman (ed.), I, 424-447.
87. Brecher, p. 224.
88. Bipan Chandra, "Jawaharlal Nehru and the Capitalist Class, 1936," *Economic and Political Weekly*, X, nos. 33-35 (August 1975), 1307-24.
89. Nehru, in Norman (ed.), I, 626.
90. Gopal, "The Formative Ideology...," p. 10.
91. Nehru, *Toward Freedom*, pp. 416-31; Norman (ed.), I, 464-72.
92. See Norman (ed.), I, 477-698, and II, 3-340.
93. Chandra, "Jawaharlal Nehru and the Capitalist Class," p. 1321.
94. Joshi, in Nanda (ed.), pp. 122-41.
95. Joshi, p. 131.
96. Gopal, "The Formative Ideology....," p. 8.
97. Joshi, p. 130.
98. Cited in B.N. Pandey, *Nehru* (New York: Stein and Day, 1976), p. 176.
99. Gopal, *Nehru*, I, p. 180.
100. Cited in Brecher, p. 218.
101. Gopal, "The Formative Ideology...," p. 12.
102. Gupta, p. 197.
103. Jawaharlal Nehru, *The Discovery of India* (New York: John Day Company, 1946), p. 400.
104. *Ibid.*, p. 402.
105. Joshi, p. 135.
106. Nehru, *Discovery*, p. 404.
107. *Ibid.*, p. 403.

108. *Ibid.*, p. 405.
109. Jawaharlal Nehru, *Selected Works,* vol. IX, pp. 373-74.
110. K.T. Shah, *National Planning, Principles & Administration* (Bombay: Vora & Co., 1948), pp. 46-66, 82.
111. Gopal, *Nehru,* I, p. 248.
112. Joshi, p. 136.
113. Gopal, *Nehru,* I, p. 126.
114. Nehru, *Toward Freedom,* p. 148; Norman (ed.), I, p. 176.
115. Norman (ed.), I, p. 333.
116. D.K. Barooah, "The Master and Mentor," in B.K. Ahluwalia (ed.), *Jawaharlal Nehru: India's Man of Destiny* (New Delhi : Newman Group of Publishers, 1978), p. 127.
117. Joshi, p. 128.
118. Chandra et al., "The Communists...," p. 733.
119. Joshi, p. 139.
120. Gupta, p. 188.
121. Rehman Sobhan and Muzaffer Ahmed, *Public Enterprise in an Intermediate Regime: A Study in the Political Economy of Bangladesh* (Dacca: Bangladesh Institute of Development Studies, 1980), pp. 8-9.
122. Bipan Chandra, "Jawaharlal Nehru...," pp. 1307-24, "The Indian Capitalist Class...," pp. 390-413, and summary of oral presentation, *Social Scientist,* XI, no. 11 (1983), pp. 32-33.
123. Aditya Mukherjee, "Indian Capitalist Class and the Public Sector 1930-1947", *Economic and Political Weekly,* XI, no. 3 (January 17, 1976), and "Indian Capitalist Class and Congress on National Planning and Public Sector 1930-47", in K.N. Panikkar (ed.), *National and Left Movements in India* (New Delhi: Vikas, 1980), pp. 45-79.
124. James O'Connor, *The Fiscal Crisis of the State* (New York: St. Martin's Press, 1973, p. 42.
125. Note: (1) "the merchant communities in their own way were quite as deeply imbued with the spirit of nationalism" as the professional and service classes. Rajat K. Ray, *Industrialization in India: Growth and Conflict in the Private Corporate Sector 1914-47* (Delhi: Oxford University Press, 1979), p. 287. (2) "...we are Indians first and merchants and industrialists afterwards..." Purshotamdas Thakurdas, cited in Aditya Mukherjee, "The Indian Capitalist Class," in Romila Thapar (ed.), *Situating Indian History* (New Delhi: Oxford University Press, forthcoming).
126. Ajit Roy, *Monopoly Capitalism in India* (Calcutta: Naya Prakash, 1976), pp. 168-75.
127. Of course, one needs to be warned against treating the Indian capitalist class as a united entity. On its heterogeneity, socially and politically, see Claude Markovits, *Indian Business and Nationalist Politics 1931-1934* (Cambridge: Cambridge University Press, 1985), esp. ch. 3.
128. Ghanshyam Das Birla, *Bapu: A Unique Association* (Bombay: Bharatiya Vidya Bhavan, 1977), vol. IV, pp. 189-91, and *In the Shadow of the Mahatma: A Personal Memoir* (Bombay: Orient Longmans, 1953), pp. 207, 305-307.
129. Bipan Chandra, "The Indian Capitalist Class and British Imperialism," in Sharma (ed.), *Indian Society,* pp. 390-413. For a different opinion that the Indian capitalist class was not even nationalist, but an ally of colonial rule though not comprador, see G.K. Lieten, "The Civil Disobedience Movement and the National Bourgeoisie," *Social Scientist,* XI, no. 5 (May 1983), pp. 32-48.
130. Cited in Chandra, "Jawaharlal Nehru...", pp. 1307-24.
131. *Ibid.* See also Ray, pp. 125-27, and Markovits, pp. 109-115.
132. It needs to be underlined here that in examining Birla's memoirs, one is struck by how minor a place Nehru had in his preoccupations; there is very little on him or about him in four huge volumes; see Birla, *Bapu.*
133. Markovits, p. 86.
134. Chandra, "The Indian Capitalist Class...", p. 410. Following in Chandra's footsteps, Mukherjee also refers to "the national movement being led under strong bourgeois hegemony, with an independent bourgeois government imminent." Panikkar (ed.), p. 57.
135. Scholars often make slippery use of the term bourgeoisie; note for example: (1) "Lawyers were among the first and most important numbers of the modern bourgeoisie to put in an appearance on the Indian scene." Barrington Moore, Jr., *The Social Origins of Democracy and Dictatorship: Lord and Peasant in the Making of the Modern World* (Boston: Beacon Press, 1966), p. 370., (2)

"The Indian bourgeoisie during the nineteenth century consisted of bankers, merchants, landlords, industrialists, plantation owners and brokers." R. Suntharalingam, *Indian Nationalism: An Historical Analysis* (New Delhi: Vikas, 1983), p. 65.
136. Chandra, "Jawaharlal Nehru...", p. 1321.
137. Chattopadhyaya, pp. 244-59.
138. Birla, *Bapu*, II, p. 206.
139. Mukherjee, in Panikkar (ed.), pp. 71, 75.
140. *Ibid.*, pp. 45-79.
141. Sachs, p. 115.
142. C.T. Kurien, *Indian Economic Crisis: A Diagnostic Study* (Bombay: Asia Publishing House, 1969), p. 85.
143. Purshotamdas Thakurdas et al., *Memorandum Outlining A Plan of Economic Development for India* (London: Penguin Books, 1945),pp. 55, 58.
144. *Ibid.*, pp. 70, 87, 90-91.
145. K.T. Shah (ed.), *Industrial Finance (Report of the Sub-Committee)* (Bombay: Vora & Co., 1948), p. 65.
146. Thakurdas et al., pp. 65-66, 91-95.
147. Girish Mishra, *Public Sector in Indian Economy* (New Delhi: Communist Party Publication, 1975), p. 7; E.M.S. Namboodiripad, *Economics and Politics of India's Socialist Pattern* (New Delhi: People's Publishing House, 1966), p. 124; G. Adhikari, *Communist Party and India's Path to National Regeneration and Socialism* (New Delhi: CPI, 1964), pp. 164-67; and Sachs, p. 118.

Chapter IV
State Entrepreneurship in the Nehru Era: Ideology vs Necessity

At independence, any Third World country is confronted with the necessity of choice as to the preferred type of economic organization. Economic development is, of course, an imperative for it. Equally, given the compulsions of economic backwardness, it is inevitable that the state must play an active role in economic development. But the precise nature of the role of the state, both in terms of the kind of society desired and also the means through which it should be attained, is a matter for choice. The political leadership has to decide whether the state shall confine itself to a regulative and promotive role or take on an entrepreneurial role as well. If the latter, should it be socialist in a thoroughgoing manner, or a mixed economy? If the choice is of a mixed economy, what are the relative roles to be accorded to the public and private sectors, and which sector is projected to be dominant? More importantly, is the mixed economy a transitional stage or a permanent one? If the former, is it intended to move toward a socialist or a capitalist society? Undoubtedly, the choice will depend on a number of factors, but fundamentally it will be related to the nature of the class basis of the political leadership, the ideological vision of decision-makers, and the struggle among political elites. These aspects are treated here systematically in relation to India under Nehru, with the discussion focused on: (1) the social background of the Congress leadership; (2) the nature of economic policy during a period of political transition and ideological stalemate; (3) the socialist thrust after consolidation of political power by Nehru; and (4) Nehru's conception of the mixed economy.

1. Social Classes and Political Power

At India's independence in 1947, the Congress party as the organizational instrument of the nationalist movement assumed full power at the national level. Subsequent Indian political development up to 1984 can be divided into two rather distinct periods as between (1) the Nehru era from 1947 to 1964, when Nehru was Prime Minister, and (2) the post-Nehru period from 1964 to 1984, which though it saw several prime ministers in office, can basically be identified with Mrs. Indira Gandhi, his daughter.

The Nehru era was remarkable in many respects, most of all in laying down the basic framework of policies in the political, economic and social arenas. A noteworthy aspect of this framework, consistent with the liberal values of the middle class leadership of the Congress party, was the making of the principle of adult franchise the cornerstone of the country's structure of political power. The adoption of this principle had the potentiality of changing the balance of power between the middle class and the peasantry in the political system, given the enormous numerical strength of the peasantry in a basically agricultural country with traditional controls over subordinate classes. The Nehru period is noteworthy as well for the basic political stability it provided, which was largely a function of the hegemony of the Congress party in the political system. The Congress party was the dominant political force in the country, indeed, the only relevant political force, quite identical and co-extensive with the political system, exercising power at both national and state levels. The Congress party, and thus the political system, commanded political legitimacy derived from massive popular mandates, based no doubt in considerable measure on the personal mass popularity of Nehru. In this period, the Congress party as a relatively well-disciplined party had considerable autonomy from society, so that it could enact major policies fairly removed from the pressure of narrowly based interest groups. The Congress could do this partly because the level of social mobilization among the population was relatively low, and therefore the intensity of demands on the state and the pressure to meet them was not too high.

To what extent did the class basis of leadership inherited from the nationalist movement change during the Nehru era? In this period, the Congress remained a multi-class party, with its basic support structure cutting across class lines, though there was an erosion of support for it among intellectuals and white-collar salaried groups.[1] In terms of leadership, one can distinguish among three levels at which it is exercised -- local, state and national. For the power structure at the local level, it is hard to generalize for a country of India's size and diversity. However, several case studies by Myron Weiner near the end of the Nehru period demonstrate that at the district level the bulk of Congress leadership, close to two-thirds or more, came from the peasant-proprietor class, largely from the dominant-caste rich peasantry owning 30 acres or more of land; this was followed by business with about one-sixth to one-fifth of the leadership, while professionals and the salaried middle class had only one-tenth of leadership positions, representing a decline from pre-independence days.[2] In the towns and cities, studies by Rosenthal and Weiner underline businessmen (for example, owners of bicycle and cloth shops, restaurants and tea stalls) as the largest single group, followed by professionals and the salaried middle class; however, public esteem went largely to leaders associated with the modern professions. Curiously, doctors were more visible than lawyers in the urban power structure.[3]

At the state level, compared to the years of the nationalist movement when the new middle class dominated the leadership, there is apparent a marked trend toward the ruralization of political elites. Although earlier longitudinal data is not available for comparison, one study of Uttar Pradesh shows that agriculturists constituted nearly 40 per cent of Congress legislators in 1952 while middle class elites (education, law, medicine) were 32.3 per cent and business (composed largely of merchants and traders) 13.8 per cent. By 1962, there had been some change but not drastic; agriculturists increased their share to 42.2 percent at the cost basically of the middle class whose share was reduced correspondingly, while business held constant.[4] It would seem that, in the alliance between the middle class and peasantry in the Congress, the balance of power was moving in favour of the peasantry. There are problems, of course, in taking the category of agriculturists as representing the rich and middle peasantry only; besides former landlords, it also includes scheduled caste representatives, who are apt to be opposed to the peasantry and to ally with the middle class.

It is likely that the council of ministers did not mirror the legislative caucus in terms of the ascendance of the peasantry, but the trend seems to be toward dominance by agriculturists. What this indicates is that state governments as the institutionalized expression of the power of the rich and middle peasantry (along with the middle class) were not only able to implement peasant demands at the state level (especially in relation to agriculture and land reform, which are state subjects) but also place constraints on what leadership at the national level could do in these areas.

At the national level, the shift in political power toward rural elites is quite manifest but, while there was some decline in the position of the middle class, the latter continued to be the dominant element. Agricultural elites increased their share from 10.8 per cent in 1950 to 18.3 per cent in 1952 and 26.1 per cent in 1962 (see Table IV.1); the sharp increase was a function of the introduction of universal franchise. On the other hand, middle class professionals declined from 50.4 per cent in 1950 to 40.3 per cent in 1962. Noteworthy here is the heavy representation of lawyers which, even though it came down considerably, still by itself rivalled that of agriculturists. These figures on middle elites cannot be treated as conclusive, because the category of "service"-- consisting of retired army and civil service officers -- could also be included in middle class elites, while the category of "public work" is rather vague, including social workers and professional politicians. At the same time, there is overlap among occupations; thus, in 1952, while 18 per cent of legislators had agriculture as their occupation, 24 per cent had income from agriculture. Business had around 10 per cent of the representatives in 1952, but this figure declined to 7.3 per cent in 1962.

The general category of "agriculture" does not indicate fully the class background of national legislators. However, research by Kochanek provides more detailed data on the Second Lok Sabha of 1957 even though one may differ with his perception of what constitutes "middle peasantry," which seems to incorporate the rich peasantry as well. While 24 per cent of Congress legislators (total sample 224) declared themselves to be agriculturists, actually some 43 per cent held land to be their major source of income; indeed, the total of all those who had some land was as high as 71 per cent. If this larger total of all those who had some income from land is taken into account, then Kochanek believes "most of the Congress members of the Second Lok Sabha who were landowners tended to have small to medium-sized holdings. Some of the smaller landholders were obviously agriculturalists only in the sense that they shared in their family's land, which they themselves did not have responsibility for cultivating; thus, they had no direct connection with land, a status probably most applicable to the 27 per cent who owned less than twenty acres. Of the 35 per cent who owned more than 20 acres, some 15 per cent were in 20-50 acre range and 11 per cent owned more than a hundred acres. These figures lead to the conclusion that the majority of the Congress landholders in the Second Lok

Table IV.1

Occupational Background of Congress Legislators in Lok Sabha

Occupational Category	1947 Prov. Parliament No.	%	1952 No.	%	1957 No.	%	1962 No.	%
Agriculture	32	10.8	62	18.3	54	24.1	93	26.1
Business	30	10.6	37	10.9	25	11.2	26	7.3
All Professionals	143	50.4	165	48.7	68	30.4	144	40.3
--Law	(93)	(32.6)	(103)	(30.4)	(52)	(23.2)	(90)	(25.2)
--Other	(50)	(17.8)	(62)	(18.3)	(16)	(7.1)	(54)	(15.1)
PublicWork	42	14.8	67	19.7	63	28.1	59	16.5
Service	28	9.9	8	2.4	7	3.1	13	3.6
Other	-	-	0	0	6	2.7	6	1.7
Not known	10	3.5	0	0	1	0.4	16	4.5
Total	285	100	339	100	224	100	357	100

Source: Rosen, p. 73; Kochanek, p. 380; W.H. Morris-Jones, *Parliament in India* (Philadelphia: University of Pennsylvania Press, 1957), p. 123; and Nandini Upreti, *Provisional Parliament of India* (Agra: Lakshmi Narain Agarwal, 1975), p. 20. The 1957 figures are apparently on the basis of a sample survey by Kochanek.

Sabha were drawn from the middle peasantry, which forms the base of the Congress leadership in the rural areas."[5]

Kochanek further supports this conclusion by looking at that group, amounting to 43 per cent, whose members owned land as their major source of income. Of this group, 33 per cent had less than 20 acres while 41 per cent had 20 to 100 acres and another 16 per cent more than 100 acres. Kochanek states: "Altogether, some 50 per cent of the Congress legislators most directly connected with the land were small or medium holders, and the majority of these were in the middle peasantry." He further sees "the growing strength of this middle peasantry at the national level" reflected in its progressively increasing presence in successive legislatures, concluding: "it is the middle peasantry in particular, rather than the agriculturalists as a whole, who are beginning to gain greater and greater representation at the national level."[6]

In terms of caste breakdown, which is usually not available, Kochanek's study shows Brahmins to have one-fifth of the Congress representation in the Second Lok Sabha, while the other higher castes (Kshatriya, Kayastha, and Vaishya) had around 36 per cent; Harijans and tribals had 26.2 per cent, and non-Hindus and anti-caste Hindus 17.6 per cent.

If agriculturists are an ascendant group, their ascendancy was, however, not manifest to the same extent during the Nehru era in the council of ministers, the apex of political power in India. Satish Arora has documented the enormous presence of middle class professionals in the council. In the years 1962-64, only 16 per cent of the ministers were agriculturists, while 55 per cent were from the professions (law alone had 33 per cent) and 18 per cent were in social and political work; only 7 per cent had a business background.[7] The higher the level in the power hierarchy in India, the more entrenched seemed to be the position of the new middle class.

What is obvious overall at the national level is a typical "intermediate regime" as an alliance between the middle class and rich and middle peasantry under the leadership of the middle class. Although there was a sharp decline in its position after 1952, still the middle class remained the dominant class in political power. Of note is the fact that the leadership of the middle class in politics is further buttressed by its domination or near monopoly in the bureaucracy, civil and military, as well as the managerial cadres in the private sector.[8]

Of course, any national administration would have to take account of other classes that are not adequately represented in the power structure. Still, it is one thing to have the political system to be responsive to one's demands and quite another to possess the actual power to respond. The fact that the middle class has a dominant position in the power structure means that its interests are most likely to be protected and advanced by the state. This would seem to have enormous implications for the growth of the public sector. Sobhan and Ahmed suggest that the middle class or the petty bourgeoisie has a special propensity for "the expansion of the state sector. The growth of the state sector provides both expanded employment opportunities and income earning opportunities through trade and industry.... the state sector gives them more direct control over the instruments of patronage."[9] Another scholar seems to agree, stating "the very nature of this class make them support state capitalism."[10] One Marxist scholar attributes the peculiar qualities of avarice and envy to the middle class in India, holding that these take "the ideological form of populist or middle class socialism," which he regards "at the moment the most important ideological force in Indian politics."[11] The comparatively stronger or dominant position of the middle class as against the bourgeoisie during the Nehru era thus meant that there could be no automatic tendency toward capitalism, which would have been the case had the bourgeoisie been dominant.

However, it does not follow that the middle class would automatically prefer to go in for a mixed economy, especially one oriented toward socialism. The relative weakness of the bourgeoisie no doubt inclines the state toward an active public sector, while the presence of labour as a subordinate partner of the middle class strengthens that tendency. However, as has been obvious, the middle class is not necessarily united on the issue of the role of the state in the economy. The establishment and expansion of the public sector as a function of the will to power of the middle class could then only be a result of contingent factors that would shift the gravity of power in the Congress party from the right wing to the left wing; it could not simply be an automatic manifestation of the domination of the middle class in political power at the centre.

While the new middle class continued to be dominant, the rich and middle peasantry, as is obvious, quickly rose to rival its power, even though not in the council of ministers at the centre. The fact that it had such enormous electoral clout compelled the power structure to pay adequate attention to its demands. Although there is no certainty which way the peasantry may look in terms of the organization of the economy, perhaps it would favour capitalism because it is itself organized around private property. It is likely to fear that any damage to the principle of private property at the level of industry may rebound on it in agriculture. On the other hand, some scholars posit that "the rich peasant sees an expanding state sector as beneficial to himself in that it increases the volume of resources and instruments of control over the poor," while he acts as an intermediary between the ruling party and the rural masses. Regardless of its stance on the public sector, the rich and middle peasantry is seen as instrumental in pushing through the "first generation of land reforms which break up the big estates and undermine the feudal power of the landlords. The beneficiaries of such land reforms have invariably been the rich peasant who now become the dominant element on the rural scene" and block further land reforms.[12]

The small share of business in the legislative caucus and the council of ministers at the centre suggests a weak position for it in the power structure. Of course, one can suggest that the entire leadership structure is a facade, and that business runs the political system from behind the scenes. On the known evidence, there is no basis for such a conspiracy theory. In any case, the question of whether the leadership responds to business desires in the way business wants it to is something to be examined, not accepted in advance. One may even raise the question as to why business is not more assertive in openly claiming a larger share in political power. But it would seem that the same forces that run counter to its claiming a greater share in power also militate against it having its way with the political leadership. The Indian bourgeoisie encounters low public esteem and an anti-business political culture. As one government commission observed, "it is not possible to ignore the reality that the widespread hatred for big business, whether based on good reasons or not is a serious consequence of concentration of economic power... This dislike and hatred are serious in themselves..."[13] On the other hand, the public sector has been seen by the electorate as being more in the national interest; this element, according to Kochanek, is "reinforced by the pervasive influence and symbolic role of socialist ideology, which is especially strong among the urban intellectuals and the urban middle class." He adds: "In short, the political culture in India is such that business is completely defenseless when the political leadership can state an issue in terms of the vested interests or can pit the haves against the have-nots."[14] This was especially so when the regime, as under Nehru, commanded political legitimacy.

Sobhan and Ahmed take the position that an intermediate regime favours the extension of the role of the state "not only because the upper bourgeoisie was identified with the colonial power but because it was intrinsically unable to perform the role of the `dynamic entrepreneur' in building capitalism on a sufficiently large scale."[15] Contrary to this position, it has already been seen that the bourgeoisie in India was not a comprador class but that it supported the nationalist movement, albeit passively. Furthermore, on the basis of evidence provided earlier, it is not easy to accept the notion of its possible lack of dynamic entrepreneurship. Indeed, the argument itself may be more a rationalization to advance the interests of the intermediate regime. Against the evidence of its performance in the 1930s, it is difficult to sustain the "incapability thesis" about the Indian bourgeoisie. Of course, the Indian bourgeoisie was not able to develop the economy during colonial rule, but that is a function of colonialism, not the Indian bourgeoisie. If therefore measures were adopted to restrict or curb opportunities for the Indian bourgeoisie after independence they emerged from sources other than the alleged incapability of the bourgeoisie, basically its lack of political power, which had been pre-empted by the middle class.

2. Post-Independence Crises and Policy Moderation

Since independence came with the partition of the country, the new state was immediately overwhelmed, especially in North India, with large-scale communal warfare and carnage, resulting in some quarter million deaths, vast destruction and loss of property; administrative breakdown and chaos, and the massive influx of some six million refugees. The unsettling encounter with masses of humanity in convulsion at the dawn of independence was to have a significant impact on Nehru in countenancing uncontrollable change and conflict in the context of a state with few resources. It was to induce caution and moderation in bringing about economic change.

Nor was this the only factor inhibiting, indeed paralyzing, the pushing through of a radical economic programme. The new state was, in fact, overwhelmed by multiple crises and what was at stake was sheer survival. As Nehru later explained, "we were really busy for about a year just to keep ourselves above water."[16] No sooner had the country emerged as an independent entity than it was involved in a 15-month war over Kashmir with the secessionist state of Pakistan. Besides, hardly had six months passed after independence that Gandhi was assassinated, raising fears of political turmoil. Again, the new state confronted the spectre of disintegration with several of the larger princely states asserting their own ambitions for independence. Foreign and local observers foresaw India going the path of China under Chiang Kai-shek.

If all this were not enough, the state was confronted by an economic crisis of severe proportions. The separation of Pakistan disrupted the integrated economy of the subcontinent, but for India this had immediate economic consequences. With food surplus areas going to Pakistan, the country was confronted with the problem of food deficits, necessitating priority to food production. Nor was this problem made any easier by the necessity for India to simultaneously expand production of jute and cotton for its factories, since the areas specializing in these products belonged to Pakistan. To top all this, the business community was seized with insecurity over its future, terrified now by the ideological predilections of Nehru as the head of the new government, despite its earlier brave conclusions about his essential sobriety. For its part, the government initially made little effort to reduce business fears.

As if in apparent implementation of Nehru's ideas, the All India Congress Committee (AICC) resolved in November 1947 to establish "an economic structure which will yield maximum production without the creation of private monopolies and the concentration of wealth" as "an alternative to the acquisitive economy of private capitalism and the regimentation of a Totalitarian State." This was followed in January 1948 by what was for the business community a bombshell by way of the report of the AICC's Economic Programme Committee, chaired by Nehru and consisting essentially of socialists and Gandhians.

With the object of establishing "a just social order", the committee -- whose radical proposals were subsequently endorsed by the AICC -- felt it necessary "to bring about equitable distribution of the existing income and wealth and prevent the growth of disparities in this respect with the progress of industrialisation." It recommended removal of intermediaries between the tiller and the state and the fixation of land ceilings, with the surplus land organized in village cooperatives. But its proposals for industry were ominous for the business class. All industries having to do with food, clothing and other consumer goods were recommended to be reserved for the cottage or small-scale sector and managed on a cooperative basis. At the same time, all "new undertakings in defence, key and public utility industries", as also those industries which were large in scale or monopolistic in nature, were to be reserved for public ownership. As for existing enterprises in these sectors, "the process of transfer from private to public ownership should commence after a period of five years" when their prices would have come down partly in response to the changed economic conditions and partly "under pressure of appropriate legislation or administrative measures". Furthermore, importantly, banking and insurance were to be nationalised. What industry there was left in the private sector was to be "subject to all such regulations and controls as are needed for the realisation of the objective of national policy in the matter of industrial development". The latter would include controls over investment, income and dividends, which were to be limited to a maximum of 5 per cent. The committee also asked for the establishment of a planning commission at the centre.[17] Congress leaders have considered the committee's report as of momentous importance, for the report "conceived in Shri Nehru's great mind laid down the blue-print for India's future path of development. All the other schemes and policies, which have since been worked and implemented in India, have their roots in this Report."[18]

The cumulative impact of the earlier 1947 "soak-the-rich" budget, the economic havoc resulting from partition, and fear over the ideological intentions of the new government was just "too much for Indian capital. It underwent what was probably the severest crisis of confidence in its history. Industrial production fell off drastically from its wartime peak in most spheres."[19] The ensuing economic crisis therefore made production, over the next more than a half-decade, the top priority for government rather than any objective of bringing about a new social order. Some observers saw in the strike of capital "the deliberate desire to cause embarrassment and thus force the Government to change its policy",[20] or a design on the part of Indian business, whose "major preoccupation in the first years was to *set the limits of state activity*" against "the Government's populist leanings and of its tendency to formulate policy from general principles."[21] However, it is not necessary to accept this conspiracy theory to comprehend the gravity of the crisis. Generally, the mass of investors simply respond to perceived opportunities of profitable investment rather than to political advice to bring the government to its knees. Similarly, any positive response on the part of government to create incentives for business to expand economically need not imply a conspiracy between business and government. For any government in the modern age, political survival depends on economic growth, for that growth is linked with increased employment and improved living standards. But if economic entrepreneurs and managers must undertake activities that subserve the end that is central to government, they have to be provided appropriate incentives. In truth, the principle extends no less to state-managed economies, except that here economic managers and decisionmakers constitute the state.

In order to cope with the economic crisis that enveloped the country after independence, the government could have taken the route of wholesale nationalization, but it would have been utopian, if not quixotic, to do so. The government had few resources, financial or physical, to cope with this aspect of economic management on top of all the other problems it confronted. However, there was, besides but no doubt critically, a decided political block against this path. For one thing, the government at the time was at its very apex one in which power was equally shared by Nehru and Patel, ideological antipodes on the question of socialism. Furthermore, not only had Patel in the past before independence led political forces inside the Congress opposed to Nehru on the issue of socialism, he was now in control of the party machine. Even though the party was about evenly divided in its loyalty to Patel and Nehru, that did not mean that half the party belonged to the left wing, for many followed Nehru out of personal loyalty to a charismatic leader rather than commitment to his ideas. Moreover, the government until around 1950 was constituted not out of a single party but a coalition, out of a sense of the need for national unity in a time of transition. Ministers from other parties or from outside politics held the important portfolios of finance, industry, transport and railways, defence, and law.[22] Any decision in relation to the economy could not therefore be an *authoritative* one and a function simply of the ideology of Nehru or the Congress party, but a result of *composite* choice reflecting the balance of forces or interests among several groups.[23]

The composite nature of the choice was manifest in the Industrial Policy Resolution brought forward by the government on April 6, 1948,[24] which as a compromise sought to satisfy the various interests within the party, government and society. Its immediate objective was to overcome the crisis of production by restoring business confidence through clarification of the new government's economic policy. This was done by the renunciation of any intention of nationalization for a period of ten years even in respect of important industries, by leaving to the private sector a vast economic arena for operation, and by making clear that increased production, rather than redistribution, was the first priority. The resolution was, in effect, a disavowal by the government headed by Nehru of the earlier report of the Congress party's Economic Programme Committee, also headed by Nehru. His split role now paralleled the pre-independence pattern of distinguishing between the articulation of ultimate goals and the recognition of what is possible within a given context.

However, the declaration of some general, if vague, principles in the resolution offered some satisfaction to socialists of Nehru's ideological persuasion. The document was, indeed, schizophrenic with its proclamation of general principles for an interventionist and entrepreneurial state, and simultaneously its intent to hold them in abeyance in view of the priority for production and the limited resources of the state. Affirming the nation's goal of establishment of "a social order where justice and equality of opportunity shall be secured to all the people," the resolution held the immediate objective to be a rapid advance in living standards through increased production and expanded employment opportunities under a regime of economic planning. However, such advance was premised on "an increase in wealth", for "a mere redistribution of existing wealth" would make no essential difference to the people and would "merely mean the distribution of poverty". Consequently, what was required was a "dynamic national policy...directed to a continuous increase in production by all possible means, side by side with measures to secure its equitable distribution." But for now, given the condition of general mass poverty, "the emphasis should be on the expansion of production."

The resolution urged viewing the role of the state in the context of the requirement of production but, apparently as an ideological concession to Nehru and the left wing, took it as given that "no doubt that the State must play a progressively active role in the development of industries". It at once qualified this by stating that "the immediate extent of State responsibility and the limits of private enterprise" were to be determined by "the ability to achieve the main objectives." Here it confessed that "under present conditions, the mechanism and the resources of the State may not permit it to function in industry as widely as may be desirable." In other words, what blocked an expanded role for the state was the compulsion of circumstances, not principle which, quite the contrary, led the other way. For the present then, while the state would seek to remedy its current situation, it would confine itself to "expanding its present activities wherever it is already operating and by concentrating on new units of production in other fields rather than on acquiring and running existing units." In this way, present business in the private sector was under no threat. But the left wing was assured that this was only an interim proposition, because of the qualifying declaration that "*meanwhile*, private enterprise, properly directed and regulated, has a valuable role to play." Apparently, the government regarded private enterprise, in Hanson's view, "as no more than a temporary developmental expedient;" he was further convinced that even at this early stage the government was "*implicitly* socialistic in outlook".[25] But the government itself refrained from openly identifying with any particular ideology. At the same time, in the actual demarcation of fields to the public and private sectors, the government took a restrictive view of the area reserved for itself, though it was careful enough to employ language that allowed it to do literally almost anything. It seems that the ideological ambiguity of the declaration meant that its operational impact would be at the mercy of the changing configuration of political power and balance of social interests rather than rigidly determined beforehand. The concession to private enterprise lay not in principle but in the then power of the right wing and the circumstance of the crisis of production.

The resolution made a tripartite division of the tasks in the industrial field. There was first (1) an exclusive monopoly sector for the state, but which in reality only reflected the existing situation. Of the only three items included under it -- arms and ammunition, railway transport and atomic energy -- the last item at the time had little meaning for an underdeveloped country while the others were already in the public sector. This exclusive state monopoly sector was supplemented by a second (2) reserved public sector, where in six fields -- coal, iron and steel, aircraft manufacture, shipbuilding, telephone and telegraph manufacture, and mineral oils -- the establishment of new undertakings was left to the state. The resolution did not exclude, if necessary in the national interest, the cooperation of the private sector even in respect of these fields (though subject to government control and regulation). While proclaiming the inherent right to nationalize, the government in effect renounced nationalization, having decided "to let existing undertakings in these fields develop for a period of ten years"; what is more it promised them over that period "all facilities for efficient working and reasonable expansion" and "fair and equitable" compensation in case of nationalization.

Barring these two sectors, (3) the rest of the industrial arena remained with private enterprise, whether individual or cooperative. But this was not without qualification, for the state reserved the right to "progressively participate" as well as "intervene" even here if performance of private enterprise was not satisfactory. An illustrative list included 18 basic industries whose "planning and regulation" by the state was considered "necessary in the national interest." Additionally, the government thought that small-scale and cottage industries had a very important role, while it expressed a cautious vagueness over foreign capital, holding it to have a useful role in development but requiring regulation in the national interest.

Notwithstanding the claim about the implicitly socialistic outlook attributed to it, it seems that several aspects of the Industrial Policy Resolution of 1948 made it a prescription for a mixed economy of the instrumental-capitalist type:[26] the rather wide arena left to the private sector, the removal of any immediate threat of nationalization, the restricted area reserved to the state, and the promise to provide a promotive environment for private enterprise through resolving infrastructural bottlenecks, arranging adequate raw material imports, organizing tariff protection against foreign competition and legislating a taxation system which encouraged savings and investment. The role of the public sector was apparently conceived of as only "supplementing and stimulating private enterprise."[27] For that reason, while business was buoyed by it, the left was dismayed with the apparent "retreat from socialism".[28] K.T. Shah, who had served as secretary of the National Planning Committee under Nehru, described the resolution as an "utter disappointment".[29] But confronted with the production crisis and the limited resources at its disposal, there seemed to be no other alternative for the government under Nehru even if it theoretically held to its goals over the long haul. The underlying considerations with the government were given lucid expression by Nehru:

> situated as we are today in India, after all that has happened in the course of the last seven or eight months, one has to be very careful of the steps one takes so as not to injure the existing structure too much. There has been destruction and injury enough, and certainly I confess.... that I am not brave and gallant enough to go about destroying any more....It seems to me that in the state of affairs in the world and in India today, any attempt to start with what might be called a clean slate, that is to say, a sweeping away of all that we have, would certainly not bring progress nearer, but might delay it tremendously. Far from bringing economic progress, it might put us so far back politically that the economic aspect itself might be lost sight of. We cannot separate these two things. We have gone through big political upheavals and cataclysms.....[30]

Despite the vague ideological concessions to the left wing, it was the context of economic crisis that underlined the thrust of the resolution -- promoting production through restoring business confidence. Still, the larger historical importance of the resolution for future economic policy lay in its being the first formal declaration by the independent government for a "mixed economy", though it did not use the term,[31] even if it was of the instrumental-capitalist type for the time being.

3. Nehru's Political Coup and Policy Change

The 1948 Industrial Policy Resolution, despite its vague phraseological concessions to the state's entrepreneurial role in industrialization, essentially confirmed reliance on the private sector. If this did not represent necessarily a final defeat for the left wing, it did signify at least the frustration of its attempt to achieve a more substantial role for the public sector, to reduce the role of the private sector through nationalization, and to curb the future growth and expansion of large-scale private enterprise. The resolution essentially codified existing practice, and seemed to set the policy framework for economic change in independent India. The profound Marxist scholar, Daniel Thorner, noted in 1950 that the quasi-socialist programme embodied in the consensus of the National Planning Committee, which Nehru had headed, was not any more an object of serious concern. He believed that land reforms had been "for the most part indefinitely deferred" and further that

> Similarly, the Congress has suspended for the foreseeable future its programme for industrial development via broad economic planning under governmental auspices. In place of the older programme, the Government has now announced that private, rather than public enterprise will govern the selection of fields and the intensity of industrial development.

He thought that this outcome was a consequence of India's conflict with Pakistan, since the resulting "martial atmosphere" had enabled "princes, landlords and other traditionally powerful groups" to retain "a rather large influence in public life."[32] In like fashion, D.R. Gadgil held: "The old socialistic programme has, however, receded more and more into the background during the last twelve months and a new pattern of economic policy, especially industrial policy, has begun clearly to emerge."[33] The retreat from socialism seemed set and, indeed, irreversible.

But change policy did, and in directions that Nehru had all along perceived desirable within the framework of his ideology. In retrospect, Hanson was to note: "How the rather dismal situation was transformed, during the course of a few years, into one of infinitely greater hopefulness, and how India's leaders not only became converted to 'a socialist pattern of society' but conceived and began to operate a programme for its achievement, is an important episode in recent history of which our knowledge is extremely deficient."[34] Fundamental to the change in the policy framework, however, was antecedent political change, partly a function of human mortality and partly a consequence of political design and struggle, but basically on the part of one man.

The outstanding feature of the power arrangement at the apex immediately after independence was the almost equal sharing of power between Nehru and Patel, who stood ideologically apart but politically united by a common interest in India's strength and loyalty to Gandhi before and after his death. The Congress was evenly divided, but Patel held final control over it by tacit agreement even though he did not formally head it. The strength of the right wing, particularly of its leader Patel, in the Congress is obvious in its ability to veto Gandhi's recommendation for the election of a socialist to the Congress presidency in 1947; subsequently, it was able to oust the socialists from the Congress through a party resolution, aimed precisely at them, that barred organized groups within the party. In 1948, when elections to the party presidency took place, while the two key candidates were identified with either Nehru or Patel, the election itself was not invested with critical importance. Even though Patel's candidate -- Tandon -- lost, the election really demonstrated that the party was evenly split between the left and right wings led respectively by Nehru and Patel, with the left wing winning 1199 votes and the other 1085.

Despite the right wing's loss, it seemed that in the states the Congress organization was already coming under its control. When the party's presidential contest came around again in August 1950, the right wing, with the conservative and religiously orthodox Tandon again as its candidate, won with 1306 votes against the 1052 votes of the left wing's candidate, certifying again the rather even division within the party. But Tandon's election "placed the right wing of the Congress in complete control of the party organization and its decision-making organs". But not only did this complete control prove to be short-lived, it led directly to a political "coup" on the part of Nehru that installed him in power in the party.

At the root of this reversal of fortunes of the right wing lay intervention by nature, for with the death of Patel on December 15, 1950, "at the height of its power, the group lost the support of its major patron".[35] There then ensued a bitter conflict between the left wing and the right wing, especially over control of the party's election committee, for what was at stake here was the allocation of party tickets in the forthcoming elections in 1951-52 and thus control over the future parliamentary wing and government. The party was pushed to the breaking point, with many dissidents seceding from the party. Finally, Nehru forced the issue in August 1951 by resigning from the party's Working Committee as also the election committee, declaring "I am convinced that I do not fit into the Working Committee and am not in tune with it." Shocked by this decision, Tandon himself resigned from the party's presidency, stating that Nehru's resignation could not be accepted, for "Nehru is not an ordinary member of the Working Committee. He represents the nation more than any other individual today". Underlying Tandon's graceful and self-abnegating action was the recognition of Nehru's indispensability for the party in the imminent general elections, due less to his economic ideology than to his personal charisma. Soon, Nehru himself was elected president of the Congress by the AICC by 295 votes to 4; "the crisis had finally been overcome. Nehru emerged as the indisputed leader of the Congress" organization[36] in addition to his unchallenged power as prime minister in the government which was now, with Patel no longer alive, essentially "a one-man show".[37]

The right wing reconciled itself to its defeat because of its perception of the indispensability of Nehru for victory in the forthcoming elections, but its inner feelings were expressed by one right wing leader, D.P. Mishra, who, in the words of Kochanek, "issued a public statement denouncing Nehru's actions as dictatorial and declaring that, if Patel had been alive, he would have placed his support squarely behind Tandon. Mishra felt that ever since the death of Patel a group very close to Nehru had begun a 'sneaking vilification of Sardar Patel' while also carrying on a campaign against the Congress President simply because he had been supported by Patel...he lashed out at Nehru's policies and tactics....Unrealistic economic controls were being enforced, and India's foreign policy had led to disaster".[38] A British observer noted that Nehru's actions had "left mixed sensations, ranging from tempered approval to condemnation of tactics employed to achieve his objective, and bitter criticism of his supposed motive." The Punjab newspaper *Tribune* saw an explanation for the actions in his perception of the business community's dominant influence in the party under Tandon.[39]

Nehru's political supremacy did not mean that the right wing was ousted or liquidated. Nor did it mean a complete overhaul of the Congress party. Such actions would have run counter to Nehru's democratic sensibilities and to the political culture of the party at the time. They would have also been reprehensible to Nehru personally, for vindictiveness and ruthlessness were foreign to his nature, as they would have seemed unwise politically, for the unity of the Congress was in his eyes indispensable for India now as it had been before independence. As a consequence, the Congress party and government continued to be a broad, heterogeneous coalition. It is remarkable that even as late as 1958, there was only one real leftist in Nehru's cabinet, along with 3 rightists, 4 centrists and 3 moderates; indeed, as Brecher underlined, "Nehru tends to surround himself with persons who do not share his ideology in any marked degree".[40] What the real consequence of the political "coup" was that it shifted the centre of gravity toward the left wing. But more importantly, not only did the clean surgical operation by itself make vivid the effective power of Nehru, but in the process it also demonstrated that there was no comparable leader around which the right wing forces could rally.

The coup's aftermath would enable Nehru to push through, after due debate and discussion, his ideas in a party which was dependent on him, in large measure, for electoral success. Composite decision-making was to be replaced more by *authoritative* decision-making, where the party accepted his largely ideologically-based policies in terms of the rationale he provided, regardless of how the groups within the party felt about the resulting impact on their interests. However, these policies did not emerge immediately. At first, there was preparation for the upcoming general elections, and then the actual campaign stretching over many months in 1951 and 1952 during which Nehru bore an extraordinary superhuman burden. After the elections, in which his party received a political mandate based for the first time on universal adult franchise, there followed a period of consolidation of his power in the party as well as government. The power consolidation was symbolized by the handing over of the party presidency in the fall of 1954 to a handpicked but relatively unknown successor, U.N. Dhebar, who "accepted Nehru as supreme leader".[41] It is around this point in time that Nehru's ideological predilections, moderate but yet consequential, were to become translated into policy.

If the earlier policies, under the compulsion of circumstances and of resistance by the right wing, could be labeled "a retreat from socialism" then the new policies which overrode them in directions desired by Nehru could, by the same token, be no less deserving of being characterized as an "advance towards socialism", at least in the sense of an active and expanding entrepreneurial role for the state. However, such advance should serve as an essential antidote to the common facile assumption of policy as basically only a function of social forces, and therefore inevitably destined to manifest itself as such. For, the whole complex of policies inaugurated in the mid-1950s that are associated with the Indian "model" of development rested on an accident of the death of Patel and the survival of Nehru. If the physical capabilities of the two leaders had been reversed, it is not certain that that complex of policies would have become installed and solidified. Regardless of whatever rhetorical concessions they may have made to salve the conscience of socialists in the party, the right wing and its leadership had all along opposed not only the initiation of those policies but their very advocacy.

The advance should serve as an essential antidote as well to the commonplace attribution by Marxian analysts of the "model" to the interests of the bourgeoisie because of the Congress party's being allegedly a bourgeois organization under the domination of the bourgeoisie. For the earlier policies, too, but with greater justification, would have also been said -- indeed, they had been so pronounced -- to serve the interests of the bourgeoisie. If the interests of the bourgeoisie can be served by contradictory sets of policies, then the notion of either bourgeoisie or interest is rendered meaningless. Nor would it be any more meaningful to suggest that Nehru somehow understood the real objective interests of the bourgeoisie better than did the right wing. Nor can that set of policies be attributed to the great strength of the left forces inside the Congress, for by the time the policies were pushed through those forces were weaker than ever before, with the departure from the Congress in 1948 and afterwards of socialists and Gandhians, who eventually formed the Praja Socialist Party. Indeed, the left forces in the Congress party had seen a consistent decline since 1939 with the withdrawal of the Forward Bloc under Subhash Chandra Bose and then the effective secession in 1942, subsequently ratified by their expulsion, of the communists. On the other hand, the imminence of the Congress party's ascent to power in government after the mid-1930s brought in forces that were largely interested in the status quo. Nehru was thus, as Karunakaran repeatedly stresses, the leader and "socialist theoretician of the most well organized conservative party of India."[42]

It goes against the grain of social science, especially Marxian social science, but there can be little doubt about the critical role of Nehru as an individual in delineating a specific path of development, regardless of the merits of that path, and launching India on it with consequential results. Nor, further, can the adoption of such a path be attributed simply to the domination of the intermediate strata, for that domination was there before the political "coup" as it was afterwards; if anything, with the departure of the socialists and with the rise of agriculturists, the role of the more state-activist component, the urban middle classes, had been weakened in the Congress party. The relative greater strength of the intermediate strata in comparison with the weakness of the bourgeoisie, no doubt, provided a social environment receptive to an activist entrepreneurial state, but until the accomplishment of Nehru's political supremacy within the party that domination did not automatically translate into policy.

4. The Push Towards "Socialism" Under Nehru

Socialism had largely receded from Nehru's concerns in the years immediately after independence. Even his speeches until around 1954 were largely innocent of any reference to it. As weighty a document as the constitution made little concession to it. In the constitution's directive principles of state policy, which were non-justiciable even though they had moral import, it was no doubt recognized that "the State shall strive to promote the welfare of the people by securing and protecting as effectively as it may a social order in which justice, social, economic and political, shall inform all the institutions of the national life" and further that the state was obliged to see that "the operation of the economic system does not result in the concentration of wealth and means of production to the common detriment". But this was really counterbalanced by the much more critical concession, in the justiciable chapter on fundamental rights, to the right to property and to compensation in the event of acquisition of property by the state -- and this on the insistence of the right wing of the Congress party. While there was thus nothing suggestive of socialism in the constitution, the principles enunciated therein were themselves compatible with any variety of policies and social systems. Furthermore, economic planning at the time was unable to move things in the direction of an activist entrepreneurial state.

An important concession that Nehru had been able to extract, much to the annoyance of the right wing, was the establishment in 1950 of the Planning Commission. But the right wing was able, on the plea that it would alienate the business community, to exclude from its terms of reference the objective originally proposed of "the progressive elimination of social, political and eocnomic exploitation and inequality, the motive of private gain in economic activity or organization of society and the anti-social concentration of wealth and means of production".[43] In the event, the substitute objective that was finally accepted instead followed the formulation included in the directive principles concerning the operation of the economic system, mentioned above.

The First Five Year Plan for the period 1951-1956 also did not register any advance towards socialism or a large public sector. The public sector, in effect continued to be viewed as simply supplementing and stimulating the private sector. If anything, there was a retreat from even the role envisaged for it in the 1948 Industrial Policy Resolution. To begin with, the First Plan focused more on production than on redistribution. While the planners were, of course, opposed to inequalities, they were equally anxious to "ensure a continuity of development without which, in fact, whatever measures, fiscal or other, might be adopted for promoting economic equality might only end up in dislocating production and even jeopardizing the prospects of ordered growth". The Plan accordingly eschewed any ideological goals. Secondly, the First Plan itself was a modest plan, with little intent at restructuring the economy. It was less an effort to initiate a process of rapid growth, but more to bring some measure of rehabilitation to an economy that had been physically exhausted by World War II and partition. It incorporated no ambitious strategy, but was rather an aggregation of ongoing projects strung together within the compass of a comprehensive review of the economy. Thirdly, the focus of the Plan was agriculture, to which the planners accorded, as they repeatedly stressed, the highest priority.[44] This was quite justified in view of the deficit in food and agricultural raw materials and the considerable international embarrassment that resulted from it for the government. As against agriculture, industry had a low priority, and here increased production was expected from better utilization of existing capacity rather than through new plant.

Fourthly, in the industrial sphere, as more generally, reliance was to be placed on the private sector. Ideally, the planners saw the state alone as "capable of satisfying the legitimate expectations of the people" and felt that this required "a progressive widening of the public sector and a re-orientation of the private sector to the needs of a planned economy." But this was more in the realm of aspiration, for the state's investment in agriculture and infrastructure left them with few resources for industry in the public sector. Providing only a minor outlay for the public sector, the planners conceded that "the initiative and responsibility for securing the necessary expansion over the bulk of the field of industry rest with private enterprise."[45] The lack of capacity on the part of the state led to an emphasis on the complementarity and lack of rivalry between the two sectors. The planners figured at the time that "the productive capital in industry and in services essential to it is so small compared to the needs of the country that in the further accumulation of it, the two sectors can well supplement each other and need not necessarily expand at the expense of the one or the other." They envisaged the two sectors not "as anything like two separate entities; they are and must function as parts of a single organism". Hanson concluded that the planners, apparently not convinced at the time of "any inherent superiority" of public enterprise, were obviously putting aside the public-private controversy as "obsolete and irrelevant", believing that the 1948 resolution's demarcation of territory between the two sectors was "not particularly helpful", and therefore now "proposed to apply strictly economic criteria, unmodified by considerations of ideology."[46]

Any conception of an instrumental-socialist orientation for a mixed economy thus seemed to be ruled out at the time. The planners pragmatically declared: "The scope and need for development are so great that it is best for the public sector to develop those industries in which private enterprise is unable or unwilling to put up the resources required and run the risks involved, leaving the rest of the field free for private enterprise".[47] There could be no clearer statement than this declaration on the conceptualization of the state's role as instrumental for the building of capitalism. However, it would not be fair to attribute the lack of emphasis on the public sector to the absence of intent on the part of Nehru, now that he had emerged as the undisputed leader of the Congress party and government; the fundamental constraint was obviously of resources. Meanwhile, the state proceeded to install a regime of controls over private enterprise under the Industries (Development and Regulation) Act of 1951 in order to see that it functioned in the public interest. The complex of discretionary controls initiated under the Act came to constitute one of the most comprehensive and stringent systems of controls over the economy and private sector anywhere in the world outside the communist countries. A key provision of this Act, though of little consequence in influencing the pattern of development during the course of the First Plan, pertained to licensing of industries by government.

The orientation of the First Plan can be justifiably ascribed, apart from the resource constraint, to the fact of its momentum having been set before Nehru achieved political supremacy within the Congress party and government. But Nehru was not about to jettison his ideological vision of the future from becoming manifest in state activity. Since the latter was to take, as already indicated, a "socialistic" direction, it is necessary to underline its autonomous roots in Nehru's ideological framework rather than responsiveness to outside pressures, whether in terms of calculation of political advantage, or in terms of imitation of or inspiration from foreign models. One particular interpretation, while pointing to the growing recognition by the Congress party of the problem of unemployment, holds that "the fact that the political challenge came from the Praja Socialists and Communists, not from the Right, determined to some degree both Congress' choice of solution -- industrialization -- and of method -- state production. Congress had to compete, on their terms". By way of evidence, there is no follow-up to the reference on communists but on behalf of the socialists it is pointed out that they, "having already doubled their parliamentary fraction through the adherence of practically all Independents, won a crucial by-election in Agra, July 1953. Thereafter, and until the Second Plan, the provision of jobs dominated politics."[48] On such slim foundations of the diverse socialist group in parliament with some 25 members, against the Congress leviathan of 364 members, and of a single by-election after the mammoth victory of the Congress a year earlier, rests the case of political calculation.

Another interpretation places the inspiration for the reorientation in Nehru's policies to his visit to China in October 1954; as one observer summarised it, "the China trip may have served to stir up his ever-impatient soul with a view to reassessing whether the purpose and direction pursued by the Congress corresponded with the declared objectives." Nehru himself firmly rejected any such interpretation; as he told one of his biographers in 1956, the new complex of policies had "absolutely nothing to do with it.... we had talked about socialism throughout (the struggle for independence) and as long as twentyfive years ago the Congress said that the chief industries should be owned and controlled by the State. After the coming of independence it developed gradually and ultimately came out. Nothing special happened last year".[49] Nehru was right. The facts are in his favour.

Not long after the 1951-52 elections, Nehru moved to strengthen the socialist forces within the government and party. The most spectacular manifestation of this larger aim was his initiative in the fall of 1952 in inviting and meeting with Jayaprakash Narayan, the eminent leader of the Praja Socialist Party, which had just been formed in September 1952 out of socialists and Gandhians who had earlier withdrawn from the Congress. The purpose was to "enlist cooperation at all levels", suggesting coalition at the centre and in the states and possible merger back into the Congress. As Nehru was to write later: "I am not satisfied, if I may say so, with the rate of our progress....I wanted to hasten it and I wanted your help."[50]

After a long series of talks extending over six months, the negotiations finally broke down in March 1953 because the socialist party rejected cooperation or merger on other than a take-it-or-leave-it basis, demanding among other things land confiscation without compensation and cooperative farming as also nationalisation of banks, insurance companies and mines, all this within four years. The socialists seemed more interested in the publicity the occasion provided for their party than in actual cooperation. Nehru, for his part, sympathized with their goals and endorsed them in principle but demurred: "But surely it is beyond me both as Prime Minister and as President of the Congress to deal with such vital matters and give assurances in regard to them....One can hardly take these things in a bunch."[51] Fundamental to Nehru's position, it would seem, was the notion that movement toward social goals should undoubtedly take place but that nothing should be done to cause political upheaval and undermine India's national unity, and that the people must be carried along in the pursuit of social change -- all values of long standing with him. Left without the assistance or cooperation of the socialists, Nehru then turned to the only political instrument available to him, the Congress party, and to push it as far as possible in directions that he considered desirable.

The focus of such action could not but be economic planning, for that was in Nehru's mind the overall tool for social and economic change. But there was little that could be done about the First Five Year Plan since that was already in midstream. Nonetheless, the First Plan was important in that it occasioned the question as to what should follow it, especially since it seemed more than successful in meeting targets, perhaps because they were so modest to begin with. The considerable improvement in agriculture, largely a result of the monsoons, price stability and the rehabilitation of the economy, indeed, gave rise to euphoria among the planners that anything could be accomplished. Consequently, the conceptualization of what should follow the First Plan in the Second Five Year Plan became the fulcrum around which the new directions, initiated and masterminded by Nehru, began to be organized. These new directions became manifest in several policy instruments and declarations, and it is the totality of these, rather than any single one of them—though each is important in and of itself— that marks a turning point in decisively shifting the balance in a mixed economy and reorienting it on an instrumental-socialist path.

4a. *Political Endorsement of Goal of "Socialism":* In January 1953 the Congress party at the Hyderabad session had endorsed the First Five Year Plan, but starting in May the party under Nehru's presidency began increasingly to express concern over the problem of unemployment, to stress that the aim of planning should be full employment, and to underline the need to expand the development of industries, particularly cottage and small-scale industries, to provide greater employment (Working Committee, May 1953; AICC Agra, July 1953; AICC Ajmer, July 1954). However, gradually this concern with unemployment and industrialization began to be set in the ideological context of opting for socialism. At first, in its resolution on "planned development" the AICC in July 1954 emphasized that "the present social structure, which still continues to be partly based on an acquisitive economy, has to be progressively changed into a *socialised economy*" -- perhaps the first ever such reference in an official resolution of the party. What form this change ought to take became obvious in another resolution on "industry", which declared: "The country already has a powerful State-owned public sector in Industry. This should be enlarged by the addition of other basic and new industries, wherever possible. Where social ownership of basic industries is not possible in the near future, effective social control should be exercised. The resources of the country should be utilised in building new State industries and not in nationalising existing private industries, except where this is considered necessary in the national interest."[52]

The opting by the Congress for an enlarged and powerful public sector or for social ownership of the basic and key industries, as symbolic of a socialised economy, stood as a morally desirable end in itself. The preference for it was not based on any consideration of its being either functional for the expansion of the private sector or as necessary to compensate for some inherent lack of capacity in the private sector. The causal basis for it was simply ideological, and it could not have been otherwise given the whole development of Nehru's thought processes from the mid-1920s onwards. But the preference for both socialism -- which had largely lain dormant since independence -- and therefore the public sector began now to be openly and emphatically stressed. Not insignificantly, the year 1953 also saw India's first major nationalisation, that of the airlines.

A sharp escalation in the publicly expressed preference for socialism by Nehru came in early November 1954, not long after his return from a visit to China. At a meeting of the National Development Council he pondered over the question of the kind of society he eventually envisioned for India. With the foundations of the economy having been strengthened, he thought, the country could now advance faster and that this involved "industrialization at a fairly rapid pace" with heavy industries receiving priority. While urging a more rapid pace of development, however, he rejected capitalism as a path, stating that "a system which is based purely on the acquisitive instinct of society is immoral" and, in any case, its days are over. This was quite in character with his pronouncements before independence. In more positive terms, he went on to say:

> The picture I have in mind is definitely and absolutely a Socialistic picture of society. I am not using the word in a dogmatic sense at all. I mean largely that the means of production should be socially owned and controlled for the benefit of society as a whole. There is plenty of room for private enterprise here, providing the main aim is kept clear.[53]

Socialism for Nehru was not only the morally desirable society to be achieved but also, considerng the context of an underdeveloped country, the strategy for development. As he was to explain subsequently in 1960, one reason for socialism as strategy was that, if not aimed at "right from the beginning" through an expanding public sector, then the consequent concentration of economic power and monopolies would encourage "a process which will come in our way badly and be harmful now and later. It will take us away very far from any kind of progress towards socialism."[54] Particularly, this required the private sector to be excluded from basic industries, because such industries would add to the economic power of the private sector.

In the two months following his speech at the National Development Council on his socialist vision, Nehru drove home the socialist theme repeatedly, though "against considerable opposition, even within his own party, to take the country to the Left,"[55] for "many highly-placed Congressmen were lukewarm about `socialism'."[56] As a signal of the ideological change that had occurred after the Council meeting, the Industrial Policy Resolution was reviewed by the cabinet, which decided that it now "had to be interpreted in terms of the socialistic objective."[57] As a more momentous step in advancing the new socialist theme was the passage, in the second half of December 1954, by the Lok Sabha of a resolution stating that the achievement of a "socialistic pattern of society" was the objective of economic policy of the state; it also urged that the tempo of economic activity, especially industrial development, "should be stepped up to the maximum possible extent." There followed then, in early January 1955, the enactment of a historic resolution at the Congress session at Avadi -- hence the Avadi resolution -- which declared that "planning should take place with a view to the establishment of a socialistic pattern of society, where the principal means of production are under social ownership or control, production is progressively speeded up and there is equitable distribution of wealth." It emphasized that "the national aim is a welfare State and a socialistic economy," and went on to demand that "the public sector must play a progressively greater part, more particularly, in the establishment of basic industries." It recognized that the private sector will continue to be important, but its view of the private sector emphasized cooperatives and small-scale industries; in any case, the private sector was to "play its part within the broad strategic controls of the Plan."[58]

The thrust for the public sector in these declarations and resolutions arose as part of the ideological preference for the "socialist pattern of society." But as criticism and opposition arose in some quarters against the government's drive toward a dominant public sector, its spokesmen began to supplement the consummatory commitment to it with an instrumental rationale in support of the public sector, asserting that it was equally essential as a means to rapid economic development.[59] One particular instrumental reason advanced was that the large investment outlays necessary for basic industries were beyond the means of private enterprise. The consummatory commitment and instrumental rationale -- that, in the words of Kurien, "the growth of the public sector and the dominating role of the State in economic affairs is not only justifiable on ideological grounds but is also necessary from an economic point of view"[60] -- made a potent combination to silence any opposition to the public sector. Socialism was thus not only morally desirable as an end but also considered more efficient as a means.

The Lok Sabha and Avadi resolutions represented official endorsement by government and party of Nehru's long-held aim of a socialist society, undoubtedly to be achieved not all at once but nonetheless steadily and insistently. They sealed the framework for government policy in the period ahead. In due course, "socialistic pattern of society" became replaced by "socialist pattern", "socialism" and, at the last Congress session that Nehru attended in January 1964 at Bhuvaneshwar, "democratic socialism". Near the end of the 1950s Nehru received a severe reverse -- indicative of the powerful political status that the rich and middle peasantry had come to occupy in the country's politics -- when the Congress refused to accept joint cooperative farming as part of a strategy for agrarian reform. As a consequence, in the final analysis, his socialist plans boiled down to ever-widening state ownership of the principal means of production in the non-agricultural sphere, rather than equality or redistribution and structural change. Not only for Nehru, but for the general public, too, "the growth of the public sector and of State activity, therefore, is usually seen to be an extension of socialism. Socialist enthusiasts welcome it and clamour for it, and those who are averse to socialism oppose it."[61] Though thus restricted in scope, Nehru's ideology nonetheless had momentous consequences for the social organization of the economy. Still, without a prior restructuring of the agrarian system, it was reduced to an attempt at industrialization and socialism "by stealth."[62]

4b. *Policy Actions and the Socialist Path:* Soon after the political endorsement of the "socialistic pattern of society" at Avadi, there followed a series of policy actions on the part of the government that marked a decided elevation in the role of the public sector and the imposition of new controls over the private sector. The budget announced in February 1955 provided for special taxes on the salaries of business executives and extended favourable treatment to cottage industries. Later, parliament approved a constitutional amendment that gave the government, rather than the courts, the authority to decide on compensation for property acquired by the state. Then there took place the nationalisation in May 1955 of the largest commercial bank, the Imperial Bank of India, and, later in January 1956, of life insurance companies -- actions which the Congress session at Amritsar in 1956 considered as "significant steps towards the evolution of a socialist structure." Earlier, in the first half of 1955, the National Development Council approved the critically important "Plan-frame" for the Second Five Year Plan, developed by Nehru's planning advisor Mahalanobis. Patterned after the Soviet model, the Plan-frame provided for a "Big Push" strategy for industrialization, with a major thrust to basic industry. Although the strategy was reportedly addressed to solving the newly emerging problem of unemployment, it was in fact not only based on goals Nehru had outlined prior to independence, but was more pointedly addressed to economic independence and national self-reliance, indeed autarky.[63]

In order to conserve resources for investment in basic industry, which was simply presumed to be the preserve of the public sector, Mahalanobis recommended withholding investment in the factory-produced consumer goods industries sector, with the demand for consumer goods to be met largely through reliance on cottage industries. Thus, major industry in the private sector was simply to be squeezed out in investment considerations. But Mahalanobis further urged expansion of the public sector as an ideological desideratum:

> The public sector must be expanded rapidly and relatively faster than the private sector for steady advance to a socialist pattern of economy. In order to make available large capital resources for investment of the Plan, Government will be prepared to enter into such activities as banking, insurance, foreign trade or internal trade in selected commodities.

Little wonder that the business community was "alarmed", not simply by the Plan-frame but also by the whole series of events in 1955.[64] Nehru attempted to pacify the community by underlining that the private sector will continue to play an important role, but the government was not deterred from the larger course it had chosen to adopt. With the adoption of the Second Five Year Plan for the period 1956-1960, "the ideological pendulum in India moved once again towards socialism."[65] The planners now underlined that what they sought was not simply "better results" within the existing social and economic framework, but rather the remoulding of society according to new values:

> These values or basic objectives have recently been summed up in the phrase 'socialist pattern of society.' Essentially, this means that the basic criterion for determining the lines of advance must not be private profit but social gain, and that the pattern of development and the structure of socio-economic relations should be so planned that they result not only in appreciable increases in national income and employment but also in greater equality in incomes and wealth. Major decisions regarding production, distribution, consumption and investment -- and in fact all significant socio-economic relationships -- must be made by agencies informed by social purpose. The benefits of economic development must accrue more and more to the relatively less privileged classes of society, and there should be a progressive reduction of the concentration of incomes, wealth and economic power.[66]

Within the larger context of the underlying philosophy emphasizing social gain rather than private profit, the planners then advanced a combination of ideological premise and functional rationality for the preference accorded to the public sector:

> The public sector has to expand rapidly. It has not only to initiate developments which the private sector is either unwilling or unable to undertake; it has to play the dominant role in shaping the entire pattern of investments in the economy...it is inevitable, if development is to proceed at the pace envisaged and to contribute effectively to the attainment of the larger social ends in view, that the public sector must grow not only absolutely but also relatively to the private sector.

While the planners emphasized, as in the First Plan, that the two sectors "have to function in unison and are to be viewed as parts of a single mechanism," they repeatedly insisted that the public sector had to grow not just rapidly and absolutely but relatively to the private sector. With the Second Plan, then, the public sector assumed "a leading role" in development, with stimulation of private enterprise reduced to secondary importance.[67] Kidron says in relation to the Second Plan that "above all, it was a plan for the public sector", aiming to increase that sector's share of investment outlays to 61 per cent from the 44 per cent achieved in the First Plan. More significantly, Sachs points out that whereas private sector investment outlays were projected to increase only by 50 per cent, those in the public sector were planned to be increased by 153 per cent so as to change the public:private ratio in investment from 0.94:1 to 1.58:1. Important as that might have been, especially striking was the differential pattern in respect of increase in investment in industry compared to the First Plan; the increase in the public sector was placed at 353 per cent as against only 148 per cent in the private sector.[68] These quantitative changes, in reality, marked a shift to the instrumental-socialist type of mixed economy.

With respect to the First Five Year Plan, the eminent economist D.R. Gadgil, later himself to head the Planning Commission, had charged: "Being wedded to private enterprise the Indian Government cannot think in terms of a central pool of savings, of centrally directed investment and of general regime of austerity imposed thorugh direct controls and fiscal devices."[69] As if to prove precisely that it was not so wedded, with the Second Plan the government resorted to most of the recommended instruments. The public sector became the agency for centrally directed investment, and for the necessary central pool of savings the government resorted to a variety of mechanisms to transfer resources to itself from savers. Apart from "direct transfer" by way of sale of savings bonds and "indirect transfer" through compulsory and quasi-compulsory measures such as taxation, designed to reduce disposable income, the state resorted to "forced transfer" by restricting investment in the private sector.[70] One mechanism of "forced transfer" was limiting private investment to levels determined by the state, in addition to excluding the private sector from specified areas of the industrial arena. Accordingly, the state came forward with a new Industrial Policy Resolution in 1956.

The Second Five Year Plan established the overall development strategy for the remainder of the Nehru era. The same basic strategy was continued in the Third Five Year Plan for 1961-1965, stressing again rapid industrialization, with special emphasis on capital and producer goods industries. The role of the public sector was boosted even further in the Third Plan, both practically and ideologically. With the public sector now projected to play an "even more dominant" role in economic development, it was expected "to grow both absolutely and in comparison and at a faster rate than the private sector." Again, the end purpose of this burden on the public sector was the achievement of a socialist society: "In an underdeveloped country, a high rate of economic progress and the development of a large public sector and a cooperative sector are among the principal means for effecting the *transition towards socialism*." Increased stress, however, came to be placed on the public sector's role as an alternative and a counter to concentration of economic power and alleged monopolistic tendencies in the private sector. This emphasis on the countervailing role of the public sector arose as a result of the faster expansion of the private sector beyond what the Second Plan had envisaged for it

-- a development that really indicated, against all suggestions to the contrary, the inherent strong capability of the private sector. To counteract what the planners saw as the growing concentration of economic power, they decided on a three-pronged strategy: (1) "extension of the public sector into fields requiring the establishment of large scale units and heavy investments"; (2) exercise of "considerable vigilance" against "the growth of large existing business" in the application of state controls and regulations, especially licensing of new industrial units; and (3) reorientation of policy toward the private sector so as "to ensure broad-based ownership in industry, diffusion of enterprise and liberal facilities for new entrants, and the growth of cooperative organisations." These measures were clearly directed against big business or upper bourgeoisie and favoured the middle sectors, including the petty bourgeoisie, that constituted the base of the intermediate regime.

The expanded role of the public sector envisaged under the Third Plan is evidenced both in the investment outlays and share in industrial output. The Third Plan provided for an investment of Rs. 18.0 billion in the public sector for industry and mining as against Rs. 11.8 billion in the private sector. In 1950-51 the public sector's share in net output of all organized manufacturing industries had been less than 2 per cent; by the end of the Second Plan it had gone up to around 10 per cent; now it was projected at 25 per cent by the end of the Third Plan, mostly concentrated in capital and producer goods. Similarly, the public sector's share in minerals production was expected to increase from less than 10 per cent in 1950-51 to over 33 per cent in 1965-66. With this rate of expected growth, the planners exulted: "As the relative share of the public sector increases, its role in economic growth will become even more strategic and the state will be in a still stronger position to determine the character and functioning of the economy as a whole."[71]

4c. *Industrial Policy Resolution of April 30, 1956:* While discussions were under way on the Second Plan, the government brought forward a new Industrial Policy Resolution in 1956, replacing the 1948 resolution, in order to adjust to the changed circumstances, especially the proclamation of the objective of "socialist pattern of society". From that ideological objective followed, as an automatic proposition not requiring further justification, the intent "to accelerate the rate of economic growth and to speed up industrialization and, in particular, to develop heavy industries and machine-making industries, to expand the public sector, and to build up a large and growing cooperative sector." Not only that, the aim of socialist pattern of society also required reduction of income disparities, and prevention of private monopolies and concentration of economic power. Given these requirements, the resolution declared: "Accordingly, the State will progressively assume a predominant and direct responsibility for setting up new industrial undertakings and for developing transport facilities. It will also undertake State trading on an increasing scale." It was not, however, the aim to have the public sector corner the entire field: "At the same time, as an agency for planned national development, in the context of the country's expanding economy, the private sector will have the opportunity to develop and expand." But here the principle of cooperation was to be extended "and a steadily increasing portion of the activities of the private sector developed along cooperative lines." Definitively drawing out the dominant role of the public sector from the goal of socialism and reinforcing it by referring to the requirement of quicker development, the resolution laid down the larger principles for the scope of public and private sectors:

> The adoption of the socialist pattern of society as the national objective, as well as the need for planned and rapid development, require that all industries of basic and strategic importance, or in the nature of public utility services, should be in the public sector. Other industries which are essential and require investment on a scale which only the State, in present circumstances, could provide, have also to be in the public sector. The State has, therefore, to assume direct responsibility for the future development of industries over a wider area.

Having thus pre-empted on principle a large field for the public sector, the resolution recognized "limiting factors" in the present state of development that made necessary a clear demarcation of industries in which the State "will play a dominant role." It then distingushed among three categories of industries. In the first category (1) were industries of basic and strategic importance and public utilities, whose development was henceforth "the exclusive responsibility" of the state, with all new units in the future reserved for it. This category encompassed 17 items, with four of them (railways, air transport, arms and ammunition, and atomic energy) required to be monopolies of the central government. Among the other items were: iron and steel, heavy plant and machinery, heavy electricals, mining of coal and several other important ores, mineral oils, mining and processing of several metals, aircraft manufacture, shipbuilding, telephones, and electric power. As a remarkable index of the government's intention to retain flexibility, it is noteworthy that inclusion of industries in this category did not "preclude" expansion of existing units in the private sector, nor the state's obtaining "the cooperation of private enterprise" even in the establishment of new units "when the national interests so require."

The second category (2) included industries which were to be "progressively State-owned and in which the State will therefore generally take the initiative in establishing new undertakings." However, here private enterprise was also "expected to supplement the effort of the State," and to have "the opportunity to develop in this field, either on its own or with State participation." The more important among the 12 items in this category were: aluminium, machine tools, basic and intermediate chemicals, antibiotics and essential drugs, fertilizers, synthetic rubber, and road and sea transport. One decided effect of the first two categories -- which was no doubt intended in the cause of the larger aim of national independence -- was to bar, with the stroke of a pen but without saying so, foreign enterprise from a broad front of important industries. If, indeed, in respect of such industries the choice was between foreign enterprise and public sector, the state had excluded foreign enterprise in the name of excluding private enterprise.

Finally, the third category (3) comprised all other idustries, and these were generally "left to the initiative and enterprise of the private sector." But, ever reluctant to give up any option, the state reserved the right "to start any industry even in this category." Nonetheless, it promised "to assist the private sector in the development of these industries" through provision of an adequate infrastructure of power, transport and finance. Further, while insisting that the private sector must fit in with the requirements of the policy framework, the resolution assured it of "fair and non-discriminatory treatment" where it co-existed with the public sector in the same industry. At the same time, there was no mention of nationalization as if either this was no longer on the agenda or that the government did not wish to tie its own hands. But the government emphasized the high importance it attached to the growth of cottage, village and small-scale industries, and pledged to continue its policy to assist them "by restricting the volume of production in the large scale sector, by differential taxation, or by direct subsidies." Besides, it promised to reduce regional disparities through the development of backward areas by means of dispersal of industry.[72]

By and large, despite the more or less precise demarcations, the resolution took a flexible and ambiguous position by qualifying them so as to enable the government to do anything it wished. For that reason apparently, it continues to serve as the foundation of industrial policy. However, the flexibility and deliberate ambiguity have made for continual controversy over whether particular actions were consistent or not with the resolution. Regardless, the overall thrust was to shift the balance between the public and private sectors in the "mixed economy" so as to reorient it to an instrumental-socialist type, given the wide arena allotted to the public sector and the intent to have the state progressively assume a dominant role. Kochanek would seem to agree with this view when he says that through the resolution "the economic future of India was taken from the hands of private capital and placed in the public sector" and that the resolution "contained a strong bias in favour of the public sector."[73] Besides, the commitment in the preamble to the socialist pattern itself constrained the search for choice in favour of the public sector. As Hanson states: "Its exact wording was less important than its `socialist' context. Although the government could claim that it licensed an `empirical' stance in the public-private controversy, the paragraphs that could be made to bear this interpretation had to be read in the light of offically-adopted 'socialist' objectives."[74] In the alternative, the government rendered itself vulnerable to attack by leftist forces on grounds of betrayal of socialism. After making allowance for all the exceptions and qualifications, Hanson concluded that the resolution contained "a strong built-in bias in favour of the public sector. Public enterprise is encouraged, private enterprise permitted on certain conditions. Hence the choice, if not absolutely predetermined, is steered in a particular direction by considerations which, while not necessarily hostile to rapid growth, are essentially irrelevant to it."

The underlying drive in the demarcation of the industrial arena and in the provision for faster expansion and domination by the public sector was clearly ideological. No instrumental purpose by way of allegedly promoting the private sector can be said to be served by the decision to exclude beforehand that sector from a broad front of vital industries. It is one thing to have the state express its readiness to undertake the establishment of industries which are perhaps beyond the capacity of the private sector -- and there could really be none which are not at the same time beyond the capacity of the state, especially one with little knowledge or experience of establishing or running industry -- but to determine a priori that the private sector is automatically barred from a group of industries that are the foundation of economic power can only be the result of an initial commitment to ideology. The notion of large investment outlays for basic industries being beyond the financial resources of the private sector is based on the state pre-empting those resources in the first place, through the socialization of savings and high taxation, and the simultaneous unwillingness to provide incentives and risk insurance to the private sector. Besides, the state itself was not so well equipped for these industries in terms of resources, for it had to resort to dependence on foreign aid. Of course, to provide assistance to private enterprise to develop across the entire industrial spectrum presupposes a political predilection of a different order, such as in Japan and South Korea, but its lack here is indicative of an alternative ideological commitment. That commitment was to socialism, or more narrowly to ownership of the principal means of production which, as Nehru expressed it, were "sacrosanct for the State."[75] Nehru's posture towards the private sector was based, not on its incapacity to undertake the job or to do it well and efficiently, but on the undesirability of it from the viewpoint of the ultimate goal of socialism.

Quite crucially, no government declaration provided for sale of industry to the private sector once the state had proven its viability; the private sector's exclusion was taken to be permanent. Thus it is unconvincing to hold the particular design formulated for the public sector in India to be simply promotive of capitalism. Of course, any public investment has an expansionary effect on the private sector, particularly if it takes place in infrastructure, but the purpose of the demarcation was to exclude precisely those areas which would add to the strength of the private sector and thus to reduce its status and power relative to the public sector controlled by the state. To be sure, the government at times was compelled under the duress of economic crises, as in 1957, to relax its requirements on exclusion, and leftist scholars have been quick to attack it for the "abridgement of the public sector" and "encroachment by the private sector."[76] But there could be no better testimony to the ideological basis of the sectoral demarcation in the industrial arena than such a relaxation, for what can be allowed to the private sector in a time of crisis could certainly be allowed it in normal circumstances. At the same time, the fact that an entire ideological framework is at times not rigorously executed does not mean that that part of it which is implemented is not ideology-based. Nor does the presence of an ideological framework always exclude the application of norms of rationality about costs and benefits; the pertinent question is whether that rationality is allowed to operate within the discipline of that framework or outside it.

In addition to the Industrial Policy Resolution and the Second Five Year Plan, the government's new posture toward the private sector was reflected in the enactment in 1956 of a new Companies Act to bring legislation into accord with the objective of the socialist pattern. With that, India adopted "one of the most detailed and stringent codes of business legislation to be found anywhere in the world."[77] The central government, rather than any independent body, acquired wide-ranging powers in regard to the management of companies, notwithstanding vociferous opposition from the private sector.

5. Mixed Economy as Transition to Socialism

Despite the emphatic thrust provided to the public sector through the complex of policies and programmes adopted in 1956, and the correspondingly constricted industrial space left to the private sector, Nehru's attitude toward the existing private sector and even to its expansion within the limits prescribed was not entirely negative, leave aside vindictive, even if it was not altogether positive. Nehru's intent, as it finally evolved over the years after independence, was not an immediate and wholesale replacement of the private sector through nationalization. This was objectionable to him since it was not the peaceful and Indian way; it was more the way of violence, and he saw no reason to cause injury when it could be avoided. As he explained in 1955: "We shall do so in our own way, and that is a peaceful way, a cooperative way and a way which always tries to carry the people with us including those who may be apprehensive or even hostile to begin with."[78] Nehru was therefore not one to favour a mechanical adoption of the Soviet model.

However, this did not mean any diminution of his commitment to socialism, narrowly understood as the public ownership of the principal means of production. Rather, in a creative adaptation, the socialist society he envisaged was to be achieved, not through displacement of existing private enterprises, but through the steady and relentless expansion of the public sector within a "mixed economy", so that over the long haul it progressively supplanted, eclipsed and subordinated the private sector. Whatever private sector continued to exist in this reduced and shrinking position would have to function according to the larger needs of society rather than merely of private profit. The public sector would thus in due course come to determine the nature of society, since it will be the dominant force in society not only because of its large absolute and relative size but more importantly because of its control over the strategic points of the economy by way of basic and key industries, "the commanding heights". The domination of the public sector in the economy was a first and unchallengeable principle with Nehru, and the private sector would have to function within that encompassing principle.

What was significant for Nehru was that colonial rule had left India with no significant base in basic and key industries. Thus that which constituted precisely the powerhouse of an industrial country was absent; it represented vacant industrial space, occupied neither by the public nor private sector. Consequently, the public sector could move into this space without injury to the existing private sector. What was important therefore to Nehru was to set the course for expansion of the public sector through reservation of basic industries for it. With that, the transition to socialism would become irreversibly determined. The present private sector could be left alone, without resort to nationalization, to be reduced to insignificance in due course. In this fashion, socialism would be brought about not through nationalisation of the existing private sector but through an expanding public sector, the result of state entrepreneurship. In 1961, Nehru saw the pattern of a socialist society emerging in India through this process in about ten years. He did not expect much resistance from the private sector to this course, because of what he felt to be the peculiarly persuasive Indian approach to things; for, after all India had abolished princely states and the rural intermediary system (*zamindari*) without much trouble.[79]

This would seem to constitute Nehru's painless and peaceful path for the transition to socialism. In addition, Nehru supplemented this organizational design for the transformation of the balance between the two sectors, within the mixed economy, by a technological dynamic which made nationalization redundant. Moved by his realization of the rapidly changing nature of technology, he thought it made little sense to acquire enterprises which were destined for obsolescence. Rather, the state ought to establish new undertakings with the most recent technology. These arguments were further buttressed by the notion that increased production, which the advance toward socialism required, necessitated the mobilization of all sources of growth, especially when the resources in the hands of the government were limited, as he repeatedly stressed. But, if the private sector was then to be given an area for operation, restricted as it was, there was no point in thwarting its growth; rather it should be given all facilities to expand within the larger framework determined by the state. Some of these themes emerge in a major speech Nehru gave in the Lok Sabha in 1956 on the Second Five Year Plan:

> May I say here that while I am for the public sector growing, I do not understand or appreciate the condemnation of the private sector? The whole philosophy underlying this Plan is to take advantage of every possible way of growth and not to do something which suits some doctrinaire theory or imagine we have grown because we have satisfied some text-book maxim of a hundred years ago. We talk about nationalization as if nationalization were some kind of a magic remedy for every ill. I believe that *ultimately all the principal means of production will be owned by the nation*, but I just do not see why I should do something today which limits our progress simply to satisfy some theoretical urge. I have no doubt that at the present stage in India the private sector has a very important task to fulfil, provided always that it works within the confines laid down, and provided always that it does not lead to the creation of monopolies and other evils that the accumulation of wealth gives rise to. I think we have enough power in our laws to keep the private sector in check. We are not afraid of nationalizing anything.
>
> The House knows we have taken some big steps even during the last few months. Only a little while ago, the House was dealing with the Bill concerning insurance. These are all mighty steps and we are not afraid of taking them, but we do not propose to take any step merely to nationalize, unless we think it is profitable to the nation. On the other hand, we would much rather build up new national industries than pay compensation to decrepit industries in order to take charge of them. Why should we, in this age of changing technology, pay to take possession of any old plant unless it happens to serve some strategic purpose? In that case I would do it because I want to hold the strategic points in our economy. I should like the House to appreciate the philosophy behind this report, namely, that the public sector and the private sector should be made to cooperate within the terms and limitations of this Plan.
>
> While the public sector must obviously grow -- and even now it has grown, both absolutely and relatively -- the private sector is not something unimpor-

tant. It will play an important role; *though gradually and ultimately it will fade away*. But the public sector will control and should control the strategic points in our economy. The private sector, as we have stated in the Industrial Policy Resolution, will be given a fairly wide field subject to the limitations that are laid down. It is for us to decide, from time to time, how to deal with that sector.

The point is that since we are an under-developed country, the scope for industrialization and advance is very vast. The field, so to speak, is occupied by nobody. Let us advance; let the public sector advance. But why should we spend time and energy over acquiring some old factory or old plant?[80]

At other moments, Nehru added to this grudgingly accorded role to the private sector more positive functions. These included a wariness, stemming from his basic liberalism, about an all powerful totalitarian, bureaucratic state; he stated in 1958:

I do not want State socialism of that extreme kind in which the State is all powerful and governs practically all activities. The State is very powerful politically. If you are going to make it very powerful economically also it would become a mere conglomeration of authority. I should, therefore, like decentralization of economic power.[81]

He also saw the private sector as a stimulus to efficiency in the public sector, stating in 1956:

I think it is advantageous for the public sector to have a competitive private sector to keep it up to the mark. The public sector will grow. But I feel that, if the private sector is not there, if it is abolished completely, there is a risk of the public sector becoming slow, not having that urge and push behind it.[82]

Considered in this light, Nehru's conception of the "mixed economy" is not simply an instrumental one as a stage of transition to socialism but rather one that shades into a consummatory view positing it as almost a stable order, with its own moral justification, even though he takes the dominant role of the public sector as a given. Such a view would be consistent with Nehru's refusal to identify completely with either of the competing global ideologies of capitalism and communism. This persistent posture of ideological autonomy at home in relation to the two rival economic systems would seem to correspond to non-alignment abroad between the two rival power blocs, even though tilted in one direction or another from time to time. It would seem at the same time to be consistent with his endeavour always to develop a creative and distinctive "middle way", "a middle path", at home as abroad in foreign policy. As he always insisted and as the Second Plan repeated, "each country has to develop according to its own genius and traditions."[83] There was then an element of national integrity, legitimacy and symbolism associated with the "mixed economy."

However, even if distinctive, the endeavour did not guarantee success. Rather than assuring a combination of the merits of the two systems of capitalism and communism -- dynamism and discipline, respectively -- it harboured the danger of combining the ills of both systems, that is, inequality and bureaucratic domination. A dominant public sector under the intermediate strata could no more guarantee social justice than private enterprise under the bourgeoisie. Perhaps Nehru was unrealistic in underestimating the vested interests of the intermediate strata in control of the state. Here, the very flexibility of his socialism constituted a key danger; the intermediate strata could interpret it so as to corner economic resources and strategic sectors to bring them under their own control as against those above, but to sabotage any plan to bring about equality and redistribution in favour of those below. Besides, the absence of prior destruction of vested interests could allow them to combine with the intermediate strata in their domination over the lower classes. The installation of democracy prior to, or simultaneously with, socialism could furthermore only facilitate such an outcome; the combination of democracy and socialism, howsoever noble and attractive intellectually, could prove fatal for both in the context of an underdeveloped society in contrast to a developed society.

But more crucially for an underdeveloped country, there could be no assurance that public enterprise would ever serve as an adequate growth mechanism. The advance towards socialism could not be assured simply by placing the principal means of production under social ownership and control. That advance was premised on efficiency of the public sector in creating genuine surpluses to become an engine of economic growth. Nehru apparently assumed that the public sector -- or for that matter, the constricted private sector in the context of a protected economy operative under administrative controls, or even the small-scale sector endowed with special privileges by the state -- would be efficient and therefore took no specific measures to ensure efficiency. If, however, the public sector were to prove inefficient in this task, it would merely represent resources misinvested and wasted, and the total endeavour a decidedly "premature" socialism. The end result, despite the good intentions, then could be neither socialism nor growth.

The main threads of Nehru's ideology and programme can now be pulled together briefly:[14]

1. Nehru rejected capitalism because it was morally unacceptable on the grounds of its being based on the spirit of acquisitiveness, private profit and inequality. Instead he favoured socialism -- since it was based on sharing social gain and equality--and democracy; the two, far from being incompatible, were considered incomplete without each other and therefore had to go together.

2. Socialism for Nehru meant social ownership and control of the means of production, state intervention through central planning, and equality. But socialism was not only a desired mode of social organization for a developed society, it was also the means for development of an underdeveloped country. Besides, he believed it necessary to undertake it now so that vested interests were not created in the process of development that would block its establishment later.

3. Nehru's conception of socialism was not fixed, rigid or doctrinaire, for he wished it to be adapted to each society and to circumstances or stages of development. For a developing society, he envisaged it as involving not immediate or complete state ownership of the means of production, but of the principal means, not violent overthrow of the existing social system or

class war but steady advance with the consent of major groups within a democratic framework, not comprehensive nationalization but only selective socialization in order to control the strategic points of the economy.

4. The appropriate form of transition to socialism was a mixed economy in which the state is economically dominant, controlling the "commanding heights" and all strategic points by virtue of its ownership and management of the basic and key industries and more generally through economic planning and regulation. In the mixed economy, while the public sector's dominance was beyond dispute, there would be room for the private sector to function as well. While over the long term the private sector will become subordinated to the public sector, in the short term the two sectors were seen as complementary to each other. A key requirement for socialism was increased production, otherwise socialism will mean only distribution of poverty; production required tapping all sources of growth, especially in view of the limited resources of the state, but within the overall framework established by the state. He thought that one should not underestimate, carried away by slogans, the enormous organizational problems involved in the state taking over the entire economy in an underdeveloped country.

5. For a developing society, there was no need for large-scale nationalization, for here the space for industrial expansion is so vast, especially in the field of basic industries, that the public sector can simply eclipse the private sector by reserving key areas for itself. Further, considering the fast pace of technological change, it is best to start with new industries in the public sector rather than nationalize obsolescent industries in the private sector.

6. By expanding absolutely and relatively, the public sector will be able to change the balance within the mixed economy and thus enact the transition to socialism. But this is not a sudden process; rather it will be a protracted and long-term process, spread perhaps over a generation or two.

7. In some sense, the end stage of socialism itself will be a kind of mixed economy, for the combination of total economic and political power in the hands of the state is undesirable since it is likely to lead to regimentation under a bureaucratic state. The ultimate purpose of socialism is the fullest development of the individual, not of the state. Besides, the continued existence of the private sector is necessary as a stimulus to efficiency in the public sector. If this model departs from the more orthodox Marxist framework, that is because that framework has to be creatively, not mechanically, applied.

Summary and Conclusions

The creation of the public sector understood as comprising a vast complex of industrial enterprises is entirely a phenomenon of the Nehru regime in independent India. While the existence of the public sector is rather generalized throughout the Third World, the actual form it takes and the intent in bringing it about varies immensely. Any adequate understanding of the public sector in India requires recognition of the fact that its creation took place in the context of an openly proclaimed ideology, even if such an ideology was not a rigorous or rigid one.

That ideology was, briefly, socialism. The public sector arose as part of a set of policies designed for the transition to socialism. As Kurien maintains, "the strongest ideological element that lies behind Indian economic policy, of course, is socialism" and that "the ideological leaning toward socialism has given to the state a prominent role in economic affairs"; equally, Bhatt confirms, "after independence, the dominant ideology has been the socialist one, evolved by Nehru himself."[85] The public sector was seen by Nehru and his planners both as an index of socialism as well as a route to socialist society.

The public sector in India did not have to be created by reference to any ideology as has happened in many Third World countries, but in this particular case it was and that has not been without important implications for the nature of the public sector. In the Indian case, the public sector came to be created not as a means of "last resort" after the failure of private enterprise to come forward,[86] or as a "substitution" for a missing or incapable bourgeoisie,[87] but as a first resort because of the compulsions of ideology. It is mistaken therefore to neglect, as Bardhan does, the role of ideology in the founding of the public sector in India and to see it as a function simply of interest-based bargaining.[88] As the eminent economist Chakravarty states, "the approach adopted by Nehru could be justified on purely pragmatic considerations alone....But there were more things involved than mere pragmatism. There was a whole theory of transition to a more humane social order which was behind the choice that Nehru made."[89]

Since the public sector in Nehru's plan was largely to be created through new entrepreneurial activity on the part of the state, rather than through thoroughgoing nationalization following a revolutionary overthrow, there was necessarily involved a transition period during which the private sector would continue to exist. But it was envisaged that the private sector was inevitably destined to be overwhelmed by the public sector. In other words, the conception that Nehru had of the mixed economy was an instrumental-socialist one, that is, as transition to socialism, and certainly not an instrumental-capitalist one. Of course, the public sector's cause was sought to be advanced by holding it to be at the same time functional for rapid economic development, but even then it is significant that this rationale was proffered as advancing economic development, never capitalist development or capitalism.

It is often claimed nonetheless that Nehru was, through the creation and expansion of the public sector, attempting to build capitalism. Although no initiator of policy can ensure how his handiwork will ultimately be utilized, it is patent that the policies Nehru undertook were not exactly keyed to the expansion of capitalism but rather to bypassing and supplanting it with a view to bringing about its eventual demise: the expansion of the public sector not only absolutely but relatively in relation to the private sector; the securing of the "commanding heights" of the economy in the hands of the public sector and in its hands alone; the exclusion of the private sector from critical sectors of the economy; and the effort to reduce the economic power, and therefore the political, power of the private sector.

If the public sector in post-independence India arose out of ideology, then it is critical to note that the party in power that came to acknowledge socialism as its guide in setting it up had until the mid-1950s not accepted that ideology as its platform.

Before independence, the Congress party would not countenance accepting socialism as its ideology; indeed, it would not even countenance Nehru propagating socialism while occupying high office in the organization. After independence, the dominant conservative trend within it became initially even more pronounced as left-oriented groups seceded from the party. With that, Nehru became virtually the sole, but most strategically located, socialist leader of any prominence in the party. While his socialist ideology did not command much support within the government, the Congress party, or the country, Nehru's popular mass appeal as a charismatic leader, who had undergone tremendous personal sacrifice for India's freedom, was nonetheless of critical importance to the Congress party from the viewpoint of its electoral prospects. This appeal, however, had little to do with the ideology of socialism though it was related to Nehru's personal identification with the underdog; another charismatic leader, Mahatma Gandhi, had had an even greater mass appeal to the same population without any pretense at socialism, indeed, with publicly expressed opposition to it.

However, the conservative forces opposed to Nehru within the party crumbled once they lost their political bulwark through intervention by fate, first in respect of Gandhi and later, more critically, Patel. Nehru then moved quickly through a political "coup" to establish his domination in the Congress party, thus bringing both government and party under his control. Relying on the leverage of his popular mass appeal -- and therefore his critical importance to the electoral prospects of the Congress party -- he was then able to move the party in directions in accord, though not entirely, with his personally preferred socialist vision. This outcome underlines the importance of the structural context -- rather than simply class configuration -- which made electoral success a political requirement for the Congress party's exercise of state power.

Nonetheless, one element that enabled Nehru to move the Congress party toward acceptance of the goal of socialist society was the fact that the Congress party was not, and had not been, under the domination of the bourgeoisie or capitalist class. The fact that the intermediate strata, comprising the new middle class and peasantry, were in command of the Congress party precluded an automatic opting for a capitalist or instrumental-capitalist economy, which would surely have been the case were the capitalist class in control. However, control by the intermediate strata did not altogether preclude the capitalist model any more than it assured the adoption of a socialist one. The fact is that, in the Indian case, the same intermediate strata had during the course of the nationalist movement refused to go along with any acceptance of socialism. Even after independence, the dominant trend in the Congress party remained conservative until the mid-1950s, and some would add that it remained ever so afterwards despite the formal acceptance of socialism.

It was only after consolidation of his power in government and party that Nehru was finally able to push the Congress to accept socialist society as a morally desirable end and also the progressive socialization of the means of production as a route to that end. The ideological change, with its commitment to measured but relentless movement toward socialism, may not have met with the approval of critics both on the right and the left, but it had critical consequences for the kind of public sector that was envisaged and installed. In bringing this change in the ideological orientation of the party and government, the role of one individual as a charismatic leader was critical. It is doubtful if in his absence the Congress party would have moved in that direction. However, one cannot attribute the acceptance of socialism by the Congress party to a genuine conversion, for such large numbers from so many different strata could not have been moved overnight to a change of faith on the basis of speeches by an individual leader. Nor was the acceptance of socialism a result of a change in the social composition of the Congress party, for none had occurred. Critical to the party's acceptance of the goal of socialism was rather political calculation in regard to the unique position of Nehru as a national leader who could act as an electoral magnet for the party. O'Connor suggests as a premise that "the capitalistic state must try to fulfil two basic and often mutually contradictory functions -- accumulation and legitimization".[90] In the Indian case, it is obvious that the Congress party in its search for legitimization through Nehru had paradoxically to concede to his ideological predilection for the socialization of accumulation, substantially if not completely. However, once having been shown the way by Nehru on the basis of ideology, literally led by the hand as it were, the intermediate strata would come to see virtue in it from the viewpoint of interest as well.

While the role of the individual leader was critical in guiding the Congress party to the acceptance of the goal of socialist society, it is equally necessary to underline the relevance of the class configuration which made that acceptance possible. It was the domination of the intermediate strata in the Congress party, rather than that of the capitalist class, that enabled Nehru to lead it in that direction. One cannot, of course, take the public sector to be a function simply of the domination of the intermediate strata, as Sobhan and Ahmed assume.[91] But that domination was a necessary, if not sufficient, condition. However, it is mistaken to believe, as Ashok Desai does,[92] that the creation of the public sector in India followed the emergence of a large army of educated children of the petty bourgeoisie some two decades after the expansion of school and college education following independence. Quite the contrary, the public sector arose simultaneously with the expansion of education and before the products of schools and colleges had come out. But, once created, it provided the intermediate strata a ready means to accommodate their children in public employment.

More importantly, constructed in the context of the domination of the intermediate strata, the public sector opened up the possibility of leading -- especially in the absence of a prior genuine ideological conversion -- not to a transition to socialism, but to perpetuating and reinforcing that domination by the same strata through their seizing the public sector, once it was created in the name of socialism, as a source of autonomous economic power to be added to their already existing political power.

NOTES

1. George Rosen, *Democracy and Economic Change in India* (Berkeley: University of California Press, 1966), p. 84.
2. Myron Weiner, *Party Building in a New Nation: The Indian National Congress* (Chicago: University of Chicago Press, 1967), pp. 120-21, 208, 273-74, 300, 467-68.
3. Donald B. Rosenthal, *The Limited Elite: Politics and Government in Two Indian Cities* (Chicago: University of Chicago Press, 1970), pp. 106-108, 129; and Weiner, pp. 409-18.

4. Stanley A. Kochanek, *The Congress Party of India* (Princeton: Princeton University Press, 1968), p. 374.
5. Kochanek, pp. 383-84.
6. *Ibid.*, p. 384.
7. Satish K. Arora, "Social Background of the Indian Cabinet," *Economic and Political Weekly*, VII, nos. 31-33 (August 1972), pp. 1523-32.
8. V. Subramaniam, *The Managerial Class of India* (New Delhi: All India Management Association, 1971), pp. 33-35.
9. Rehman Sobhan and Muzaffer Ahmed, *Public Enterprise in an Intermediate Regime: A Study in the Political Economy of Bangladesh* (Dacca: Bangladesh Institute of Development Studies, 1980), p. 10.
10. Susheela Kaushik, *Election in India: Its Social Basis* (Calcutta: K.P. Bagchi & Co., 1982), p. 49.
11. P.C. Joshi, "Reflections on Marxism and Social Revolution," in K.N. Panikkar, *National and Left Movements in India,* (New Delhi: Vikas, 1980), pp. 196-97, 200.
12. Sobhan and Ahmed, pp. 10-11.
13. India, *Report of the Monopolies Inquiry Commission,* (New Delhi: 1965), p. 135.
14. Stanley A. Kochanek, *Business and Politics in India* (Berkeley: University of California Press, 1974), pp. 50, 198-202, 331.
15. Sobhan and Ahmed, p. 9.
16. Cited in Ignacy Sachs, *Patterns of Public Sector in Underdeveloped Economies* (Bombay: Asia Publishing House, 1964), p. 129.
17. Indian National Congress, *Resolutions on Economic Policy, Programme and Allied Matters (1924-1969)* (New Delhi: All India Congress Committee, 1969), pp. 18-32.
18. H.D. Malaviya, *Socialist Ideology of Congress* (New Delhi: Socialist Congressman, 1966), p. 27.
19. Michael Kidron, *Foreign Investments in India* (London: Oxford University Press, 1965), pp. 78-79.
20. *Economic Weekly,* January 1949, cited in Kidron, p. 83.
21. Kidron, p. 83.
22. See W.H. Morris-Jones, *Parliament in India* (University of Pennsylvania Press, 1957), pp. 156-57.
23. Solo differentiates these as the two forms of organizational choice. *Authoritative* decisions refer to those made "by individuals in authority *for* the organization." Since an authoritative decision, even though made by an individual, must be acceptable to the organization, it "will be made by reference to that which can be explained, rationalized, or justified" and therefore "surely the criteria of evaluation and the boundaries of the acceptable are determined by an ideology shared by those who exercise and those who accept and ultimately support the exercise of authority." For Solo, "ideology will have the greatest weight when decision is authoritative, whereas expediency is more likely to prevail when choice is composite." Robert A. Solo, *The Political Authority and the Market System* (Cincinnati: South-Western Publishing Co., 1974), pp. 36-37.
24. For the text, see M.L. Trivedi, *Government and Business* (Bombay: Multi-Tech Publishing Co., 1980), pp. 613-21; emphasis added.
25. A.H. Hanson, *The Process of Planning* (London: Oxford University Press, 1966), p. 449.
26. See Sachs, pp. 121-22.
27. Clarence J. Dias, "Public Corporations in India," in International Legal Center, *Law and Public Enterprise in Asia* (New York: Praeger, 1976), p. 52.
28. Michael Brecher, *Nehru: A Political Biography* (London: Oxford University Press, 1959), p. 511.
29. Kidron, p. 85.
30. Jawaharlal Nehru, *Speeches: Volume One: September 1946-May 1949* (New Delhi: Publications Division, 1967), p. 123.
31. The Congress party did characterize the resolution the same year as "favouring a Mixed Economy." See Congress, *Resolutions,* p. 35.
32. Cited in A.H. Hanson, *Public Enterprise and Economic Development* (2nd ed.; London: Routledge, Kegan Paul, 1965), pp. 153-54.
33. Cited in Hanson, *Process,* p. 236.
34. Hanson, *Public Enterprise,* p. 154.

State Entrepreneurship in the Nehru Era: Ideology vs. Necessity 211

35. Kochanek, *Congress*, p. 34.
36. *Ibid.*, pp. 45, 49, 527.
37. Sarvepalli Gopal, *Jawaharlal Nehru: A Biography: Volume Two : 1947-1956* (Cambridge, Mass.: Harvard University Press, 1979), p.304.
38. Kochanek, *Congress*, p. 50.
39. Cited in Brecher, pp. 435-36.
40. Brecher, p. 463.
41. Kochanek, *Congress*, p. 61.
42. K.P. Karunakaran, *The Phenomenon of Nehru* (New Delhi: Gitanjali Prakashan, 1979), pp. 7, 22.
43. Kochanek, *Congress*, p. 141.
44. India, Planning Commission, *The First Five Year Plan* (New Delhi: 1953), pp. 31, 44, 71, 420, 425.
45. *Ibid.*, pp. 32-33, 44, 71, 425, 430.
46. Hanson, *Process*, p. 452.
47. *The First Five Year Plan*, p. 422.
48. Kidron, p. 128.
49. Brecher, p. 529.
50. Brecher, pp. 464-65.
51. *Ibid.* Many years later, Narayan rejected his own position for fear of totalitarianism. See R.C. Dutt, *Socialism of Jawaharlal Nehru* (New Delhi: Abhinav Publications, 1981), p. 211.
52. *Congress, Resolutions*, pp. 71-84; emphasis added.
53. Jawaharlal Nehru, *Planning and Development: Speeches of Jawaharlal Nehru, 1952-56* (New Delhi: Publications Division, n.d.), p. 17.
54. Jawaharlal Nehru, *Speeches: Volume Four: September 1957-April 1963* (New Delhi: Publications Division, 1964), pp. 139-40.
55. Brecher, p. 528.
56. Hanson, *Process*, p. 460.
57. Kochanek, *Congress*, p. 175.
58. *Congress, Resolutions*, pp. 86-89.
59. Note Nehru's comment in 1959 : "We have accepted socialism as our goal not only because it seems to us right and beneficial but because there is no other way for the solution of our economic problems." Nehru, *Speeches : Volume Four*, p. 5.
60. C.T. Kurien, *Indian Economic Crisis: A Diagnostic Study* (Bombay: Asia Publishing House, 1969), p. 20.
61. *Ibid.*, p. 12.
62. Nigel Harris, *India-China: Underdevelopment and Revolution* (New Delhi: Vikas, 1974), p. 10.
63. See Baldev Raj Nayar, *The Modernization Imperative and Indian Planning* (New Delhi: Vikas, 1972), pp. 42-44, 132-33.
64. Brecher, pp. 531, 533.
65. Kurien, p. 87. Note also Rosen's comment: "Much of the reasoning behind the detailed plans is ideological. It rests on the grounds that such planning is part of `democratic socialism,' and that it is scientific." Rosen, p. 240.
66. India, Planning Commission, *Second Five Year Plan* (New Delhi: 1956), pp. 22-23, 28, 30, 136, 393.
67. Dias, p. 52.
68. Kidron, p. 136; Sachs, pp. 131-32.
69. Cited in Kurien, p. 91.
70. *Ibid.*, pp. 96-97.
71. India, Planning Commission, *Third Five Year Plan* (New Delhi: 1962), pp. 7, 10, 13-14, 50, 64, 459. Emphasis added.
72. *Second Five Year Plan*, pp. 43-50.
73. Kochanek, *Business*, p. 79.
74. Hanson, *Process*, pp. 464-473.
75. Jawaharlal Nehru, *Speeches: Volume Three: March 1953-August 1957* (New Delhi: Publications Division, 1958), pp. 101-103.
76. Kidron, pp. 141, 143.

77. Hanson, *Process*, p. 486.
78. Cited in Brecher, p. 532.
79. Sarvepalli Gopal, *Jawaharlal Nehru: A Biography: Volume Three: 1956-1964* (London: Jonathan Cape, 1984), pp. 165, 283.
80. Nehru, *Speeches: Volume Three*, pp. 101-103.
81. Cited in Brecher, p. 532.
82. Cited in Sachs, p. 129.
83. *Second Five Year Plan*, p. 23.
84. I was persuaded to summarize thus by the essay by Julius Silverman, "The Ultimate Objective of Nehru's Socialism," in Ashok V. Bhuleshkar (ed.), *Towards Socialist Transformation of Indian Economy* (Bombay: Popular Prakashan, 1972) and V.K.R.V. Rao and P.C. Joshi, "Some Fundamental Aspects of Socialist Transformation in India," *Man & Development*, IV, no. 2 (1982), pp. 9-22.
85. Kurien, pp. 7, 19; and V.V. Bhatt, *Development Perspectives* (Oxford, U.K.: Pergamon Press, 1980), p. 58.
86. Werner Baer, Richard Newfarmer and Thomas Trebat, "On State Capitalism in Brazil: Some New Issues and Questions," *Inter-American Economic Affairs*, XXX, no. 1 (Summer 1976), pp. 69-91; and Werner Baer, "The Role of Government Enterprises in Latin America's Industrialization," in David T. Geithman (ed.), *Fiscal Policies for Industrialization and Development in Latin America* (Gainesville, Florida: University Presses of Florida, 1974), pp. 263-281.
87. Alexander Gerschenkron, *Economic Backwardness in Historical Perspective* (Cambridge, Mass.: Belknap Press, 1966).
88. Pranab Bardhan, *The Political Economy of Development in India* (Oxford, UK: Basil Blackwell, 1984).
89. S. Chakravarty, "Nehru and the Public Sector," *Eastern Economist*, vol. 75, no. 22 (November 28, 1980), pp. 1182-85.
90. James O'Connor, *The Fiscal Crisis of the State* (New York: St. Martin's Press, 1973), p. 6.
91. Sobhan and Ahmed, pp. 8-9.
92. Ashok V. Desai, "Factors Underlying the Slow Growth of Indian Economy," *Economic and Political Weekly*, vol. XVI, nos. 10-12 (March 1981), pp. 381-392.

Chapter V

Nehru's Socialism and Group Interests

No matter what particular ideology may motivate decision-makers in the determination of their policies, those policies are likely to favour the interests of some groups as against others. An important question for investigation then ought to be as to how given policies have affected different interests. However, that question lies outside the scope of the present study. But there is a prior question specifically in relation to the Nehru model which is equally of importance. It concerns the perception and attitudes of groups and classes in regard to the policies that constituted the Nehru model when they were proposed and enacted, regardless of the subsequent consequences.

Three large classes or groups seem relevant to any investigation of perception and posture in relation to Nehru's socialist model: (1) the upper classes comprising the bourgeoisie and landed aristocracy; (2) the intermediate strata consisting of the urban middle class (the professionals, the salaried group, and small businessmen and traders or petty bourgeoisie) and the rural rich and middle peasantry (or kulaks); and (3) the lower classes composed of industrial and agricultural labour and the poor peasantry. Given their strategic position in the political system, the intermediate strata are obviously of key importance.

On the other hand, among the upper classes, the landed aristocracy is not of particular importance. It was, of course, the target of agricultural reform, even if not a very radical one, but it was otherwise little affected by Nehru's socialist pattern. That pattern affected more directly the industrial and commercial classes, and consequently their position is of central concern.

As for the lower classes, a large bulk of them -- the poor peasantry, agricultural labour, and workers in the unorganized or informal urban sector -- remained relatively unmobilized during the 1950s and therefore were not part of the relevant political strata that had to be taken into account in decision-making. Regardless of how their interests were likely to be affected by the Congress model of "mixed economy", they had little to say about economic policy. To the extent these classes were involved in the political process, their support could be obtained by either traditional ties of loyalty to the higher classes or by the slogan of socialism, allegedly intended for the advance of the masses. On the other hand, organized labour in the corporate sector and public sector was in a different situation. Kalecki himself took it to be a part of the intermediate strata, while Myrdal referred to "the urban workers in large-scale organized industry, but the latter constitute a privileged middle class in a country where over 80 per cent of the people live in villages and even the majority of the poorer urban dwellers are outside the organized sector."[1] The spokesman for one component of this working class has been the Indian National Trade Union Congress, which was founded by the Congress party and has been sympathetic to it. Its position requires no special consideration since it basically supported the Nehru model. The other major component of the working class has been organized into the All India Trade Union Congress, a front organization of the Communist Party of India (CPI). Besides, the CPI has held itself to be the historical revolutionary agent of the proletariat. The position of CPI and its front organization on the Nehru model is of particular interest since it provides insight into the perspective of ideologically revolutionary groups. In summary, then, this chapter looks at (1) the relationship of the intermediate strata to the Nehru model; (2) the posture of business; and (3) the reaction of the Communist party. The central argument of the chapter is that, even though his project for an "instrumental-socialist" mixed economy was rooted in ideology, (1) Nehru built a wide-ranging consensus among the intermediate strata, centered around "countervailing power" against the capitalist class; (2) the Communist Party of India was, despite its opposition to the allegedly capitalist state, part of this consensus, and (3) business was, contrary to most versions of the radical argument, opposed to the project.

1. Intermediate Strata, Ruling Group and Countervailing Power

The critical importance of the intermediate strata in the political system is obvious from their dominance of most political parties, especially during the 1950s. However, since the intermediate strata are composed of rather varied and disparate elements, it is difficult to attribute inherently common interests to them. But they can be taken to share a common interest in avoiding domination by the business class and therefore of favouring a dominant and rapidly growing public sector run by a state which they dominated. Interestingly, India's foreign policy of non-alignment is indicative of the way that a state dominated by the intermediate strata tended to strengthen its own position as against big business above and the lower classes below; an alignment with the United States would have strengthened big business, while alignment with the Soviet bloc may have bolstered the working class.

Among the intermediate strata, it would seem that the urban middle class had the greatest to gain from a dominant public sector and indeed a wider socialism, for not only would these provide it a source of economic power independent of the business class but also greater economic opportunities of employment. This would be especially so since Nehru's "mixed economy" model was based on execution by bureaucrats and technocrats rather than by a party of mass mobilization. Politically, however, the urban middle class was not organized in a single party but was fragmented among several parties. Still, during the Nehru era, at least until the end of the 1950s, there prevailed on the one hand the overwhelming dominance of the Congress party in the political system and, on the other, a considerable consensus among the important political parties around the Nehru model. Differences pertained to the degree and pace of change rather than the basic structure of the model. Indeed, the consensus constituted a challenge for some opposition parties, such as the Praja Socialist party and even the Communist party, as to how to support the consensus and at the same time maintain an identity distinct from the Congress party.

The central coalition partner of the urban middle class was the rich and middle peasantry. While apparently willing to go along with the notion of a dominant public sector, this social segment would have nothing to do with any element of the socialist project that would involve the restructuring of rural society beyond the initial reduction in power of the landed aristocracy, of which it was the chief beneficiary. Politically powerful in state governments and represented in the Congress caucus, it was able to scuttle effectively any plans to have either cooperative farming or land redistribution. Aside from that, however, it was willing to let the government under Nehru proceed with his plans for industrialization and the public sector. Subsequently, after Nehru, it could press the government, because of its enormous electoral power, to provide subsidized inputs and support prices for agricultural output. A large part of the rich and middle peasantry became disaffected with the Congress party after Nehru, in spite of having been a major beneficiary of its economic programmes, and segments of it even opposed the public sector. But during the Nehru era the peasantry can be said to have concentrated more on its narrower economic interests, which could be effectively pursued at the state level, and gone along with the "mixed economy" model. Indeed, one could say that there existed a tacit compact between the urban middle class dominant at the centre under Nehru and the rich and middle peasantry dominant or ascendant in the states to allow each to pursue its project or interests rather autonomously within its sphere without interference from the other.[2]

Insofar as small-scale business is concerned, it can be differentiated as between traders and industrialists. While the former would have nothing to do with state trading, because their very livelihood would be affected, the latter may have been more forthcoming in support of the public sector because of the opportunities that would thus be made available to them through limiting the economic power of large-scale industry. As Shirokov says: "Since the small and even the middle entrepreneurs were powerless to develop heavy industry on their own, they welcomed the increased involvement of the government in this field as a counterbalance to the activities of national and foreign monopolies. It was probably pressure from the lower ranks of the national bourgeoisie that was responsible for the state-capitalist measures in the mid-1950s."[3] No evidence is offered by Shirokov on the pressure that he refers to as a probability, but the small-scale entrepreneurs could not have been opposed to the particular "mixed economy" model since the public sector did not touch their interests directly while the model incorporated special measures to promote the small-scale sector; besides, they may have also felt that subsidized inputs would be made available to them by large-scale industry in the public sector, under the notion of social gain, than by the private sector.

Beyond social strata and classes, there is the question of the interests of the ruling group of political leaders in the Congress party in supporting the socialist model. To be sure, Nehru personally favoured the model because of his long-term ideological commitment to it. The same may have been true of some of his more ideologically committed supporters. But neither he nor they, and certainly not others, could be considered to have been immune to political considerations in expanding the entrepreneurial role of the state and limiting that of the capitalist class. The fundamental fact is that in the context of colonial rule the new middle class initiated and dominated the nationalist movement, to which other classes later rallied. Having moved into positions of power vacated by the colonial rulers, the middle-class nationalist leadership, which derived its power from sources independent of the capitalist class, could not but look askance at the possibility of being subordinated to the capitalist class if that class were to expand its economic power. Put simply, the new ruling group at independence found itself in possession of political power but lacked economic power while the capitalist class had economic power within limits but no political power. In this situation of incongruence between economic power and political power, the ruling group confronted a stark choice: either develop its own autonomous economic power or see the capitalist class emerge into political power. The ruling group apparently decided to develop its own autonomous economic power while limiting the economic power of the capitalist class. In this fashion, it attempted to bring a correspondence between economic power and political power. The choice may perhaps have been less productive from the viewpoint of economic growth, but the aim was clearly political. Hanson underlines how considerations of political calculation may not have been removed from ideology:

> Ideology can, of course, also act as a screen for power calculations which have nothing to do with economic development and which may be inimical to it. When, for instance, a ruling group has achieved political monopoly, it may well use 'socialist' criteria to justify a balance between public and private enterprise which is really intended to serve no other end than the consolidation of that monopoly. The essence of such a policy is *the progressive elimination of such rival (or potentially rival) power centres* as remain. One of these may be identified in the private sector of the economy, which consequently has to be decimated by wholesale nationalisation.[4]

Hanson here, of course, is talking in theoretical terms in general, rather than specifically in relation to India. But Venkatasubbiah, the official historian of the pre-eminent peak business association, would seem to be thoroughly in agreement; pointing to business opposition to the public sector on grounds of economic efficiency, he underlines the fundamental point that business patently misunderstood the motivation of the political leadership:

> But the Federation, whether by accident or design, overlooked the springs of the Government's thinking. To the Government and the planners economy and efficiency in the use of capital and other factors of production

were only desirable objectives. They were not determinants of policy. The determinant was the building up of a parallel state sector of manufacturing and service industries as *a countervailing economic power to that of private enterprise.* The latter's power was perhaps illusory to a large extent but the Congress Party believed it was real. If economy and efficiency were the sole criteria, the state could lay claim to very little investible resources; whereas the objective, whether or not explicitly stated, was eventually to dominate total investment in the economy directly -- its so-called commanding heights. It must have been obvious to anybody on the eve of the Third Plan that *Indian planning was no mere economic planning. It was also political and social planning.*

Again, he states:

A strict demarcation of industry as between state and private enterprise would mean that the Government would rather forgo some economic growth -- which could be substantial -- than get it through the private sector. *The basic difference in approach between the two centred on economic power.* Private industry looked upon its development simply as part of national development. The Government looked upon it as increasing economic power in private hands, not merely as increasing private profit. And this was believed not to be in the social interest.[5]

These are, of course, matters where motivation is attributed but for which there is little evidence; not only that, there is little likelihood for evidence to be forthcoming because it would tarnish the more noble image that is conveyed by reference to ideology. However, Nehru was highly sensitive to issues of power and, by extension, to the necessity of countervailing power. When attacked for excluding the private sector from basic industries, he sharply remarked that, while there were some basic industries already in the private sector, that sector wanted more because "not only they might prove to be very profitable but because it gives them economic power." For his part, Nehru found it "highly objectionable that economic power should be in the hands of a small group of persons, however able or good they might be," and he was emphatic that "such a thing must be prevented."[6] It is instructive that Nehru did *not* say to the business community in respect of basic industries: "Look, here is a field which is not profitable for business, you should therefore let the state carry the burden on your behalf." Rather, he said that this is a field which is apt to add to the economic power of business, and precisely for that reason he meant to see that business stayed deprived of it. Interestingly, three years after Nehru's death, the Congress party in its 1967 election manifesto made explicit the political intent underlying its economic programme:

In an economically under-developed society like ours, the very structure of political power and its interlinking with command over economic resources make it necessary that the commanding heights of the economy shall not be in private hands. For, they who hold the levers of economic power will also ultimately run the political apparatus. The free exercise of the democratic process demands therefore the intervention of the state in the running of the economy of the country.[7]

One may take the motive of countervailing power as the exclusive cause for the thrust toward a dominant public sector rather than that sector flowing out of action intended for other purposes. Or, one may consider countervailing power as only a subsidiary motive and thus simply a reinforcing element in the larger design for the public sector. Regardless, it is obvious that the intent in either case is no less anti-capitalist in terms of drastically restricting the sphere of operation of private entrepreneurs -- at least the larger ones, who are precisely the powerful ones and thus presumed to constitute the ruling class -- than it would have been if emergent out of ideology as articulated by Nehru. Nor could the political consequences, apart from shrinking the arena for private accumulation, be less meaningful to the private sector, for -- if some Marxist scholars like Poulantzas are correct -- while willing to make economic concessions to the working class, the capitalist class is determined not to concede on sharing state power with others.[8] And here we have a political ruling group, emerging out of an anti-colonial movement, determined precisely not only to control state power but to have an exclusive monopoly on it, and equally determined to see that the capitalist class does not assume control of the state even though it may be allowed to perform certain economic functions within the confines established by the state.

Though anti-capitalist in its determination to limit both the economic and political power of the capitalist class, the political ruling group -- drawn from the intermediate strata -- was not eager for a social revolution in behalf of the lower classes, which would only have served to undercut its own power. Indeed, the limits on the economic and political power of the capitalist class could be rationalized to that class precisely in terms of avoiding a revolutionary upheaval, even while proclaiming them as assuring a smoother transition to a socialist society. That would seem to be the implication of Hanson's comment -- who though would have rather preferred a revolution as "a consequence of growth than one which is a consequence of stagnation", which he saw as the likely result of a growth-constraining public sector:

> we have to remember that 'socialism' has widespread support from the politically more articulate groups and that the private industrialist, generally an unpopular figure, rarely behaves in a way likely to improve his social standing. 'Socialism,' therefore, besides being a morally respectable ideal, has its role to play as a slogan. Perhaps an even more important consideration is the danger, of which at least the more socialistically-minded members of the government are constantly aware, of bringing into existence a private enterprise Frankenstein which will use its economic power to dominate the public authorities in such a way as to make the eventual realization of socialist purposes impossible, except at the cost of revolu-

tion.... As for the Frankenstein-risk, this must be squarely faced. It is possible, although not inevitable, that as a result of a type of economic development where the criterion of choice between public and private is simply rate of growth, the 'capitalists' will come to dominate society, which will become set in a grossly inegalitarian mould, characterized by universal materialism and self-seeking.[9]

Apparently, for the same reason, Nehru appealed to the self-interest of the capitalist class in justifying the national project for a dominant public sector, warning that there would be no private sector left if the public sector were prevented from coming into existence.[10] But the calculations of the ruling group in supporting the Nehru model undoubtedly encompassed as well the bringing under their control of vast opportunities for patronage and mobilization of funds for their own benefit and that of their supporters.

If the ruling group could be said to have had reasons of self-interest in rallying behind Nehru's project for a socialist society, the same may be said of the bureaucracy. After all, the controls placed on the private sector would enable it to exercise enormous authoirty over the economic future of particular entrepreneurs, if not the entrepreneurial class as a whole. Used as it was to administering economic controls during World War II, not only could it not be averse to such controls but would have positively endorsed them. Besides, the erection, commissioning and running of vast economic enterprises would add immensely to its power. At that time, with Nehru's political power at its zenith and the political system with an enormous store of legitimacy, it would not have mattered much even if the bureaucracy had not been favourably disposed towards the project, but it was perhaps important to Nehru that one more centre of power was not opposed to it.

Indeed, it could be said that it was Nehru's particular genius to suggest a course of economic policy around which the widest possible consensus could be developed among social classes and institutional groups. Such a consensus could also be said to encompass even the capitalist class, for while the policy excluded -- for reasons of ideology and perhaps power -- certain parts of the economic arena from that class it did not only, by and large, not hurt its existing economic position but allowed it to expand considerably within the restricted area assigned to it. As Maddison expressed it:

> For him this industrial strategy was a surrogate for social change, although it did nothing to promote equality or economic growth. It won support from the bureaucratic establishment because it added to their power, it was supported by politicians because it increased their patronage, it met no opposition from established industry because it did not interfere with vested interests and it was supported by intellectuals who generally identified capitalism with colonialism, and who assumed that government control would contribute to social justice. It aroused no opposition because it conflicted with no vested interests.

Karunakaran would seem to agree: "Nehru had some understanding of the power structure in the country...he took all these centres of power with him. [11]

In investigating the consensus around Nehru's economic policy on the basis of calculation of interest on the part of different social classes and groups, attention needs to be drawn to the limits of mere intellectual analysis. The lack of hard evidence is an obvious shortcoming. No less so are the imponderables of the historical process. The pushing through by Nehru of his socialist model in the 1950s invests it with an aura of inevitability and consequently persuades the student to search for the configuration of interests that made it possible, but primarily through attributing interests and motivations on grounds of logic.

However, if one concedes that there was no necessary inevitability about Nehru's emergence to political supremacy within the Congress party one could envision, in a counter-factual alternative, that another leadership -- headed by Patel, Tandon or some other conservative leader or set of leaders -- could have become either dominant in the party and government or could have at least checkmated Nehru. That may have solidified into a more permanent framework the economic pattern that had prevailed until around 1955 under both the Industrial Policy Resolution of 1948 and the First Five Year Plan. In that case, it could have been as easily argued that, of course, such a state of affairs was but the natural outcome of the patent domination of the Congress party by the bourgeoisie. In this argument, the political ruling group would simply have been taken to be an agent of the bourgeoisie, performing according to instructions. Further, it would have been asserted that nothing different than selling out to the bourgeoisie could have been expected from the double-faced, unstable and unreliable middle class, petty bourgeoisie and the rich and middle peasantry. And if the working class went along with it, it would have been accused of being bought off as a labour aristocracy through economism and thus misled from its true interests.

An analysis along these lines is not so fanciful as it may seem; such interpretations were put forward by Marxists prior to the mid-1950s. Indeed, shades of it, particularly in relation to the domination of the state by the bourgeoisie, have continued to feature in the programmes of one or another Marxist group, notwithstanding the pushing forward of a dominant public sector. That only highlights the centrality of considering the reaction of the capitalist class and of the Communist party during the 1950s to Nehru's plans for a socialist pattern of society.

2. Business and the Socialist Pattern

A general argument exists in the literature that the public sector serves the capitalist class by providing externalities, and therefore that class is not only not opposed to the public sector but requires and demands it.[12] Specifically, in relation to South Asia, Myrdal has argued:

> the interests of private business do not normally conflict with expansion of the public sector of big industry. Government investment is meant to be

concentrated in heavy industries where little private initiative is forthcoming. To the extent that these investments create external economies or provide goods that would otherwise be scarce owing to the strained foreign exchange situation, there should be, on the contrary, a harmony of interest. Much construction will also be done by private contractors and the purchase of various supplies will increase demand in private industry. More important, the investment and pricing policies pursued by public enterprises are usually such that, by holding down prices, they swell the profits of the private sector. Thus, when put into practice, the vaguely socialist notion that public enterprises must render services at low prices in fact boosts considerably the returns on private capital. Instead of being used to supplement government revenue and help to mop up purchasing power, the public sector functions to inflate private profit.

The phrase "public *versus* private" is, therefore, misleading. It conveys an impression of competitiveness between the public and the private sectors whereas the two are, in fact, mostly complementary. Successful operation of the public sector would normally increase opportunities for private enterprises.

Given such a perspective on the alleged harmony and complementarity of interests, it would be natural for business to support the maintenance and expansion of the public sector, especially in relation to basic industries. However, "successful operation" is the decisive reservation. But Myrdal also enters a note of ambiguity with another qualification: "Only in one sense is there real competition: in regard to funds and, in particular, foreign exchange. The enlargement of the public sector must to some degree decrease the investment opportunities for private entrepreneurs. Yet even this is not an invariable rule....But undoubtedly it is this possible competition for available funds that causes private Indian business organizations sometimes to demand less investment in heavy industry and greater consideration for the consumption industries, which belong more naturally to the private sector."[13]

Others are reluctant to make even this concession. The Communist Party of India (Marxist) asserts that the bourgeoisie lacked the technical and financial resources to build capitalism and consequently used state power for the purpose.[14] Similarly, even a spokesman for the Communist Party of India, which otherwise has a different position, maintains that the public sector has been "used to build capitalism in India. The public sector has acted as a catalyst and helped the growth of capitalism and monopolies in this country. It has provided infrastructure, cheap raw materials, equipment and financial resources to the private sector and thus led to the creation of favourable conditions for the emergence of monopoly houses in this country."[15] Again, though his position is otherwise more complex and nuanced, the eminent Communist leader Namboodiripad averred "that, though our ruling classes claim to be building a socialist society, they are, in reality, helping private capitalism to grow and expand at a very rapid rate." In like manner, political scientist Karunakaran states: "Nehru created a large public sector which satisfied the leftists and other nationalists but the public sector was so organized as to serve largely the independent big business. This satisfied the capitalists. The rate of growth of big business during Nehru's time was faster than that of any other group in any other country for a comparable period." And the Trotskyite scholar Harris maintains, "within the organs of the State, the national bourgeoisie was able to protect and enhance its position, limiting the State to those industrial activities directly of need to private capital."[16]

All this reflects a rather strong position on the purposes of the public sector. However, a different case can be made though it is not often made. As a first step in making this case, one could distinguish between (1) infrastructure, both physical and financial; and (2) basic industries. Many would concede that the economic infrastructure has promotive implications for the building of capitalism because of the externalities it provides. Others would maintain the same, perhaps less vigorously, for basic industries, too. However, even in the case of infrastructure, should it be assumed that the public sector is somehow a natural order of things for providing externalities? A negative response can be made not only by taking different countries of the world into account but also by looking at India itself. Before the thrust for public sector was inaugurated in the mid-1950s, the private sector was prominent in the field, for example, of electricity generation and supply. Even today, private enterprise owns and manages the Calcutta Electric Supply Company in Calcutta and the Tata Electric Company in Bombay, and what is more it does so efficiently and profitably. If private enterprise can do this in respect of electric supply in this restricted sphere it can do so also on a larger scale, and it can do it in other areas as well. Equally, in relation to basic industries, the Tatas built and continue to run a modern steel plant, and do so again efficiently and profitably; its products serve as inputs for the rest of the economy no less or no more than do the products of the public sector steel plants. It is misleading therefore to suggest that the public sector is somehow essential for providing private enterprise with externalities and subsidised inputs.

Furthermore, the notion that the government is again somehow especially capable of aggregating large amounts of capital to provide an expanded thrust for economic growth through the public sector rests on the prior assumption of socialization of the community's savings by way of heavy taxation and monopoly in insurance and banking. These savings could be as effectively utilized -- and certainly so in retrospect -- by private enterprise as by government, if the government were oriented to encouraging private enterprise. Besides, it is a misconception that market constraints can somehow be miraculously overcome by government intervention. Often, market failures are simply transformed into government failures. In the final analysis, the case that the public sector was in the interest of business and that business somehow demanded it rests on assertion rather than on evidence.

A contrary argument on why the private sector would oppose the expansion of the public sector has been made effectively by Hanson:

A rapidly developing public sector involves the denial of certain fields to private capital -- fields which it would *eventually* wish to occupy; and because the occupation of these by the state can be effected only through the mobilisation of resources which would otherwise remain in private hands, opportunities for both private capital accumulation and private consumption have to be limited, via taxation and controls. Such restrictions are rarely hailed with joy by the upper classes.[17]

But, supposing one concedes that the public sector was in the interest of business and that business demanded it, then one should expect business to support it or even remain quiet about it. If business should do that, then the radical hypothesis on the public sector as a function of business interest and demand would be confirmed. The requirement here, of course, is that evidence should not be based on some odd statement of some odd businessman, for from the total universe of businessmen it would be easy to find a statement here or there in support of the hypothesis. A more rigorous investigation of the problem is, however, somewhat eased by the fact that there has existed for over sixty years now a peak business association, the Federation of Indian Chambers of Commerce and Industry (FICCI), whose position on various questions touching business interests should be taken as authoritative; the other peak association, the Associated Chambers of Commerce, was for much of its history tied up with foreign capital in India though there is not much to distinguish it now from FICCI. The stand of FICCI on various issues relating to the public sector should be decisive, though it would need to be supplemented by reference to two other organizations in order to fully appreciate the position of business -- one to build public opinion favourable to business interests (Forum for Free Enterprise) and the other to pursue business interests directly in the electoral arena with the hope of capturing state power (the Swatantra Party).

FICCI and the Public Sector

FICCI had been an ally of the nationalist movement. However, the relationship between the two underwent a change after independence. The political leadership of the nationalist party came to occupy the seats of power vacated by the departing colonial rulers, while FICCI was, as before, cast in the position of a supplicant before authority though initially, while Patel was alive, it had access within the cabinet. Indicative of the distance that developed between the two after independence was the attitude of the government on the question of foreign capital. Both business and nationalist movement had before independence opposed foreign capital but when business protested the government's welcome to foreign capital in 1949, the government responded: "we can't wait forever for Indian capital to come out."[18]

Leaving aside questions of substance, FICCI was soon disenchanted with the value that government placed on consultation with it. FICCI found that while, no doubt, the government consulted with it, or perhaps went through the motions of consultation, "Government's final decisions have no relevance to discussions which take place and the advice that is tendered."[19] In its view,[20] the government also tended to confront business with *fait accomplis*. What business, however, wanted was not simply consultation but that somehow its advice on matters that concerned it should be binding on government. In other words, business wanted the government to act precisely as a committee of the bourgeoisie that Marx envisaged. Government, on the other hand, also entered into consultations with labour and other concerned groups, and made up its own mind. Kochanek, in his thorough study of business and politics in India, maintains that government consulted with business on a non-committal basis, that it was not deflected from its course where policy principles were involved, and that business influence on government tended to be more individual than collective.[21] The anguished sense of helplessness, while government pursued its own course, was manifest in the words of an eminent businessman in 1955: "Can nothing be done in the matter."[22] In some sense, the business complaint was of long standing. As early as 1931 the then president of FICCI, Shri Ram, had protested on the occasion of the Karachi resolution that the Congress party made unilateral decisions on economic matters without consultation with business.[23]

Angry as business might have been over the government posture toward it, there was not much it could do about it, since government derived its power not from business but from the electorate. Fundamental to the weakness of business in relation to government was the fact that business functioned in an environment that was not only not favourable to business but hostile to it; there was widespread public contempt for the business community for the failings of its members. What is more important is that business was aware of its own vulnerability in this respect. At several annual meetings of FICCI, members acknowledged the existence of the public perception of business and, in desperation, asked if there was not something that the organization could do about it. At the 1956 session, one business representative stated: "We have no sympathetic consumer, no sympathetic Government, our labour is not sympathetic to us. We start friendless and end friendless.... We are living in hard times. We cannot ignore the writing on the wall. We must face facts. The country is moving towards a socialistic pattern of society. Even if you and I want to check it, it cannot be held or checked; it will go on." A year later, another businessman, expressing his dismay at private enterprise being "looked upon with contempt, hatred and suspicion in our country," similarly commented: "It is therefore no surprise that no political party in India ventured to advocate and support our cause in the general elections. As things are now moving, we cannot deny that our days appear to be numbered and our system is on its way out in India. And, if the end comes, I am afraid it may go unwept, unsung and unhonoured. Whatever may be our own contention, we have failed to win the confidence of the people."[24]

More substantively, three issues in relation to government were critical to business: (1) controls over business and economy; (2) nationalization; and (3) sectoral demarcation of economic activity as between business and government. On these issues, three periods can be distinguished in respect of the business posture toward government and its economic policy during the Nehru era. First, there is the period from 1947 to 1953 when, despite a critical stance, business was fairly comfortable with government policy which reflected a stalemate between the right and left wings and therefore a perpetuation of the status quo. Second, the period from 1954 to 1962, when FICCI, and certainly the business community as a whole, was involved in a two-pronged posture of protest against government policy and simultaneously resignation to its assigned fate. Finally, starting from 1963 business adopted a more activist posture of protest against government policy and displayed little reluctance at being in conflict with government.[25]

(a) 1947-1953: Even before independence arrived, business had become "a little nervous" at the certainty of Nehru becoming prime minister, for "it knew Nehru's convictions." Then it was struck a heavy blow with the budget by Finance Minister Liaquat Ali Khan, who took advantage of Nehru's weakness for socialism to inflict heavy punishment on the business class which was largely Hindu and therefore fair game for the chief representative of the Muslim League in the government. Business also found other policies of the interim government to be restrictive. With that background, Nehru's address to the annual meeting of FICCI in 1947 could provide comfort to business only in negative terms; business felt reassured that "we won't leave this hall with the feeling, which we entertained up to this morning, that the industrial future of India was doomed." At the same session, business left no doubt as to what its preferred model of economic organization was; it held forth in a resolution that private enterprise was the "best and speediest means" for the country's industrial development, and accordingly wanted it encouraged in every way. It was willing to concede on the regulation of private enterprise by the state in the "larger national economic interest", and even to accept basic industries in the public sector but only if private enterprise was unwilling to undertake them.[26] Thus, at this early stage, just prior to independence, business was in no way asking for its own exclusion from any sector; nationalization for it was simply contingent, not on principle, but on failure of private enterprise.

The turmoil of partition and uncertainty over the future economic policy of the Indian government then threw business into a state of crisis. But business fears about government policy apparently ended, temporarily it would seem in retrospect, with the Industrial Policy Resolution of 1948; business, according to FICCI's official historian, "reacted guardedly."[27] It correctly perceived that the resolution certified the status quo.

A major issue over which the business community became highly agitated soon after independence, and has remained to be so exercised ever since, pertained to economic controls by the state. In early March 1951, when it wrote the government on its own initiative on the Colombo Plan, FICCI opposed the "pervasive controls," holding them to have been "a serious drag on production and have been helpful neither to the consumers nor to the country at large." The Federation warned that "regimentation of the economic life is not going to work in this country and the sooner it is realised the better for all concerned. The Committee are convinced that planning and freedom of enterprise can go together and should go together." Later that month, the annual session passed a resolution attacking controls for their failure in increasing production, their resulting in administrative abuse and corruption, and their causing dissatisfaction everywhere; it asked for their removal in favour of "more free play for the laws of supply and demand," and asked that government "declare in unequivocal terms that their ultimate aim is to remove all controls." The mover of the resolution held that, despite the economic deterioration that had arisen as a result, "Government seems to be wedded to the policy of controls more on ideological grounds than on factors that justify the practical necessity for the same."[28]

In another resolution at the same session, the FICCI let it be known what kind of economic planning and policy it deemed desirable -- not one that sponsored state enterprise but one "mainly directed to assist and encourage industrial development by private enterprise, inasmuch as such development will be the cheapest and best and will also ensure early improvement in the standard of living of the people." Toward that end, it asked for provision of tax relief as an incentive, revision of labour legislation on a realistic basis, and removal of controls.[29]

First Five Year Plan: When the government initiated economic planning, FICCI's opposition to controls and an extensive public sector was carried over into its critique of specific plans. In relation to the Draft Outline of the First Five Year Plan, a FICCI memorandum proclaimed its commitment to the mixed economy and planning and even criticized the lack of ambition on the part of planners in setting targets. But it charged that "in their doctrinaire adherence to the efficacy of controls and with scant consideration to the problems of incentive in an economy they have proceeded to recommend policies which make a mockery of the mixed economy." In the FICCI view, the Planning Commission's version of mixed economy was "a very detailed and comprehensive system of controls." The Federation opposed the Industries (Development and Regulation) Bill, for it gave the government powers in respect of industry that were "sweeping and obnoxious in character" for which there was little justification. Equally, it opposed the Commission's opinions on state trading. It was critical of government for being primarily concerned with the public sector, and sharply reminded it to address "the basic issue as to what Government should do to create a suitable atmosphere for investment in industries." Indeed, if government were to do that, FICCI remarked, it was bound to "reorientate their whole approach to the whole question of industrial development vis-a-vis private enterprise." FICCI was distressed that the result of the recommendations of the planners would be "to regulate and stifle the ability, initiative and enterprise at every turn."[30]

(b) *1954-1962:* Although the Federation made its preferences clear, the issue of public versus private sector was not really very critical at the time of the First Five Year Plan, for that Plan was modest in both size and thrust. Business concern centered more on controls over existing commerce and industry -- and this concern would intensify as time went along -- rather than any constriction of industrial space in the future. However, this situation was not to last too long, for planning would result in "a virtual cold war" between business and government over sectoral division.[31] The results of the 1951-52 elections, no doubt, assured political stability and must have come as a relief to business, but the consequent consolidation of political power by Nehru in both government and party could not but have been unwelcome to it, at least in retrospect, for that would enable Nehru to shift economic policy according to his ideological commitments. With the declaration by Nehru in 1954 to launch India on a socialist course there developed a major divergence between business and government. The year 1954 constitutes a decisive divide, initiating a new period in their mutual relationship. But it is in 1956 that the full implications of the new course became manifest for business.

The nationalisation of life insurance in January 1956 came as a tremendous shock to the business community and caused extreme bitterness as no other policy of the government had by then.[32] Even though business was equally opposed to state entry into trading, at least that did not involve ouster of the private sector from trading. But in insurance, the government had suddenly transformed a major service developed successfully by the private sector into a state monopoly, ostensibly for removing malpractices but really to place larger financial resources within its own easy reach. The government had evidently brushed aside the FICCI plea that "there did not exist any ground for nationalising insurance as it was unlikely that such nationalization would in any way serve public interest." Business was simply left to deplore government's action but otherwise remained helpless to do anything about it. The very first substantive resolution of the FICCI session following nationalization recorded "its disappointment and dissatisfaction"; it pointed out that "almost overnight the final decision of Government was conveyed to the public as a *fait accompli* without previous consultation or discussion with the interests concerned" and that the government had failed to provide a "strong and convincing" case for nationalisation. The resolution's mover, incoming president Lakshmipat Singhania, stated that the government's motivation was "not to leave economic power in the hands of private enterprise, or also [for] political reasons."[33]

The business community was no less in uproar against state trading. Here, too, FICCI made a case against state trading on the grounds that it will lead to the dislocation of small traders who worked on very low profit margins and that "competition, not monopoly, was the current way of safeguarding the interests of the consumers and the general socio-economic objectives." But the government went ahead to establish in May 1956 the State Trading Corporation and progressively gave it exclusive responsibility for foreign trade in several important commodities, such as manganese, iron ore and cement. Nor was the government bothered by business opposition to the reform of company law and amendment of the constitution facilitating state acquisition of private property.[34]

Again, when the government announced the Industrial Policy Resolution in 1956, the business community was simply left to react to an accomplished fact. It is curious though that scholars differ as to what the business reaction was. Kochanek states that business "welcomed" the resolution because it clearly demarcated the sphere allocated to the private sector and removed the threat of nationalisation. Hanson takes the position that the resolution satisfied no one, and that the business community was "as critical of its radicalism as the

Socialists were of its moderation." And Brecher apparently believes it satisfied everyone, for the Left "could take comfort from the shift towards some form of socialism. Businessmen could find compensation in the loopholes."[35] The truth of the matter is that business had no alternative but to accept it. FICCI expressed agreement only with the stated objectives of rapid economic growth and industrialization and, while it saw some flexibility in the resolution, it correctly concluded that "it is intended primarily to operate in favour of the public sector in that the State may enter any industry or trade it chooses". In the circumstance, FICCI could only plead for a modification in government's approach so as to ensure the expansion of both public and private sectors "without prejudice to one another." It particularly opposed the state "displacing normal trade channels." It was emphatic, however, about what it saw to be the proper place for the public sector -- certainly not in industry, rather "the energies of Government must be devoted mainly to improving and augmenting social capital" and to "nation-building activities," while "there are broad sectors of any economy in which private enterprise is the most effective agent in furthering economic development". It did not feel assured on nationalization and wanted a specific guarantee to that effect. Nor was it in favour of the state having equity capital in private enterprises, for it would amount to direct control.[36] There is nothing in the record to suggest that FICCI welcomed -- and it would be preposterous to so expect -- the exclusion of the private sector from a broad industrial front which had been reserved for the public sector; it was willing to accept an active role for the state in building economic infrastructure but that is as far as it would go.

Second Five Year Plan : The controversy over the respective roles of the private and public sectors was carried over into the debate over the Second Five Year Plan. Even before government's thinking was apparent, FICCI had on its own initiative published an ambitious draft plan intended to restructure the economy through rapid industrialization, with emphasis on heavy industry; its aim was autarky and self-sufficiency. To that extent, FICCI's thinking was close to that of the government. However, once the government planners disclosed their intentions, FICCI was visibly upset at the implications of the Second Plan for the private sector; indeed, Venkatasubbiah believes the Second Plan marked the point of divergence between business and the planners.[37]

Interestingly, the Marxist scholar Sachs suggests that the Second Plan, by concentrating on infrastructure and steel, became "by and large acceptable both to the left-wing opinion and to important sectors of the bourgeoisie not excluding big business. The former saw in it a step towards national economic development, while the latter realized in particular, that the 'multiplier effect' of public investments would provide them with new possibilities and increased profits." Even though he cites G.D. Birla in support of the point, business reaction was not limited to that "realization", however. Sachs himself adds: "The bourgeoisie recognized, moreover, that the acceptance of the goals established in the Plan was a necessary concession to public will -- concession easier to accept, as it did not challenge the foundations of capitalist production."[38] None of this could be considered a tribute to business strength, and the lack of challenge to the foundations of capitalism was more a consequence of self-limitation on the part of the state, not of business muscle. Brecher more accurately assessed that the business community was "alarmed."[39]

In the first place, FICCI found the plan-frame document that preceded the Second Plan draft to be disturbing because of "differences of a basic character" in approaches to planning by government and business, and it took umbrage at "total comprehensive planning.". It questioned the compatibility of "total planning" with democracy, and underlined that "the dangers of centralized planning involving regimentation of the economy are great." It was dubious about comprehensive planning assuring rapid progress and equally about the capacity of the administration to bear the required burden. Instead, it demanded that "where there is a well-established private sector, it would be enough for the state to lay down only a few strategic targets." More specifically, FICCI's concern was centered on the plan-frame's near-exclusion of the private sector from large-scale modern industry through the strategy of reliance on cottage and small-scale industry for consumer goods and on the public sector for producer goods. It raised a profound point about the prospective use of the products of basic industry if no development was to be allowed of large-scale consumer goods industry. It also doubted the organizational capacity of government to mobilize small-scale and village industries, and predicted failure of government policy. Fot its part, FICCI advocated pushing forward with large-scale consumer goods industry, refused to be persuaded by any Gandhian sentiment on behalf of village industries, and called on history as witness for the inevitable elimination of village industries. It rejected the artificial propping up of village industries "*at the cost of* organized industry."

FICCI was dismayed at the diminished role given to the private sector. It was adamant that, first, the public sector did not have the capacity to absorb the accelerated pace of investment set for it and, second, that it would be less productive in terms of income and employment generation than a similar investment in the private sector. It challenged the assumption that the private sector did not have the resources for the kinds of projects envisaged in the public sector, for that was to take a static view of things and ignored the performance of the private sector over the First Plan. FICCI then presumed to lecture government on the appropriate approach:

> In fact, it is an unwarranted interpretation of the recently announced objective of a socialistic pattern of society which leads some to advocate a growing share for the public sector in the total resources of the economy. But this is by no means necessary. What is essential and deserves to be emphasized is the need to increase the total wealth of the community without which a socialistic pattern of society can have no meaning.

FICCI found the pre-eminent role given to the public sector in the plan-frame objectionable not merely for what it meant in terms of a restricted development of the private sector, but also for the implications it had for resource mobilization, making for "unrealistic increases in taxation and public borrowing." It warned the government that "a certain preoccupation with doctrinaire objectives may hinder the process of growth itself." FICCI advised the government on what should be its real concern:

> The proper spheres of activity for the public sector in the future are the formation and maintenance of Social Capital. In an underdeveloped country like India, the task of providing the basic requirements, such as power, or from a long term point of view, education, health, etc., as also the institutional framework within which private enterprise is to work, is in itself sufficient for any government administration to be fully occupied with.

At the end, FICCI asked the planners "to avoid a doctrinaire approach embodying the prejudices of a few and resulting in viewing the private sector with suspicion or antipathy."[40]

After the publication of the Draft Outline of the Second Plan by the government, FICCI submitted another memorandum in which it basically reiterated its earlier position, expressing satisfaction with some parts and disappointment with others. It asked that "an empirical rather than a rigid doctrinaire approach" be employed in economic planning. It was again critical of the expansion of the public sector at the cost of the private sector, holding that government's preference for the public sector was "ideological not economic." The Federation opposed state ownership of manufacturing enterprises, especially on the scale planned, from the perspective of both consumer interests and government capabilities. It commented: "From the point of view of broader social policy, it is not imperative for the public sector to own manufacturing units nor is it feasible in practice for the State to do so on the scale contemplated. Apart from delays associated with and inherent in Government administration, the question of cost to the consumer is also important. It is by no means proven that the state as an agency for the conduct of business is efficient in terms of results from the point of view of the community". FICCI underlined "the ideological character" of government's allocations by pointing to the coal sector. It asked government to confine itself, for the sake of rapid economic development, to infrastructure. It continued to oppose the "obscurantist" reliance on village and small-scale industries.[41]

The Second Five Year Plan, along with the Industrial Policy Resolution of 1956, marked a critical stage in the development of the state's entrepreneurial role in the Indian economy, for it concretized the implications of the pledge for a socialist society. FICCI reacted to the evolving programme, but what is striking, in the abundant documentation that is available on its reactions, is that there is absolutely no warrant for the position that it endorsed -- leave aside initiated or encouraged -- the expansion of the public sector into manufacturing industry in order to avail of some assumed externalities. Quite the contrary, at every step it asked the government to stay out of industry and to confine itself to providing good government and an adequate social and economic infrastructure. On this point, its difference with the government was fundamental. It is likely that even its endorsement of the state's role in power and transportation stemmed out of a recognition that no one was likely to listen to a contrary opinion. If FICCI's official position did not truly represent its real sentiment, for which there is no evidence, it is amazing that it could sustain such a monumental hoax over so many years, not only in numerous documents but also at its annual sessions in which hundreds of key representatives of business participated.

In the end, however, FICCI had to resign itself to the policies adopted by government; it could not shake the government out of them. It could influence their application in individual cases, but it could not shift the policies as such. To be sure, the government did not hurt existing business, though even that is not entirely correct, given the nationalisation of life insurance and state entry into trading. Furthermore, business expanded vigorously during the Second and Third Five Year Plans in the Nehru era, for state activity in the economic arena and the wall of economic protection thrown up around India by the government had an expansionary effect on the private sector. Some of the industrial houses saw rapid growth, a fact that evoked a significant political reaction in the post-Nehru era. Besides, within the larger framework of policies derived from Nehru's ideological commitments the government at times displayed some pragmatism and flexibility in recognition of having overextended itself, even though it did not do so always (as Hanson argues with considerable force in relation to "the triumph of ideology" in respect of coal and steel).[42] At least, the government was not vindictive toward business even while it would not swerve from its chosen path. When all is said and done, however, notwithstanding the expansion of private enterprise, in a larger sense the private sector's economic and political fate was being sealed insofar as its power relative to the public sector was concerned and equally in respect of government control over the private sector.

It is obvious that as the government launched on a change of course in its economic policy in the mid-1950s, there developed serious divergence in perspective between government and business on the desirable model for development and operation of the economy. It is equally obvious that the government was determined to follow its own course, basically untrammelled by business views; indeed, it constantly confronted business with *fait accomplis* on major policy issues. Given the manifest divergence in perspectives and the determination of the Nehru government to proceed with its own priorities, what course of action did business adopt in relation to government? Here, two major schools in the business community surfaced in the mid-1950s. One school, reflecting a minority opinion in practical terms, advocated the adoption of a more aggressive path of protest against the government. Identified with a group of Bombay businessmen, including the business house of the Tatas, the school's basic thrust stemmed from the perception that "influence is at work in New Delhi which has no faith in private enterprise and which may succeed in putting the economic clock back" and that "Government policies today are being dictated by a lack of trust in the business community."[43] The "Bombay school" wanted the adoption of a more activist approach to bring about change in public opinion on private enterprise, directly through business-organized campaigns, as a means to influence the government, if not to change the political configuration.

The other school was identified with FICCI more generally (which, in turn, in terms of business communities drew largely from Marwari and Gujarati businessmen) and with the house of the Birlas more particularly, but especially G.D. Birla, a disciple of Mahatma Gandhi. Reflecting majority opinion in the business community, this school was eager to avoid alienating the government and tended to go along with what the government decided. It did so not because it approved of government policy -- the record speaks to its strong disapproval -- but out of a spirit of resignation. It was deeply conscious of how that policy ran counter to business interests, but it was persuaded to reconcile itself to it out of recognition of the compulsion of circumstances. Critical in that assessment was the premise that there was no alternative to the existing government -- or rather the alternatives were worse -- in the situation in which the country was placed.[44] The political stability that the Nehru government provided was a crucial element in the FICCI posture, despite business disapproval of its policies. It was conscious of where power lay; it pointed out to a member of the Bombay group: "The rank and file of the political parties and the general mass of people of the country, however juvenile or unreasonable their thinking may seem, are the most potent process to be reckoned with." In a larger sense, FICCI was aware of the limitations on what could be accomplished through a direct appeal to the public, because of public distrust and suspicion of business. Pertinent on this point was its comment: "First and foremost there is no denying that because of the notoriety earned by a few during the war years and thereafter, it must be lived down by the business community as a whole. It is very unfortunate, but it has to be faced."[45]

The Bombay school, however, started a new organization in 1956 known as the Forum for Free Enterprise, which engaged in sponsoring lectures and publishing pamphlets, addressed to the English educated public, that sought to underline the virtues of private enterprise and the economic pitfalls and political dangers of state enterprise. Later, in 1959, the strategy of the Bombay group led to the organization of a new political party, the Swatantra party (Freedom party), which was devoted to private property and free enterprise and vigorously opposed the entire framework of economic policy fostered by Nehru and the Congress party.[46] Noteworthy is the Janus-faced attitude of the business community toward the different strategies of the Bombay group and FICCI. Based on an understanding of what was ideally desirable from the viewpoint of business interests, the bulk of FICCI membership was in sympathy with the Bombay strategy.[47] On the other hand, in recognition of what was realistically possible under the given circumstances, only a minority openly identified with the Bombay strategy, while the business community at large considered it the better part of discretion to operate within the existing political system with all its constraints on business. However, the wider subterranean sentiment was available for eruption under appropriate circumstances. A few years hence, alienation drove even FICCI to the Bombay strategy, only to be taught the virtues of its earlier posture.

Third Five Year Plan: As the Third Five Year Plan began to be formulated, FICCI attempted to reopen the issue of demarcation of roles between the public and private sectors. At the annual session in 1959, even before the government had come to any firm decisions on the Third Plan, a spate of resolutions was passed on questions of economic planning. It was followed later the same year by an 82-page document which put forward the business community's vision of the appropriate economic strategy.[48] Emphasizing consumption as the motor for economic growth, FICCI demanded limitations on growth of the public sector, reversal of the heavy industry strategy, expansion of consumer goods industries, and a new taxation policy that would encourage savings and investments. FICCI did not have much quarrel over being ambitious in terms of the size of the plan, but its major grievance was policy, for in the absence of the right policy measures, "it would be unrealistic to think in terms of a large Plan." Basically, what FICCI seemed to want now, without explicitly saying so, was a return to the strategy incorporated in the First Plan -- little emphasis on heavy industry, a weak public sector, few restraints on the private sector, and an emphasis on consumer goods and agriculture.

With its premise of the importance of demand and investment of resources where they would yield maximum return, FICCI rejected the notion that "heavy or basic industry has an overriding priority". Instead, medium-scale industries should receive due emphasis, because of their greater employment potential and their direct relation to demand for consumer goods. Of major concern to FICCI was that no restraints be placed on factory production of consumer goods, whose expansion should be determined solely by the market rather than state protection of cottage and village industries.

More sharp was the Federation's position on the sectoral issue. It would not countenance the "school of thought which has come to increasingly emphasize that the first priority on the available resources should be assigned to the State, and further that the share of the State must continually increase as a matter of course, not only absolutely but also relatively". FICCI believed that such an approach had implications that extended beyond the economic sphere; rather they "impinge upon the structures and viability of social and political institutions which have been accepted as forming an integral part of our political democracy and national life". The Federation, disenchanted with the pattern of the Second Plan, wanted "an almost equal division of investment resources" between the two sectors as in the First Plan. It found unacceptable the enforcement of "any rigid or doctrinaire division between the Public and Private Sectors with respect to their scopes of industrial development and rights of entry". For it, what was important in the assigning of projects was the establishment of "an efficient, modern and competitive industrial structure". Contrary to the assumption of business having favoured state enterprise in basic industry, FICCI specifically stated that "there is no reason why the private sector should be excluded from participating in the development of iron and steel industry or the exploration and exploitation of mineral fuels."[49] At the same time, in pointed reference to state trading, land reforms and cooperative farming, FICCI opposed "forcing the pace of institutional changes on doctrinaire grounds."

FICCI's recommendations on resource mobilization reinforced its position -- or were indeed the basis of it to begin with -- on a reduced role for the public sector, de-emphasis of heavy industry, a more prominent role for the private sector, and greater emphasis on consumer goods industry. It argued for this position on grounds of economic efficiency and maximum return, but carefully noted that these principles were meant to serve the public interest through faster economic growth.[50]

FICCI had published its long document with a view to having an impact on government thinking, but when the government published the Draft Outline of the Third Plan, it was sorely disappointed on this score and made its concern known in sharper terms.[51] Its difference with the government was minimal on Plan size, but substantial on policy. FICCI criticized the tendency of government "to enlarge their industrial and business activities" which were of little benefit to the consumer because of the high cost of state enterprises; it found that "while social capital has to be augmented, resources are verily being frittered away in other directions." The Federation asked government to attend to its proper functions by coping with regionalism and disintegrative pressures rather than dissipate its energies in economic activity. While acknowledging that the government had handled the sectoral issue with some flexibility, it left no doubt about its dissatisfaction. It found that the place accorded to the private sector in investment outlays "cannot be considered to be very satisfactory." It made particular grievance of the fact that the planners ignored its superior performance in the first two Plans, but in this it seemed to overlook the fact, underlined by Venkatasubbiah, that the central issue with government was relative power rather than efficiency. Nonetheless, FICCI took umbrage at "the emphasis placed on ensuring the primacy of investment in the public sector," stating that it "cannot be viewed with equanimity," given the earlier performance of the public sector, and regretted government's "doctrinaire" approach to the matter. It warned that it was contrary to national interest "to hold back the rate of growth of the private sector and to insist on assigning the pride of place to the public sector." It seemed miffed by the short shrift given to business by government despite "the interest which private sector has exhibited in setting up steel plants," and equally by the role allotted to the private sector in coal, for which "the allocation does not seem to be satisfactory if looked in the light of the performance of the two sectors in the Second Plan."

After the government published the final version, and the formulation of the Third Plan was an accomplished fact, FICCI perforce gave it the customary welcome, emphasizing that there was little difference between the two on goals but only on approach, policy and details. In the end, FICCI's "repeated attempts to get the government to reopen the sectoral issue" between 1959 and 1962 met with failure.[52]

(c) *After 1962:* While the private sector was up to this time aggrieved at the limited arena left open to it and at the increasing controls on its operation, relations between government and business had not been marked by overt conflict. But this situation did not last beyond 1962. The subsidiary trend of disenchantment with government policy that had existed among businessmen since at least the mid-1950s now became more generalized and affected FICCI as well. Already in early 1962, the Swatantra party, as an outspoken supporter of free enterprise and a staunch opponent of economic planning and public sector, had done well in the third general elections; even though it was a relative newcomer on the political scene, it emerged as the second largest opposition party in the lower house of Parliament.

But what triggered off the conflict between business and government were the consequences of the 1962 India-China war. Heavy taxation coupled with continued and more intensified controls, along with inflation and economic slowdown, angered the business community even though at first in a fit of nationalism business, like the rest of the country, rallied around the government. The accumulated frustrations of the busines community with government policy burst out in a strong attack on the government. Business felt that, on the one hand, "policy seems to be hampered by old ideology rather than meet speedily and effectively the exigencies of the situation" and, on the other, government was making "a scapegoat of private enterprise" for its own policy failures. As one resolution declared, "the doctrinaire approach...pursued over a period of years now, is chiefly responsible for the fall in the rate of economic growth." Business alienation was further fed by anger at what was perceived to be a new bout of hostility toward it through the establishment of the Monopolies Inquiry Commission and escalation of the ideological impulse in policy-making as a result of the adoption of the resolution on "Democracy and Socialism" by the Congress party in early 1964. It saw the spectre of another series of nationalisations, overwhelmed by the thought that "a new superstition is growing that the competence and resources of governmental organisations are infinite."[53]

The new aggressive approach on the part of business toward government was to emerge in its fullness only after Nehru's death in 1964. But the review here of the business posture toward Nehru's socialist model leaves no doubt that his model with a dominant state sector in the industrial arena as its centrepiece, far from being desired by business, was forthrightly and actively opposed by business. Business would adjust to the *fait accompli* but in a spirit of resignation and helplessness. Years later FICCI officials would state, not that they accept policies flowing from the model, but that they are only "reconciled" to them.[54] The public sector expanded subsequently but it was really the Nehru era that was critical, for it was then that the economic-political framework centering on the public sector was imposed, a coalition in support of it organized, and a momentum for it established. The reasons for its imposition were, as the abundant evidence amply demonstrates, clearly ideological, having to do with the political values of Nehru, who occupied a strategic position in the political system. Even if they were narrowly political -- in the sense of advancing the interests of a political ruling group, drawn from the intermediate strata, through the creation of countervailing power under its command -- their consequences ran counter to the interests of business, and were so perceived by business. Precisely for that reason they were opposed by the business class.

3. "State Capitalism" and the Proletarian Vanguard

Despite the radical assumption that the establishment and expansion of the public sector is meant to facilitate the building of capitalism, it is manifest that the supposed beneficiaries of this policy -- that is, the capitalist class -- were opposed to it. But what kind of response did this attempt on the part of Nehru to build an "instrumental-socialist" mixed economy evoke on the part of the Communist Party of India (CPI) as the vanguard of the Indian proletariat. Since such an attempt was obviously meant, in CPI eyes, to advance the interests of capitalism on behalf of the capitalist class which had captured state power, then the logical conclusion ought to be that the CPI would have been opposed to this building of "state capitalism". However, CPI's ideological position on this issue was not quite that simple, and was marked by considerable complexity, indeed contradiction.

Notwithstanding its claim to be the party of the working class, the CPI has been, in terms of its leadership and cadres, essentially a party of the intermediate strata, more especially of the urban middle class, particularly the intelligentsia. What has distinguished it, however, is a thoroughgoing political commitment to an ideology which speaks on behalf of the interests of the urban proletariat. This ideology, when reduced to its essentials, took the Indian state in the Nehru era to be the state of the bourgeoisie and landlords, but too feeble to perform the essential tasks of national reconstruction. Since CPI's position on state power has already been examined in Chapter II, it is not necessary to repeat the tortuous history of its changing posture on the Indian state. However, attention needs to be directed at its stance toward the Second Five Year Plan, whose central thrust was not only the emphasis on heavy industry but building it through the public sector.

The CC Compromise in 1955

Despite its having given up the path of insurrection and violence in 1951, CPI remained determined to oppose the Nehru government, which it considered to be an anti-popular and anti-democratic government under the domination of landlords and the reactionary big bourgeoisie, collaborating with British imperialism. However, by the mid-1950s several events conspired to press the CPI to rethink its posture toward the Nehru government and its policies. Among these were: the continued endorsement of India's foreign policy by the Soviet bloc, the severe reverses suffered by CPI in elections in Andhra Pradesh in early 1955, and the concretization of the Congress party's commitment to socialism in the shape of proposals for the Second Five Year Plan. In the middle of 1955 there then occurred a marked shift -- as manifest in a resolution of the party's central committee (CC) -- in the CPI posture toward the government in contrast to the earlier singular oppositional stance.[55]

The CC resolution was obviously a compromise among several views, some suggesting forthright cooperation and coalition with the government and Congress party as a whole, others of combining with the progressive wing of the Congress party to oppose and defeat the reactionary wing, and still others of continuing its earlier course of opposition to the government with the aim of replacing it with a coalition of other democratic parties.[56] In this resolution, the

CPI continued to maintain that since the big bourgeoisie, which held the leading position in the Indian state, was in alliance with landlords and compromised with imperialism, the Indian bourgeoisie was incapable of completing the required tasks of the bourgeois-democratic stage of the country's revolution. What was essential therefore was a people's democracy, not only for the efficient accomplishment of these tasks but also to place the country firmly on the path of socialism. However, while adhering to this view of government and its own long-term strategy, the CPI was ready to acknowledge the reality of new developments in the government's external and internal policies and to make corresponding shifts in its own posture toward them.

In the CPI's new assessment, it was not merely some specific measures of the government in the international arena that were worthy of support but foreign policy as a whole. The CPI recognized Nehru's personal role in developing this foreign policy. But since the policy was, in its view, threatened by powerful forces in the ruling class that were linked to imperialism, the CPI pledged not simply to confine itself to welcoming and supporting it but to campaign to expose and isolate its opponents and "to strengthen the broad mass movement of people of all parties" in support of it.

Despite the newly-acquired positive view of foreign policy, the resolution maintained that the foundation of India's freedom remained insecure, because a government under the domination of the big bourgeoisie refused to complete the anti-imperialist and anti-feudal tasks. Still, there was some modification of the earlier totally negative appraisal of the government's internal policies. The resolution acknowledged that, even though the First Five Year Plan had not laid the foundations of a strong economy, increased agricultural production and foreign exchange reserves "have resulted in partial improvement of certain aspects of Indian economy, as compared to earlier years, and have strengthened the position of the Indian bourgeoisie." But these gains rested on "a precarious economic and political basis," for the government was attempting to build capitalism, in the epoch of general crisis of capitalism, without first completing the anti-imperialist and anti-feudal tasks of confiscating foreign capital and abolishing landlordism. As a consequence, the mass movement for the completion of these tasks necessarily had "to develop in opposition to the government's general internal policies."

If this was all there was to the party's revised perspective on domestic policy, it represented no departure from its earlier stand. However, the resolution underlined the party's new assessment that the presence of a socialist bloc and the balance of class forces in India made it "possible even today for the democratic movement to secure a limited advance in the direction of economic development of the country." With that, the party then endorsed the key features of the proposals for the Second Five Year Plan as embodied in the "plan-frame":

> The proposals to build basic industries, if implemented, would reduce the dependence of India on foreign countries in respect of capital goods, strengthen the relative position of industry inside India and strengthen our economic position and national independence. The party, therefore, sup-

ports these proposals and also the proposal that these industries should be mainly developed in the public sector. It supports the proposal that the demand for consumer goods should be met, as far as possible, by better utilisation of the existing capacity and by development of small - scale and cottage industries so that jobs are provided for an increasing number of people and maximum possible resources are available for the development of basic industries. The party not only supports these proposals but will expose and combat those who want them to be modified in a reactionary direction.

The party advanced some shrewd reasoning for its support of the plan-frame proposals. It believed that, if implemented, the proposals would result in conflict between the Indian bourgeoisie on the one hand and British capital and feudal elements on the other; furthermore, they would even intensify conflict within the bourgeoisie itself, "facilitating the weakening and isolation of the most reactionary elements." All this could, moreover, mean the further "adoption of limited measures by the government against imperialist, feudal and reactionary monopolist interests." How the CPI had come to differentiate these interests from the government, and why a government of these very interests who supposedly dominated the government allowed it to proceed thus, was not explained.

The CPI, however, did not bank on any automatic curbing of imperialism and feudalism as a result of the plan-frame proposals, especially since it considered the proposals not to be bold or progressive enough, indeed, it believed them to be in some respects reactionary. The proposals did not, in its view, place any control on British capital or restrict monopoly profits, while the economy remained linked largely to the crisis-ridden capitalist world. For the CPI, these reactionary features and inadequacies stemmed naturally from the class character of the government.

What the CPI thus seemed to do was to develop a policy which enabled it to support those government policies that had popular appeal, indeed claim credit for them by its vigorous endorsement of them, but at the same time to retain the option of mobilizing opposition to oust the govenment in order to achieve power itself. In this fashion, it attempted to counter being overwhelmed by government policy that claimed to be socialist both internally and externally, a claim the CPI could not concede without undermining its own existence.

However, what the CPI did not do in relation to the "plan-frame" proposals, especially with regard to the public sector, also needs to be underlined. It did not say that, since the state was a state of the bourgeoisie and landlords, with its leading force the monopoly bourgeoisie collaborating with imperialism, any public sector under its auspices would fortify the position of the bourgeoisie and, therefore, it should not be endorsed but rather must be exposed as a bourgeois manoeuvre to strengthen its hold over the economy. Some elements within CPI did exactly offer such reservations, but not the party as such. Rather, the party calculated that a large and expanding public sector would make for conflict between the Indian bourgeoisie and British capital, and also among various sections of the bourgeoisie but mostly to the detriment of the monopoly bourgeoisie. What the CPI held as potential here, however, was by and large precisely the intent of Nehru, if not of his government as a whole. For the same reason, of course, the CPI demanded that the government go further and have the public sector encompass a much wider chunk of the economy.[57] But this raises basic questions about CPI's concept of the leading force in state power, for if the monopoly bourgeoisie could not prevent Nehru from building and expanding the public sector, it did not testify much to its potency as a sharer in state power, leave aside as the leading force. And if the monopoly bourgeoisie was not in the position in state power that had been attributed to it, then it raises doubts as well about the appropriate relationship that the CPI envisaged with the government.

Perhaps on account of that very reason, opposition within the party could not be pacified by the new stance developed in the CC resolution. This opposition featured a strong yearning for a coalition with the Congress party,[58] at least with its progressive wing. It was persuaded to do so by the declared commitment of that party under Nehru to achieve socialism, by the desire to consolidate and push further this progressive change through altering the balance of forces within the government, and by a sense of external danger to India stemming from the country being besieged and encircled by imperialism through the arming of Pakistan and the establishment of SEATO. In response, the following year in 1956 at the Fourth Party Congress at Palghat, the CPI arrived at a still newer understanding of the Indian bourgeoisie and its activities;[59] this understanding represented a notable shift in the party's position on critical questions.

Policy Shift at Palghat in 1956

To begin with, the Palghat resolution shed all reference to the *big* bourgeoisie being the leading force in state power as also to its lack of motivation to choose policies favouring independence and industrialization. Rather, India was now considered to have a bourgeois-landlord government, with the bourgeoisie as its leading force, which was motivated by "the desire to develop India along independent capitalist lines." What was striking about the new assessment was its discovery of a progressive aspect to the bourgeoisie. This was not simply in the realm of foreign policy where the CPI found that "the present foreign policy of the government of India conforms to the interests of the entire Indian people (including the national bourgeoisie)," but also more generally in its plans to build an independent capitalism, which "inevitably bring the government into conflict with imperialism, with feudalism and sometimes with the narrow interests of sections of the bourgeoisie."

However, CPI's assessment was not entirely positive; it could not be if the CPI itself were to survive as a separate party. Its articulation of the progressive aspect was complemented by the underlining of a reactionary aspect in what was CPI's conception of the dual nature of the bourgeoisie. For the CPI, the manifest result of the dual nature of the bourgeoisie was contradiction in government policies:

Therefore, while opposing imperialism and attempting to weaken its grip over national economy, the bourgeoisie simultaneously maintains its links with British capital and gives facilities for further inflow of foreign capital. While striving to curb and weaken feudalism, it simultaneously maintains its alliance with landlords, against the democratic forces and makes concessions to the landlords. While striving to industrialise the country, it seeks to place the burdens of economic development mainly on the common people. While extending the public sector, it simultaneously pursues policies of support to monopolists in their attacks on the working people and adopts many measures which enrich the monopolists and thus help them to strengthen their position in important spheres of our life.

Corresponding to the progressive and reactionary aspects of the bourgeoisie, the CPI saw within the ruling party a developing political differentiation with the "growth of radical and democratic sentiments inside the Congress and among masses following the Congress." Accordingly, the CPI now adopted a two-pronged policy of unity and struggle in relation to the Congress party. Unity with the left wing of the Congress party, even if not with the entire party, was meant to consolidate and push further the progressive policies of the government. At the same time, struggle against the Congress party had as its objective the overthrow of the Congress party from power in order not only to complete effectively and expeditiously the necessary tasks of the bourgeois-democratic revolution but also to place the country firmly on the path to socialism.

CPI's position on the Second Five Year Plan was a more specific application of its understanding of the dual character of the bourgeoisie and of its strategy of unity and struggle. The Palghat resolution acknowledged that the Indian bourgeoisie and government was launched, not on a course of economic subservience to imperialism, but on the path of independent capitalist development. This was reflected in the Second Plan's fundamental objective of rapid industrialization with concentration on basic and heavy industries. The reasons for the adoption of this course now were "the aspirations of the Indian bourgeoisie to develop India as an independent capitalist country" and "the relative strengthening of the national economy as also the position of the Indian bourgeoisie," apart from the popular desire for national reconstruction.

The Palghat resolution admitted that, while the original plan-frame proposals had been changed "in a reactionary direction due to the presence of big business," nonetheless the achievement of the targets in industry would "contribute to the reduction of India's dependence on foreign countries in respect of capital goods, particularly for light industries, and strengthen the relative position of industry in our country." All of this represented a positive appreciation of the progressive aspect of the bourgeoisie on the part of CPI. Besides, the increase in the proposed investment outlays for village and small-scale industries was considered to be "also a welcome feature." However, the CPI was opposed to the modifications that had been made in the plan-frame, stating that "the restrictions on the growth of the public sector on the one hand and the most unduly increased allocations for the large-scale industries in the private sector, would lead to the strengthening of the monopolists and retard planned reconstruction of our economy." What the CPI was driven to, however, was not any novel or radical formulation of its own in economic planning, but to demand of the allegedly bourgeois government to return to its earlier proposals. It declared: "What is essential is to fight for the restoration of the original proposals in the plan-frame, fight for their full implementation." Urging the organization of popular intervention, it found the restoration a feasible objective "in view of the conflicts over policies which have grown and are growing inside the Congress." Meanwhile, the resolution asked the party to participate actively in "schemes and projects sponsored or run by the government."

The Underlying Motivations in Policy Shift

The Palghat resolution of 1956 is critical in any examination of the posture of Communist leaders in relation to the public sector. Regardless of differences among them, they all supported the establishment of the public sector by the bourgeois government, rather they wanted it to encompass an even larger sphere of economic activity. The split in the party was still more than a half-dozen years away, and no one would have had an inkling of it at the time, for it was a prospect unheard of in communist parties until the Sino-Soviet break. All the major figures in the Communist movement were a party to that resolution. Indeed, subsequent party positions after Palghat but before the party split went even further in support of the progressive wing of the government and its policies and of cooperation with it, largely because the bulk of the Communist leadership saw economic planning and public sector under attack from a rising tide of the reactionary monopoly bourgeoisie. But all that can be ignored. What is critical though is that the leaders who seceded from the party in the early 1960s, and later expressed a different position on the functions of the public sector, had at Palghat joined in a common party stand on the public sector; besides apart from a general oppositional and confrontational posture toward the government on their part subsequently, their attitudes on the public sector in concrete terms did not differ that much from the mother party.

If the public sector was, indeed, an instrument to serve the interests of the capitalist class and capitalism, then there can be no doubt that the Communist Party of India emerges as a full partner to the scheme. If the CPI supported the public sector in the expectation that it would lead to socialism even while under the hegemony of the capitalist class, then it displayed poor understanding of class dynamics as understood in Marxism. On the other hand, if it supported the public sector under the auspices of an allegedly capitalist state because it served the larger interests of the country beyond those of the capitalists, then it ran counter to the Marxian notion of the zero-sum nature of conflict between capitalists and workers. But it seems that the CPI supported the public sector fundamentally because it served to reduce the power of big business or monopoly bourgeoisie -- an interest rightly to be attributed to the intermediate strata. Exactly as the Nehru government, the CPI wanted to stop as far as possible the growth of the private sector in large-scale industry. It endeavoured to see that all future industrial growth was in the hands of either the state sector or village and small-scale sector. For that reason, it welcomed the Second Plan's objectives in these two sectors. At the same time, exaggerating the modifications made to the plan-frame proposals in the Second Plan, the CPI denounced them for their sectoral implications, stating that "the restrictions on the growth of the public sector on the one hand, and the most unduly increased allocations for the large-scale industries in the private sector, would lead to the strengthening of the monopolists and retard planned reconstruction of our economy."

A deeper insight into this essential motive of curbing the power of the private sector is provided by an examination of the position of CPI's trade union arm, the All India Trade Union Congress (AITUC). Especially important in any assessment of the position of AITUC is the strategy developed at its silver jubilee session at Ernakulam in December 1959.[60] That session underlined several features of the Second Five Year Plan as progressive, the key ones being the pre-eminent position of the public sector (particularly with new units in the vital industries being allocated to that sector) and the establishment of socialist pattern of society. On the one hand, AITUC doubted whether a government under the influence of monopoly capital could really either achieve the Plan's objectives or abolish capitalism in favour of socialism. But, on the other, since any advance towards industrialisation and economic independence "weakens imperialism and strengthens democracy and peace and ultimately world socialism," it urged that "the country and the working class can demand that these features be maintained and fulfilled. He who breaks these features can be denounced as an opponent of the people's interests." It was especially pushed into this course of support because the Plan was also under attack from "right-wing reaction," which was opposed precisely to those progressive features that AITUC welcomed. AITUC therefore reminded radical critics to see through the designs of the right wing and act to protect the Plan, even if it came from a government dominated by the bourgeoisie:

> Some trade unions only denounce either the whole Plan as merely a conspiracy of the bourgeoisie to defraud the people. Such a one-sided and unreal view, though put in radical and Left phraseology only helps the Right-wing reactionaries and frustrates or misleads mass initiative from achieving positive improvements and gains.
>
> The trade unions of the AITUC should reject both these positions. We must shoulder the responsibility to educate the masses on the need to fulfil the Plan and to defend it against the three main disruptors, namely, the foreign monopoly capitalists, the Indian monopolists, and their agents in the state.

A year later at Bangalore, AITUC warned the working class of "the serious offensive against the industries in the public sector that has been launched by some of the big businessmen in the private sector." While acknowledging that the private sector needed opportunities for development within limits since it was not possible to eliminate it under existing conditions, AITUC pointedly underlined at that time the reasons that were fundamental to supporting the public sector by the communists:

> But *what the monopolists want is not such development but total power over the economy and politics of the country*, their self-aggrandisement at the expense of the working class through low wages and high profits, monopoly control and high prices, and above all, an alliance with Anglo-American finance. This must be fought by the working class, because they hit our direct interests and also those of the people.[61]

It seems that a key function of the public sector in the eyes of the communists, even though often not expressed as such explicitly, was the limits it placed on the economic and political power of the capitalist class. This was so notwithstanding their frequent commentary that the public sector was being established by a state under the domination of the bourgeoisie and was meant to build capitalism and thus enhance the power of the bourgeoisie. Insofar as limiting the power of the bourgeoisie was a key consideration with Nehru in order not to block the path to socialism, Communist thinking was totally in line with that of Nehru's. Furthermore, there were as well other more positive functions to the public sector that were acknowledged by Communists at Ernakulam and which therefore necessitated support:

> We must give efficient work according to the terms of the service, particularly so in the state sector. State-sector economy is not yet a socialist economy. But it is built by taxing the people, out of public money. The surplus in the state sector does not become wholly the private dividends of the rich few, though a part no doubt is used for that purpose and for harnessing the toiling masses in the service of capital, through state force, whenever they show signs of protest or revolt. Even then, the state sector, when made amenable to Parliamentary control can be made to use the surplus for tax relief, education, amelioration of conditions for all people, etc.[62]

This view of the public sector then led AITUC at Ernakulam to basically the path of economism and moderation in the trade union field. It adopted "a two-pillar policy -- to help in the development of the economy and to defend the interests of the working masses in that economy." AITUC realized that evolving the tactics to be employed under the "two-pillar policy" was difficult since that policy could not mean either no strikes at all or a strike for every demand. It therefore decided that unions must now "negotiate and settle; if that fails, strike peacefully and as a last resort," avoiding both disruption and inaction.

Summary and Conclusions

The public sector in the industrial arena has often been viewed as a project of the capitalist class in order to advance its interests through building capitalism under the auspices of a capitalist state. It is ironic though that, in the Indian case, the overwhelming burden of the empirical evidence is that the capitalist class opposed the building of the public sector and the corresponding constraints that it placed on the private sector. Not only is this actually the case, it was so recognised by leaders of the Communist Party of India[63] though they did not stop to ask as to why something that was in the interest of the capitalist class was opposed by that class or, more basically, how a state controlled by the capitalist class did things that were unacceptable to that class.[64] On the other hand, it is equally ironic that those opposed to the capitalist class and capitalism, whether socialists or communists, have been the most vociferous in supporting it and having it be pushed further. Among this group were also radical scholars -- foreign and Indian, inside and outside government -- who played an active role in the formulation and propagation of proposals for the creation and expansion of the public sector, and accordingly must share the credit or blame for its subsequent performance.

The reasons for opposition to the public sector by the capitalist class and for support by the socialists and communists were fundamentally the same, however. It seems specious to suggest that it was opposition to imperialism and desire for economic independence on the part of the capitalist class that was critical to the creation of the public sector. Any capitalist class that controlled the state, and was sufficiently motivated to counter imperialism and achieve economic independence, could as well have done it through private entrepreneurship after the pattern followed under Meiji Japan than through the kind of mixed economy installed in India that barred private enterprise from the "commanding heights." If for some reason, however, the capitalist class really believed that this kind of mixed economy was a more effective instrument for countering imperialism and achieving economic independence, then it did not have to openly oppose its creation. Equally, on the other hand, if the public sector does nothing but advance capitalism, what difference does it make to the proletariat whether it is private or state capitalism? Why then such vociferous condemnation of those who opposed the public sector?

In fact, whatever its alleged role in the advancement of capitalism, "the instrumental-socialist." public sector was seen by opponents and supporters alike as excluding important arenas of the economy from private accumulation. More fundamentally, both saw it as reducing the economic and political power of the capitalist class. Again, both saw it as effectively facilitating -- to be dreaded by one side and welcomed by the other -- the prospect eventually of abolition of capitalism and the capitalist class altogether. This would occur either in a unilinear but peaceful fashion through the ever-increasing absolute and relative role of the public sector (as envisaged by Nehru and socialists of his persuasion), or in a dialectical fashion through the eruption of violent conflict between a weakened capitalist class and the strengthened socialist forces cohering around a potent public sector (as envisaged by the communists). It is striking that, despite the difference in rhetoric and even more over the question of the mode of transition to socialism, Nehru and the communists essentially shared the same vision of the end of capitalism and the same calculations about the role of an expanding public sector in achieving it.[65]

In the actual process of history, the fears and hopes of capitalists and socialists or communists, respectively, may or may not turn out to be justified. It may well be that, despite them, those who held the public sector to be an instrument of building capitalism may in the ultimate analysis be proven right. On the other hand, the intermediate strata as a whole, some of whose segments constitute the socialist and communist groups, may continue to favour a social and economic system in which they continue to exercise domination with the public sector as a critical source of economic and political power, precluding hegemony by or on behalf of either the bourgeoisie or the proletariat. In that sense, the communist may well have, through their vigorous support for the creation and expansion of the public sector, acted as the Trojan Horse of the intermediate strata.[66]

NOTES

1. Gunnar Myrdal, *Asian Drama* (New York : Pantheon, 1968), vol.II, p. 762.
2. Lloyd I. Rudolph and Susanne H. Rudolph, in Meghnad Desai, et. al. (ed.), *Agrarian Power and Agricultural Productivity in South Asia* (Delhi: Oxford University Press, 1984), pp. 336-37.
3. G.K. Shirokov, *State and Industrialization* (Moscow: Progress Publishers, 1973), p. 59.
4. A.H. Hanson, *Public Enterprise and Economic Development* (2nd ed.; London: Routledge & Kegan Paul, 1965), p. xxv. Emphasis added.
5. H. Venkatasubbiah, *Enterprise and Economic Change: 50 Years of FICCI* (New Delhi: Vikas, 1977), pp. 101, 104.
6. Jawaharlal Nehru, *Speeches: Volume Five: March 1963-May 1964* (New Delhi: Publications Division 1968), p. 130.
7. R. Chandidas, et al., *India Votes* (New York: Humanities Press, 1968), pp. 5-11.
8. "In the case of the capitalist state, the autonomy of the political can allow the satisfaction of the economic interests of certain dominated classes, even to the extent of occasionally limiting the economic power of the dominant classes...but on the one condition...that their political power and the state apparatus remain intact." See Nicos Poulantzas, *Political Power and Social Classes* (London: NLB, 1973), pp.191-92, 188-90, 284-89.
9. A.H. Hanson, *The Process of Planning* (London: Oxford University Press, 1966), pp. 474-75.
10. He warned : If private enterprise has full play, one of the first casualties in this country will be private enterprise itself. Jawaharlal Nehru, *Speeches : Volume Two : August 1949 February 1953* (New Delhi : Publications Division, 1954), p. 49.
11. Angus Maddison, *Class Structure and Economic Growth* (London : George Allen & Unwin, 1971), pp. 89-90; K.P. Karunakaran, *The Phenomenon of Nehru* (New Delhi: Gitanjali Prakashan, 1979), p. 31.
12. Raymond D. Duvall and John R. Freeman, "The State and Dependent Capitalism," *International Studies Quarterly*, XXV, no. 1 (1981), pp. 99-118.
13. Myrdal, II, p. 819.
14. Communist Party of India (Marxist), *Programme: Adopted...1964* (New Delhi : 1979), pp. 4, 7.
15. Girish Mishra, *Public Sector in Indian Economy* (New Delhi: Communist Party Publication, 1975), p. 39.
16. E.M.S. Namboodiripad, *Economics and Politics of India's Socialist Pattern* (New Delhi: People's Publishing House, 1966), p. 194; Karunakaran, p. 43; Nigel Harris, *India-China : Underdevelopment and Revolution* (New Delhi : Vikas, 1974), p. 309.
17. Hanson, *Public Enterprise*, pp. 177-78
18. Venkatasubbiah, p.86.

19. FICCI, *Correspondence*...1951, p. 161.
20. Venkatasubbiah, pp. 78, 97, 99, 104, 105, 133.
21. Stanley A. Kochanek, *Business and Politics in India* (Berkeley: University of California Press, 1974), p. 63.
22. R.G. Saraiya, cited in Venkatasubbiah, p. 79.
23. *Ibid.*, p. 4.
24. Shantinarayan, in FICCI, *Proceedings...1956*, p. 109; and S.S. Kanoria, in FICCI *Proceedings...1957*, pp. 15-16. Many years later, in 1970, Kanoria reiterated: "Our weakness is that there is not broadbased, informed public opinion in our favour. On the contrary, criticism of private business, however irrational and ill-informed, seems to evoke wide public approval." See Venkatasubbiah, p. 136.
25. Although this periodization departs somewhat from Kochanek's demarcation of periods and description, I found his discussion of the different periods eminently useful. Kochanek, pp. 216-25.
26. Venkatasubbiah, pp. 70-72.
27. Venkatasubbiah, p. 78.
28. FICCI, *Correspondence...1951*, pp. 147-50; and *Proceedings...1951*, pp. 34-35.
29. *Ibid.*, p. 98.
30. FICCI, *Correspondence...1951*, pp. 162, 170, 186, 187, 189. See also FICCI, *Proceedings...1952*, p. 87 ff.
31. V.D. Divekar, *Planning Process in Indian Polity* (Bombay: Popular Prakashan, 1978), p. 256.
32. Venkatasubbiah, p. 98.
33. FICCI, *Report...of the Executive Committee...1955*, pp. 44 -46, and *Proceedings...1956*, pp. 45-48
34. FICCI, *Report...Executive Committee...1955*, pp. 23-24, 30, 33-34, and *Proceedings...1956*, pp. 63-64
35. Kochanek, p. 79; Hanson, *Process*, p. 464; and Michael Brecher, *Nehru: A Political Biography* (London: Oxford University Press, 1959), p. 538.
36. FICCI, *Correspondence...1956*, pp. 582-85.
37. Venkatasubbiah, p. 85.
38. Ignacy Sachs, *Problems of Public Sector in Underdeveloped Economies* (Bombay: Asia Publishing House, 1964), pp. 132-33.
39. Brecher, p. 533.
40. FICCI, *Second Five Year Plan: A Comparative Study of the Objectives and Techniques of the Tentative Plan-Frame* (New Delhi: 1955).
41. FICCI, *Correspondence...1956*, pp. 354-86.
42. Hanson, *Process*, pp. 464-73.
43. R.G. Saraiya's letters in 1955, cited in Venkatasubbiah, p. 79.
44. Kochanek, *Business*, p. 37.
45. FICCI letter in 1955, cited in Venkatasubbiah, p. 80.
46. See Howard L. Erdman, *The Swatantra Party and Indian Conservatism* (Cambridge, Great Britain : Cambridge University Press, 1967).
47. Kochanek, pp. 39-40, 221.
48. FICCI, *The Third Five Year Plan : A Tentative Outline* (New Delhi : 1959).
49. *Ibid.*, pp. 7, 25, 39-40, 78.
50. *Ibid.*, p. 22.
51. FICCI, *Draft Outline of the Third Five-Year Plan : An Analysis* (New Delhi : 1960), pp. 3, 13-15, 30-31
52. Venkatasubbiah, pp. 102-03.
53. FICCI, *Correspondence...1963*, pp. 94-99; and *Proceedings...1964*, pp. 4, 28-29.
54. Interviews with FICCI officials in 1978 and 1982, including the late R.G. Aggarwal.
55. "Communist Party in the Struggle for Peace, Democracy and National Advance," in Mohit Sen (ed.), *Documents of the History of the Communist Party of India* (New Delhi : People's Publishing House, 1977), vol. VIII (1951-1956), pp. 416-440, but especially pp. 418-28, 434-35, 438.

56. Victor M. Fic, *Peaceful Transition to Communism in India* (Bombay: Nachiketa Publications, 1969), pp. 138, 150-51.
57. Communist Party and Problems of National Reconstruction, in Mohit Sen, *Documents*, VIII, pp. 441-63
58. E.M.S. Namboodiripad, *Indian Planning in Crisis* (New Delhi:National Book Centre, 1982), p. 53.
59. Political Resolution, in Mohit Sen (ed.). *Documents*, VIII, pp. 523-71, especially pp. 546-47.
60. S. A. Dange, *General Report at Ernakulam: Silver Jubilee Session, All India Trade Union Congress, December 25 to 29, 1957* (New Delhi: AITUC Publication, 1958), pp. 11-13, 19-20, 89-90.
61. S. A. Dange, *Crisis and Workers: Report to AITUC General Council* (Bangalore Session, 14-18 January 1959) (New Delhi : All India Trade Union Congress, 1959), pp. 1, 3, 45. In the original, the entire paragraph is italicised.
62. Dange, *General Report*, p. 20
63. Referring to the proposals for the public sector and regulation of the private sector as well as for ceilings on landholdings, the Communist leader Namboodiripad commented in the mid-1960s : "Neither of these policies were to the liking of the spokesmen of Big Business. They were afraid that, if these two policies were pursued by the government, there would be far less opportunities for them to make profit...The publication of the Plan Frame was therefore greeted by them with loud protests...against the 'pro-communist' policies of the planning authorities." Namboodiripad, *Economics and Politics,* p. 125
64. Note the interesting rationalization: "Big Business, however, knows that it would be extremely unrealistic to expect the policy of nationalization to be completely given up. After all, this is an epoch in which the influence of socialism as an ideology is being felt throughout the world; India therefore cannot be an exception to it." *Ibid.*, p. 195.
65. One radical argument acknowledges business opposition but endows Nehru and his advisors with the extraordinary capacity of discerning and pursuing the real interests of the bourgeoisie: "how to explain the Government of India adopting a strategy which turned out to be one for building State Capitalism in the name of Socialism but which was initially opposed by the bourgeoisie itself?...This is one more instance in history of leaders of a ruling class being much more farsighted than individual members of the same class. The distance between the understanding of ruling class interests as perceived by ordinary individual members of the class and their representatives in the state can be so big that the former may actually oppose the actions of the state until they come to understand the real motive behind the state policies....Nehru and his closest cabinet colleagues were alone crystal clear about what was happening -- they alone did not suffer from any delusions." Ashok Rudra, "Planning in India : An Evaluation in Terms of Its Models," *Economic and Political Weekly,* XX, no. 17 (1985), pp. 758-64. Framed in this manner, of course, the statement lifts the argument beyond empirical social science.
66. With reference to the intermediate class, Jha notes : "the left has unwittingly played the role of a stalking horse in achievement of its designs. It has done this by invoking the sanction of Marxism-Leninism and thereby cloaking the policy aims of this class in a mantle of intellectual respectability and humanitarian concern." See Prem Shankar Jha, *India: A Political Economy of Stagnation* (Bombay : Oxford University Press, 1980), p. 125.

Chapter VI
Economic-Political Crisis and the Transition to Radicalism (1964 - 1969)

Nehru's socialist vision envisaged a comprehensive attack on India's social and economic problems. But, despite his endeavours, there could be no radical reconstruction of the country's agrarian system because of opposition from the middle and rich peasantry represented in the Congress party. In the final analysis, his socialist programme in practice was reduced to a "big-push" strategy for industrialization -- with emphasis on heavy industry rather than consumer goods -- carried out largely in the public sector under a regime of centralized economic planning, with accompanying restrictions on the private sector. Even this programme was difficult to implement because of the tremendous demands it placed on resource mobilization. Indeed, it was inconsistent with the country's democratic political framework insofar as it required the population to undergo sacrifices through taxation and forced savings while postponing economic gratification into the future.

Despite the difficulties encountered, Nehru was able to implement this programme because of a special set of circumstances. First and foremost, there was Nehru's own charismatic leadership which inspired faith in his chosen goals and confidence in the future. Second, partly as a result of that and also as a legacy from the nationalist movement, the Congress as an umbrella party was dominant politically at the centre and in virtually all the states, without much challenge from a fragmented opposition. Third, notwithstanding the political consciousness aroused by the nationalist movement, social mobilization of the population was low during the Nehru period. Universal adult franchise was too recent a development to have made for any drastic change in this respect. As a result, class differentiation did not manifest itself sharply in politics. Fourth, relatedly, the system did not face a sharp political polarization. The Congress party drew diffuse political support from across the entire spectrum of social and ethnic groups. Fifth, there existed a basic national consensus on Nehru's programme; the socialists and communists, then the key groups in opposition, basically endorsed his goals. Sixth, Nehru's government benefited from a favourable international environment where superpower competition brought considerable economic aid while friendship with China enabled the country to keep defence expenditures at a low level.

This special set of circumstances could not be expected to continue for long, however. In a changed situation, Nehru's successors would have to confront extremely daunting challenges. They would, in addition, have to cope with the consequences of Nehru's own policies in respect of the neglect of agriculture and consumer goods. And they would have to manage these without a comparable political stature and level of support. As conflict sharpened over policy after Nehru's death, his socialist project, even in the restricted sense of a vast and expanding public sector, could not remain unaffected; indeed, it had come under question in the last years of his own life. Accordingly, this chapter addresses itself to: (1) Nehru's struggle to save and advance his socialist project near the end of his political career; (2) the bypassing of Nehru's programme in coping with new problems under Prime Minister Lal Bahadur Shastri; (3) the economic crisis during Prime Minister Indira Gandhi's first year in office, the adoption of unpopular decisions, and the resulting electoral punishment; and (4) the ensuing political crisis which generated a phase of political restructuring and radicalism under Mrs. Gandhi.

1. The Last Days of Nehru : Protecting the Socialist Project

By the end of the Second Five Year Plan (1956-61), Nehru faced a rising conservative trend, both inside and outside the Congress party. Under pressure from the lobby for the middle and rich peasantry, Nehru had to retreat in 1959 from his plan for drastic land ceilings and joint cooperative farming in the rural areas. Meanwhile, the Swatantra party, launched in 1959 on behalf of the same lobby as well as of big business, challenged the whole philosophy of socialism, economic planning and the public sector. Surprisingly, this party did well in the 1962 elections, emerging as the second largest opposition party, with the Communist party continuing, as before, in first place. Hitherto, criticism from the Communist party had centered on the Congress party not advancing rapidly or consistently enough in the direction of socialism, but now there was, in an emerging ideological polarization, attack on the very premises of government policy. The Congress party itself had lost some support in the 1962 elections compared to the 1957 elections, with its share of the national vote falling from 47.8 per cent to 46.0 per cent; its losses would perhaps have been severer were it not for patriotic fervour in the wake of the Indian liberation of Goa from Portuguese colonialism. For the time being, though, despite the attack from the right, the Congress party seemed firmly in control.

The economy's performance during the first two years of the Third Five Year Plan (1961-1966) was sharply inadequate; even as performance fell behind targets, pouring cold water over expectations of economic advance under the plan, inflation was creating serious social stress. But the coup de grace to the Third Plan was delivered in the mountain ranges of the Himalayas by the Sino-Indian war of 1962. With that war, both Nehru's foreign policy of non-alignment and his development strategy lay in ruins. The war resulted in the ouster of his most ardent socialist colleagues from the cabinet, notably Krishna Menon. Moreover, as defence expenditures literally doubled in a year, they necessitated the imposition of enormous new tax burdens over and above the immense resource mobilization effort for the Third Plan. Business, thoroughly alienated, was up in arms; its alienation now resulted in an attack on the very fundamentals of economic planning and socialism, and on government's preference for the public sector and constraints on the private sector. In the political arena, the attack against government policy was spearheaded by the rightist parties, Swatantra and Jan Sangh. Reeling under the impact of the reverses in war, the Congress party's hand was further weakened politically by losses in three key by-election contests in 1963. The acclaimed national consensus thus had broken down during Nehru's tenure; his successors were left with the task of coping with the consequences of the breakdown.

Nehru responded not by changing policy but to protect it by seeking to change the balance of power within the government through forcing the resignation of several key conservative ministers, such as Morarji Desai -- under the cover of the Kamaraj plan of August 1963 to revitalize the Congress -- and by obtaining a reaffirmation of his socialist policies from a party that was divided on the central issues. This reaffirmation of socialist ideology came at the annual Congress session at Bhuvaneshwar in January 1964, five months before Nehru's death. The resolution on "Democracy and Socialism" at Bhuvaneshwar was a historic declaration, culminating a lifelong mission on the part of Nehru to have the Congress party commit itself to the socialist path and vision. It constituted his last political testament, attempting to remove ideological ambiguity and clarifying and codifying the party's ideology.

Through the Bhuvaneshwar resolution, the Congress party sought to pre-empt a highminded position distinct from the opposed but exclusive advocates of either democracy or socialism, the former rejecting state intervention at the cost of social justice, the latter pushing it at the expense of negating individual freedom. Viewing history in a broad sweep, it expressed pride in "its role as the emancipator of the country from alien rule". But, along with that, the party believed that it had always had a social objective which after independence had assumed the form of the aim of "the establishment of a socialistic pattern of society". It took its goal of "planning through socialism" to have been endorsed by the nation since the Congress had been returned to power on the basis of that programme in two elections. The party laid the blame for the slow economic advance not on any structural defect in its goal or strategy but on faulty implementation.

With the Bhuvaneshwar resolution, the Congress party now laid claim to "working for a revolution in the economic and social relationships in Indian society" whereby "privilege, disparities and exploitation should be eliminated". Of course, it pledged to achieve all this through peaceful means and popular consent, and therefore summed up its ideology as "democratic socialism". The party considered the removal of poverty as fundamental to its aim of a socialist society, which would replace "the present day acquisitive structure". For it, that required rapid development in agriculture and industry. The party at this time committed itself to providing "a national minimum comprising the essential requirements in respect of food, clothing, housing, education and health" by the end of the Fifth Five Year Plan. In the endeavour to reduce disparities, the Congress pledged, in what could only be viewed by the propertied classes as a serious threat, "to bring about a limitation of incomes and property in private hands," with that limitation applying especially to "inherited wealth and urban property". Accordingly, it resolved that "the State should secure a large share of capital gains and appropriate a much larger proportion of unearned income". Even though the party did not commit itself to bank nationalization, it asked the government to assume a more effective role in directing "the means of credit and investible resources of the country along the lines of national priorities and our social purposes". In an apparent attempt to build countervailing power against big business, the Congress was particularly concerned over the disadvantages suffered by small entrepreneurs and new entrants in obtaining finance.

As for economic organization, the party reiterated through the resolution what was by then established policy: "The public sector has to play a strategic and predominant role in the field of trade and industry. The public sector must grow progressively in large scale industry and trade, particularly in the field of heavy and basic industry as well as trade in essential commodities". This open-ended call for expansion of the public sector could not have caused any cheer in business and trade circles. The Congress, no doubt, acknowledged that there was an important role for the private sector, but it had to be "within the broad strategy of the national plan of development". Further, it held that even in such a circumscribed private sector "the cooperative method of organization will occupy an increasingly important place, especially in the field of agriculture, small-scale and processing industries and retail trade". The party was emphatic that the "processing of agricultural produce especially paddy should not remain in private hands"; the state was asked to take over the task until such time as it could be done on a cooperative basis. On land reform the party expressed itself in favour of -- what by then had become only a vain hope -- placing a limit on size of landholdings, assuring a minimum wage for agricultural labour, and eventually organizing a cooperative rural economy.[1]

Although the Congress party thus clarified and codified its ideology and programme in broad strokes through the Bhuvaneshwar resolution, some of the more left-inclined Congressmen insisted that the party commit itself to rapid and vigorous but specific steps towards establishing socialism, such as nationalization of banks, state trading in foodgrains, controls on monopolies, and public sector entry into profitable lines in consumer industries like textiles.[2] Indeed, the Bhuvaneshwar session was dominated by the demand for nationalization of banks, but the Congress leadership was able to reject, despite the ideological discontent in the party, what seemed to it as an extreme measure.[3] Perhaps Nehru was more interested in protecting the existing public sector against the rising attack from conservative forces than in assuring any further advance if that opened up schisms within his broadbased umbrella party. However, business could not look with equanimity at the ringing reaffirmation of socialism as the Congress party's goal. Already angry at the tax burden and harrassment through stringent administrative controls, business dismay mounted steeply at what it saw as an onslaught on the private sector : "The latest Resolution has raised a chorus for nationalization and establishment of cooperatives in respect of all conceivable economic activities -- from banks to rice mills and wholesale and retail trade." It angrily noted the Congress tendency "to make a scapegoat of private enterprise" and "to chastise, with reckless confidence and without acquaintance with facts, all actions of businessmen."

The Bhuvaneshwar resolution with all its radical professions of socialism was less a reflection of the true overall ideological inclinations of the Congress party than a sentimental concession to the ideological wishes of a dying political hero. The same party's stalwarts had provided a mirror of their real intent by scuttling the joint cooperative farming plan pushed by Nehru and by thwarting genuine land reform. The Congress party had never been a radical party even as it included some radical elements, and Nehru's individual role had been critical to what had been pushed through in terms of the public sector. However, the party had become progressively more conservative as political change in the period since independence shifted the balance of forces within the party.

The interaction of the modern state and traditional society through the instrumentality of adult franchise resulted in a system of political clientelism where local notables mobilized votes for the party in exhange for benefits. At the state level there arose party bosses who knew how to manage the party's electoral machine. These party bosses were, in the main, conservative, committed largely to the status quo, and basically interested in winning elections so as to control the government which then provided them political patronage and economic largesse. Vernacular-educated and rural-based, they stood in contrast to the English-educated and urban-based government leadership with a socialist orientation. Since the Congress party at the national level was dependent on the party electoral machine at the state level, the rise of party bosses marked a shift in power from the centre to the states and also from government to the party. The state party bosses gradually became critically important decisionmakers in the party at the centre. The shift in power was highlighted by the election to the party's presidency in 1964 of Kamaraj, a leader with an independent political base in South India. Earlier in 1963, several of the party bosses had met in a secret conclave and decided as to who shall succeed Nehru. The label "Syndicate", with its reactionary and mafia-like associations, came to be applied to them and became part of the local debate and parlance. Interestingly, the Congress Working Committee's very first post-Nehru resolution of 3000 words, passed two days after Nehru's death, excluded altogether the word socialism even as it listed many of his various goals as deserving of continued support by the peoples.[5] This could not have been an accident, considering the elemental role that the term occupied in his philosophy and political career.

Even as political change within the Congress party was giving it a more conservative cast, Nehru's legacy could not be easily dispensed with. Howsoever limited his policies may have been in radical content and scope, by sustaining them over a decade Nehru as a political hero gave them the aura of nationally established policies and goals. It would become sufficient later simply to declare that they were the handiwork of the national hero to stop debate, without questioning their functionality. Besides, the sheer existence of a vast public sector and the momentum developed for it created pillars of support in its behalf. Leaders associated with the development of the public sector during the Nehru era could not easily forsake it without having to acknowledge that they had been wrong in the past. Within the party, at the same time, there gathered a group of radicals committed to Nehru's policies and programmes. Outside the party, regardless of the criticism in terms of the allegedly narrow scope of the public sector and inadequacies in implementation, the socialists and especially the communists were staunch and vehement defenders of the public sector. They attacked any revision in respect of the public sector -- except to enlarge it rapidly and vigorously, and correspondingly constrain and reduce the private sector -- as a reversal and betrayal of *national* policies and a function of an alliance of local reactionary traitors with capitalist imperialism. While all these elements favoured continuity of his legacy, Nehru also left the country in the grip of a developing crisis, which was rooted in his policies, howsoever compelling their need may have been. His successors had thus not only to cope with the crisis -- and do so without his commanding political stature -- but also with the question of persistence with his vision.

2. Shastri as PM: Policy Adaptation Amidst Crisis

In what seemed like a marvel of political management, the Congress party's political bosses were able to ensure a smooth succession after the death of Jawaharlal Nehru on May 27, 1964. The selection of Lal Bahadur Shastri as Prime Minister also represented popular opinion, and most likely was Nehru's preferred choice as well. Through a consensual process, the Syndicate was able to outflank Morarji Desai, the right wing but rigid and self-righteous leader, as also the leftists within the party who distrusted Shastri's credentials as a socialist and did not believe him to be the appropriate vehicle for protecting Nehru's legacy. Unlike his predecessor, Shastri did not tower above his colleagues; as a moderate, a pragmatist and a natural conciliator, he was inclined to work with them in a collective leadership; these very qualities had attracted him to his colleagues in the first place.

The new prime minister did not have the long-term ideological vision and commitment that was characteristic of his predecessor; his faith in socialism had developed as part of the acceptance of Nehru's leadership, who was his political mentor and patron. For some, "Shastri was no socialist....Essentially, he was a Gandhian in thinking and even in his way of life."[6] But Shastri repeatedly affirmed his faith in socialism and the mixed economy; on his formal election as prime minister, he proclaimed: "the Government of India will continue to follow the policy of Nehruji in international matters and democratic socialism will continue to be our objective in our domestic policy." He took as axiomatic the expansion of the public sector: "Our objective...is socialism. This does mean an immense growth of the public sector".[7] At the same time, he did not carry the animus against the private sector that Nehru and his leftist colleagues did; nor did he give speeches castigating private profit as evil and immoral, or extolling public ownership because it was based on social gain; for these reasons he commanded the respect of the business community.[8]

What was more characteristic of his attitude was flexibility and quiet pragmatism; he did not see any inherent conflict between the public and private sectors, and wanted them to work in a complementary way, rather than at cross-purposes, within a mixed economy for the sake of national production and employment. As he told one business group: "It is not possible to have some kind of free economy in our country. We also cannot have a regimented type of economy. Our objective, as you know, is socialism. We will have to evolve our own pattern....There has to be mixed economy; that is both the public and the private sectors will have to function in India. I would suggest that the private sector and the public sector should not contradict each other. The public sector naturally will have a bigger charge in our allocations, in our Plans. But the private sector will also be fairly big...."[9] Consistent with his own approach, his cabinet had a balanced, centrist orientation.[10]

But there was another characteristic to his professed belief in socialism which distinguished him from his predecessor. As a person of humble origins who had gone through not only political struggle -- with his prison record of nine years[11] matching that of Nehru's -- but also economic austerity, indeed suffering, Shastri had an instinctive empathy for the common man. He made the test of socialism and planning not necessarily some abstract ideological criteria but how it served the concrete daily needs of the masses in terms of food, clothing and shelter. He recognized the important role of industry, not excluding heavy industry, because: "Only industry can relieve the pressure on land and thus make possible a real improvement in the standard of living of the people. Indeed, even agricultural development depends to a great extent on the supply of fertilizers and insecticides, steel and cement, which are the products of industry. There is no getting away from the fact that the industrial base has got to be widened in India". But for him the central question still concerned the common man and his needs: "To my mind, socialism in India must mean a better deal for the great mass of our people who are engaged in agriculture, the large number of workers who are engaged in the various factories and the middle classes who have suffered much during the period of rising prices. These are what I call the common men of my country. As the head of the Government, it would be my continuous endeavour to see that these objectives are realized and that a social and economic order is established in which the welfare of our people is assured".[12] Again, he stated: "One has to make allowance for the situation such as it develops. The strains that have shown up in recent months cannot be ignored. I believe the first task is to provide food, clothing, shelter and medical aid to India's millions. I have therefore suggested that planning should be geared to meeting these primary needs at the same time as we pursue other goals".[13]

Perhaps this question may not have been of much consequence if Shastri had succeeded to his office in a time of normalcy, even if one of continuing poverty. But Shastri became prime minister in a situation of economic crisis, in part a function of earlier policies. Stagnant food production and heavy defence expenditures had released severe inflationary forces in the economy, so that on top of an increase of over 6 per cent in 1963-64 prices rose by 10 per cent in 1964-65 and still another 7.6 per cent in 1965-66. The food shortages in 1964 necessitated the hurried diversion to India of food ships destined for other countries. The year 1965-66 saw an unprecedentedly calamitous drought, resulting in a drop in agricultural production of some 20 per cent. Food imports resulted in a severe balance of payments problem, with foreign exchange reserves cut to the bare bone. The crisis would deepen with the suspension of foreign aid in the wake of the India-Pakistan war in September 1965. But the Shastri administration had to rethink its economic policy even before then as the economy was bedevilled by food shortages and, relatedly, price rises and humiliating dependence abroad. Confronted with these realities, pragmatism came to have "precedence over ideology" under Shastri, while the administrative custodian of Nehru's ideology, the Planning Commission, was downgraded.[14]

Nehru's economic strategy had centered on rapid industrialisation with emphasis on heavy industry, which implied in the context of limited resources a neglect of agriculture and consumer goods industries; it also implied a bias in favour of the public sector, for the private sector had been excluded from the field of heavy industry. The consequences of the strategy were, as some had warned, a stagnant agriculture and food shortages as well as scarcity of consumer goods. This had caused public alienation and disenchantment with government planning, which also manifested itself in public disorder, strikes and violence. Shastri felt that a reorientation of planning was necessary "if a sense of hope was to be sustained among the people. For the new prime minister the central concern was the impact on the general public, and here he felt that "there must be something wrong with planning if it achieved so little for the common man", and therefore was eager for concentration on "certain practical issues" of critical importance, especially food and prices. In any case, unlike Nehru, Shastri was not fascinated by gigantic projects; he told his first cabinet meeting: "I am a small man and believe in small projects with small expenditure so that we get quick results".[15]

The logic of the situation then led to placing "the highest priority" on agriculture; and with his earthy realism, he shunned the path of food mobilization from the farmer through compulsory levies under a regime of state trading in food, with controlled prices and a rationing system. He and his advisors concluded that the fundamental problem was inadequate production, and here they initiated the "new agricultural strategy" whose central premise was that the Indian peasant was a rational economic man who would respond to incentives in the improvement of productivity. These incentives were provided in an open market system through sufficiently attractive support prices, which were to be set by a newly established Agricultural Prices Commission; Shastri rightly regarded the fixing of renumerative food prices as "a revolutionary step".[16] The improvement in productivity was to be facilitated through the provision of modern scientific inputs, such as high-yield or miracle seeds and fertilizer. This was a momentous set of decisions, innovative and courageous. It set the subsequent framework for management of the agricultural economy, which was not shaken or changed by later administrations even when in a radical phase. The self-sufficiency in food that India was to attain later was intimately related to policy change pushed through by the government under Shastri, a man of quiet determination though outwardly modest and seemingly diffident. Some have been critical of Shastri for having opted for a technocratic approach to agriculture rather than structural change in the rural areas.[17] But it is plainly unrealistic to expect Shastri, without an independent political base during his brief 19-month tenure, to have undertaken radical agrarian reform which neither Nehru as a political giant nor his daughter with a massive political mandate could undertake during their long terms in office.

Shastri had intended no attack on the public sector; indeed, he was eager for a large Fourth Five-Year Plan even as he was persuaded of the need for a period of consolidation, concerned at the lack of pay-offs from the massive investments made in the public sector. However, policy change in relation to agriculture could not leave the whole structure of policy inherited from Nehru without impact. First of all, it meant a diversion of resources from heavy industry to agriculture, which really was to favour a sphere within the jurisdiction of the private sector as against that which was virtually a monopoly of the public sector. Again, it meant reliance on the market rather than physical controls under government planning. Besides, because of the recognized inefficiency in public sector fertilizer plants and the priority for agriculture, the government felt compelled to downgrade the public sector through opening the fertilizer field, hitherto restricted to the public sector, to the private sector, indeed to foreign investors. Furthermore, the stress on consumer goods also detracted from the public sector, for that was a sphere largely intended for the private sector. Apparently, in realization of these changes, Mrs. Gandhi told an eminent journalist privately that her "father's policies had been eroded and that the government was dominated by capitalists, Indian and foreign".[18]

The same thinking no doubt underlay the Communist attack on Shastri for "deviation" from Nehru's policies. His response was that the notion of deviation was foreign to democratic practice, for " in a democracy there is every opportunity for re-thinking and freedom for the formation of new schemes and policies", and that "in the political field situations change, men change, conditions change, environments change and the real leader must respond to the changing conditions".[19] Shastri's policy adaptation, if not the precise policies, was situationally determined; it was, in a larger sense, a function of the situation not of his own making but one bequeathed to him.

3. Economic Crisis, Controversial Policies and Electoral Disaster

Lal Bahadur Shastri died suddenly in the Soviet city of Tashkent on January 11, 1966 when he was at the height of his popularity and political power during his brief tenure in office, largely because of his firm leadership during the India-Pakistan war of 1965. There had already occurred some erosion in the influence of the Syndicate, but the new succession process marked a further deterioration in its influence; internally divided, the Syndicate had to accommodate itself to the course of events determined by other actors, primarily Congress president Kamaraj and the Chief Ministers, who were able to push successfully Mrs. Indira Gandhi at age 48 as India's third prime minister. Most members of the Syndicate were initially opposed to Mrs. Gandhi for reasons of her "access to the Left", her youth, and her spirit of independence;[20] in this they were highly prescient, for precisely these qualities would before long become implicated in a collision course between the Syndicate and Mrs. Gandhi. What weighed with Kamaraj and the Chief Ministers, on the other hand, were the forthcoming elections within around a year in 1967 where--as Nehru's daughter and being from the populous Hindi region as well as being liked in the South and by the minorities -- she carried decided assets to assure a victory for the party. Their initiative built the momentum which overwhelmed the Syndicate and compelled its members to join in the final decision, but "one astute observer cautioned that they would remain together to fight another day";[21] on the other hand, Mrs. Gandhi did not owe her office, and therefore any loyalty, to the Syndicate. She was formally elected by the Congress Parliamentary Party where she defeated by 355 votes to 169 the only other candidate, Morarji Desai, whom most of the top leadership, including the Syndicate, was anxious to see lose because of the perceived impression of him as a liability in the forthcoming elections.

The Ideological Universe of Indira Gandhi

Mrs. Gandhi was no intellectual after the model of her father. Unlike him, she had not articulated any coherent and systematic ideology before entering office.

But that did not mean that she was devoid of an ideological orientation with critical implications for politics and policy. To make a systematized, well-articulated and intellectualized expression of ideological preferences the sole possible source of implicating ideology in politics and policy is to arrogate the right of rulership to intellectuals only. But as for ideological orientation, there is contemporary information as to where Mrs. Gandhi's preferences lay. Writing in 1966, soon after her assumption of office and therefore uninfluenced by the course of events that followed in the subsequent decade, Brecher had observed: "Mrs. Gandhi shares her father's outlook; indeed, she is probably further to the Left. But like Nehru from the 1930s onwards, she will tread carefully lest an overt commitment isolate her from the Centre". Indeed, earlier, Welles Hangen had noted that "She has the reputation of being more radical and incisive than her father".[22]

That Mrs. Gandhi should share her father's ideological orientation should be no surprise, for after all early in her life she received a Marxian education through a series of letters written by him while in jail that were subsequently published as *Glimpses of World History*. Her hostel room at Tagore's university was abundantly furnished with books on socialism, leading her to be characterized as "the Red Lady of Shantiniketan". Later, while studying in England in the mid-1930s after the death of her mother, she was not particularly active in politics and was by nature reticent. But she was politically a radical and "dabbled in the activities of various Left-wing organizations in London " under the influence of her future husband, himself a "firebrand" among the socialist students at the London School of Economics; he also brought her into contact with other revolutionary students from India, chief among them Bhupesh Gupta and Mohan Kumaramangalam, who later became prominent in the Communist movement. During this period a significant influence on her was the leftist leader Krishna Menon.[23]

Returning to India soon after World War II broke out but without finishing her education, she married outside her religion and caste in defiance of the advice of her father and friends, served an eight-month prison term for the nationalist struggle, gave birth to two sons, and became her father's official hostess after independence. In this last role, she became conversant with the inner workings of Indian politics and international diplomacy. In 1955, she was more formally inducted into politics when she was made a member of the Congress Working Committee. Although not attached to any group, she had affinity as a reputed left-winger with the Congress Socialist Forum. Four years later, she served as the president of the Congress party, a role that she did not find particularly fulfilling but one that is known for her recommendation to dismiss the Communist ministry in Kerala, notwithstanding her stated general position favouring cooperation with the Communists for purposes of national reconstruction. After Nehru's death, Shastri made her a minister in his cabinet, an appointment that was endorsed by the leftist group in the Congress in the belief that "she will ensure that Nehru's policies are carried forward and that India will uninterruptedly march to the socialist goal".[24] Mrs. Gandhi soon became critical of both party and government for what she took to be a rightward shift in policy.

Inquiring simply into the ideological preferences of Mrs. Gandhi on a left-right dimension is likely to be seriously misleading, however, insofar as it neglects a singularly important element that apparently overrode that dimension -- a thoroughgoing commitment to India and its national interests. Opinion on this point is likely to be governed by the partisan passion of domestic conflict, especially as it relates to a woman of indomitable will and courage. It is best therefore to refer here to those who can view the working of political forces as a whole rather than piecemeal, that is, people who can detach themselves from divisive domestic issues, usually people on the outside. Even though a potent adversary and one who was worsted by her in conflict, Henry Kissinger made the most penetrating comment in this regard. In his memoirs, which are overly critical of her posture as it relates to the United States, he nonetheless acknowledged: "Mrs. Gandhi was a strong personality relentlessly pursuing India's national interest". Ironically, in the context of the Bangladesh crisis in 1971, Kissinger felt that the United States could with impunity apply unrestrained pressure on India in the sure knowledge that Mrs. Gandhi would not, because of her nationalist commitment, subordinate the country's interests to the Soviet Union. He calculated in terms of "what the next turn of the screw might be" on the part of the United States, and told the secret meetings of the Washington Special Action Group that "we cannot afford to ease India's state of mind. 'The Lady' is cold blooded and tough and will not turn into a Soviet satellite merely because of pique. We should not ease her mind".[25] Accordingly, operating within that mind-set the United States felt free to despatch against India a naval task force headed by the nuclear-armed USS *Enterprise*. It is in this commitment to India's national interest that her policy of national self-reliance -- an abiding theme with her -- was rooted. But, of course, with this commitment was seemingly joined the conviction, rightly or wrongly, that that interest was safe only in her and her family's hands.

The commitment to national interest did not stand in contradiction to a commitment to one or another version of socialism, but neither was it coequal with it. The prior commitment, in the circumstances of a poor, underdeveloped country, was -- as in the case of the Chinese leadership -- to national development, not necessarily to any particular brand of socialism. If national development and socialism were ever to be in conflict, it is the commitment to socialism that would give way rather than the other way round.

Even in relation to socialism, she was clear in her mind that she was neither revolutionary nor bound by dogma: "I have never claimed to be a doctrinaire socialist. I have my own version of socialism and my own vision of what Indian society should be like, and I have been working towards that vision steadfastly. It is a slow movement, but it is a surer way." Besides her gradualism, her primary point of reference, as she insistently maintained, was the centre, albeit left-of-centre, not the left as such. She believed it important "to steer clear of the extremes of ideologies, whether capitalist or any other". She further held: "In no country is socialism interpreted in the same way. Our major aim is not socialism but how to remove the poverty of the people. Socialism is a tool or a path because it seems necessary for what we have to do; and in any case, the development of any system has to be by trial and error. We have to see what actually solves problems".[26] If path and goal were in conflict, it is the goal that would have priority. One may or may not agree with her version of socialism, but that does not make her devoid of preference for a socialist society. Her stand underlines the basic point that in the Third World the choice of socialism has come out of its perceived functionality in respect of development.

Furthermore, the commitment to socialism on her part was combined with a commitment to democracy. When pressed to move more rapidly in relation to social and economic change, she repeatedly underlined -- whether out of genuine belief or as rationalization -- the inherent worth of democracy even as she manifested awareness of its limitations:

> we cannot go very much faster without changing the whole system. And it is up to us to weigh whether the price of changing the system will not ultimately come to the same cost, the same as slightly slower progress along the path of democracy.
> I believe there is no short-cut. The only sure real short-cut is more efficient functioning, a clear cut path, a clear cut ideology.... That is the only short-cut. But there is no short-cut of bypassing democracy.[27]

Such expressions are, in retrospect, likely to be deemed questionable because of the authoritarian regime she subsequently installed during the emergency in 1975. That event turned most intellectuals, especially in the mass media, against her politically. However, personal preferences of political leaders can not always be an adequate guide to their political behaviour in the unstable and at times explosive context of Third World societies, including that of the Indian subcontinent. Quite importantly, the one non-partisan study by a foreign scholar that examined the emergency thoroughly, not only in relation to her actions but also to those of her opponents, came to the conclusion that the declaration of emergency was not the perverse manifestation of an authoritarian personality. Rather, it reflected not only the legitimate but responsible exercise of executive judgement that a threat to the polity existed.[28] More seriously, one cannot write off, in terms of any fair examination of her democratic credentials, the fact that within twenty months she subjected herself to a genuine election and, when defeated, gracefully resigned.

Given this simultaneous commitment to socialism and democracy, it is understandable that Mrs. Gandhi's approach to socialism was, quite astutely, gradual and reformist. She could proceed only with that which the traffic -- the particular conjuncture of political forces -- could bear. As she expressed it, "you can't go faster than your people unless you want to cut off their heads. Now all you can do is try to educate them and push them along". Further, she perceived that "the public as a whole -- whether it is the political parties (and I include all the political parties, including Marxist) -- are not really for radical changes."[29] It is unrealistic then to expect, as some did,[30] that she could proceed with the goal of socialism as if the constraints of democracy were not operative. Quite the contrary, as a person holding the supreme political office in the land, she had to take into account several commitments -- nationalism, socialism, democracy, secularism -- and, since their demands often stood in contradiction, the goals were in tension. Consequently, there necessarily had to be trade-offs among the various elements, and commitment to any one element in the ideological complex could only be tentative. It is this which led observers to call her either pragmatic or opportunist, depending on their own preferences, and to question her ideological commitment to socialism.

The Trials of Apprenticeship in Power

Whatever the strength of her socialist orientation, her first year in office did not show any bent toward socialist policies but was rather marked by continuity with the previous administration, including basically the same cabinet team that Shastri had assembled. Apparently, she was eager to correct what she saw as deviations from her father's policies under Shastri but, with her own political position insecure and elections due in a year, she was believed to be reluctant to displease the Syndicate.[31]

Most critical to policy making in her first year in office was not only the continuation of the economic crisis which existed under Shastri but its deepening. To the already serious dimensions of that crisis had been added: another massive failure of the monsoons, leading to tremendous food shortages and a humiliating dependence on the only possible source of concessional food supplies, the United States; the inflationary consequences of food shortages and the India-Pakistan war of 1965, which reverberated in society and polity through public protest and mass violence; and the suspension of foreign aid following the 1965 war, which created a budgetary crisis for the government and an economic slowdown of vast proportions, and effectively jettisoned the Fourth Five Year Plan (1966-1971). All this took place in the context of: a resurgence of regional and religious conflict; the country's drastically diminished status in the international arena; and a prime minister who worked with a team that the Syndicate had imposed on her, who was distrusted by the party organization under the control of the Syndicate, and who was essentially on probation until the next elections in early 1967. Social unrest and political strife threatened the very fabric of the country and, as one of her biographers pointedly underlined, "it was over this scene of decline, confusion and incipient collapse that Indira Gandhi had been called upon to preside".[32]

Inexperienced and new to her office, Mrs. Gandhi came to rely in her administration on an inherited "economic" team -- consisting of Finance Minister S. Chaudhuri, Food and Agriculture Minister C. Subramaniam, Planning Minister Asoka Mehta, and her principal secretary L.K. Jha. The team was without any independent political standing but was, following a technocratic approach, oriented to modernization through reliance on the market; it was not persuaded of the utility of the public sector. Its immediate design for pulling the country out of the economic crisis centered on a policy package of (1) a projected currency devaluation in order to obtain renewal of foreign aid; (2) continued and intensified support to the "new agricultural strategy" of the Green Revolution as a long term solution to the food problem; (3) liberalization of controls to promote economic growth in the private sector; and (4) an active population control programme. This technocratic design was, of course, far removed from any pretense at socialism. After her visit to Washington, undertaken obviously to secure American aid to relieve the food crisis, Mrs. Gandhi came under severe attack for "selling out" to the United States. The left group within the Congress led the attack against her, accusing her of deviation from Nehru's policies. Increasingly isolated, she responded in a spirit of defiance and independence characteristic of her: "If it is necessary to deviate from the past policies, I would not hesitate to do so. I must pursue the policies which are in the best interests of the country as a whole. If you do not agree with these policies, you have every right to remove me and have your own leader".[33]

If criticism of Mrs. Gandhi's government was not severe enough by then, the floodgates opened with what was a bombshell announcement in early June 1966 of the devaluation of the rupee. Although she demonstrated great courage in taking what was undoubtedly an unpopular action, her decision was perceived as an abject surrender to the United States. It also alienated Congress president Kamaraj who had not been consulted; already irked by her penchant for independent decision making, he moved back into an alliance with the Syndicate. As prices rose in reaction to the devaluation, criticism mounted within the party and the country. Demands rose for the resignation of the government, with the Communist party leading a massive march on parliament. With the slogans of Bombay Bandh and Bharat Bandh, the Communists also organized strikes to shut down India's economy. To mollify the Left, Mrs. Gandhi made some placatory gestures by emphasizing the anti-imperialist aspect of non-alignment and sticking to an ambitious Fourth Five-Year Plan. In cooperation with the left group, she was able to assure a left-of-centre manifesto for the Congress party for the forthcoming elections, which asked for the public sector to secure "the commanding heights" of the economy and to place the banking system under "social control". However, her attempts to reshuffle the cabinet were blocked by the Syndicate, which also demonstrated its power by denying Krishna Menon the party's election ticket to run from his constituency in Bombay.[34] The denial of the ticket to Menon was but an illustration of the Syndicate ignoring Mrs. Gandhi in the selection of nominees for the 1967 elections. Mrs. Gandhi was in power thus only on the Syndicate's sufferance, and was on notice that she would have to go after the elections unless she subordinated herself to it.

The Congress party went into the elections in February 1967 in an enormously hostile political environment. The preceding year had been one of particularly acute economic distress as a consequence of the second successive unprecedentedly severe drought, the devaluation, and economic recession; besides the physical shortages of food and consumer goods, with the rationing system breaking down, the incessant inflation hit the public hard. During 1966-67 alone prices rose by nearly 14 per cent, a record high since the inauguration of economic planning. Indeed, since the preceding elections in 1962 prices had risen every single year for a total increase of some 50 per cent. The economic slowdown and stagnation over the period is evident from the fact that national income grew over the entire period of the Third Five Year Plan by only 15 per cent and in 1966-67 by 1.9 per cent; per capita income increased over the Third Plan by a bare 2.7 per cent and declined by 0.5 per cent in 1966-67. Agriculture was in a crisis with severe declines; the index numbers for agricultural production (base 1950-51) for the years 1961-62 to 1966-67 was 151.5, 146.0, 149.7, 166.7, 139.2 and 137.7. The consequent economic distress led to an increase in social disorder, crime and violence.[35] It also made for political alienation of the public, including the intermediate strata of the urban and rural areas which had been the pillars of support for the Congress. The resulting anti-Congress environment made for open divisions within the party, with many groups seceding from it. Party-splitting became established in the Congress at the state level. The opposition parties, sensing a unique opportunity to knock the Congress off its pedestal, agreed on electoral cooperation. Business, too, alienated from the government, changed its traditional Birla strategy of support to the Congress as a bastion of political stability, notwithstanding whatever grievances it had against it, and now adopted the Bombay strategy of strengthening the opposition in order to humble and shock the Congress.[36]

Electoral Disaster

In its first election after the death of Nehru, the Congress party, even though returning to power at the centre, met with an electoral disaster which put an effective end to the one-party dominant system that had been characteristic of the Indian polity. Until then used to being the natural party of government, the Congress was ousted from power in eight states. This made for increased political instability in the states in the subsequent years as party members defected back and forth and coalition governments fell continually; meanwhile, with governments of different political hues in power in the states, centre-state tensions grew. The political system underwent a drastic change; under Nehru, it had been only formally pluralist; now it became really pluralist. At the centre, the Congress party's majority was reduced from 361 seats to 283, and it now faced a more self-confident and increasingly obstreperous opposition. Unlike before, when the opposition was tilted to the left, the opposition in the new parliament was ideologically polarized, with the rightists having an edge; the right-wing Swatantra party was the largest opposition party (44 seats), followed by the Jan Sangh (35); the left wing included the CPI (23), Samyukta Socialist Party (23), CPI-M (19) and Praja Socialist Party (13), but all these were preceded in strength by a regional party, the DMK (25).

Many of the top political leaders of the Congress party were defeated in the elections, among them Congress president Kamaraj, West Bengal political boss Atulya Ghosh, and the right-wing leader from Bombay, S.K. Patil. The resulting weakening of the Syndicate, however, was of enormous strategic advantage to the prime minister, for it undermined any plans it may have had for the removal of Mrs. Gandhi, who was supported by the Left. Still, the Syndicate was able to impose on Mrs. Gandhi the inclusion of the other contender, Morarji Desai, in her cabinet as Finance Minister and Deputy Prime Minister. Mrs. Gandhi compromised in the event because of the party's slim majority in parliament, where the defection of only 25 members could bring about the downfall of the government. But she regarded Desai as part of a Syndicate design to keep a check on her and to build a rival centre of power; for his part, Desai thought that without him in the cabinet "the woman would sell the country to the Communists."[37] At the same time, with the conviction that she carried support in the country, Mrs. Gandhi was determined not to subordinate herself to the party organization, whose conservative leadership had apparently been discredited at the polls. Although the party seemed to have momentarily closed its ranks in the face of political adversity, the situation augured ill for continued unity, given the domination of the government and party by different and opposed groups, a division which soon assumed an ideological form.

The 1967 elections also marked a rise in the political power of the rural intermediate class and by the same token some decline in that of the urban middle class. This was not simply in terms of the rise in the proportion of representatives from the rural areas,[38] but also of change in the texture of politics. Previously, the rich and middle peasantry may have only exercised power at the state level and also blocked land reform which affected adversely on its privileged economic position. But now it became a more potent actor in politics at the centre, vociferously demanding a positive response to its demands. Chief among these demands were: support prices for food, subsidized inputs for agriculture, reservations in the bureaucracy and, importantly, a greater share in political power. These demands would be pressed with greater intensity as time went on. For now, the serious consequences of the election results related more to what happened overall to the Congress party.

4. The Radical Thrust Amidst Political Realignment

The Congress party's electoral disaster in 1967 led to an internecine conflict within the party which finally eventuated in a split in 1969, about two and a half years after the elections. The same period saw a radical thrust imparted to government policy, which was manifest in (1) a reversal of the trend of economic liberalization -- which had started under Shastri, if not earlier -- managed through the adoption of restrictive legislation in relation to the private corporate sector, and (2) a push toward nationalization, most dramatically in banking. These radical measures were not always the result of calm and deliberate inquiry; rather, they were pushed through in the heat of factional and party conflict, indeed of brinkmanship. However, the radical thrust was a consequence not of any single factor, but of a political conjuncture of several elements.

First and foremost among these elements was the insistent pressure on the part of the left group in the Congress party, which identified itself fulsomely with Nehru's goal of a socialist society for India. Organizationally differentiated as the Congress Forum for Socialist Action, the Left had been strengthened in recent years with the entry into the Congress of former socialists and communists. Some of the latter had entered the Congress party as part of a deliberate infiltration design to achieve socialism through internal pressure.[39] If a fundamental assumption of Nehru had been that a socialist society could be brought about by a relatively faster rate of expansion of the public sector until it reduced the private sector to a mere appendage of the national economy, that assumption had been dealt a disastrous blow by the economic crisis that struck the Third Five Year Plan and the diversion of public investment into defence and agriculture. Meanwhile, even during the Second Five Year Plan the private sector had shown greater buoyancy than the role that had been envisaged for it, whereas the public sector fell behind the plan targets. The Left was also angered by direct business participation in the 1967 elections which it saw as a blatant "attempt to convert private sector economic power into political power, in order to gain control of the government and change the direction of the Indian economy".[40] Furthermore, the downgrading of the position of the Left in government and party that followed the Sino-Indian border war, the greater liberalization of the economy after Nehru which served to accord a more important and freer role to the private sector, the increasing self-assertion on the part of the business community, the electoral reverses administered to the Congress in the 1967 elections evidently because of failure on the economic front, the rise of right-wing political parties to pre-eminence in the opposition instead of the communist and socialist parties, and the continued economic distress with inflation increasing by another 12 per cent during 1967-68 -- all these developments made the Left in the Congress desperate to save the socialist project of Nehru and determined to reverse the course that events had taken since the Sino-Indian border war.

The Left consequently launched an attack on large-scale business in the private sector, in order to have restrictions placed on its operation, and pressed for increased nationalisation which would serve the double purpose of reducing the economic power of the private sector and advancing that of the public sector. For the accomplishment of these purposes, it was ready to seek the cooperation of other like-minded or sympathetic parties. Given its ideological thrust, the Left was bound to run into collision with the Syndicate with its conservative cast. The Left regarded the Syndicate as a gang of old and exhausted men whose power base was constructed on a pyramid of bogus membership; it viewed these party bosses as interested in nothing more than personal power and patronage; it held them responsible for undermining the Nehru design for the achievement of a socialist society, and for bringing the party to its present pass. Between the Left, often proclaimed as the "Young Turks", and the Syndicate as the Old Guard there prevailed also a generational conflict which reinforced the ideological gap. Notwithstanding its ability to make considerable political noise, however, the Left could hardly succeed if the rest of the party was united.

Critical to the success of the Left therefore was the alliance with it eventually, perhaps always implicitly, of Mrs. Gandhi. No doubt, Mrs. Gandhi had ideological affinity with the socialist goals of the Left, for those were ultimately her and her father's goals. But, placed as she was at the head of government, she might have been more prudent -- as her father had been and as she herself had been prior to the 1967 elections -- in regard to further radical measures in recognition of the limited administrative capabilities of the state. Indeed, she was inclined toward standing above factions as her father had done, though she did not as yet command a comparable independent political base; there was a prolonged initial refusal on her part after the 1967 elections to be identified with any of the factions. However, two factors finally impelled an open alliance with the Left.

First, Mrs. Gandhi, like most Congressmen, identified the national interests of the country as being safe only in the custody of the Congress party. But the 1967 election results obviously served a warning to the Congress party that it would have no political future unless it worked out a radical programme and offered a party leadership that appealed to the electorate. This message of the decline and downward slide of the Congress party in the affections of her countrymen was reinforced by the party's poor performance in the mid-term elections of 1969 in four states. The serious potentialities of public alienation from the party, and the consequent political instability, were earlier demonstrated by the rise of a violent insurrectionary Communist movement, known as the Naxalite movement, in different parts of India, but more particularly in West Bengal. With these developments, the spectre of electoral defeat stared the party in its face. Mrs. Gandhi felt strongly that, as she expressed it later, "the party was really getting so far from the people, that I could not see the Congress Party surviving even till 1972, till the next elections".[41] She fervently believed that in this situation the Congress could survive politically only through moving to the left. There is considerable testimony on this fundamental point. One scholar notes: "Indira Gandhi had become convinced that the country as a whole was shifting towards the left. Therefore a bold leftist policy not only was the one hope to preserve the absolute majority of the party at the Centre, but, failing this, it would make possible a coalition Government with the leftist parties". Another commentator remarks: "She believed that the Congress Party was doomed to extinction in the not very distant future if it did not make a serious effort to square its practice with its professions. The party must prove what it claimed to be, the party of the common man, bent on banishing the poverty that still haunted millions of lives".[42] Several years after the event, Mrs. Gandhi herself spelt out the essential relationship between her radical change and political survival:

> There was a group in our Party which did not like the way we wanted to go. They created a situation in which it was my very considered view, that if their way would have been followed, Congress would have been completely finished. As you saw, in the next election, that part of the Congress was practically finished. Had we been with them, we would also have finished.

Even when in the political wilderness, after having been ousted from power, she maintained: "After the 1967 debacle, serious thinking led to the conclusion that a progressive and radical programme was essential to revitalise the Congress and regain the confidence of the masses. I took the initiative". Much earlier, Jagjivan Ram had stated the dilemma sharply: "The Congress has reached a stage when it must pursue radical policies or disintegrate".[43]

Ironically, even though she was to enter into a tacit alliance with the CPI to maintain herself in power and to implement and legitimize her radical thrust, her swing to the left was aimed no less at the Communists outside than it was at the conservatives inside; she told one journalist: "If I don't do anything to take wind out of the sails of the Communists, the entire country will go Red". In another context, she stated: "I dislike the word 'contain' but if only people paused to think, I am containing communism in India. The Government cannot afford to move even the slightest bit to the Right, because this will in no time begin a move on the part of the communists to take over, interpreting my actions as anti-people. I am no communist....but I do believe in trying to achieve a socialistic pattern of society in India". Again, she told another journalist: "these measures are aimed at reducing disparities. If we do not go ahead with them, the poor will rise in revolt. And I am violently against a violent upheaval....We have no option. Either we do these things peacefully ourselves or we will be overtaken by a violent revolution." Thus, hers was a pre-emptive political strategy of reform. A move to the left-of-centre was essential to undermine the Communists, but she perceived that the Syndicate, as a front of big business and landlords, stood in the way of reform and therefore had to go. Mrs. Gandhi also blamed the conservatives in her party for helping, through their policy stance, to bring the communists to power, as in Kerala and West Bengal".[44]

Despite her position as the head of government, curiously, Mrs. Gandhi was viewed neither by the public nor by the parliamentary party as having been responsible for the 1967 electoral debacle.[45] Rather, it was the party leadership of the Old Guard -- as indicated even by the treatment meted out to it by the electorate -- that was believed to be culpable on this account. Given the conservative orientation of the Syndicate, any reorientation of the party in a leftist direction was bound to set Mrs. Gandhi on a course of conflict with the Syndicate. On the other hand, in a country whose political culture had a strong strain of resentment against the wealthy class, any move to the left was likely to be popular. But for a considerable period she suppressed her political divergence with the Syndicate for fear that the consequent split would result in the collapse of the government because of its tenuous majority.

Mrs. Gandhi's larger design was rooted in a strategic calculation that the political mood of the country demanded change in a leftist direction. The perception of the direction of change demanded by the public mood was decisive in bringing about political change. In the final analysis, it was the nature of the political context that required and facilitated implementation of ideological preferences of the leadership. If the perception had been that the public mood demanded change in an opposite direction, perhaps the leadership would have accommodated it policy-wise. In a political system where holding of office was a function of an electoral mandate, Mrs. Gandhi could not do otherwise. In this sense, again, the nature of the political system, requiring a periodic resort to electoral validation, is absolutely fundamental in understanding her developing political stance. That the stance coincided with her own reputed ideological orientation lent conviction to her struggle with the Syndicate. The lack of a radical policy orientation on the part of her government before the elections could be attributed to the domination of the Syndicate, just like Nehru's own policy moderation in the half decade after independence had been a function of the control of the party machine by Sardar Patel. Furthermore, Mrs. Gandhi took the Congress party to be electorally safe only under her own personal leadership, just as Nehru had felt at the time of his political coup against Congress president Tandon in 1951.

Second, it was not simply a question of Mrs. Gandhi being at ideological odds with the Syndicate in the light of her conception of what the political situation required; it also became a question of sheer political survival in view of the threat of her removal from office by the Syndicate. Regardless of whether it was because of her "access to the Left" or her Nehru arrogance, the Syndicate had been initially opposed to her appointment as prime minister and, when outwitted, took it to be only a transitional appointment until the 1967 elections. Several moves by the Syndicate, now joined by Morarji Desai and reunited with Kamaraj, after the 1967 elections convinced Mrs. Gandhi of a political conspiracy to oust her, a conspiracy that extended to encompass cooperation and collusion with opposition parties that had in the past been anathema to the Congress party.[46] By way of a psychological explanation of Mrs. Gandhi's political behaviour, it has often been claimed that she acted out of a sense of personal insecurity that was bred into her by an unstable childhood;[47] it is not often admitted, however, though the facts are not disputed, that her political survival was, indeed, in jeopardy by virtue of the activities of her erstwhile political colleagues. With political survival at stake, the clash with the Syndicate assumed titanic proportions, in which the alliance of Mrs. Gandhi with the Left was forged, and led to the split of the Congress party.

It is and will perhaps continue to be a matter of intellectual dispute whether the conflict was a mere factional struggle to which was cynically added an ideological dimension,[48] or whether it was basically an ideological conflict to begin with which was necessarily expressed through a factional form. Perhaps the intellectual dispute lends itself to no easy resolution, for both factional and ideological elements were inextricably intertwined in ways that cannot be satisfactorily disentangled. Perhaps it is sufficient for the purpose here to recognize that the parties to the conflict gave expression to their differences in an ideological form, with one of them insistently proclaiming that the issue in conflict was fundamentally ideological, and further that political parties outside ranged themselves on one side or the other along ideological lines as events took shape. To be sure, not all individuals in either group thought exactly alike, but it is significant that ultimately each group was to enter into alliances with political parties outside which left no doubt about their overall ideological bent. And it was in the heat of this political clash between the Syndicate and the Left led by Mrs. Gandhi that a radical thrust was initiated in government policy.

Politics, Ideology and the Course of Conflict

The conflict within the Congress manifested itself during the post-mortem on the results of the 1967 elections. It was taken as given by Congress leaders that the party's disaster at the polls was related to the disparity between the Congress profession of socialism and acceptance of socialist policy, on the one hand, and its actual practice and policy implementation on the other. But there was division over who was to blame, with one side casting blame on the government, eager to subordinate it to direction by the party; the other side held the party responsible on account of its weak organization and its failure to mobilize support for the government adequately. The latter side obviously wanted a radical reform of the party organization and its subordination to the government as had been the case under Nehru during much of his tenure as prime minister.[49]

At first sight, the conflict seemed like a conventional one between party organization and government that had been a pervasive feature at the state level under the umbrella of the one-party dominant system. The party-government conflict is understandable, given the party's desire to protect its power and influence, only recently acquired in the last declining years of Nehru; earlier, its power was manifest more indirectly in undermining his policies in practice at the state and local levels. The conflict is understandable also given Mrs. Gandhi's uncertainty about her political future and her simultaneous penchant for independent decision-making. However, it would be a superficial conclusion to take the conflict as simply a party-government conflict. Such conflict is understandable in the context of a normal ongoing political process, but the present cleavage arose or was intensified in the context of an electoral disaster. The question in its aftermath was what ought to be done in relation to the patently obvious, not just an apparent, decline in the fortunes of the Congress. Where such a question arises and policy changes have to be undertaken to cope with the decline, ideological orientation or preferences necessarily come into play in the aggregation and organization of policy options. This does not mean that opposed positions have to be the extreme ends of the ideological spectrum, particularly in the case of the Congress where party members subscribed, at least formally, to an ideological consensus and where the party was known for, and had prided itself in, its ability to work on the basis of consensus. Rather, in this case, it meant that the centre of gravity of the opposed groups fell on either side of the centre, one to the right and the other to the left of it. It is not without significance that on her election as prime minister after the 1967 elections, Mrs. Gandhi made no claim to being a fiery radical but rather expressed the view that, in the light of the problem of poverty of the masses, the government would need to proceed more expeditiously with the policy of democratic socialism and that her government would have to be "left of centre".[50] On account of this essential moderation, the consensus orientation of the Congress, and the party's heritage of the nationalist movement, the present cleavage did not necessarily imply an immediate break-up. Neither side wanted to take the blame for splitting a historic organization; furthermore, each had to be cautious as to how the power position of the Congress party, but more so of each faction, would be affected by a split.

By May 1967, Mrs. Gandhi had reason to be flush with success after getting Zakir Hussain elected president of India against the wishes of Congress president Kamaraj. This was indicative of her broad appeal encompassing other leftist parties and parties of regional and minority religious groups. But she was not eager to plunge the party into a bitter division at the Working Committee meeting that month. Indeed, with both sides apparently attempting to shun ideological polarization of the party, and keen to have the party continue as a broadbased national force, the executive was able to work out a consensus on a package of future policy reforms, known as the Ten-Point Programme, to hasten the advance towards socialism.

Although to a considerable extent a restatement of past policy positions, the Ten Points would later become the springboard for intensification of ideological conflict in the party and subsequently a driving force behind new policy measures. There were reservations on the part of Morarji Desai and others sympathetic to his viewpoint, and they aimed to block or delay their execution, but for now the different sides agreed to the consensus. The Ten Point Programme covered: (1) social control of banks; (2) nationalization of general insurance; (3) progressive expansion of state trading in imports and exports; (4) public distribution of food grains; (5) the organization of consumer cooperatives in rural and urban areas for supplying essential commodities at fair prices, and the processing and manufacture of such commodities on an extensive scale under the auspices of the state and cooperatives; (6) effective curbs on monopolies and concentration of power; (7) provision of minimum needs for everyone; (8) limits on urban income and urban property; (9) rural works programme for providing employment to landless labour, implementation of land reforms, provision of credit to agricultural labour against assets that would be created, and supply of drinking water; and (10) removal of the privileges of former princes. In the opinion of one scholar, in reference to Mrs. Gandhi, "it was largely at her insistence that the first Congress Working Committee after the general elections adopted a new radical ten-point programme".[51]

When the larger AICC met in New Delhi in June 1967, the Congress Left attacked the party's reluctance to go in for a more radical programme, including the nationalization of banks. But the AICC endorsed the Ten-Point Programme, reflecting the leadership's determination to present a united front. The Congress Left renewed its attack on the leadership at the AICC meeting in Jabalpur in late October 1967, and pressed the party and government to implement the Ten Point Programme on a time-bound basis or "perish", and demanded immediate bank nationalization. Finance Minister Morarji Desai determinedly opposed bank nationalization until social control over banks had been tried. Although Mrs. Gandhi herself favoured bank nationalization, she refused to be a partner to public disruption of the party. Behind her refusal was an eagerness to avoid a challenge from Desai "before she had consolidated her position," indeed, to detach Desai from the Syndicate in order for her to get a new Congress president more to her liking.[52] Morarji Desai was thus able to have his way on this issue as also to stall on the remainder of the Ten Point Programme. Jabalpur demonstrated that, for all its vocal aggressiveness, the Left could not succeed in the face of a united leadership. However, despite the surface unity, real unity escaped the leadership. Even the surface unity had been obtained by Mrs. Gandhi at the cost of conceding to Morarji Desai the primary decision-making role in economic affairs, but this situation could not continue without eventually undermining her position as prime minister.

At the same time, "however conciliatory the Prime Minister's attitude, it was obvious that she could not hope to play a decisive role in the party affairs so long as she lacked a grip on the party machinery". Mrs. Gandhi could not get her way on the new Congress president and had to settle for a compromise choice. But the new Congress president, S. Nijalingappa, joined forces with the Syndicate and Desai, and "soon made it clear that the Prime Minister could not expect to meddle in party affairs". Indeed, the party leadership at the AICC meeting in January 1968 at Hyderabad ignored her in constituting the Working Committee; except for one member, none others on the committee were her supporters. Mrs. Gandhi thus stood isolated in the party. Masani comments: "The Hyderabad session marked a watershed in the evolution of Indira Gandhi's political strategy. It indicated that so long as Congress politics turned on narrow factional and regional alignments, the Prime Minister, lacking a provincial or an organizational base in the party, was bound to come off badly....So far Mrs. Gandhi had sought to maintain her authority and the stability of her government by eschewing ideological controversy. At Jabalpur and again at Hyderabad she had avoided raising economic issues for fear of polarising the party. But in doing so she had surrendered the political initiative...."[53]

Faced by a hostile party, Mrs. Gandhi came increasingly to rely on close political confidantes, and remained silent when members of the Congress Left criticized conservative leaders of the Syndicate. The breach widened, and by August 1968 the bulk of the Syndicate had decided to remove Mrs. Gandhi, but were held back by the threat of the resulting break-up of the party.[54] Meanwhile, the verbal warfare between the Congress Left and the Syndicate continued; in December 1968, S.

Chandrashekhar, a member of the Congress Left, along with Communist members accused Morarji Desai in parliament of favouritism toward the Birlas, one of the country's two largest business houses. Mrs. Gandhi angered the Syndicate by not joining a move for disciplinary action against Chandrashekhar. Congress president Nijalingappa concluded: "I am not sure if she deserves to continue as P.M", while Desai "discussed the necessity of the Prime Minister being removed."[55] The results of the mid-term elections in four critical states in early 1969 -- in none of which the Congress party could achieve a majority -- confirmed the trend of declining fortunes of the Congress party revealed in the 1967 elections. The party leaders were in a panic, for the prospects for the next general elections in 1972 were now held to be possibly worse than had been earlier imagined. In a chastened mood, they raised the possibility of alliances with other parties to stay in power. But with which parties? On which side of the ideological divide? The election results thus made critical the question of the appropriate party strategy if it were not to be defeated. However, Mrs. Gandhi, who had been denied a role in the selection of candidates, still remained remarkably restrained. During April 1969 she reiterated her faith in democratic socialism but as a middle-of-the-road policy, and opposed ideological polarization.[56]

But ideological polarization could no longer be stemmed, and became manifest at the AICC meeting in April 1969 at Faridabad. Here came "the first public indication that the Syndicate was prepared for a direct confrontation with Mrs. Gandhi," with the Congress president hurling an "open challenge to the prime minister's authority in the key area of government's economic policy".[57] The session was marked by heated exchanges between the Left and the conservative leaders, particularly Desai, and there was deadlock over most issues. Even as leaders expressed their opposition to polarization and splitting the party, that was precisely what was under way. Apart from the stalemate on policy issues, the Faridabad session was noteworthy for Nijalingappa's speech, which one author characterized as "incendiary".[58] Even if what he said was factually correct, his speech attacked the very heart of Nehru model to which the Left was strategically or tactically devoted, for he made utility, not sacredness, the test for the fostering of public or private sectors:

> It is a fact that monopolies in industry are growing. It is no use our crying hoarse against these monopolistic tendencies. What we have to consider is how best we can control the growth of monopolies without detriment to industrial development....We should also see that rapid industrial progress is maintained at all costs. I believe that industries, by whomsoever established, should be encouraged. If private industries misbehaved or made undue profits, they can both be punished and controlled through fiscal measures. While we should encourage the development of industries, we must see that big industries are not established in the public sector without due regard to demand and the capacity to produce.

He then went on to sharply question the functioning of the public sector and the entire system of controlled economy, and instead urged encouragement of the private sector:

> Some industries are so badly managed that the full capacity is not utilised and that they are run on very unscientific methods. The labour employed in some of them is so large that the result can only be loss and no profit. I am of the view that the public sector too should yield profits. Simply because an industry is in the public sector it does not mean that we should be saddled with losses and be content with whatever is produced. If production of articles in the private sector can be achieved more economically we can even encourage the private sector. There is a case for reviewing this public sector attempt at establishing large scale industries.
>
> There are lots of complaints about the delays in granting licenses, resulting in corruption. Where there are controls and licensing, there is always corruption and *the sooner we do away with licensing and controls the better* it would be unless there is compelling necessity.[59]

Such views had thus far come from the opposition parties of the right, not from the Congress leadership, especially at the very top. The Left was in an uproar, but was thwarted from challenging the president. However, Mrs. Gandhi now did what she had avoided so far, identifying herself with the cause of the Left against the conservative leadership; in a fighting speech she defended the public sector:

> If there was delay in arriving at quick decisions it was warranted by the paramountcy of India's political and economic interests. The Government had to consider all the aspects involved when the question of issue of licenses came up.
>
> Some people criticised the public sector for its failure to make profits. It was true that some of these undertakings did not earn profits. But profit was not the sole motive in organising this sector of Industry. The public sector was conceived as the base of Indian industry so that the country might have more machines, more steel. It also ensured India's freedom. It catered to defence and agricultural needs. To the extent India depended upon imports, its independence was compromised, encroached upon.[60]

She declared that "the public sector must continue to occupy a special place in our policies and our outlook". While acknowledging that there were legitimate grounds for dissatisfaction with the performance of some parts of the public sector, she pointedly underlined: "we should not forget that a great deal of fire is deliberately directed against the public sector by those who are ideologically committed to the system of market economy".[61] The ideological battle lines were thus being sharply drawn. Notwithstanding her strong defence, Mrs. Gandhi still did not precipitate matters on issues critical to the Left, such as bank nationalization and new elections in the party. This was despite her awareness of "the plots being hatched to dislodge her from office". Nonetheless, the electoral disaster of the 1967 elections had steadily brought about an ideological polarization and intensification of the inner-party struggle despite the sentiment against a party break-up. Masani insightfully concluded: "Till Faridabad she had been moving cautiously, feeling the ground and testing reactions. The favourable response of the party's rank and file to her Left-of-Centre stance had been encouraging and appears to have convinced her that only by identifying herself with a radical, socialist programme could she assert her leadership over the Congress and rehabilitate the party in the eyes of the masses. In the following months she would cast aside her initial hesitation and pursue this strategy with a political sagacity that was as ruthless as it was unexpected".[62]

The occasion for that strategy arose in the context of the election of a new president for India, when there was a tussle between the Syndicate and Mrs. Gandhi as to who should be nominated. The Syndicate was anxious to have one of its own members in the presidency in order to assist in the ouster of Mrs. Gandhi, while Mrs. Gandhi wanted someone who would act on her advice in case members of parliament defected from her party. But this struggle became intertwined with the ideological conflict. With a design to split the Syndicate ideologically, Mrs. Gandhi submitted a memorandum, which she characterized as "stray thoughts", to the Working Committee for the AICC meeting at Bangalore in July 1969. The memorandum incorporated into it a list of radical demands, among them: nationalization of banks, establishment of a commission on monopolies, barring of big business from most consumer goods industries, and restrictions on foreign investment in sectors where local technology was available. She had thus irreversibly chosen to link her political fortunes with the Left, which for its part stood ready to rally behind her and to mobilize popular forces for her. But the Syndicate did not want to be diverted from its main path; despite strong opposition from its conservative members, the Syndicate routinely endorsed the contents of the memorandum and then went on, in what was the first step in its strategy for replacing Mrs. Gandhi, to nominate Sanjiva Reddy for the presidency against her opposition. Congress president Nijalingappa noted that "she has been taught a lesson now". Not only that, a protege of Desai's let it be known publicly that, after Reddy was ensconced in the presidency, it was planned to ease Mrs. Gandhi out of office.[63]

Publicly humiliated and with political death staring her in the face, Mrs. Gandhi responded with surprise and celerity; she dismissed Morarji Desai from the Finance Ministry, took over the portfolio herself, and nationalized the fourteen largest banks through an ordinance. The measure met with great public acclaim and enthusiasm, and Mrs. Gandhi's popularity soared with her decisive actions. Rejected by her own party, she had turned to a different constituency, the national one, and demonstrated that she was a political force to be reckoned with there. As the Syndicate feared for the election of Sanjiva Reddy, it began to negotiate with the right-wing parties, Swatantra and Jan Sangh, to assure his election. Provided thus with evidence of the Syndicate's alliance with what she considered reactionary forces as also its intent to remove her, Mrs. Gandhi rallied support from within the Congress party and leftist and regional parties to have V.V. Giri, the candidate favoured by her, elected. The election of Giri drove home to everyone that, whatever her problems with the Syndicate, Mrs. Gandhi commanded an all-India popularity that reached beyond her own party; this phenomenon was comparable to her father's presumptive popularity at the time of his struggle with Congress president Tandon in 1951. This broad support indicated that, in case a coalition government became necessary, Mrs. Gandhi was the leader that would be most acceptable to the opposition and the nation. Subsequently, in the manoeuvres to censure and discipline her and her supporters, and the attempts of rival factions to retain or obtain control of the party organization, the party split into two in November 1969, an event which "shook Indian politics to its roots".[64] Prior to the break-up, Mrs. Gandhi gave expression to her view of the prospective split:

> What we witness today is not a mere clash of personalities, and certainly not a fight for power. It is not as simple as a conflict between the parliamentary and organizational wings. It is a conflict between two outlooks and attitudes in regard to the objectives of the Congress and the method in which the Congress itself should function. It is a conflict between those who are for socialism, for change and for the fullest internal democracy and debate in the organisation on the one hand, and those who are for *status quo*, for conformism and for less than full discussion inside the Congress....This group is not a new phenomenon. It has existed in our party throughout the last 22 years and even before. I know that this group constantly tried to check and frustrate my father's attempt to bring about far-reaching economic and social changes....My own experience even before the Fourth General Elections was that the forces of *status quo*, with close links with powerful economic interests, were ranged against me.[65]

On the other hand, Congress president Nijalingappa held on behalf of the Syndicate that the entire party was for the policies that Mrs. Gandhi had cleverly appropriated to herself, but charged: "You seem to have made personal loyalty to you the test of loyalty to the Congress and the country. All those who glorify you are progressives....It appears that everything is pemissible and pardonable to those who are recognised supporters of the personality cult that is threatening democracy in the organisation....I have my apprehension that this pattern may lead us to one-man rule in the organization and the government, and not to the strengthening of democratic socialism".[66] Although the Congress president in this way attributed the conflict to Mrs. Gandhi's hunger for personal power or domination, it is noteworthy that soon afterwards, indeed even earlier, the Syndicate accused her of being pro-Communist,[67] even a crypto-Communist, and thus, ironically if not entirely accurately, agreed with her that the conflict had been implicitly an ideological one even from the viewpoint of the Syndicate. Furthermore, within less than a year and a half, the Syndicate entered into an alliance with the Swatantra party and Jan Sangh, thus demonstrating the ideological differentiation between the Syndicate and the Congress under Mrs. Gandhi and confirming the political evaluation that Mrs. Gandhi had placed on the Syndicate. She herself underlined that "it is not without point that the Opposition they sat with is the extreme rightist Opposition -- the two parties which are in different ways entirely opposed to anything which Mahatma Gandhi stood for.[68]

With the split, the new Congress under Mrs. Gandhi was able to carry with it some 60 per cent of the AICC delegates. In the parliamentary wing, all but some 60 MPs remained with Mrs. Gandhi; while this placed her government in a minority position there was no threat to its existence, because of the support offered by the CPI and the DMK along with a large contingent of independent members. Mrs. Gandhi had thus been able to enact a political coup on the pattern of her father's in 1951. But it had been much more of an uphill task for her, considering the seasoned leadership that was ranged against her, until then regarded a transient political figure.

Although the Congress had been purged of the Syndicate, it had not been made over into a new organization; it could not have been, given the fact that the bulk of it, especially in the parliamentary wing, remained with Mrs. Gandhi. That much was clear to the Left itself: "We must accept the painful reality that many who do not subscribe to the socialist objectives have taken position on our side because of political expediency".[69] To be sure, some members of the Left, including several ex-communists, achieved greater prominence, while the CPI could exert greater pressure on the government through its leverage in parliament. Also, former Congressmen with socialist and Communist leanings sought to return to the new Congress. Besides, the new party's pronouncements came to be dominated by Marxist phraseology and epithets; not only did Congress documents bristle with phrases such as reactionaries, fascists, right reaction and left adventurists, but they also came to employ the Marxian mode of analysis, distinguishing between base and superstructure.[70] Even more moderate Congressmen such as the new Congress president, Jagjivan Ram, spoke emphatically that "as a corrective to the existing trends" of monopoly and concentration of economic power, "which are disturbing, the public sector must expand. More of the commanding heights of economy must be manned by the nation".[71] But the overwhelming characteristic of the new Congress was the personal dominance of Mrs. Gandhi. The new Congress was therefore not inherently radical, it would be so only as long as she wished it to be so.

Summary and Conclusions

Near the end of Nehru's tenure, India was faced with a serious economic crisis. Nehru's successors inherited that crisis, they did not create it. The crisis arose as a result of Nehru's economic strategy, and it was worsened by the Sino-Indian war of 1962 and further aggravated by the India-Pakistan war of 1965. Confronted by food and consumer goods shortages, inflation and balance of payments deficits, and a breakdown in the national consensus on policy, Nehru's successors first attempted to shift away from Nehru's economic strategy. That strategy had centered on socialism, essentially consisting of heavy industry, a strong public sector, physical controls on the private sector, and distrust of the market. Instead, his successors sought to emphasize agriculture, restraint on expansion of the public sector, relaxation of controls on the private sector, and greater reliance on the market.

Soon, however, the crisis and the electoral consequences it generated were followed by a titanic struggle within the Congress between Mrs. Gandhi and a Left group that supported her, on the one hand, and the party caucus on the other. Innately reticent, diminutive, retiring, young and untested, Mrs. Gandhi was, as she was to say of India, "always undervalued, underestimated, not believed".[72] Socialist leader Rammanohar Lohia called her *gungi gudiya* (dumb doll) and Morarji Desai dismissed her as a *chhokri* (a slip of a girl). However, she surprised everyone with her shrewd political skills and, by the end of 1969, emerged as the most powerful political leader on the Indian scene, triumphing over the collectivity of party stalwarts from the Old Guard that was ranged against her. In the process she split the Congress party, a historic national institution. This later led to her being characterized as the dismantler of institutions for the sake of personal power. It seems, however, that if she succeeded as a single individual in destroying institutions, as charged, then it is no great tribute to the strength of the institutions in the first place.

With her political struggle against the Old Guard was associated a marked shift to a more radical stance, especially the dramatic measure of bank nationalisation which she pushed through. Because of this association of her radical stance with the factional struggle within the Congress, many have questioned the ideological basis of her policymaking. It would seem, though, that often such questioning is rooted in the "saintly idiom" that W.H. Morris-Jones isolated in an insightful and seminal contribution in the 1960s.[73] In the Indian political tradition, it is the purity of motives that is critical in any examination of policy rather than the context or the consequences; thus in the case of Jayaprakash Narayan who, over the course of his political career, changed several times from one ideology to another -- not just from one version of the same ideology to another -- his actions are beyond question because of a perception of the selfless nature of his motives. On the other hand, in the case of Mrs. Gandhi, her actions are often considered base because they allegedly proceeded out of self-regarding motives. However, since the purpose of analysis is to understand and explain political phenomena, rather than render moral judgement, the issue deserves more serious attention.

In the questioning of Mrs. Gandhi's actions, a striking contrast is drawn between her and her father. For example, one eminent commentator proclaims that, no doubt, "Mrs. Gandhi had a leftist inheritance from her father but she was far from being an ideologist. She was too pragmatic for that. She was a politician more than anything else. Ideology was her weapon; she was not its weapon".[74] Even more strongly, another commentator states: "Mrs. Gandhi's socialism was of the populist variety. In other words, she put through measures largely to impress the masses that she was more 'radical' than other leaders in the country"; as against this, he finds that to Nehru the "achievement of socialism was a matter of conviction." Further, "basically, Mrs. Gandhi was a political creature. She did not pursue economic goals for their own sake. She had use for them only if, in her calculation, they were capable of yielding political dividends".[75] Such a position is taken despite the fact that the results of policy, at least in the economic realm, are found to point to a more favourable evaluation of Mrs. Gandhi in comparison to her father. As an otherwise critical observer points out, though others never even deign to acknowledge:

> But he ended with many of his policies in tatters. The humiliation of the Chinese invasion exposed thoroughly the shortcomings of his foreign policy. He swore by self-reliance but in practice made the economy so dependent on foreign aid that it collapsed when aid was cut off after the 1965 war. His heart went out to the starving masses but his neglect of agriculture worsened starvation and made the country abjectly dependent on PL480 wheat, the full dimensions of which became apparent in the two drought years following his

death.

Mrs. Gandhi reversed all these trends decisively. When she first became Prime Minister in 1966 the country was living a ship-to-mouth existence, begging the US for more food. On her death 18 years later India enjoys record food stocks of 22 million tonnes and has become a net exporter of grain. In 1966 India had run out of foreign exchange and had to devalue the rupee to get succour from the International Monetary Fund.....net reserves were actually negative at minus Rs. 19 crore. On her death, the country's reserves (net of IMF loans) stand at a healthy Rs. 2,500 crore, having risen by Rs. 600 crore in the last six months. Her victory in the 1971 liberation of Bangladesh made India the dominant power on the subcontinent and wiped out memories of the Chinese humiliation. During her rule, India built up the third largest technical workforce in the world which sent satellites into space and mastered nuclear explosions....

Her father operated in relatively easy economic times. He inherited huge sterling balances on Independence; he ruled when the world economy was surging along in the postwar boom, he enjoyed unprecedented infusions of foreign aid; and he lived in the era of cheap oil. Yet with all these advantages he could not raise the economy's growth rate above 3.5 per cent, a rate which persisted for three decades after Independence and was dubbed 'the Hindu rate of growth'....

Mrs. Gandhi had to work in an infinitely more difficult environment. She inherited no sterling balances, only debts to the IMF. The world economy sagged and stagnated in the last decade of her life, creating problems for all developing nations. She saw the flow of foreign aid slow down to a trickle. In 1966-67 when she came to power, aid accounted for 4.5 per cent of GNP. In the year of her death it will be just a fraction over one per cent. She had to deal with two oil shocks which devastated most developing nations, but helped India overcome these more effectively than almost any oil-importing country in the world. Indeed, in these very adverse circumstances she saw India finally break out of the Hindu rate of growth, and average 4.4 per cent in the last decade. The Indira era witnessed a new dynamism combined with self-reliance that did not exist in Nehru's time...

It can be argued that her greatest single achievement was the conquest of starvation. Public memory is short, and many have forgotten that mass starvation was a normal fact of life in every drought till a decade ago.... In the 1965-67 drought India was invaded by photographers taking snaps of dying babies, and some learned scholars wrote books proving that India was non-viable, should be left to starve, and PL480 wheat should be diverted to more deserving recipients. In 1982 there was another drought of similar intensity, but the world barely knew about it since foreign photographers and journalists could not discover dying babies any more. There was no food aid or ship-to-mouth existence, and India fended for itself. This was possible not just because of the green revolution over which Mrs. Gandhi presided. It was due as much to the creation of a public distribution system and food-for-work schemes. These ensured that grain moved where it was really needed, and purchasing power was put in the hands of the really needy. The conquest of starvation has been, in human terms, the most significant achievement since Independence.

In the public mind she will for long be associated with the nationalization of banks, but this is a misleading association.[76]

There may be merit in the position that Mrs. Gandhi pursued policies for reasons of politics rather than ideological conviction, but it is misleading on several counts. First, Mrs. Gandhi could not avoid being a politician, for she did not inherit office as a leader above the political fray because of unquestioned and unchallenged eminence in the nationalist movement. In contrast, Nehru arrived to office as a function of the nationalist movement; he could, besides, implement his ideological preferences as if by edict because of the low level of political conflict at the time and because of his personal charisma. That path was denied to Mrs. Gandhi by history. Her ideological preferences, regardless of the strength with which they were held, could be implemented only in the vortex of politics and bitter conflict. This was necessarily so given the fact that she held office at the sufferance of the party, the bulk of whose top leadership was opposed to such preferences, as also because of the greater and sharper social differentiation in her time, and the higher and more rapid rate of social mobilization. Eventually, she was victorious by resorting to an appeal outside the party on the basis of her professed socialist credentials, an appeal that was always taken for granted in respect of her father. In the process, the Congress was split, something that her father also, albeit implicitly, threatened but did not have to carry through since his already established popular appeal was sufficient to bring the party leaders to heel. However, in both cases the operation had the character of a political coup in relation to the party organization.

Second, it is illogical to hold against Mrs. Gandhi her endeavour to stay in office. That is the first priority for anyone seriously in politics, especially one with a vision for society that needs to be implemented. Supposing she did have ideological convictions, how else would she have implemented them other than by staying in office? And could it be argued that her opponents were more self-abnegating in this regard? And no one did or can seriously claim that in the situation of serious economic crisis the opposition -- regardless of party and ideology -- behaved with any particular restraint or discipline in its effort to overthrow the Congress party, whether before or after the 1967 elections. Indeed, its behaviour was nothing short of reckless.

Partly underlying the sentiment against Mrs. Gandhi on this question of staying in power is the fact that the associated struggle resulted in splitting the party. But that outcome also lay in the logic of the situation. Supposing, again, that she had pursued the policy measures that she did but out of ideological conviction rather than the alleged thirst for power, would the struggle have been any less bitter? Except for an odd leader here and there, the two sides did lie on the left and right of the centre, and the respective outlooks were vividly embodied in the opposed ideological preferences of the arch antagonists, Morarji Desai and Mrs. Gandhi.

Third, it is true that she sensed that the public wanted change in a radical direction following the electoral disaster of 1967, and she reached out to respond to it. But that again is precisely the requirement that the nature of the political system which India adopted in 1950 -- a pluralist democracy -- imposed on her. Here, again, however, it cannot be suggested that what the public desired, or what she thought the public desired, ran counter to her ideological preferences. To suggest that would be to stand reality entirely on its head, given what we know about her background and ideological development. It is correct that in her first year in office she did not implement radical policies and, indeed, she undertook modifications in the framework inherited from Nehru that ran counter to Nehru's intentions. That, however, may well have been a function of the executive team she inherited from the previous administration and which the Syndicate, indeed, imposed on her. It may have been, more importantly, a result of the very momentum of policies that had been conceived and initiated in the predecessor regime; this is not unknown in public policy, witness Kennedy's Bay of Pigs invasion which had been planned by the Eisenhower administration. Untested in office, she may have also deferred to expert opinion and also pressure from the United States and international financial institutions. Furthermore, it is noteworthy that certainly Nehru, whatever his ideological convictions, did not act in a radical manner for almost his entire first decade in office. On the other hand, Mrs. Gandhi's turn to a radical stance -- and in much more difficult circumstances -- came at the end of a much shorter period. What is more, Mrs. Gandhi undertook radical measures to which Nehru was ideologically committed -- whether land reform, bank nationalization, coal nationalization or nationalization of insurance -- but did not pursue or enter into political conflict over them. The reason for such inaction on his part was apparently the recognition of the constraints that politics and the administrative capabilities of the state placed on him. On a broader level, the responsiveness of Mrs. Gandhi to society's desire for political change within a pluralist democracy suggests that some of the cruder versions of the theory of autonomy of the state, especially those patterned after Marx's notion of the Asiatic mode of production, do not apply to the contemporary Indian state.

Fourth, the entire notion of separation of ideology and politics is a flawed one. Politics and ideology are not entirely dichotomous and exclusive elements; in actual life, they are combined, and political leaders do not necessarily follow one or the other. Indeed, it is difficult to separate politics and ideology; ideology issues out' of politics and the political condition, while ideological issues are clarified in the crucible of political conflict; in turn, ideological preferences become implicated in politics and policymaking. What is important for analysis is the nature of the policies and the total context in which they were generated. Until man is able to take psychic X-rays it is best to suspend judgement on the personal motives behind policy adoption.

Finally, on the issue of the relationship of ideology and politics to policy, it is instructive to note that Mrs. Gandhi's radical posture was not exhausted by what took place in the course of her conflict with the Syndicate. Once secure in office, she did not stop the enactment of radical measures. The radical course continued long after the Syndicate had been completely shattered and scattered. The results of this subsequent radical course also compel attention in any comprehensive evaluation of the relationship of ideology to policy under Mrs. Gandhi.

NOTES

1. Indian National Congress, *Resolutions on Economic Policy, Programme and Allied Matters (1924-1969)* (New Delhi: All India Congress Committee, 1969), pp. 145-54.
2. H.D. Malaviya, *Socialist Ideology of Congress: A Study in its Evolution* (New Delhi: 1966), pp. 49, 54, 58.
3. Dilip Mukerjee, From Crisis to Confidence in the Economy, in Sunil Shastri and Chander M. Bhalla (eds.), *Lal Bahadur Shastri: Commemoration Volume* (New Delhi: 1970).
4. FICCI, *Proceedings of the Thirty-Seventh Annual Session 1964* (New Delhi: 1964), p. 4.
5. Kuldip Nayar, *India After Nehru* (New Delhi: Vikas, 1975), p.13.
6. *Ibid.*, p. 15.
7. Lal Bahadur Shastri, *Selected Speeches of Lal Bahadur Shastri* (June 11 1964 to January 10, 1966) (New Delhi: Publications Division, 1974), pp. 16, 65.
8. D.R. Mankekar, *Lal Bahadur Shastri* (New Delhi: Publications Division, 1973), pp. 112, 175.
9. Shastri, *Speeches*, p. 79.
10. Mankekar, p. 175, and Michael Brecher, *Succession in India* (London: Oxford University Press, 1966), p. 114.
11. Ram Chandra Gupta, *Lal Bahadur Shastri: The Man and His Ideas* (Delhi: Sterling Publishers, 1966), p. 7.
12. Shastri, *Speeches*, pp. 17, 72.
13. Mukerjee, *op.cit.*
14. K. Nayar, *India After Nehru*, pp. 15, 21, 23, 63.
15. *Ibid.*; Brecher, p. 119; Mukerjee, *op.cit.*
16. Shastri, *Speeches*, pp. 12, 85.
17. Francine Frankel, *India's Political Economy, 1947-1977* (Princeton: Princeton University Press, 1978), ch.7.
18. K. Nayar, *India After Nehru*, p. 27
19. Shastri, *Speeches*, pp. 16-17.
20. Brecher, p. 204.
21. *Ibid.*, p. 237.
22. Brecher, pp. 113-14; Welles Hangen, *After Nehru Who?*, cited in Khushwant Singh, *Indira Gandhi Returns* (New Delhi: Vision Books, 1979), p. 20.
23. Zareer Masani, *Indira Gandhi: A Biography* (London: Hamish Hamilton, 1975), pp. 41, 47-48, 100, 111.
24. Cited in Masani, p. 131.
25. Henry Kissinger, *White House Years* (Boston: Little, Brown, 1979), pp. 879-80; Vinod Gupta, *Anderson Papers: A Study of Nixon's Blackmail of India* (Delhi: Indian School Supply Depot, 1972), pp. 149, 151 (the book reprints the various memoranda on the secret meetings originally published in the *Washington Post*).
26. A Moin Zaidi, *Full Circle, 1972-1975* (New Delhi: Michiko and Panjathan, 1975), p. 24; Mary C. Carras, *Indira Gandhi: In the Crucible of Leadership: A Political Biography* (Boston: Beacon Press, 1979), pp. 156, 251.
27. Cited in Frankel, p. 443.
28. Carras, p. 214.
29. Carras, pp. 236-37.
30. Carras, ch. 5.
31. K. Nayar, *India After Nehru*, p. 67.
32. Masani, p. 148.
33. Masani, pp. 155-160.
34. Masani, pp. 163-66, 169.
35. Baldev Raj Nayar, *Violence and Crime in India: A Quantitative Study* (New Delhi: Macmillan, 1975), chs. 3 and 4.
36. Stanley A Kochanek, *Business and Politics in India* (Berkeley: University of California Press, 1974), p. 223; H. Venkatasubbiah, *Enterprise and Economic Change: 50 Years of FICCI* (New Delhi: Vikas, 1977), p. 134.

37. Kuldip Nayar, *India: The Critical Years* (New Delhi: Vikas, 1971), p. 29.
38. Ratna Dutta, "The Party Representative in Fourth Lok Sabha," *EPW*, IV (Annual Number, January 1969), pp. 179-89.
39. See Satindra Singh, *Communists in Congress: Kumaramangalam's Thesis* (Delhi: D.K. Publishing House, 1973), pp. xx-xxiii.
40. Kochanek, p. 223.
41. Indira Gandhi, *The Speeches and Reminiscences of Indira Gandhi* (Calcutta: Rupa & Co., 1975), p. 85.
42. Michelguglielmo Torri, "Factional Politics and Economic Policy: The Case of India's Bank Nationalization," *Asian Survey*, XV, no. 12 (December 1975), pp. 1077-96; Trevor Drieberg, *Indira Gandhi: A Profile in Courage* (New Delhi: Vikas, 1972), p. 78. See also Kuldip Nayar, *India: The Critical Years* (New Delhi: Vikas, 1971), p. 80.
43. Zaidi, *Full Circle*, p. 202; K. Singh, p. 122; A. Moin Zaidi, *The Great Upheaval 1969-1972* (New Delhi: Orientalia, 1972), p. 44.
44. K. Nayar, *India: The Critical Years*, pp. 48, 65-66, 106; Masani, p. 285; K. Singh, p. 49.
45. Michael Brecher, *Political Leadership in India: An Analysis of Elite Attitudes* (New York: Praeger, 1969), pp. 57-62.
46. K. Nayar, *India: The Critical Years*, pp. 24-41.
47. Henry Hart (ed.), *Indira's India: A Political System Reappraised* (Boulder, Colorado: Westview Press, 1976), ch. 9.
48. Frankel maintains: the Congress split originated in a power struggle during which Mrs. Gandhi's faction made an expedient alliance with party radicals as a means of evoking popular support and that she turned "the power struggle into an ideological confrontation in an appeal for popular support over the heads of the party leaders" (Frankel, pp. 419, 429). Of course, earlier, Rao had stated: "Through her great leap Mrs. Gandhi tried to give the power struggle within the party the colour of a conflict of policy and ideology." R.P. Rao, *The Congress Splits* (Bombay: Lalvani Publishing House, 1971), p. 109. A similar position is manifest in M.M. Rahman, *The Congress Crisis* (New Delhi: Associated Publishing House, 1970). It deserves mention that this stance coincides largely with the posture of the Syndicate itself; see Atulya Ghosh, *The Split in Indian National Congress* (Calcutta: Jayanti, 1970).
49. Masani, p. 185; Frankel, pp. 395-97.
50. Rao, p. 43.
51. Torri, p. 1078.
52. K. Nayar, *India After Nehru*, p. 120; Torri, p. 1081.
53. Masani, pp. 187-88.
54. Frankel, p. 401; Masani, pp. 192-193; K. Nayar, *India After Nehru*, p. 137; K. Nayar, *India: The Critical Years*, pp. 24-25.
55. K. Nayar, *India: The Critical Years*, p. 26.
56. Rao, pp. 78-79.
57. Frankel, p. 401.
58. Rao, p. 85.
59. See Zaidi, *Great Upheaval*, pp. 67-71; emphasis added.
60. *Ibid.*, pp. 71-73.
61. Zaidi, *Great Upheaval*, p. 268.
62. Masani, p. 195.
63. K. Nayar, *India: The Critical Years*, pp. 35-36, 41; Carras, p. 139.
64. Carras, p. 3.
65. Zaidi, *Great Upheaval*, pp. 35-40.
66. *Ibid*, pp. 162-71.
67. Rao, pp. 17, 110-11, 115, 135, 139, 143-48, 209, 220-21; see also K. Nayar, *India: The Critical Years*, pp. 1, 29, 48, 54, 65, 250-51.
68. Indira Gandhi, *Speeches*, p. 85.
69. Zaidi, *Great Upheaval*, pp. 256-65.
70. Note, for example: "The political problem is thus to resolve the dichotomy between a highly evolved social and political consciousness of the broad masses and an institutional structure of

the economy which thwarts a correspondingly rising level of social and economic development. The democratic consciousness of the Indian people has advanced by leaps and bounds, thanks to the unique contribution of the Prime Minister since 1969. But the simultaneous expansion of the productive base of the economy has not taken place to the required extent." See Zaidi, *Full Circle*, p. 219.
71. Zaidi, *Great Upheaval*, p. 338.
72. Carras, p. 36.
73. W.H. Morris-Jones, *The Government and Politics of India* (London: Hutchinson University Library, 1964), pp. 52-61. The other two idioms delineated were the "traditional" and "modern".
74. K. Nayar, *India After Nehru*, p. 142.
75. K.N. Subramaniam, "A Balance Sheet," *Seminar*, No. 304 (December 1984), pp. 36-38. Similarly, a radical journal comments: "Nehru's fascination for some sort of socialism -- of course, some milk and water variety of it -- was genuine and longstanding. This was never so with Indira Gandhi who started making deliberate use of radical slogans, measures, etc., for immediate tactical gains. "Cynical Cynicism," *EPW*, XX, No. 18 (May 1985), p. 770.
76. Swaminathan S. Aiyar, "The Legacy," *Seminar*, No. 304 (December 1984), pp. 19-21.

Chapter VII

The Reign of Ideology :
The Grand Era of Nationalization (1969-1973)

There is a popular impression that the public sector in India is basically the result of entrepreneurial activity by the state rather than of nationalization, To a large extent the impression is correct. But nationalization also looms large in the creation and expansion of the public sector. Of course, Nehru had undertaken several measures of nationalization, but by and large he had shown great restraint in pushing nationalization in his pursuit of socialism. He seemed especially reluctant to nationalize industry already established in the private sector, primarily because of his belief that there was plenty to do for the state in any case. However, it was under Mrs. Gandhi's regime that nationalisation became an important instrument of state policy.

The split in the Congress party in 1969 had transformed the Congress government under Mrs. Gandhi into a minority government. But the minority government was able to continue in office for over a year from the time of the split in November 1969 to December 1970 on the basis of support of other political parties, primarily the CPI and DMK. This period was marked by the enactment or attempt at enactment of several radical measures, the target of which was the corporate business sector and the upper classes. Indeed, the failure of one such measure to receive support by a narrow margin in the upper house of Parliament led to the dissolution of the lower house by the government and to the quest for a new political mandate from the electorate. A resounding victory in the election was followed by numerous acts of nationalization of sectors and individual enterprises. However, the most dramatic measure of nationalization remained what had preceded the split of the Congress -- bank nationalization.

In evaluating the elements of ideology and interest in the various radical measures undertaken by the government under Mrs. Gandhi, this chapter systematically examines: (1) bank nationalization; (2) restrictions placed on the functioning of the private corporate sector; and (3) the adoption of new measures of nationalization following the 1971 electoral m

1. Bank Nationalisation

Was bank nationalization undertaken for reasons of ideology or interest? This is, no doubt, a complex question and its complexity is deepened by the fact that political actors themselves at times claim multiple bases for their actions. Like her father in relation to the public sector, Mrs. Gandhi advanced both consummatory and instrumental reasons for bank nationalization; she told the Lok Sabha in support of the ratifying legislation on July 29, 1969: "The nationalization of fourteen banks is totally justified in strictly economic terms as well as in terms of the broad objectives which we have pursued and shall continue to pursue so as to ensure that the hopes and aspirations of millions of our people are not sacrificed." She then added as if to underline the element of ideology contained in the notion of "broad objectives": "As early as 1954, the objective of a socialist pattern of society was adopted by Parliament."[2] Understandably, she did not allude to factional conflict as a factor. In respect of its foundations in ideology, however, bank nationalization had roots that lay in a much earlier period.

Even though Nehru during World War II could be said to have been in his moderate phase, he envisaged the nationalisation of banks as part of his vision to establish a socialist structure in India. His closest associate on the National Planning Committee produced a document as early as 1943 which assumed that banking and "insurance of all kinds" were to be "conducted as public monopolies".[3] After independence, the radical report in January 1948 of the AICC's Economic Programme Committee, under Nehru's chairmanship, included a recommendation that "Banking and Insurance should be nationalized" as part of its total package for establishing "a just social order"; the report was endorsed by the AICC at Bombay in April 1948 and by the annual Congress session at Jaipur in December 1948.[4] Earlier, the government had brought forward legislation to nationalize the country's central bank, the Reserve Bank of India, as part of the effort to exercise effective and direct control over monetary policy; the Bank stood nationalized from January 1, 1949. During discussion on the legislation in Parliament, demands were made for the nationalization of all commercial banks. Despite the recommendation of the AICC's Economic Programme Committee and the insistent pressure of others, the government made no effort for nationalization of commercial banking for seven years, when its target became the Imperial Bank of India.

In 1950, the Rural Banking Enquiry Committee referred to the popular feeling that the Imperial Bank worked to the disadvantage of local commercial banks because of government patronage, despite the fact that it was foreign-managed and bureaucratically-run; the committee went on to recommend that "the people can, therefore, legitimately look forward to this institution being developed as a national organization". Subsequently, in 1954, the committee on All India Rural Credit Survey recommended the nationalization of the Imperial Bank. This recommendation coincided with the ideological turn of the Congress and government to socialism, and it was accordingly hurriedly accepted before the Avadi session of the Congress. In the Lok Sabha in 1955, Finance Minister Deshmukh stated that the step "is not any doctrinaire plunge into nationalisation. It merely seeks to give control of a sector of commercial banking, in order to facilitate a comprehensive development of banking and the extension of credit facilities to important sectors of the economy at present not adequately served."[5]

On the other hand, the Congress at its Amritsar session in February 1956 proclaimed that "the conversion of the Imperial Bank of India into a public-owned and public managed State Bank, and the recent nationalization of Life Insurance are significant steps towards the evolution of a socialist structure." Similarly, the 1957 election manifesto of the party noted: "This is another step towards a socialist pattern and it gives a greater measure of strategic control for planning and other purposes of the State." Later, eight banks of the formerly princely states were made subsidiaries of the State Bank of India. But the Nehru government resisted the nationalization of private banks, and his grounds in this specific instance in 1955 paralleled his general posture on the private sector: "There is so much scope for State banking to expand, then why should I worry about the private banks."[6]

The demand for bank nationalization picked up in India again after the take-over by the Ceylon government of the Bank of Ceylon in 1961 and the nationalization by Burma of its entire commercial banking system; also of consequence in this demand was the Sino-Indian war of 1962 which stimulated the need for greater resources for defence. Two resolutions were moved in Parliament during 1963 for bank nationalization.[7] At the time, Finance Minister Krishnamachari maintained that the case for bank nationalization was not very strong since "the direct or immediate gain to Government in the form of an addition to its income cannot be very great"; he was more interested in expanding the entire pool of resources than merely diverting "existing resources from one sector to another."[8] The Congress Left intensified the demand at the Bhuvaneshwar session in 1964, but the leadership was able to deflect it with a promise to have credit policy serve national priorities. After Nehru's death, precisely because the new leadership veered toward greater reliance on the market and liberalization of the economy, the Left pressed more vigorously for bank nationalisation among other measures of socialist reform. Apparently, Finance Minister Krishnamachari now recommended bank nationalization but Shastri did not act positively on his recommendation.

In 1966, the issue was discussed in the cabinet under Mrs. Gandhi, but nationalization was rejected because the new finance minister was opposed to it.[10] After the devaluation fiasco and in view of the upcoming general elections, however, Mrs. Gandhi and the leadership became eager to pacify the Congress Left. Accordingly, the Congress party's 1967 manifesto pledged "to bring most of these banking institutions under social control in order to serve the cause of economic growth and fulfil our social purposes more effectively and to make credit available to the producer in all fields where it is needed." This pledge needs to be read in conjunction with the manifesto's initial reaffirmation of the party's commitment to "the goal of a democratic socialist society", and more significantly also with its aim to control the commanding heights in order to prevent the private sector converting economic power into political power.[11] Subsequent to the disastrous results of the 1967 elections, the Congress incorporated social control of banks as the first item in its new Ten-Point Programme. The Congress Left, however, refused to make any distinction between social control and nationalization, interpreting the former to mean the latter, and continued to place insistent pressure for nationalization. The case of the Left against private commercial banks rested on three major grounds: (1) the concentration of economic power in the hands of a few big business houses; (2) the misuse of bank resources by the business houses; and (3) the failure to meet national priorities in terms of credit supply to rural areas, backward areas, new entrepreneurs and exports.[12]

Although well-argued cases were presented in behalf of these points, others suspected these to be *post-hoc* justifications of an *a priori* ideological position.[13]

Firstly, although commercial banking was not new to India, its real expansion had come in the period after independence. The cause of this expansion was the extremely active role that business houses had played, with nearly every large business house identified with a major bank. Rather than being made a matter of political grievance, it is the dynamic entrepreneurship of the business houses behind the impressive performance of commercial banking that needed to be appreciated. Besides, public sector banking through the State Bank of India already controlled nearly one-third of the country's deposits while the other two-thirds was spread among at least a dozen major banks. The private commercial banks had performed a useful service to the public, which continued to have faith in them by entrusting their deposits with them. Indeed, deposits at the five largest private banks had during 1960-68 grown at a much higher rate than those with the public sector banks. The fact that there was competition between the public and private sectors had also provided some incentive for efficiency in public sector banking.

Secondly, while directors of banks could divert credit to their business houses, the extent of this diversion was exaggerated and could not have been more than 20 per cent even at the upper limit; ironically, the State Bank of India was more guilty in this respect. Furthermore, if there had been misuse the fault for it lay with the Reserve Bank of India for its failure to exercise effectively its far-reaching supervisory and directing powers over banking. Thirdly, private commercial banking had largely served the country's planning objectives in support of rapid industrialisation, for its advances to industry had increased from 34 per cent in 1951 to 63 per cent in 1965 while those to commerce had declined from 53 per cent to 26 per cent.[14] More critically, in relation to the other desired objectives, the real failure lay with the government in not evolving a national credit policy for any of the first three Five-Year Plans; otherwise the government could have directed the banks in regard to credit allocation.

What is surprising is that the private banks were being blamed for neglect of rural areas, when in fact they had earlier been discouraged from entering this sector which had been reserved for cooperatives and public sector banking. Indeed, even public sector banking had abdicated its role in the rural sector on the assumption that cooperatives were the state's chosen instrument for the purpose. Also, private commercial banks had not neglected small-scale industry; they had provided considerable credit to this sector: for example, of the Rs. 908 million outstanding as advances to small-scale industry on March 31, 1966, the share of the private sector was 56 per cent while that of the public sector was 44 per cent.[15] Furthermore, banks were custodians of the deposits of the public and could not dispense with proper security for loans. Thus, the banks were being unfairly held responsible for a situation which was a result of the government's own creation. Again, the continued existence of private commercial banking seemed essential if the notion of mixed economy had any meaning.

However, the issue was not to be decided by rational argument, but in political conflict. Because of the resistance of Morarji Desai to bank nationalization, on the ground that it "would severely strain the administrative resources of the Government while leaving the basic issues untouched," the Congress for the moment eschewed nationalization and opted instead for a scheme of social control. The scheme was announced in Parliament in December 1967 and, after a lengthy legislative process, came into effect in February 1969. It snapped the linkage between big business houses and private commercial banks through changes in the composition of boards of directors, and vested the management of banks in professional bankers. But its most fundamental feature was the establishment of a National Credit Council, as a high-powered and broadbased body headed by the Finance Minister, to decide on policies and priorities for loans and investments among the different sectors of the Indian economy. Even before the new scheme came into effect, commercial banks had undertaken a shift in credit allocations, and they actually exceeded the targets for agriculture and small-scale industries that had been fixed for these sectors.[16] However, the Congress Left considered social control of banks a failure even though it had hardly been in operation for six months.

What hastened the end of the social control experiment was Mrs. Gandhi's decision to make bank nationalization a central issue in her conflict with the Syndicate. That, in the immediate context of this conflict, not much thought or preparation had gone into the subject is evident from her note to the party's Working Committee in Bangalore in July 1969, which she significantly described as consisting of "just some stray thoughts rather hurriedly dictated". In respect of banks, she was apparently still undecided: "There is a great feeling in the country regarding the nationalization of private commercial banks. We had taken a decision at an earlier AICC but perhaps we may review it. Either we can consider the nationalization of the top five or six banks or issue directions that the resources of banks should be reserved to a larger extent for public purposes." She went on to suggest that greater resources for the public sector could be obtained by asking banks to increase their investment in government securities from the existing 25 per cent to 30 per cent; as for the adverse effect this may have on credit for trade and industry and on bank profitability, she cavalierly dismissed the issue, stating "in times of credit squeeze, private industry somehow adjusts itself." In another note to the Working Committee, Finance Minister Desai averred that social control was having a significant impact on the working of banks and that nationalization was therefore not necessary.[17]

Despite the apparent lack of preparation, Mrs. Gandhi felt compelled to act precipitously by the political rebuff administered to her by the Syndicate on the question of the party's nomination for the country's presidency. She then dismissed Finance Minister Desai, overruled the reservations and opposition of senior bureaucrats,[18] and issued an ordinance in July 1969 nationalizing fourteen of the largest commercial banks with deposits of over Rs. 500 million, thus bringing under public ownership commercial banking covering over 85 per cent of the country's deposits. In a radio broadcast, she viewed bank nationalization as "a continuation of the process which has long been under way" to adopt policies to achieve a socialist pattern of society and to obtain "control over the commanding heights of the economy." She regarded nationalization as being necessary for the expeditious achievement of the objectives that had been set for social control, more especially making available sufficient credit to agriculture and small industry and to encourage new classes of entrepreneurs. In parliament, she again linked bank nationalization to the goal of socialist pattern of society, stating "public ownership and the control of the commanding heights of national economy and of its strategic sectors are essential and important aspects of the new social order which we are trying to build in this country."[19]

The government's haste in nationalizing the banks is further evident from the fact that the ordinance was issued only a few days prior to the opening of parliament, that the subsequent confirming legislation had to be modified by government itself, and that the final enactment was declared invalid by the Supreme Court. The government then had to re-nationalize the banks through a new ordinance and another legislative enactment. Regardless of these difficulties, the nationalization was greeted instantly with great popular enthusiasm, especially among small businessmen, small industrialists, and taxi drivers. A Marxist scholar viewed the enthusiasm as the result of bank nationalization having been "identified in public mind with an attack on concentration of power in few hands" and believed: "Above all, this is the first economic step, perhaps the boldest since Independence, which has been taken in defiance of the Big Business interests and their political representatives and in the background of vast social discontent against economic disparities persisting and even growing in Independent India." He further asserted: "Bank nationalization itself reflects a significant though not yet decisive shift of the class basis of political power from Big Business to the petty bourgeois in the urban sector and specially to the new and far more numerous class of rich peasants and landlord-farmers in the rural areas." He was also certain that these rural classes would be the principal beneficiaries of nationalization. [20]

Similarly, an eminent economist argued that "the political pressure for nationalization has got built up to a very significant extent on the grievances of the smaller industrialists and business men, more particularly in the less-developed regions of the country against what they consider the privileged positions of the 'big business houses' with larger resources and operating on a national scale." Another economist referred to the predominance of rural representation in state and central legislatures, and stated: "There is hardly any identity of interest between them and the big business which controls the commercial banks. It is this factor more than anything else, which has made bank nationalization a political feasibility."[21] Understood in this sense, bank nationalization was a manifestation not simply of political conflict among elites, but also of the greater crystallization of class consciousness of the intermediate rural and urban strata as a result of economic development and social mobilization over the period since independence.

If so, the Communist parties were no less partners than the Congress in this enterprise of serving the interests of the intermediate strata, nor did that enterprise seem to reflect much potency on the part of the bourgeoisie in state power. The CPI was, of course, to enter into a tacit alliance with Mrs. Gandhi's Congress party in sustaining her minority government. It enthusiastically supported bank nationalization as a progressive measure, calling it "an important victory in the struggle against monopolists and against concentration of wealth and economic power in their hands." This was quite consistent with its 1967 election manifesto where, besides berating the Congress for following the "capitalist path" and demanding its overthrow, the CPI urged: "The entire financial capital, accumulated in the banks and now used according to the will of the millionaires for their private gains, must be brought under the most effective state control and planning. For this purpose all banks must be nationalized." Similarly, the CPI(M) had, while accusing the Congress party of "building capitalism", supported the nationalization of banks; it now welcomed the measure "as a step in the right direction" even though it was not by itself to be taken as ushering in socialism.[22]

On the other hand, the Swatantra party attacked bank nationalization as did the Jan Sangh. In accord with its overall free-enterprise orientation, the former party had declared in its 1967 election manifesto: "The Swatantra Party is opposed to the nationalization of banks contemplated by the Congress Party which is utterly irrelevant to the country's problems and would retard development besides being fatal to monetary stablility, security and saving by placing the savings of lakhs of small depositors at the mercy of a government seeking to lay its hands on all available resources." The Jan Sangh simply considered "the proposal to nationalize banks as improper"; more generally, it believed that "the public sector has expanded so much that it needs consolidation."[23]

The FICCI as the spokesman of big business had been opposed to bank nationalization but it was left no recourse except to reconcile itself to the inevitable. Its president declared it to be "a hasty step, especially when social control on banks was working successfully"; he felt it was not "a well considered decision", believed it would affect the economy adversely, and saw no "economic justification or pressure of circumstances for such a drastic measure."[24] Its private reaction was of course much stronger, but it realized that it was the better part of discretion to accept the inevitable. Even though it was not responsible for the 1967 election disaster for the Congress, business had learnt to regret its "pyrrhic victory" in weakening the Congress because of the political crisis, instability, regional tensions and violence it had unleashed. The Birla strategy that the Congress was the only alternative available had been proven right. But the weak stance of big business on bank nationalization led to internal dissension and to the secession of more strongly opposed elements. Its tame reaction was due to "fear of putting itself out of court with the Government" and reflected its powerlessness in the face of government determination. Its plight at the time as "a home of lost causes" was effectively conveyed by its historian:

> It had set out to contain the public sector's thrust on both fronts, namely, establishment of an increasing number of state-owned industries and nationalization of existing financial institutions, or industries in some cases. It has lost on both fronts. There was no stemming the advance of the public sector whenever and wherever Government felt able to advance. Nationalization of banks in 1969 extinguished whatever hope had survived the nationalization of life insurance.[25]

Two progressive economists seemed to agree when they drew the implications of the measure: "Nationalization of banks weakens the stranglehold of the private sector, particularly of big business, over the economy. Big busiess can no longer use the banks to exercise control over the economy....To the extent the private sector is weakened to that extent the forces of socialism may be relatively strengthened."[26]

Bank nationalization in 1969, just as industrial policy under Nehru, demonstrated the power of a state under a political elite representative of the intermediate strata. The decision was fundamentally rooted in ideology, traceable to the groundwork laid by Nehru and built upon by his supporters. The measure was perceived as instrumental-socialist, not as making possible a more effective working of capitalism. But the decision was precipitated by political calculations in respect of, firstly, larger social forces as they impinged on assuring electoral support in the future and, secondly, more immediately building a political coalition to sustain a radicalised regime in power. Thus, the roots were ideological but the enabling conditions and the immediate motives were political.

2. Constraining the Corporate Private Sector

The perception that electoral success in the future required a radical stance on the part of the Congress did not cease with bank nationalization and the party split. That consideration continued to hold until new elections were held. In sustaining a radical posture, the tacit cooperation between the CPI and the Congress was useful to Mrs. Gandhi, for it testified to her credentials as a radical leader. At the same time, even as she thus stood distinguished from the rightist groups (Syndicate, Swatantra, and Jan Sangh), cooperation with the CPI enabled her also to meet the threat from the extreme left in the form of CPI(M) and the Naxalite group, particularly in West Bengal and Kerala.[27]

On the other hand, cooperation with the Congress was attractive for CPI in attempting to meet the challenge of CPI(M) in these two states. Beyond that, the CPI could also hope of success in its strategy of establishing the National Democratic Front through a coalition among CPI, the progressive wing of the Congress under Mrs. Gandhi, and other left groups. Out of these considerations there evolved political and electoral cooperation between the Congress and CPI in Kerala which resulted, firstly, in the overthrow of the CPI(M)-led government and its replacement by a CPI-led coalition and, secondly, in their jointly defeating CPI(M) in the mid-term elections there in June 1970. This confirmed the utility of cooperation to both the Congress and the CPI. Furthermore, former Communists within the Congress Left sought to advance the CPI cause by initiating and supporting a radical programme in order to push Mrs. Gandhi in the direction of more radical measures but, more critically and strategically — regardless of her particular actions -- by identifying themselves completely and wholeheartedly with her personal leadership in order to more effectively infiltrate government and party.[28]

Mrs. Gandhi had achieved her victory over the Syndicate by aligning herself with the Congress Left, but the latter now pressured her to undertake a more radical programme, especially the nationalization of the big industrial houses. On the other hand, having just taken over fourteen of the largest commercial banks, Mrs. Gandhi apparently felt that that was burden enough for the state for the present. However, the Congress party was more responsive to other demands of the Left designed to place curbs on the expansion of big business. For many of these potential constraints on the private sector there had already, even before the split of the Congress, developed a considerable momentum through various inquiry committees, established at times at the behest of the Left. The suggested constraints soon manifested themselves in new legislation on big business and new administrative policy on licensing, discriminatory treatment against private sector, and threat of backdoor nationalization.

a. *Monopolies and Restrictive Trade Practices Act, 1969*: This legislation was passed in December 1969, and came into effect in June 1970, but it was the result of cumulative pressure built up over the preceding decade against big business. While Nehru was insistent on the eventual domination of the public sector in the economy, he was not particularly eager to block the growth of existing private sector enterprises, in view of the vast needs of the country for industrial development. However, in order to deal with charges that concentration of economic power and economic disparities were increasing in India, he appointed an expert committee in 1960 under the chairmanship of the veteran economic planner, P.C. Mahalanobis, a great enthusiast of the Soviet economic model. Although the committee in its report of 1964 acknowledged that the "statistics do not show any definite and significant trend in concentration ratios" during the planning period, it pointed to the existent reality of concentration, indeed, that "concentration of economic power in the private sector is more than what could be justified as necessary on functional grounds, and it exists both in generalised and in the specific forms."[29] Recognizing the limitations of its data, however, the committee asked for a more comprehensive and detailed assessment.

Following this recommendation, the government established in 1964 a Monopolies Inquiry Commission. While urging that monopolistic and restrictive trade practices be curbed, the Monopolies Commission in its report of 1965 refused to recommend striking at the concentration of economic power as such. However, the draft legislation which the Commission had prepared [30] was significantly modified in select committee of parliament in response to leftist attacks on big business which had grown especially clamorous in the late 1960s, particularly with the tacit or explicit approval of Mrs. Gandhi. Mere business size was made a suitable target for control, regardless of whether or not it resulted in monopoly or restrictive trade practices.[31] Accordingly, all business houses, not simply individual firms, with assets of more than Rs. 200 million, or controlling one-third of production or distribution of any goods or services, were barred from expansion or diversification except with prior and specific approval of the government; these restrictions were over and above the usual licensing requirements. An ongoing Monopolies and Restrictive Trade Practices Commission was set up, from which the government would, if necessary, seek advisory opinion. The intent was clearly to curb the growth of large industrial houses, and not in some hidden way meant to assist in their expansion; the restrictions could not surely be said to have been undertaken at the urging or behest of big business, but rather they were strenuously opposed by such business. The state in this instance was no committee of the big bourgeoisie. Rather, as a check against the big bourgeoisie, the state made it a policy to rely on new entrepreneurs for future economic growth. On the other hand, big business prophetically warned that the consequence of these and other similar measures to block its growth would be industrial stagnation.

b. *Modification of Industrial Licensing Policy, 1970*: Big business was also a target in terms of inquiries into licensing procedures. The Mahalanobis committee itself had held economic planning, in part, responsible for encouraging concentration of economic power. Echoing it, the Monopolies Inquiry Commission also recognized that "industrial licensing, however necessary from other points of view, has restricted the freedom of entry into industry and so helped to produce concentration."[32] In view of the centrality of industrial licensing, the government also appointed R.K. Hazari, already renowned as a resolute critic of industrial concentration and big business, to review its working. While acknowledging severe limitations in the data, Hazari in his report in 1967, nonetheless concluded that large business houses, more particularly the Birla group, had pre-empted licensable capacity in many industries even as the system deterred new entrants. Not surprisingly, his recommendations were directed against the large industrial houses, which were sought to be prohibited from entering or expanding in traditional industrial activities.[33]

A heated and acrimonious debate followed in the upper house of parliament in May 1967, and resulted in the government's setting up in 1967 a larger three-man Industrial Licensing Policy Inquiry Committee. Known as the Dutt committee after its chairman, S. Dutt, its composition was weighted against the private sector, with two of its members (H.K. Paranjape and S. Mohan Kumaramangalam) committed to the public sector, the latter having been not too long before a member of the CPI. The recommendations in its report of 1969 for a more vigorous and expanded role for the public sector, restrictions on large industrial houses, and reservation of certain industries exclusively for the small-scale sector, were accepted by the government and found expression in substantial change in industrial licensing policy by the government on February 18, 1970.[34] FICCI's elaborate protest against what it clearly considered biased and unwise recommendations of the committee, and its warnings against the possible retardation of industrial growth as a result of their implementation,[35] went unheeded.

The basic intent of the new licensing policy, which "entirely reversed the trend" toward economic liberalization inaugurated in the mid-1960s,[36] was to restrict the future growth and economic power of the large industrial houses by strengthening the countervailing power of the public sector and of the medium and small-scale sectors. Henceforth, twenty large industrial houses together with their individual firms were to be restricted to core industries (those that were basic, critical and strategic) and to the heavy "investment" sector (new projects of over Rs. 50 million) where these had not already been reserved for the public sector. In order to provide for further government control in regard to such private investment, the new policy accepted the principle of "joint sector", combining private and public investment. The medium industries, with investments ranging between Rs. 10 million and Rs. 50 million, were to be open only to entrepreneurs other than the large industrial houses except where efficiency and exports demanded it. The list of industries reserved for the small-scale sector was also enlarged. The new policy, in addition, envisaged a substantial expansion of the scope of the public sector over and above the fields allocated to it in the Industrial Policy Resolution of 1956.

Furthermore, the new policy raised the threat of backdoor nationalization of private industries through the government's finance-lending institutions converting their loans into equity in the future. A government announcement in July 1970 stipulated that, where the combined assistance from public financial institutions exceeded Rs. 5 million, the inclusion of a "convertibility clause" in agreements was mandatory, giving government the right to convert its loans and debentures into equity shares. For assistance below Rs. 5 million the insertion of such a clause was discretionary. It was believed that "penetration of the private sector by the State through the operation of the conversion process can be initially significant, eventually decisive."[37] But one consequence was to inhibit growth of industrial enterprises for fear that reliance on assistance from public financial institutions may lead to control passing out of the hands of owners.

The hostility to the private corporate sector manifest in the new policy, and in the various committee reports on which it was based, reflected the attitudes of the intermediate strata, especially the intellectuals, that dominated the state. Of course, even these measures did not sufficiently appease the Congress Left, which wanted nothing less than the outright nationalization of the large industrial houses.

c. *Business Sense of Disquiet* : It was not only the actual measures that the government adopted to restrict big business, but also the fear of what government may do further in the future -- and the business community's patent helplessness in relation to it-- that caused despair in the corporate private sector. Bank nationalization had already deprived the private sector of a huge segment of its sphere of operations. What is more, this additional measure of socialization of savings, over and above that of life insurance, made the private sector excessively dependent on government for finance. The cumulative effect of state policies since the mid-1950s had been that "the private sector had been encircled not only by a wide range of legislation but by a variety of countervailing power." The business community was thoroughly "disenchanted with the system," and it perceived the government's licensing policies as the sheer pursuit of ideology rather than an attempt to genuinely work a mixed economy. Further, the prime minister's statements that bank nationalization was only the opening shot in the bitter struggle between the poor and the few rich caused "a sense of deep disquiet among businessmen."[38]

Nor was this feeling helped by what the business community perceived to be an assault on the fundamental right to property and the government's going back on its contractual commitments. Even though the immediate issue did not concern the business community, the government's attempt to withdraw, under the pressure of the Congress Left, the privy purses of the former princes did not enhance trust in the government's word. The legislation abolishing the princely privileges and privy purses failed to get the necessary two-thirds majority in the upper house by only a fraction of a vote, but the government then adopted an ingenious manoeuvre to achieve the same end -- an executive order cancelling their recognition as princes. However, the Supreme Court on December 15, 1970 ruled the order unconstitutional and an invasion of the right to property. Within twelve days of the court's decision, the government dissolved the lower house and sought a new electoral mandate.

Significantly, despite the constraints imposed on business and the blows administered to it, the business community remained divided. One part, following the Birla strategy, supported Mrs. Gandhi as the best hope for the restoration of political stability and law and order that had been wrecked by the 1967 elections. Another part supported the opposition, with even businessmen running for parliament. [39] Regardless, business had encountered serious blows during the course of the radicalization of the Congress. However, it had escaped further nationalization beyond commercial banking. Yet that provided no immunity from nationalization after the 1971 elections.

3. Renewed Popular Mandate and the Revived Nationalization Thrust

Soon after the government order on the derecognition of princes was turned down by the Supreme Court, Mrs. Gandhi called for new elections. In part, this action reflected her feeling that the government had been forced to make compromises because of dependence on other parties for support, and that it was therefore necessary to seek a new mandate. However, observers did not foresee much success for her in this endeavour, given the parlous condition of her party organization after the split. In the event, she resorted to an election strategy that was novel on the Indian political scene -- appealing directly to the masses, over the heads of intermediary elites, with the radical populist slogan "Remove Poverty". Her party also arrived at an understanding on electoral cooperation with the CPI. But the election was really centered on her personally, with a four-party opposition alliance that included the Syndicate, Swatantra party and the Jan Sangh ranged against her with the slogan "Remove Indira." The results of the national elections in March 1971, for the first time delinked from state elections, however, astounded everyone. The Congress with 352 seats won more than a two-thirds majority in the Lok Sabha, with no need thus to rely on other parties, not even for constitutional amendments. The results confounded most observers, who had expected coalition politics as the likely outcome of the elections. The landslide victory was a personal triumph for Mrs. Gandhi, and it initiated a process of personalization of power in the political system.

Mrs. Gandhi's election victory had been built on a strategy of a broad aggregation of support: from the poor, especially the scheduled castes, on the promise of poverty removal; from the minorities, on the basis of her established secularism; from the rural middle and rich peasantry, on the assurance of easier credit after bank nationalization; and from the urban middle classes and business on the prospect of political stability. [40] The social base of her support was thus a broad one though more concentrated among the poor and the minorities. The significance of the vote of the poor should not be downplayed as simply so much electoral fodder. Without the recognition of its strategic political importance, it would be difficult to understand fully the basis of government policy, regardless of its inadequacies, in respect of a public distribution system for essential commodities for the weaker sections of the population, rural employment programme, allotment of house sites for the poor, and provision of some basic needs.

The most significant change was not in respect of social classes as it was in the configuration of political tendencies. As Morris-Jones pointed out, the "progressives are in a sense the chief beneficiaries of the election." [41] The CPI maintained its stregth of 23 seats while CPI(M) increased it from 19 to 25, becoming the largest opposition party. Among the right-wing parties, Swatantra, which had earlier been the largest opposition party in the Lok Sabha, was badly mauled, being cut down from 44 to 8 seats; the Jan Sangh lost more than one-third of its seats, retaining only 22 compared to its earlier 35 seats. The Syndicate suffered an ignominous defeat, winning only 16 seats as against its pre-election strength of 60. What had seemed like a major break-away segment had been reduced to a rump; Mrs Gandhi had judged the electoral mood better than the Old Guard of the party and thus restored the Congress to a dominant position once again. The other constituent of the rightist Grand Alliance, the Samyukta Socialist Party, was virtually wiped out, winning only 3 seats compared with its earlier 23. Within the Congress, too, the Left gained critical strength, with some sixty to eighty members of the party's new contingent constituting a determined hard core of the Left. [42] Significantly, within the Congress Left, ex-Communists had a strong and strategic presence. Their leader, S. Mohan Kumaramangalam, became the Minister for Steel and Mines and emerged as a key advisor to Mrs. Gandhi until his death in a plane crash in 1973.

While the social base of the Congress as such was a broad one, in the Lok Sabha the agriculturists consolidated their ascendant position, emerging as the single largest occupational group (see Table VII.1). While the position of the middle classes

Table VII.1
Occupational Background of Ruling Party MPs:
1971, 1977 and 1980

Occupation	1967 %	1971 No.	1971 %	1977 No.	1977 %	1980 No.	1980 %
1. Agriculture	36.8	135	39.6	105	36.3	138	40.1
2. Business	5.1	22	6.4	11	3.8	24	7.0
3. All professionals	35.9	129	37.8	93	32.2	104	30.2
-- Law	(22.2)	(82)	(24.0)	(58)	(20.1)	(72)	(20.9)
-- Other	(13.7)	(47)	(13.8)	(35)	(12.1)	(32)	(9.3)
4. Public Work*	17.0	42	12.3	73	25.3	69	20.1
5. Service**	2.9	7	2.1	3	1.0	6	1.7
6. Others	2.3	6	1.8	4	1.4	3	0.9
	100.0	341	100.0	289	100.0	344	100.0

* Includes political, social and trade union workers
** Military and civil services

Source: Adapted from V.A. Pai Panandiker and Arun Sud, *Changing Political Representation in India* (New Delhi: Uppal Publishing House, 1983), pp.55-56. Data for 1967 is from Ratna Dutta, "Party Representative in Fourth Lok Sabha," *EPW*, IV (Annual Number, January 1969), pp. 179-89; since the total in the original added to 102 I have reduced the "others" category by 2.

declined, they continued to be a leading group; more particularly, the middle classes were hegemonic in the cabinet. It is startling to note the much higher degree of middle class hegemony in the executive branch after the elections compared to earlier councils. Professionals constituted around 60 per cent of the membership of the council of ministers, with lawyers alone providing 37 per cent (higher by around 5 percentage points than in earlier councils) while agriculturists were only 15 per cent (lower by about 3 per cent). More critically, in the cabinet there were no agriculturists at all whereas lawyers were 60 per cent, and educators, journalists and engineers were 15, 4 and 4 per cent, respectively.[43] Policy initiatives therefore emerged in the dynamic interaction of: a broad support-base but especially concentrated among the poor and minorities; agriculturist and middle class dominance in the national legislature; middle class hegemony in the executive; and a strategic presence of the Left in the party and government.

Notwithstanding the consolidation of the Congress Left, critical to its strength was the personal position of Mrs. Gandhi. The left programme could be implemented only insofar as it was acceptable to her. For the next three years after the 1971 elections, Mrs. Gandhi continued in her radical phase that had begun with bank nationalization. apparently convinced that she had a public mandate for it. As the party's general secretaries noted:

> The people's verdict in the Lok Sabha elections is therefore an unambiguous mandate to the Congress to go ahead towards radical socio-economic changes to break the stranglehold of the monopoly on our economy, to take steps to raise the living standards of the downtrodden sections of society, to remove constitutional bottlenecks in the way of implementing the directive principles enshrined in our Constitution and thus to take India forward towards a progressive, socialist, self-reliant economy.[44]

Initially, however, Mrs. Gandhi's government was paralyzed from radical action as its energies came to be concentrated on the crisis over Bangladesh, which erupted following the military crackdown there by Pakistan. Coping with ten million refugees was burden enough, but India had to prepare for the prospect of war. With China and the U.S. siding with Pakistan, India endeavored to assure protection through the Indo-Soviet treaty in August 1971. Mrs. Gandhi was credited with the success in the war that occurred in December. She emerged even more powerful on the political scene, a pre-eminence that was confirmed in the March 1972 elections to the state legislatures. With the state elections out of the way, the party was even more forthright and specific about what its mandate required; at the AICC meeting at Gandhinagar (Gujarat) in October 1972, it declared:

> there can be no escape from a massive taxation effort to finance rising levels of public investment. The resource base of the public sector has to be expanded to generate surpluses for investment....Rapid economic growth in the context of our socio-political objectives requires vigorous expansion of the public sector in industry and other vital spheres of national economy....The economic situation in the country demands that the private sector adopt

unambiguously criteria imposed by national priorities and discard those of mere profitability in a sheltered market. The Government shall exercise appropriate control over private sector....Policies for reducing concentration of economic power and for reducing social and economic inequalities should continue to be implemented with vigour. The most significant steps in this direction would have to be the expansion of public sector and curbing of economic power by Monopolies which is prejudicial to social objectives.[45]

Notwithstanding government's preoccupation with the Bangladesh crisis, there had already taken place some movement on radical change, more really to facilitate legally the takeover of private sector enterprises by the public sector. One constitutional amendment (twenty-fourth) sought to preclude any judicial challenge by making parliament supreme in its amending powers, even if it infringed fundamental rights, including the right to property. This extension of parliament's amending powers to cover fundamental rights was for the purpose of pushing social engineering, and was based on the premise that the judicial branch had become a perverse bulwark against progressive legislation. Another constitutional amendment (twenty-fifth) enabled the government to provide, not compensation, but "an amount" that it fixed by law for acquisition of private property. Business was naturally opposed to it, with J.R.D. Tata declaring that the amendment had opened an era of "downright confiscation and expropriation."[46] Even before the amendment was enacted, the government had launched on what amounted to a nationalization spree. Significantly, most of such nationalization centered on the cabinet portfolio of Steel and Mines, held by Kumaramangalam. But nationalization was not restricted to steel and mines. The overall ideological mood of the party and its thrust was vividly expressed by Congress president Shanker Dayal Sharma before the AICC in New Delhi in June 1972:

> The defeat of the Syndicate was proof positive of the fact that the pro-people and anti-capitalist forces in India can alter the character of the State power in India and carry through basic socio-economic changes. Indeed, the Congress, under Indiraji's leadership has successfully changed the direction of free India's political development and has moved along the path of ending the influence of big money and monopolists on affairs of the State.
>
> ...Internally we have to accomplish agrarian reforms...; establishing a growing State sector in industry which must gradually attain a dominating position in the sphere of industrial production; through nationalization establish State control over financial and banking business....The object, among others, is to ban big money from becoming the dominant social and political force....For the economic policies to which the Congress is committed, the monopolies are a hindrance....the public sector is increasingly becoming predominant, but the dominating heights held by monopolies have to be captured. A curb on monopolies is today a national demand....Nationalization is not necessarily socialism, but without nationalisation there cannot be any socialism. Nationalization on a sweeping scale can transform the socio-political structure.[47]

a. *Nationalization of General Insurance* :   The government had nationalized life insurance in 1956 and the major commercial banks in 1969, both measures having been claimed as part of the movement towards a socialist society. General insurance as another site of financial power and community's savings had remained outside government ownership even though subject to government regulation. More than ten years after the nationalization of life insurance, the Congress in 1967, after the stunning reverses in that year's elections, accepted -- as part of its consensus formula of the Ten-Point Programme -- the nationalization of general insurance even though it agreed only on social control of commercial banking. However, it was the nationalization of banks as a more dramatic measure that took precedence in the political struggle between Mrs. Gandhi and the Syndicate. Nothing was done in relation to general insurance, apparently perhaps because the government had sufficient business on its hands with bank nationalization.

After the Congress party split, several of the Young Turks demanded nationalization of insurance companies. Their intent was clearly not to build capitalism or to assist the private sector but to reduce the power of the corporate sector: "the operations of the insurance companies have a direct bearing on the question of 'concentration of economic power.' Most of the general insurance companies are used as instruments of promoting companies of the controlling business houses."[48] Soon, electoral considerations moved the Congress to make pledges of more radical change.

The Congress party's manifesto for the 1971 elections, after referring to the Directive Principles in the Indian Constitution, went on to ask for a mandate to "continue the advance to socialism through democratic process" and to "enlarge the role of the public sector". More specifically, it proposed

> the expansion of the public sector by taking over general insurance; increasing state participation in the import-export trade; greater role of the state in industries where substantial public funds have been invested; and expansion of the activities of the Food Corporation of India, coordinated with cooperatives, which can ensure the implementation of a national policy for the distribution of food and fair prices for the farmers.[49]

This was a large order but, determined to show that it meant business, the government decided within two months after the elections, even in the midst of the Bangladesh crisis, to nationalize general insurance. After bank nationalization, this seemed "merely in the nature of a mopping-up operation"[50] in the drive for socialization of the community's savings. As a first step the government, through an ordinance in May 1971, took over the management of 106 general insurance companies and then nationalized them in August 1972. Finance Minister Chavan, in justification, simply stated that there had been a growing demand for such nationalization, that "certainly, it was one of our major commitments to the electorate," and that with it "we have fulfilled at least one of the important promises that were given to our electorate." A Congress member saw it as "a measure which will put the nation on the march towards socialism and for ushering in the great socialist society."[51]

In this endorsement of the measure, the government was joined by CPI and CPI(M) members. A veteran CPI(M) member welcomed the bill "as a small step in the hope that it will check the power of individual monopolists in this country. As we are opposed to all forms of private monopoly, we welcome the measure as a step in the right direction." Similarly, a CPI member added: "I, no doubt, support the Government and I must congratulate them on bringing forward the legislation for nationalizing..."[52] Both CPI and CPI(M) were angry at the government, however, for providing any amount whatsoever by way of compensation to the owners and shareholders. As expected, the Swatantra party thought the measure to be both unnecessary and undesirable. The Jan Sangh dismissed it as "essentially a political decision and has got nothing to do with the economic considerations", holding nationalization to be no panacea for the country's economic ills.[53] One business leader asserted that the measure was "another instance of government action without any economic justification." The major peak business association seemed to agree, stating that there was "no doubt, nationalization is a political decision," and asked for proper compensation if the investment climate was not to be adversely affected.[54]

b. *Nationalization of Coal Industry*: Among the most important and dramatic measures of nationalization after bank nationalization was that of coal. The nationalization of general insurance could have been no surprise, for after all it had been part of the Ten-Point Programme and the 1971 Congress party manifesto. But coal nationalization came without much warning; indeed, the Minister for Steel and Mines, S. Mohan Kumaramangalam, had provided assurances only a month prior to nationalization that no such action was contemplated, only the amalgamation of small units.[55]

Coal had been included in Schedule A of the Industrial Policy Resolution of 1956 where, in pursuit of the goal of socialist pattern of society, future development of such a basic and strategic industry was made the exclusive responsibility of the state. But for nearly 25 years after independence the state had shunned nationalization of the coal industry. This had not, however, meant the lack of ideology on the part of government. Indeed, Hanson in 1966 underlined "the triumph of ideology" in decision-making on coal. He demonstrated how, on the basis of performance, the private sector deserved a larger role in coal production, but that "the 'socialist pattern' stood in the way"; economic rationality was pushed aside to favour the public sector despite its poor performance.[56] It is noteworthy that in 1957, in line with the Industrial Policy Resolution, the government -- through the Coal Bearing Areas (Development and Regulation) Act -- had reserved all virgin coal areas for development by the public sector while the private sector was restricted to the already worked areas. Thus circumscribed, the private sector could only survive until the exhaustion of the existing collieries; it was not a situation that could inspire great confidence among colliery owners.

The coal nationalization that took place in the early 1970s was the result of the initiative, enterprise and determination of the Minister for Steel and Mines, S. Mohan Kumaramangalam. How he arrived at that high office provides in itself an interesting background to the nationalization of coal. Kumaramangalam was one of the revolutionary students in Mrs. Gandhi's circle of acquaintances during her

The Reign of Ideology : The Grand Era of Nationalization (1969-1973)

student days in London in the mid-1930s. Like Bhupesh Gupta and Rajni Patel in that circle, he subsequently became a member of the CPI. A committed Marxist, he offered the CPI in 1964 a new tactic for the achievement of its goal. He was critical of the CPI's negative attitude and opposition toward the Congress which, in his view, had only driven that vanguard of India's national liberation movement into the arms of the more rightist elements. Instead, he urged that CPI adopt a two-fold tactic of supporting the "national aim" of non-capitalist path of development, to which the Congress party was committed by its various historic resolutions, and of putting pressure from the outside on the Congress to implement its own accepted policies. The former was to be accomplished through communists combining with the progressive sections of the Congress party -- that is, infiltration and coalition -- and the latter through building a mass movement, the two together then leading to the formation of a government of national democracy. He underlined that the Congress encompassed not only the bourgeoisie but also other major classes of the nation -- the petty bourgeoisie, the working class, and the peasantry -- and that its progressive national aims, even in rhetoric, enabled it to secure wide support: "These sections follow the bourgeois Congress leadership not because they support that leadership's real programme of developing capitalism in India, but because they are supporters of the proclaimed programme of the Congress -- the Jaipur, Avadi and Bhubaneshwar decisions of the Congress to build 'Socialism' ". Condemning the CPI's obsession with working for a governmental alternative to the Congress, he advised that the CPI work with, not against, the Congress:

> our tactical line should have been to have restated the national programme and shown how the Congress had failed to implement it; and then called for a Government of National Unity based on Congress-Communist collaboration *in order to* implement this programme. Behind this slogan a powerful mass movement should have been built and as the movement developed in strength and the masses behind the Congress were rallied in action in support of this slogan, then a real possibility would arise of the establishment of such a national government for national aims.

He believed this tactic to be better suited to India's concrete situation than the more orthodox communist preference for the Chinese or Russian models, which he referred to as "almost a disease with us, Indian Communists." He wanted, along with pressure, cooperation with the Congress on specific tasks in its national programme of non-capitalist path of development. Accordingly, he disapproved of CPI's generalized opposition to Congress though he was mindful of the fact that there had been some improvement in this regard on the part of the CPI at Palghat in 1956, including the call for nationalization of "banking, general insurance, coal-mining as a whole."[57]

Kumaramangalam's thesis, called "A Review of the Party Policy Since 1947," was ignored when first submitted to the CPI in 1964, but when resubmitted in 1969 with a postscript it was apparently accepted readily by the party. Whether because of that or the party's own autonomous shift in that direction, the CPI entered into a tacit alliance with the Congress party in 1969 and for several years thereafter. Interestingly, after the 1971 elections the CPI in a resolution stated:

> it is the *mass people's upsurge, a left upsurge, a left upsurge based on expectations of rapid fulfilment of radical programme that brought about the rout of the reactionary combine and a landslide victory for the Congress(R).* The same upsurge can be led forward and the new government forced to implement its promises if a proper lead is given by the left and democratic forces inside and outside the Congress acting jointly. The post election situation does not, therefore, signify a return to the past, but affords *immense possibilities for shifting the country's political life towards the left provided unity is forged between the progressive sections inside the Congress(R) and those outside and the path of mass mobilization, mass campaigns and mass struggles adopted.*

At the Ninth Congress in 1971, the CPI reiterated:

> The most urgent need of the hour is to build left and democratic unity which must necessarily include the progressive and democratic sections within the Congress(R) .... The aim of left and democratic unity in the new situation is to carry forward the broad popular forces including those within the ruling Congress into joint mass movements and mass struggles against imperialism, monopoly capitalism and feudalism and for revolutionary democracy and basic structural changes, including changes in the state apparatus. This makes it imperative not only to unite the left parties, but *also to find common ground as well as to develop common actions with the democratic forces within the ruling Congress.* Such common actions may even sometimes include the Congress organization as a whole....the overall policy of the CPI towards the ruling Congress is one of unity and struggle.
>
> The progressive forces in the ruling Congress today are far stronger than before....The aspect which yet dominates the postelection period, one which will have decisive effect if the party seizes it and struggles to carry it forward, is the fact that the masses as well as the progressive sections in the ruling Congress are in a far better position than before to fight back the forces of reaction inside and outside and force a shift to the left. The policies of the government can be shifted to the left provided these masses are constantly united and mobilised in joint actions and struggles on concrete issues.

Even in 1975, when the government seemed to have adopted a more rightward course, the CPI stuck to its policy; the political resolution at its Tenth Congress maintained:

> Our line will continue to be one of building up the widest unity between the left and democratic forces inside the Congress and those outside, between the masses following the Congress and the masses following the left parties, in the struggle against the forces of imperialism, monopoly capital and feudalism

and for revolutionary democracy and basic structural changes, including changes in the state apparatus....the overall policies of the party towards the ruling Congress will continue to be one of unity and struggle, taking into account the dual nature of the national bourgeoisie."[58]

Along with others of his persuasion, Kumaramangalam had not renewed his CPI membership in the mid-1960s and in due course entered the Congress party. In 1966 he became Advocate General under the Congress government in Madras, later named Tamilnadu. Subsequently, with the defeat of the Congress government in Madras in 1967, the central government appointed him Chairman of the Indian Airlines Corporation. That same year he was also made a member of the Industrial Licensing Policy Inquiry Committee, the thrust of whose recommendations in 1969 was directed against the large industrial houses. After 1969 he became the key strategist for the Congress Left, especially its ex-CPI component. In what was a spectacular rise to power following the 1971 elections, he emerged as the minister responsible for the important portfolio of steel and mines and also as the chief theoretician for Mrs. Gandhi's attempts at social and constitutional engineering.

Kumaramangalam, however, was not alone as a former CPI member inside the Congress party. Satindra Singh, a former comrade who brought the Kumaramangalam thesis to public notice, pointed out in 1973 that "nearly 70 ex-Communists and fellow travellers" were in the Congress Parliamentary Party. Among the more prominent ex-Communists in the party, apart from S. Mohan Kumaramangalam (Minister for Steel and Mines), there were: D.P. Dhar (Minister for Planning); K.R. Ganesh (Minister of State for Revenue and Expenditure); Nurul Hasan (Minister of State for Education and Social Welfare); K.V. Raghunath Reddy (Minister of State for Labour and Rehabilitation); R.K. Khadilkar (Minister of State for Health and Family Planning); Chandrajit Yadav (General Secretary, Indian National Congress); Rajni Patel (President, Bombay Pradesh Congress Committee). Singh further named the following as former Communists and fellow travellers who became "the spirit" behind the Congress Forum for Socialist Action: Arjun Arora, Amrit Nahata, R.K. Sinha and Shashi Bhushan.[59]

Given Kumaramangalam's background it should occasion no surprise that, as his wife put it posthumously, "it was the firm conviction of my husband, Mohan Kumaramangalam, that public ownership and control should be extended over basic industries such as steel and coal."[60] Indeed, coal nationalization has been considered to be "Kumaramangalam's single-handed achievement because the majority in the cabinet was opposed to the measure."[61] Curiously, however, in justifying it, he scrupulously avoided reference to ideology or socialism, advancing instead only strictly instrumental reasons. Perhaps his truer views were more simply and more forthrightly expressed by a CPI(M) spokesman in parliament: "We are wedded to the policy that all the means of production, specially of vital and essential commodities should be under the public ownership and control."[62]

The nationalization of coal under Kumaramangalam was not accomplished in one full swoop but was rather done in two parts. At first, through an ordinance on October 16, 1971, simply the management of 214 coking coal mines and coke oven plants was taken over, but with the clear intent of subsequent nationalization

This set of mines was nationalized in May 1972, for which a cogent rationale was provided. The very nationalization of coking coal mines created the conditions for a similar fate for the non-coking coal mines. Following the earlier pattern, the management of 464 non-coking coal mines was first taken over in January 1973 under an ordinance, and later these mines were nationalized in May 1973. With that the entire coal industry, outside mines attached to private sector steel plants, came under public ownership and operation.

The rationale for the nationalization of the coking coal mines centered on two major points, the first of which had to do with the conservation of this type of coal since India was not abundantly supplied with it by nature while the private sector, interested only in profit, was allegedly resorting to indiscriminate slaughter mining.[63] Kumaramangalam referred to the work of many committees in this regard, going as far back as 1937. However, in a devastating rebuttal, a critic underlined that it was meaningless to seek justification for nationalization by reference to the condition of the industry before World War II when it was under the control of British owners. Significantly, as a result of post-independence legislation on coal conservation (Mines Act 1952 and Coal Mines Regulation Act 1957), which vested the government with abundant powers for the control and direction of the industry, "coal mining in India was for all practical purposes already a nationalized industry in all its aspects."[64] Indeed, if things had not been set right, it was the fault of the government, which now used its own failure at implementation to nationalize the industry.

The second point in justification for the nationalization of coking coal was the need to expand production to meet the growing needs of the iron and steel industry. What was curious about this rationale was that steel production had not been held back because of lack of sufficient supply of coking coal. Indeed, there was overproduction of such coal which was then used for purposes other than manufacture of iron and steel.

In respect of non-coking coal, Kumaramangalam had assured that there would be no nationalization, but then suddenly sprung it on the industry,[65] justifying it on two grounds.[66] One ground pertained to the need for increased production of coal, which required enormous investment that was allegedly beyond the capabilities of the private sector. The minister charged that the private sector collieries were not making adequate investments, but he made a grievance of the fact that "the investment from their own funds has been meagre," that the required heavy investment of Rs. 1 billion was "far beyond the financial capabilities and organizational competence of the private coal mining industry in India," and that they "wanted the funds to be provided largely by public financial institutions and partly by guaranteed price increases." In saying all this, he ignored how the public sector had failed to rely on internally generated sources, but more importantly how the government through its own policies had brought about precisely the situation he criticized: the socialization of savings forced the private sector -- whether in coal or elsewhere -- to go to public financial institutions for funds; the government either directly controlled prices, or its railways as almost a monopoly buyer dictated prices, which determined the coal industry's profits; and the threat of nationalization that constantly hung over the industry left little incentive for investment. But during the 1950s and 1960s the private sector had provided a more than respectable record of performance, especially compared to the public sector, so much so that the left-oriented Minister for Mines, K.D. Malaviya, was compelled to say in parliament in April 1963:

> Yes, it is a fact that the private sector coal industry has gone ahead and they have overstepped their targeted production. We are all very glad, and have congratulated them several times. We want coal, and we badly want coal, and therefore, whosoever produces it, is most welcome to do it. [67]

The business community argued that "No case whatsoever has been made out for taking over the entire coal mining industry. Collieries have been meeting the requirements of the country fully. There is no coal shortage anywhere."[68] Ironically, shortages and poor quality were to become the order of the day after nationalization.

There was a sounder basis to the second ground that Kumarangalam offered, that is, the exploitation of labour; he spoke with great indignation: "Anybody today who raises his voice in defence of the private sector in the coal-mines is guilty of being the spokesman of the most reactionary and the most, I would say, ruthlessly, shamelessly exploiting section of the capitalist classes of our country." Mrs. Gandhi also seems to have had a similar assessment: "nationalization of coal was an essential step because the owners of the mines were not looking after the workers and were exploiting mines in such a way that workers suffered and mines themselves were deteriorating."[69] However, it should be noted that government data demonstrated that the recommendations of the wage board had been fully implemented in respect of 69 per cent of the miners, and partially implemented for 28 per cent of the remaining miners.[70]

The government used the exploitation of labour by a part of the industry to nationalize the entire sector, the real basis for which lay in ideology and, on that account, coal nationalization was acclaimed by the Communist parties who at the same time wanted no compensation paid. The government did make an exception in the case of the captive collieries owned by the Tata Iron and Steel Company, not only because labour's conditions there corresponded to those in the public sector but also, as Kumaramangalam explained, "We have got enough on our hands"[71] — precisely the reason Nehru had advanced all along for not nationalizing industry in the private sector.

c. *Nationalization of Indian Iron and Steel Co.*: The steel industry in the form of major integrated steel plants began in India in the first decade of the 20th century, when Indian entrepreneurship established the highly successful plant of Tata Iron and Steel Co. (TISCO) at Jamshedpur. By the end of World War II, this plant had an annual production of 0.75 million tonnes of saleable steel. Another major plant was with the Indian Iron and Steel Co. (IISCO), which had been set up at the end of World War I largely through British investment. By World War II end, it had a yearly production of 0.20 million tonnes.[72] The production of the two plants constituted almost the entire locally manufactured steel of about 1 million tonnes at the time. The Industrial Policy Resolution of 1948 made the establishment of new steel plants a prerogative of the state while it held the threat of possible nationalization over existing plants. The Industrial Policy Resolution of 1956 continued with the policy of vesting in the state responsibility for future development of the industry. Consequently, when in 1954 the house of the Birlas reached an agreement with a British consortium for a steel plant the government disallowed it. The government was reluctant to violate its industrial policy, especially when the financial contribution organized under the auspices of the Birlas amounted to only 10 per cent of the Rs. 1 billion cost, while the remainder was expected to come from the government or government-underwritten loans. In the event, the government proceeded essentially with the same deal but in the public sector.[73]

The period of the Second Five Year Plan (1956-1961) saw a major dramatic entry by the state in steel, with three one-million tonne plants undertaken in the public sector at Bhilai, Durgapur and Rourkela. Subsequently, in the mid-1960s the government launched another 1.7 million tonne steel plant at Bokaro, apart from proceeding with the expansion of the three plants. Even though the government was negative toward the private sector's establishing new steel plants, it was willing to allow expansion of the two existing steel plants in the private sector. However, in practice, it was not always very forthcoming because of its ideological preference for the public sector. In the case of the Third Five Year Plan (1961-1966), the government simply "refused to allow further growth of the private sector plants."[74] It did not agree to a TISCO proposal in 1961 for expansion, on the assumption that public sector expansion would suffice in meeting plan targets; similarly, IISCO encountered problems in getting approval for expansion. The government approved expansion by TISCO and IISCO only when that assumption proved questionable and, indeed, resulted in shortages.

Thus, the private sector's role was seen as merely supplemental when the public sector failed to meet targets, but even then "usually only after a crisis is imminent and it is too late for the private sector to contribute effectively to forestalling this crisis."[75] Hanson believed that the government ignored the performance of the private sector in determining its role; although he did not exclude economic rationality, he held that "there is at least a strong *prima facie* case that the basis of judgement was essentially political and ideological" and that economic rationality had "to force itself on reluctant ideologists." The result of government policy, then, was to slow down the growth of the steel industry in India. Besides, the fixing by government of "unduly low steel prices", while it benefited middlemen and users, deprived the firms of internal savings and made them excessively dependent upon public financial institutions. Even there the government role was perceived as unhelpful by the firms. For example, having provided loans to the two firms in the 1950s for expansion on the understanding that higher steel prices later would help in repayment, the government in 1963 insisted on repayment "by new share issues or a sacrifice in dividends and conferred upon itself the unilateral right to convert these advances into company shares," that is, partial nationalization.[76]

Whatever the limitations, TISCO expanded its capacity by the end of the Second Five Year Plan to 2 million tonnes and IISCO to 1 million tonnes. By 1963-64 IISCO's saleable steel production reached its peak at 0.810 million tonnes, and then began to decline consistently year after year. On July 14, 1972, through a presidential ordinance, the government took over the management of IISCO from Martin Burn Company which managed the firm. The government's rationale for the takeover, as offered by the Minister of Steel and Mines, S. Mohan Kumaramangalam, was mismanagement; it charged that neglect of maintenance and modernization of the plant had resulted in such serious deterioration as to bring down production to about 60 per cent of capacity.[77] Two years subsequent to the initial take-over, Kumaramangalam's successor, K.D. Malaviya, explained more sharply: "there was a rapidly deteriorating condition of the plant...this deterioration was the direct result of ineffectiveness....The attitude of the management too was unresponsive to the grave and very urgent problems which needed attention....So, the rehabilitation was needed. Capital was not there. And nobody bothered about investing money. Everybody at that time wanted to make money out of it. Therefore, rehabilitation was very gravely neglected....it was considered appropriate that the plant which was doing very well only a few years back should not be allowed to deteriorate further...that the management of the company should be taken over." Interestingly, Kumaramangalam made no reference to ideology or socialist pattern of society in the initial takeover. Malaviya himself said, in reference to the government's decision, that the plant's situation had "deteriorated to such a dangerous level that the action was a more of the nature of a penal action than an ideological one."[78]

There can be no doubt about the mismanagement of the company, evident in the consistent fall in production after 1963-64. However, the business community disagreed, maintaining that "the problem is not one of fall in the standard of efficiency of the Management, but is entirely of the labour situation prevailing in the area."[79] This was in apparent reference to the turbulent labour conditions in West Bengal because of the activities of CPI(M), whether inside or outside government. An economic geographer seemed to agree in part: "The decline in output in recent years has been due to shortage in raw materials supply and labour trouble among other causes."[80] But Kumaramagalam demurred, blaming management. However, the business community found the government's rationale unacceptable, pointing out that IISCO's capacity utilization at 61.7 per cent was far higher than the public sector plant's at Durgapur where it was only 43.8 per cent. Prophetically, it did not see the government take-over as a solution, saying that if it were so then the Durgapur plant "would have been doing much better than is the case."[81]

There were several factors of an ideological and instrumental nature, but at times unstated, that were responsible for the state take-over of IISCO. To begin with, one could say that the company was already half-nationalized, leaving no alternative but to complete the process. Here in steel, as elsewhere in the private sector in India, the socialization of savings had made IISCO heavily dependent on the public financial institutions. Some critics charge that in this manner the public sector has subsidized the private sector. But others hold that through this mechanism the state has used its control of the financial and banking institutions to bring about what is essentially backdoor nationalization, to be made open at the right strategic moment. In the case of IISCO, through the Life Insurance Corporation, Unit Trust of India (government-controlled mutual fund), banks and the more recently nationalized general insurance, the government held almost 50 per cent, more precisely 49.34 per cent, of the shares. With such a huge stake in the company's equity, the government could not just let what some considered a national asset go to waste by letting it be shut down. However, the takeover proved extremely costly subsequently in terms of rehabilitation, with some saying that it would have been preferable to have opened a new plant with the same investment.

With such a large share in the equity, the government really could not easily let some one else take over the company either. The initial takeover was meant for only two years after which the government was to make up its mind as to what to do with the company. However, a permanent takeover was built into the very extensive financial penetration of the company by the government. Kumaramangalam could not have been more explicit even while avoiding the term nationalization: "there is absolutely no question of the management of this company going out of the hands of the Government and back into the hands of either of the erstwhile management or of any other future private management that may rest its greedy eyes on IISCO. It is merely a question of time in order to be able to decide what would be the most appropriate form."[82] Apparently the decision came sooner than expected, even though for strategic reasons it was not announced. When the initial two-year takeover was renewed for another two years in 1974, Steel Minister Malaviya declared that the government found immediately after the takeover that the plant's condition was much worse than had been earlier imagined and so it "thought of nationalizing the plant at that time."[83]

Aiding in the resort to takeover by the government was the ideological commitment of its key decisionmakers even though, for whatever reasons, Kumaramangalam refrained from referring to it. Two years later, while emphatic that there could be no question of returning the plant, Malaviya asked for, but only in an aside in Hindi, the forebearance of the House in not discussing such matters; nonetheless, he said there was also the issue of the government getting hold of the commanding heights of the economy.[84] In truth, the real ideological intent was expressed by a CPI spokesman who welcomed bringing "this Company finally and conclusively under the umbrella of the public sector and I hope the day is not far distant when the Tata Iron and Steel Company also, in which already a very sizeable chunk of shareholding is held by public financial institutions, will join the rest of the family. There is no reason why one plant should remain outside in the private sector." CPI(M), too, while declaring that "within the capitalist system no basic change is possible by mere taking over or even by nationalizing a few industries. Nonetheless, we support this step."[85] What the episode demonstrated was the element of vulnerability of firms to government takeover when they become excessively dependent for finance on government, especially in the context of the particular ideological orientation of political elites amidst an anti-business environment, at least during the 1970s. On the other hand, with the nationalization of savings and heavy taxation the government left the private sector few other alternative channels to obtain resources.

A third factor that was evidently critical in the takeover, but one that was not expressed in the initial rationale, was that of the interests of labour. Critics often allege that the nationalization of sick enterprises is to bail out capitalists who have drained them of their resources -- the sinking sands theory. That may or may not be true. But in a representative system where political leaders have to get elected, it is really labour, especially when it is in the organized sector and exists in huge conglomerations, that is more critical than a few capitalists. In the case of IISCO, the jobs of some 40,000 employees in the steel plant and associated mines were at stake. In 1976, when the government bought out the remaining equity in the company, the then Minister of Steel and Mines (1974-77), Chandrajit Yadav -- who had also come over to the Congress party from the CPI and was part of the ex-CPI faction -- acknowledged that "the Company was almost going to be closed and it would have rendered thousands of workers jobless".[86]

Thus, a mixture of ideological and instrumental reasons made for the decision to take over IISCO. The instrumental reasons themselves had, in considerable measure, to do with the legacy of the cumulative consequences of earlier decisions over the years, in turn inspired by ideology. However, it is important to underline that in the absence of a prior commitment to ideology on the part of key decision-makers, the government may well have left the enterprise to the workings of the market.

d. *Copper and Refractories*: It is a tribute to Kumaramangalam's dynamism and dedication to his cause that in the two years from March 1971 to May 1973, when he headed the ministry of steel and mines, he was able to bring the entire coal industry into the public sector as also a major steel company. But that was not all. Among the other cases of nationalization that he successfully pushed through related to copper and refractories. Indeed, in the area of nationalization in those two years, Kumaramangalam seemed to dominate the scene.

In the realm of refractories, nationalization involved two firms -- Asian Refractories Limited and Assam Sillimanite Co. Limited. The justification for the nationalization of Asian Refractories was exceedingly summary -- it was to augment supplies of refractories "to meet the requirements of the iron and steel industry" It was thought "it would be better if we could take over the refractories company, the Asian Refractories, inside our own steel complex as it were," and that "we have to take over Asian Refractories to help us in Bokaro. That is the real object of taking this over."[87] The company, which had shut down in 1968, had been auctioned off to one of the concerns of the Birla business house, a particular bete noire of the Left at the time. Before the government could make a decision on the Birla's application for takeover under the monopolies legislation, however, Kumaramangalam intervened to nationalize it simply on the ground that the end products were useful for his steel industry, that "it is a good deal", and that "the equipment is good equipment, very modern." As for Assam Sillimanite, even before it could be put on auction, the government took over its management in 1972, only to nationalize it in 1976.[88]

In the case of the copper industry, it seemed like the government was anxious to grab a well-functioning company, the Indian Copper Corporation, with a 10,000 tonne capacity. This was especially intriguing since Kumaramangalam himself acknowledged that the public sector's own efforts to build a copper industry had "run into a number of difficulties in the past" even though he was, with this takeover in 1972 and subsequent complete acquisition the same year, more confident about achieving production targets. Though registered in England, foreign shareholding in the company was insignificant, about 2.5 per cent, while the major shareholders were India's public financial institutions, more particularly the Life Insurance Corporation, with about 20 per cent of the shares. The company mined some of the country's richest copper ore.

Kumaramangalam advanced several reasons for taking over the company. Firstly, he declared that the country urgently needed greater production of copper to avoid heavy foreign exchange outlays on imports, and accordingly this had to be done on "a priority basis using the existing industries, having the Indian Copper Corporation as the foundation for that stepping up." Secondly, he felt that an integrated development of the industry required "avoiding the artificial lines of demarcation...between the private and the public sectors." With this rationale, of course, there could be no bar to taking over any and all private enterprise. Thirdly, he believed that the takeover would enable making the best use possible of existing managerial and technical capabilities for the development of the copper industry. Fourthly, he argued that the takeover would put an end to monopoly control,[89] and apparently for that reason both CPI and CPI(M) heartily endorsed the measure. However, it should be noted that local copper production formed only a small part of the country's needs.

Before the FICCI, Kumaramangalam acknowledged that the public sector had "made a mess of Khetri", the government venture in copper production, but he made known his determination to act unilaterally through what seemed like forced reasoning:

> The Indian Copper Corporation is a flourishing concern no doubt. But it has not got the resources because it is profit conscious like all private sector organisations necessarily have to be. They live after all for profit, whether I like it or you like it. That one can say is a historical fact. Well, we in the government do not live necessarily for profit. We look at the national interest as a whole. And looking at it from the viewpoint of the national interest, we found no alternative except to take over the Copper Corporation and that is why we have done it....I can only say that ultimately that it is we, in Government, who have to make the decision.[90]

The various measures of nationalization that Kumaramangalam's ministry pushed through caused great consternation in the business community, which was "unhappy that of late Government have been over-anxious to take over this unit or that on one ground or the other" [91] More fully, it stated:

> There is much of sophism in the arguments that are put forward in the case of these takeovers: In the case of the coking coal industry, the rationale was making available larger supplies to the steel industry. As regards Indian Copper Corporation which was declared to be efficiently operated, the reason advanced was that the takeover would help the loss making Hindustan Copper Corporation, a public sector enterprise, to function better. In respect of the

takeover of Indian Iron and Steel Co., the move was mainly to stem the alleged deterioration that had set in the plant's working. We are now asked to look upon the takeover of the non-coking coal industry as a tool of development.

It is regrettable and unjustified that government should follow a policy of takeover and nationalization, especially when there is no evidence of any breakthrough in the performance of public sector enterprises. Moreover, it is most unfortunate that not only different but contrary reasons are given for pursuing this policy. [92]

Furthermore, the business community was upset that compensation was "so negligible that nationalization is confiscatory in character." However, no matter what it may say after the event the community was powerless before the *fait accomplis* that the government confronted it with. Moreover, with one nationalization after another, every private business lived in fear of being the next victim.

e. *Nationalization of Cotton Textile Mills:* The cotton textile industry is India's oldest and perhaps largest industry, contributing some one-fifth of the country's national income and providing employment to some one million people. The industry's plant was overworked during World War II and the period soon after, for it failed then to obtain new textile machinery. Managing to survive during the 1950s, the industry entered a period of serious difficulty in the early 1960s, which exploded into a crisis by the late 1960s. Closures of cotton textile mills followed, which brought in government intervention in the form first of takeover of management of over a hundred mills — which constituted about a sixth of the country's textile mills in the organized sector — and then their nationalization. Though the government blamed the closure of the mills and their industrial "sickness" on mismanagement by the mill-owners, the phenomenon of the crippling of such a large part of the industry has to be properly seen as the result of a highly unfavorable economic environment, which itself was largely the creation of government's own policies.

The failure of a large proportion of India's cotton textile mills is attributed to the neglect of modernization of machinery. That, in turn, is considered to be fundamentally a function of low profitability, which prevented the industry to mobilize capital and plough back resources into modernization. It is in relation to low profitability that the impact of government policy was critical.

Firstly, in the 1950s the government froze, with the aim of assuring greater employment, the weaving capacity of (1) the organized or mill sector of the textile industry, while entrusting the meeting of the additional demand for cloth in the future to (2) the decentralized or small-scale sector consisting of (a) powerlooms and (b) handlooms. If at the time India's national bourgeoisie could be equated essentially with the textile industry, then this blockage by the state of its growth in what then was a reasonably profitable economic activity cannot be considered to be a particularly significant index of its domination over the state. The impact of government's discriminatory policy, however, was stagnation in cloth production in the mill sector and a phenomenal expansion of that in the decentralized sector, as Table VII.2 vividly demonstrates. From contributing around a fifth of the total cotton cloth production in 1951, the decentralized sector had by 1972 come to almost equal the organized sector, straining to outstrip it soon. While production in the organized sector had seen a decline since the mid-1950s, that in the diffuse and decentralized sector, employing some 7 to 10 million people, had grown at a rate of around 13 per cent.[93]

Table VII.2
Cloth Production in India

Year	Total Cloth Production (million metres)	Mill Cloth Million Metres	Mill Cloth %age	Decentralized Cloth Million Metres	Decentralized Cloth %age
1951	4740	3727	78.6	1013	21.4
1955	6278	4658	74.2	1620	25.8
1960	6629	4616	69.6	2013	30.4
1965	7643	4587	60.0	3056	40.0
1970	7849	4157	53.0	3692	47.0
1972	8022	4245	52.9	3777	47.1

Source: Adapted from R.D. Mohota, *Textile Industry and Modernisation* (Bombay: Current Book House, 1976), pp. 28-29.

Secondly, not only was the cotton mill sector barred from expansion, it was prohibited from manufacturing certain kinds of products which were particularly profitable, such as dhotis, saris and some special fabrics, whose production was instead reserved for the decentralized sector.[94] Thirdly, the government treated the mill sector in a discriminatory manner by imposing on it disproportionately high excise duties compared to the powerloom sector, which furthermore did not have to bear the larger overhead costs, including a comparatively higher wage bill because of government's labour legislation.[95] Besides, the government provided the handloom sector with special concessions and subsidies. Because of the inability of the government to police such a large and diffuse sector, the powerloom sector was also able to obtain subsidies by falsely passing off its production as handloom cloth. As a consequence, it became "the persistent complaint of the mill sector that the crisis of the industry was partially due to the competition from the decentralized sector."[96]

Fourthly, the government imposed on cotton mills from 1964 to 1978 the obligation to produce coarse cloth for sale at a controlled price for sale to the poor or weaker sections of the population. The quantity of controlled cloth required to be manufactured as a proportion of total mill production varied over the years, starting out at 50 per cent in 1964 but reduced to 25 per cent in 1968 and about 14 per cent in 1973. Two things are significant about this controlled cloth scheme. One is that the sale price of the cloth was much below the actual cost,[97] thus directly reducing the industry's profitability and the availability of internal financial resources for reinvestment. The other was that it also affected adversely the price of uncontrolled cloth by the forced competition with controlled cloth. Both ways,

The Reign of Ideology : The Grand Era of Nationalization (1969-1973)

one sector of the industry was made to bear the cost what should have been more fairly borne by the national community. But the government at one time adopted a stern attitude in the matter; during the emergency, Mrs. Gandhi declared: "we are going to impose definite obligations on the textile mills to produce common cloth of coarse and medium varieties. This policy will be vigorously imposed, and all those who come in the way of effective implementation of this policy will be dealt with ruthlessly....If it defaults in the fulfilment of these obligations, the full force of Government will be brought to bear on the defaulting mills."[98] What should be obvious is the ubiquitous imposition of a plethora of restrictions on the textile industry by the government. Once a strong advocate of controls, Hazari comments in retrospect: "the tale of industry....is full of regulation, on the whole, positive till the mid-sixties, generally negative since then, with textiles bearing the full brunt of negativism right through."[99]

Fifthly, from 1962 onwards except intermittently, shortages of cotton developed so that only about 80 per cent of the capacity, which had expanded particularly in the decentralized sector, could be utilized. The shortages, of course, led to lower capacity utilization but the intense competition to procure raw material supplies also sent cotton prices soaring. This added tremendously to the input costs since cotton constituted about half the cost of cotton cloth. On the other hand, because of competition from the decentralized sector and from synthetic fabrics the prices of cotton manufactures did not rise corresponding to the increase in prices of raw cotton, as Table VII.3 indicates. Upto 1962-63 the index numbers of cotton manufactures kept consistently ahead of those for raw cotton, but in 1963-64 a reversal took place in this pattern in which there was no let-up. This new trend became especially pronounced in 1967-68, with the gap between the index prices for the two products becoming particularly wide in 1971-72. The crunch between high costs and low returns led to low profitability, from which ensued a vicious cycle to perpetuate the untenable position of the industry. Low profitability meant diversion of resources to more profitable activities. It also meant inability to mobilize share capital, which resulted in a change in the capital structure of the industry through excessive dependence on banks for loans. The high interest charges on loans, in turn, further reduced profitability, thus driving the marginal units to sickness. Some saw the solution in modernization of the industry which would make it more efficient and therefore more profitable.[100] Others saw that as a clever ploy on the part of the industry to obtain concessional finance from the government. But more importantly, they saw modernization as the source of the crisis in the first place as it made some units more efficient in the context of excess capacity and raw material shortages, and therefore created a crisis for marginal units.

Nor is government takeover seen by them as a solution since it merely transfers the losses of individual mills to the government except that the public is made to bear the losses; furthermore, the modernization of the mills taken over under government auspices has the same impact as modernization under private auspices, with the government then forced to take over more of the resulting sick mills in a never-ending process. The solution rather is seen in removing shortages of raw cotton, where the government has a critical role to play but has failed to perform adequately.[101] But this may very well only transform the crisis from that of high-cost

Table VII.3
Comparative Wholesale Prices
(Index Numbers; Base 1952-53=100)

Year	Raw Cotton	Cotton Manufactures
1954-55	102	107
1960-61	112	128
1962-63	113	131
1963-64	111	108
1964-65	117	110
1965-66	119	114
1966-67	127	122
1967-68	142	126
1968-69	155	129
1969-70	171	134
1970-71	192	141
1971-72	234	162

Source: Adapted from R.D. Mohota, *Textile Industry and Modernization* (Bombay: Current Book House, 1976), p.63.

inputs into a glut of cloth on the market and then forced unemployment. Meanwhile, in forging an adequate policy the government has been caught between the cross-pressures of different interest groups:

> Cotton growers want a higher price, which is no doubt in the national interest since the industry has to obtain all or almost all its cotton requirement from domestic sources. On the other hand, the Government wishes to ensure that the handloom and powerloom weavers get their yarn cheap. Then there is the third factor, namely labour, which must be enabled to improve its living standards. And the exchequer...must garner sufficient revenues from this industry, which is the country's largest, to keep the wheels of administration running smoothly.[102]

Regardless of which might be the optimum solution, the pertinent fact is that many cotton textile mills had by the late 1960s closed down or were faced with closure, unable to meet their liabilities. The government felt obliged then to intervene under the Industries (Development and Regulation) Act of 1951; it took over 57 textile mills with the clear intent to nurse them back to health and return them to their owners, and in 1968 it set up the National Textile Corporation for the purpose. There was a 15-year limit under the Act for the government to return the mills. However, the crisis refused to go away. Particularly after the 1971 elections, business felt insecure about the intentions of the new government and many mills, if they had not already closed down, began to cannibalize their assets rather than nurture them. In the event, the government through an ordinance in October 1972 took over the management of 46 textile mills but with the explicit intention of nationalization, which was accomplished in 1974 not only for these 46 mills but the earlier 57 mills as well. This was a turning point in the attitude of the government, for hitherto it had largely stayed out of the consumer goods field. In 1976 two more mills were added to the list of textile mills nationalized by the government under Mrs. Gandhi before her defeat in the 1977 elections.

The government's justification for the takeover or nationalization centered on prevention of the "fall in production of textiles and resultant unemployment arising in certain textile undertakings on account of mismanagement and other financial and technical difficulties."[103] It would seem, however, that the real purpose pertained to employment of labour. The fall in production could not have been a serious consideration since this could have been easily picked up by the decentralized sector, especially its powerloom component. Although the availability of cheaper controlled cloth may have been a factor, it is curious that the government a few years later in 1976 removed this obligation from the mills under the National Textile Corporation as also from other mills in a weak financial position.[104] At any rate, even if applicable, this factor demonstrated government's responsiveness to the interests of the weaker sections through nationalization rather than those of the capitalists as radical critics have at times charged.

To be sure, capitalists may have been aided earlier in the making of profits through government developmental assistance in the setting up and running of enterprises, but nationalization was not required to aid them. Indeed, such businessmen opposed the government taking over their enterprises, representatives sympathetic to their interests in parliament criticized government action,[105] and they went to court, though fruitlessly, to prevent the government from nationalizing them. Any compensation paid by government was meaningless since their liabilities in many cases were more than the compensation paid, which really went to the public financial institutions; compensation was simply a transfer of funds from one part of the government to another. Nor could nationalization be seen as a step toward the general modernization of the industry, for having nationalized these mills the government became more cautious about taking over other mills as it faced an endless task. As Minister of Commerce D.P. Chattopadhyaya explained in 1976: "I would like to make it very clear in the beginning that it is not the Government's policy to take over the sick mills one after another indiscriminately without going into the question of the individual mills in question....This is not a very ideal way of solving the problem. That is why the Government had a very new and I must say a deeper look at the matter....Central government takeover is not the only solution. It is not necessarily the best solution."[106] Among other possibilities he suggested was the merger of sick units with healthy units in the private sector.

The real agent for the nationalization and its intended beneficiary, it would seem, was labour. Organized labour was in the forefront in putting pressure on the government for nationalization. As one CPI spokesman said in parliament in 1974: "the workers had to start a movement and a campaign that the government should not return these mills to the erstwhile owners but nationalize them and run them. It is good that the Government had done it and we welcome it." Another CPI spokesman was grateful to the government on behalf of labour in 1976: "I therefore congratulate not only the Minister but the entire Cabinet and I also congratulate the determined workers, 7500 to 8000 of them, and their family members for remaining determined even facing starvation and even after starvation their slogan was that this mill should be taken over by the Government....The day the news was broken in Kanpur, I saw the amount of jubilation among the starving workers and their family members....There was 'Zindabad' not only for Shrimati Indira Gandhi but for the entire Cabinet Members....This was one of the greatest dreams of my life and this has now been realised."[107] CPI and CPI(M) leaders pressed for nationalization of more mills, indeed, of the entire mill sector. Commerce Minister Chattopadhyaya responded that the action "shows unmistakably the Government's concern that the interest of the unemployed people should be looked after....The very fact that this is taken over reflects Government's deep concern for the welfare of the workers."[108] However, government concern extended beyond unemployment; it was the fear of the breakdown of law and order that often seemed persuasive. The *Eastern Economist* speculated in the case of the 1972 nationalization that "it is perhaps the decision of the Apollo Mill's workers to take out a *morcha* on the day the Prime Minister was due to arrive in Bombay (November 1, 1972) that clinched the issue."[109]

The nationalization of textile mills between 1972 and 1976, though it initially took place in an environment of radical fervour following the 1971 electoral mandate, came increasingly to focus on specific interests, concretely those of labour, rather than spring from a diffuse ideological orientation. Even though the government seemed to lose its ideological fervour for nationalization after 1973, specific interests, particularly of labour, continued to play a part in individual measures of nationalization, especially relating to the textile industry. In such action, political and electoral considerations played a considerable role. Indeed, CPI(M) spokesmen charged that government tended to favour nationalization of those textile mills where labour was organised under its own auspices or agreed to move over to it.[110]

f. *Nationalization of the Wholesale Wheat Trade* : Since the Great Bengal Famine of 1943, in which three to five million people died, the Government of India has largely followed a pragmatic policy in relation to the partial mobilization of the marketed food surplus and its distribution under official channels to sections of the public. This pragmatic policy has consisted of public procurement of about 10 per cent of the foodgrains production and distributing it to urban centres at government-fixed prices, while letting the market operate for the rest. In the partial procurement of the marketed surplus, the government has relied on a variety of instruments rather than any single one, in consideration of the fact that India is a large and diverse continent-size country. These instruments have included levies on producers and traders, market operations, monopoly purchases, support price arrangements, and exchange for subsidized input supplies. Emphasis on one or more of these has varied according to circumstance, but essentially India has operated with "a mixed dual-market arrangement of private trade and the public distribution system working side by side, under some prescribed rules of the game."[111]

The Reign of Ideology : The Grand Era of Nationalization (1969-1973) 315

More often than not for the three decades after World War II India has had to supplement its surplus procurement by food imports, which at times led to a feeling of humiliating dependence abroad. For the distribution of the food procured or imported, the government has relied on a network of fair price shops which sell the food at fixed prices to segments of the population, largely in urban and metropolitan centres, while counting on the rest of the non-producing population to fend for itself. In other words, the government has been reluctant to assume the responsibility for feeding the entire population or to take the blame for not doing so when food was scarce or prices rose. This has largely been in recognition of its own limited capacity to handle such a gigantic task.

To this generally pragmatic policy - under which the government has stuck to "not giving itself over to ideological 'solutions' to the problem" - there have been two great exceptions, inspired by opposite ends of the ideological spectrum. One was in late 1947 when, under the urging and pressure of Mahatma Gandhi, the government responded to the public irritation with irksome wartime food rationing and controls by lifting controls altogether and relying on the market alone. As food shortages occurred consequent on hoarding in an economy of scarcity and as prices rose dramatically, the government ended the experiment with food decontrol within nine months and returned to a regime of controls in respect of the procurement and distribution of food. The other exception was in 1973 when the government resorted to the nationalization of the wholesale wheat trade, in the words of a leading authority on the subject, "under some leftist ideological compulsions" and "on ideological and political grounds, rather than practical."[112]

In the aftermath of the Bengal famine the government had established a Foodgrains Policy Committee in 1943, which raised the possibility of total control over the food economy with a gigantic staff as "the greatest single experiment in the history of the government" for feeding several millions of people. However, the committee dismissed the idea of a foodgrains monopoly by the central government as "wholly impracticable under present conditions".[113] Instead, on the recommendations of the committee, the government adopted a mix of policy instruments that included stabilization of market prices through statutory control, monopoly procurement of foodgrains in some areas, planned movement of foodgrains from surplus to deficit areas, and distribution through partial rationing.

Following the disastrous experience with food decontrol in 1948, the government set up in 1950 the Foodgrains Procurement Committee, which plumped for a thoroughgoing policy of controls as a means to price stability. However, the government chose to ignore the committee's recommendation as not being practical and instead, without saying so, followed the advice of the member who dissented from the committee's report. The heart of his dissent was

> The ideal system of procurement and distribution for each area would...be one that sets out deliberately to take only a part of the surplus from the surplus producer and to feed only a part of the deficit population, leaving the balance of production to be marketed freely. This system would maximize availabilities because arrivals in a free market are always more than under controls. It would maximize production because the higher price obtained by the

producer for that part of his produce which is sold at uncontrolled rates would induce him to grow more. It would keep the general level of prices from soaring without restraint by meeting a part of the demand at controlled rates....Intensification of controls in all areas, but particularly in self-sufficient or surplus areas, leads to a progressive stiffening of opposition.[114]

The government came to reject the option of either complete control or complete decontrol, and chose a middle path that combined partial mobilization of the surplus, partial distribution through both statutory and informal rationing, and partial operation of the free market. With some modifications this set the policy for the government in the future decades.

In the context of increased food production during the period 1952 to 1954, however, the government moved to the relaxation and dismantlement of controls, only to bring them back into more rigorous operation in 1957 as the food situation deteriorated. The government then established the Foodgrains Enquiry Committee in 1957 under the chairmanship of the socialist leader Asoka Mehta to advise on the new food situation. Basically, the committee hewed to a middle way in coping with the food problem, but it suggested several institutional innovations. While most of its recommendations followed the middle path, the committee departed from it in one important area in the cause of price stability:

> We should like to emphasize here that until there is social control over the wholesale trade, we shall not be in a position to bring about stabilization of foodgrains' prices. Our policy should, therefore, be that of progressive and planned socialization of wholesale trade in foodgrains.[115]

However, in the event, the government headed by Prime Minister Nehru brushed aside this recommendation even though it would have coincided well with his ideological preferences. Understandably, Shastri tried to curb the enthusiasm of "the zealots of state trading", warning: "We should be very careful in resorting to state trading, particularly in foodgrains. Unless the Government was ready to cope with the complexities of the problem, state trading would only increase corruption, besides adding to the difficulties of the common man."[116]

The Foodgrains Policy Committee, set up in 1966, was moderate in its recommendations, stating that in the cause of food self-reliance the government had to get hold of a large share of the country's food production and therefore had "to strengthen its own machinery for the procurement, transport and distribution of foodgrains for the surplus as well as deficit areas." Earlier in 1965, as part of its plan to assure greater food security and price stability, the government established the Food Corporation of India to purchase and build food stocks; thus, the government endeavoured to control the food economy, not through displacing the traders, but through the Corporation's power to intervene in the market. By 1971 the Corporation had achieved "a commanding position in the foodgrains trade.."[117]

As the political scene became increasingly radicalized in the late 1960s, however, solutions that had been pragmatically ignored on mature consideration of the total context came to be accepted or advocated in the heat of partisan conflict. After the split, the Congress party under Mrs. Gandhi at the AICC meeting in Bombay in December 1969 resolved that "wholesale trade procurement of major agricultural commodities should be done in the public sector." The party's manifesto for the 1971 elections promised an expanded role for the Food Corporation of India in the food economy in order to "ensure the implementation of a national policy for the distribution of food and fair prices for the farmers."[118] After the elections the issue was pushed aside, perhaps because of the Bangladesh crisis, but at the AICC meeting in October 1972 at Gandhinagar (Gujarat) the party made known its clear and unequivocal commitment to establish a public distribution system to provide essential commodities (such as cereals, sugar, edible oils, kerosene, and cloth) at fair prices to vulnerable sections of the population. But it felt that "a public distribution system can be a success only if government has control over the wholesale trade" in these goods. Accordingly, as a start it decided to "reaffirm the promise made to the people on taking over the wholesale trade in foodgrains by the State. Such a takeover achieved through monopoly procurement should be dovetailed into an effective system of retail distribution, holding of buffer stocks and appropriate price policy."[119]

The 74th annual session of the party at Calcutta in December 1972 urged the implementation of this decision right away in respect of rice and wheat.[120] The same month the State Food Ministers accepted nationalization of the wholesale foodgrains trade, and the policy was endorsed by the Chief Ministers in February 1973. It seems some of the Chief Ministers opposed the move but were silenced by the central leadership.[121] Their opposition stemmed from the fact that, being close to the ground, they did not want to alienate both the rich and middle peasantry and the trading community to whom they would have to turn for political support at election time. Surprisingly, some of Mrs. Gandhi's radical advisers, such as P.N. Haksar and Kumaramangalam, were also opposed to the takeover of the foodgrains trade while others, especially D.P. Dhar, heading the Planning Commission, were insistent on it.[122] Initially it was decided to take over the wholesale trade in wheat -- to be followed later by that in rice -- and it came into effect in April 1973. The Food Corporation of India was the agency entrusted with the responsibility of the wholesale wheat trade. Even though the government justified the policy on the ground of building a buffer stock for an effective public distribution system, its stance was marked, in the words of the correspondent of a radical journal, "with ideological overtones, justifying the takeover as being among the first steps towards the desired socialistic changes in the social and economic structure." The same correspondent stated that "the ultimate aim of the policy is to restructure the social fabric of our society."[123]

For the trading community the socialization of the wholesale wheat trade was an anathema. Traders in foodgrains decided to halt trade in all foodgrains markets as a protest and in order to wreck the policy. They opposed the policy, arguing that it could not have been aimed at any undue concentration of power as hundreds of thousands of traders were involved, nor against any monopolistic practices as hundreds of markets were in competition, nor against mismanagement as the traders assured the most efficient and cheapest supply of foodgrains.[124] Whatever interests the policy was designed to serve, they could not have been those of the business class, for not only was the livelihood of half a million wholesale traders with some two and a half million dependents at risk, but the trading community also let its stern opposition be known vocally and practically. Nor for that matter could the policy have been for the purpose of serving the interests of the farmers who thoroughly opposed it and acted to foil it. Instead, the policy demonstrated at this time a state acting autonomously of important economic interests.[125]

This policy -- under which the Government now assumed the responsibility to feed the entire population while it could no longer make scapegoats of the trading community for shortages and rise in food prices -- was pregnant with great risks for the government. These risks were driven home by a business journal, associated with the house of the Birlas, as it ventured into prophecy:

> Of all the ideological decisions that the government of India has taken, or is to take, it is its decision to take over the wholesale trade in wheat throughout the country from the next rabi season that will prove to be its economic Waterloo. This is not the vindictive verdict of those engaged in the wholesale trade but of the common consumer, who will judge the success or failure of this primarily ideological pursuit solely from the criterion set by the authorities themselves.. It is the government's contention that the takeover will help bring down the prices of foodgrains. Far from this happening, the consumer apprehends, he will be made to pay a higher price and for an inferior quality of foodgrains too.[126]

Unfortunately for the government, its decision, as in the case of the 1966 devaluation, was badly timed. After 1967 there had been a series of bumper harvests, partly as a result of good monsoons and party because of the Green Revolution strategy; this plus imports had made for a comfortable food situation. As a former food official noted in respect of 1971: "For the first time since Independence domestic procurement of foodgrains had exceeded the total needs of the public distribution system. Stocks at the end of the year were 7.9 million tonnes....Wheat procurement -- and production -- both were going up by leaps and bounds as it were; there was no need of any compulsory levy on producers of wheat in any State."[127] The building up of buffer stocks enabled India to cope with the food demands of 10 million refugees from Bangladesh, even as it defied the United States in the international crisis over the formation of that state. Carried away by the new abundance, the government ended imports of concessional food from January 1972, but it did so partly also in a show of defiance of the United States. However, the situation did not last. In 1971-72 there was a decline of over 3 million tonnes in foodgrains production and another one of 8 million tonnes in 1972-73.

It was in the context of a long drought and two successive bad harvests and food scarcity that the policy of the wholesale wheat trade takeover came into effect. It may or may not have been inefficient or ineffective in any other year, but it ended in failure in the existential situation. Food procurement from the 1972-73 crop was only 4.5 million tonnes compared to 8.8 million and 6.6 million tonnes for the crops of 1970-71 and 1969-70. Over and above the inherent food shortages in a drought year, the psychology of scarcity and the expectation of higher prices in the future led to hoarding by farmers. Besides, as in the case of the follow-up of devaluation, the goverment had not been adequately prepared for the experiment in socialization of trade,[128] which involved not just a single factory or a few coal mines but a complex market and trade network that included some 8 to 10 million traders, of whom wholesale traders were about half a million, apart from the hundreds of millions of farmers. The paradox is that, when there is adequate food production, compulsory public procurement is unnecessary because the private agencies perform the job more efficiently, but when there is a deficit then compulsory procurement does not work.

As it encountered resistance in mobilizing the food surplus, the government saw a conspiracy at work against it. The southern regional conference of the AICC declared at Bangalore in early May 1973: "The real threat to our democratic institutions arises from this combination of historically retrograde and opportunist elements. This unholy combination, which had become increasingly desperate on account of progressive measures like land reforms, nationalization of coking coal and non-coking coal mines, takeover of the wholesale trade in foodgrains and the long overdue changes in the judiciary, had to be fought and defeated. It must not be allowed to block our steady advance towards socialism."[129] But party resolutions did not help in food procurement which was only 4.5 million tonnes as against the target of over 8 million tonnes. On the other hand, the pressures on the public distribution system increased as government rapidly depleted its stocks: "in actual fact, the system had just about broken down in some parts of the country where the fair price shops existed only on paper, causing riots and bandhs on a minor scale."[130] In the event, prices of foodgrains, especially of coarse grains that were outside the public distribution system, shot high. The attempt at socialization had turned into a veritable disaster. In desperation, the government resorted to costly food imports but it also beat a hasty retreat in the socialization of the wheat trade; rather than extend the nationalization of wholesale trade to rice, the government in effect withdrew from that of wheat in January 1974.

It was a turning point, if not in itself, but certainly as part of a general economic crisis which the trade takeover aggravated. The failure of the experiment in socialization of the wholesale wheat trade was significant for the change it sparked in government's economic policy as it had evolved since the 1967 general elections and its attitude to the public sector. As the failure of the wheat trade takeover manifested itself, the *Economic and Political Weekly* noted concerning the Prime Minister's stance: "From all accounts, she is disenchanted with the idea and has been telling all and sundry that she had been placed in an unnecessary predicament by overenthusiastic radicals."[131] But earlier, Jayaprakash Narayan had warned: "Indeed, this policy is sure to discredit socialism in the eyes, not of the rich, but of the masses and it will hurt the ruling party itself."[132]

The failure of a rightward or market-oriented solution in the form of devaluation in 1966 had led to a leftward swing in economic policy, both in terms of a restrictive posture toward the private sector and an expansive thrust for the public sector. Similarly, now, the consequences of the leftward course — particularly the failure of the socialization of wholesale wheat trade but more generally an economic crisis resulting from a hostile attitude to business and an expansionary state interventionism -- brought the government to call a halt to that course, and instead initiate a policy of economic restraint and greater reliance on the market mechanism and a favourable attitude toward the private sector. The year 1974 thus heralded, as it were, an end to the reign of ideology that had prevailed since 1969 and an end to the generalized attack on the private sector as well as to the nationalization spree. Little wonder that FICCI president Charat Ram told Mrs. Gandhi: "The government indeed needs to be complimented on the modifications recently announced in its wheat procurement policy of last year; it indeed demonstrates the highest level of political and economic confidence. Such a perceptive step to meet the requirements of the economy was possible only under a fearless and national minded leadership such as yours."[133]

Summary and Conclusions

A series of nationalization measures was pushed through by the government between 1969 and 1973 against the opposition of big business. This set of nationalization measures served to further contract the sphere of operations of the private sector. FICCI president Mangaldas correctly underlined in 1973 that: "Banking, insurance, coal, major part of electricity and many other industries and trades are already in the hands of the State. In less than two decades, the capital assets of industrial enterprises in the public sector are as much as, if not more than in the private sector, particularly after the recent spate of nationalization. The growth of the public sector has been phenomenal. And the private sector now is under strict control of the Government." From the viewpoint of the functioning of the private sector, he added: "There are ever so many countervailing forces that have emerged." Similarly, a year earlier FICCI president S.S. Kanoria concluded, "today the levers of power over every sector of the economy, including the financial and credit institutions, are securely in the hands of the government. In these concerns no major policy decision can any longer be taken without the concurrence of some governmental agency."[134]

Business opposition to nationalization would have been logical since such nationalization served to reduce its economic power both absolutely and relatively but, more importantly, it removed from the private sector vast areas for making profits. But business opposition was also manifest, and did not just remain latent, as the empirical record shows; it was thus not simply rationally logical but also empirically obvious. Notwithstanding its opposition to nationalization, however, business could not stop the government from its chosen path. Rather, the government freely cut business to size, and its ability to do so was a momentous defeat for the private sector. Business had been highly aggressive in its criticism of government policy since the Sino-Indian war of 1962, and a segment of it had even supported and joined with right-wing parties, but the 1971 elections "disabused a section of the business mind of some of the illusions about the fabric of political power."[135]

The reasons for this state of affairs are not hard to fathom. It is often claimed by Marxist scholars that the Indian bourgeoisie lacks the dynamism to perform the historic role of transformation of the economy and that therefore a socialist regime is necessary to perform that role.[136] The relationship seems to be quite the reverse. It is rather the installation of precisely a regime that was committed to the execution of radical measures — even if not the entire programme in a single package all at once — demanded by socialist ideology that thwarted the corporate bourgeoisie from performing its transformational role. The corporate bourgeoisie — regardless of the prospects of its success — was thwarted through hemming it with multifold restrictions and through contracting severely and drastically its sphere of operations by nationalization and expansion of the public sector.

That phenomenon stemmed fundamentally from the fact that, whatever its economic power, the bourgeoisie was not the ruling class. Rather, the role of ruling class had been pre-empted, due to the particular nature of the nationalist movement, by the intermediate strata, more particularly by the new middle class. Utilizing his unique position in the Indian political system, Nehru as a representative of the new middle class had fostered an instrumental-socialist mixed economy with a vast public sector, whose very existence then threw up strong supports for its maintenance and expansion. Later, a changed political environment provided Mrs. Gandhi the opportunity and the incentive to push further in this direction. Not only was she oriented ideologically to that path, but so also was significantly an active group around her — which, though undoubtedly paying homage to Nehru, had received its ideological training and socialization earlier inside the Communist Party of India. Furthermore, not only did the CPI look with favour on the presence of these former Communists in the corridors of power as an internal pressure group pushing the Congress party toward socialism, but the CPI itself endorsed the policy measures of the government taken in the name of socialism.

The presence of former Communists inside the Congress party, at the highest levels of party and government, is a matter of no little significance. It is broadly indicative of the acceptability and legitimacy of Marxist or socialist ideology in the Congress party of the time. Its significance can be more sharply underlined by raising a counterfactual question. Suppose a similarly numerous contingent had come into the party and government from the Jan Sangh, what position would the intellectual Left have taken on it? Would it not have immediately and automatically condemned the Congress party for having gone over to fascism and communalism? Suppose, again, the Swatantra party had sent as large a group of businessmen into the Congress party and government. Would not the Left have claimed confirmation of the Marxist proposition that the state is but an executive committee of the bourgeoisie? In truth, the presence of the ex-Communists in party and government underscored the ideological leanings of those in control of the Congress party. In Nehru's time, the CPI had endorsed his programme because it largely coincided with its own programme but now it had, besides, created a physical presence inside the Congress party.

For some two decades from the mid-1950s to the mid-1970s, regardless of whether they were appropriate or not, the ruling ideas in India were socialist, not bourgeois. It was precisely because they were socialist that they were supported by the Left, including the CPI. That the ruling ideas were socialist should occasion no surprise, given the Marxist proposition that the ruling ideas of a society are the ideas of the ruling class. The ruling class in India for at least the three decades after independence has been, as documented earlier, the new middle class, especially the intellectuals, not the bourgeoisie. It is no accident that socialism has been described as the particular ideology of the intellectuals. This ruling class was able to push through its economic schemes with impunity, with no reactionary classes or military gatekeepers overthrowing it. To be sure, other ideas existed too, but they were on the defensive over these three decades, and even afterwards. The overarching economic framework was provided by the ideas of socialism, and the bourgeoisie had to labour within that framework. A vivid indicator of the hegemony of socialist ideas over this period and their feeding into the economic framework is the fact that subsequent attempts to alter that framework in the direction of liberalization have been automatically and fervently assailed by Marxist scholars and publicists, notwithstanding the repeated denigration of the framework by them earlier as having been a project of the capitalist class.[137]

Over and above what Nehru had done, the regime under Mrs. Gandhi pushed the expansion of the public sector in a dramatic and substantial way. In this lay also a hazard. Nehru had been against nationalization of existing industry in the private sector on the ground that its obsolescing technology would mean a waste of public resources and that the government already had enough on its hands. To the extent that the government under Mrs. Gandhi nationalized existing industry on a large scale, whether in textiles or coal, it added to its subsequent problems with the apparent lack of profitability in the public sector. But in no way can the case be sustained that the rapid-pace nationalization thrust between 1969 and 1973 took place at the behest of the corporate bourgeoisie. Rather, the various measures of nationalization were either demanded by labour or undertaken by the new middle class in behalf of labour; in any case, they were enthusiastically endorsed by labour. Thus, it was not capital, but labour, that was instrumental in nationalization. What is remarkable is that, regardless of whether or not Mrs. Gandhi was engaged in a reformist strategy, former Communists played a critical role -- with the public acclaim of both CPI and CPI(M) -- in pushing through that general strategy. They did so not to build capitalism, but as part of a larger design -- theirs and the CPI's -- to bring about socialism. Little wonder that Asoka Mehta could claim that "over the years the weight of the ideological factor has grown."[138]

NOTES

1. Note the comment of a longstanding staunch advocate of the public sector: "...one unique feature.... It is this that the public sector in India is almost entirely the creation of entrepreneurial efforts of the State." Raj K. Nigam, *Public Sector for Public Welfare* (lecture delivered in Patna) (New Delhi: Documentation Centre, 1983), p. 4. Note also: "A very important feature is that the public sector in India has grown not through a process of nationalisation of private companies but out of the entrepreneurship of the State." B.C. Tandon, *Management of Public Enterprises* (Allahabad: Chaitanya Publishing House 1978, p. 56.
2. India (Republic), *Nationalisation of Banks*: A *Symposium* (New Delhi: Publications Division, 1970), p.1.
3. K.T. Shah, *National Planning, Principles & Administration* (Bombay: Vora & Co., 1948), pp. 52-53.
4. Indian National Congress, *Resolutions on Economic Policy, Programme and Allied Matters (1924-1969)* (New Delhi: All India Congress Committee, 1969), pp. 20, 31, 33.

5. Narendra Kumar (ed.), *Bank Nationalisation in India*: A *Symposium* (Bombay: Lalvani Publishing House, 1969), pp. 26, 28.
6. Congress, *Resolutions*, p. 93.; Kumar (ed.), p. 33; Harish C. Sharma, *Nationalization of Banks in India* (Agra: Sahitya Bhawan, 1970), p. 26.
7. D.N. Ghosh, *Banking Policy in India : An Evaluation* (Bombay: Allied Publishers, 1979), p. 371.
8. *Lok Sabha Debates* (September 6, 1963), pp. 4909-4912; see also Deputy Finance Minister Tarkeshwari Sinha's similar response, in *Rajya Sabha Debates* (December 20, 1963), pp. 4463-66.
9. Kuldip Nayar, *India After Nehru* (New Delhi: Vikas, 1975), p. 60.
10. Kuldip Nayar, *India: The Critical Years* (New Delhi: Vikas, 1971), p. 37.
11. R. Chandidas, et al., *India Votes* (New York: Humanities Press, 1968), pp. 5-11.
12. Kumar (ed.), pp. 35, 83-90, 170-71, 199-206.
13. Sharma, pp. 6-16, 49; Kumar (ed.), pp. 60-102; K. Rangachari, in India (R), *Nationalization*, pp. 62-66; Joseph K. Alexander, in *ibid.*, pp. 99-103; and G.T. Huchappa, in *ibid.*, pp. 123-128.
14. Kumar (ed.), p. 71.
15. Kumar (ed.), pp. 74-77, 219-220; see also A.N. Agarwal, in India (R), *Nationalization of Banks*, p. 73
16. Sharma, pp. 27, 37-39.
17. A. Moin Zaidi, *The Great Upheaval 1969-1972* (New Delhi: Orientalia, 1972), pp. 81-84; Kumar (ed.), p. ix.
18. K. Nayar, *India: The Critical Years*, p. 67
19. Zaidi, *Great Upheaval*, pp. 103-106; *Lok Sabha Debates* (July 21, 1969), p. 275.
20. P.C. Joshi, in India (R), *Nationalization of Banks*, pp. 6-9. Note also the words of the veteran CPI leader S.A. Dange: "one lobby which takes the side for the present, for certain reasons, of the agriculturalists, of the rich peasants, of the medium capitalists, of the small capitalists, the real interests of the national bourgeoisie. That interest has today decided to attack the stronghold of monopoly, that is, the banking companies." *Lok Sabha Debates* (July 28, 1969), p. 334
21. K.N. Raj, in *Nationalisation of Banks*, pp. 94-98; and C.H. Hanumantha Rao, in *ibid.*, pp. 81-84
22. See statement by S.A. Dange, in *Lok Sabha Debates* (July 28, 1969), pp, 333-35; CPI, *Documents of the Ninth Congress of the Communist Party of India...1971* (New Delhi: 1972), p. 102; Chandidas, et al., pp. 37-78; and statement by P. Ramamurti, in *Lok Sabha Debates* (July 29, 1969), pp. 295, 298
23. Chandidas, et al., pp. 11-32. See also P. Sood, *Indira Gandhi and the Constitution* (New Delhi: Marwah Publications, 1985), p. 114.
24. FICCI, *Correspondence...1969* (New Delhi: 1970), pp. 108-110.
25. H. Venkatasubbiah, *Enterprise and Economic Change: 50 Years of FICCI* (New Delhi: Vikas, 1977), pp. 135, 139
26. Kamal Nayan Kabra and Rama Rao Suresh, *Public Sector Banking* (New Delhi: People's Publishing House, 1970), p. 160.
27. Francine Frankel, *India's Political Economy, 1947-1977* (Princetion: Princeton University Press, 1978), pp. 434-35.
28. Frankel, p. 446.
29. India, Planning Commission, *Report of the Committee on Distribution of Income and Levels of Living* (New Delhi: 1964), pp. 53-54.
30. India, *Report of the Monopolies Inquiry Commission* (New Delhi: 1965), pp. 159, 166-84.
31. H.V.R. Iengar, "Role of the Private Sector," in C. N. Vakil (ed.) *Industrial Development of India: Policy and Problems* (New Delhi: Orient Longman, 1973), pp. 28-41.
32. India, *Report of the Monopolies Inquiry Commission*, p. 8.
33. R.K. Hazari, *Industrial Planning and Licensing Policy: Final Report* (New Delhi: Planning Commission, 1967), pp. 6-9, 25
34. India, Ministry of Industrial Development and Company Affairs, *Report of the Industrial Licensing Policy Inquiry Committee* (New Delhi: 1969), pp. 183-197; FICCI, *Correspondence...1970* (New Delhi: 1971), pp. 307-312
35. FICCI, *Correspondence...1969* (New Delhi: 1970), pp. 324-336
36. Frankel, p. 437.

37. M.L. Trivedi, *Government and Business* (Bombay: Multi-tech Pub., 1980), pp. 273-74; Asoka Mehta, "Growth of the Public Sector as the Dominant Sector," in Vakil (ed.), *Industrial Development of India: Policy and Problems*, p. 22.
38. Venkatasubbiah, p. 149; A. Dasgupta and N.K. Sengupta, *Government and Business in India* (Calcutta: Allied Book Agency, 1978), pp. 61, 65; FICCI President Poddar's letter to the Prime Minster, in FICCI, *Correspondence...1969*, p. 113
39. Stanley A. Kochanek, *Business and Politics in India* (Berkeley: University of California Press, 1974), p. 225.
40. Weiner states: "Mrs. Gandhi's Congress won back a large portion of the modern sector that it had begun to lose in the 1967 elections -- the urban middle classes, the bureaucracy, the business community -- while strengthening its electoral base with the scheduled castes and tribes, the religious minorities (especially the Muslims), and other low-income groups." Myron Weiner, *India at the Polls, 1980* (Washington, D.C.: American Enterprise Institute for Public Policy Research, 1983), p. 7. He understandably excludes the rich and middle peasantry because of the support from it received by other parties, but the Congress party could not have won such a massive majority without support from large segments of this peasantry. Morris-Jones saw no new political behaviour in response to Mrs.Gandhi's populist politics but an old-type behaviour where "Muslims, scheduled castes and 'backward classes' voted to an unprecedented degree in solid blocs for the PM." See W.H. Morris-Jones, "India Elects for Change -- and Stability." *Asian Survey*, XI, no. 2 (August 1971), pp. 719-41. The newness really consisted in how the support was mobilized, that is, through direct appeal rather than by way of the traditional party organization operating through patron-client networks. The business community, despite the radicalism of the Congress party, supported it because of its anxiety over political stability. See comments by S.S. Kanoria, in FICCI *Proceedings of the 44th Annual Session...1971* (New Delhi: 1971), p. 53.
41. Morris-Jones, p. 740.
42. Frankel, p. 463.
43. Satish K. Arora, "Social Background of the Indian Cabinet," *EPW*, VII nos. 31-33 (1972), pp. 1523-32
44. AICC, *Congress Marches Ahead*, V (1972), p. 103
45. A Moin Zaidi, *Full Circle 1972-1975* (New Delhi: Michiko and Panjathan, 1975), pp. 58-67.
46. FICCI, *Correspondence...1972* (New Delhi: 1973), p. 534.
47. AICC, *Congress Marches Ahead*, VI (1972), pp. 40-47.
48. Zaidi, *Great Upheaval*, pp. 261-62.
49. Zaidi, *Great Upheaval*, pp. 418-29.
50. Venkatasubbiah, p. 136.
51. *Lok Sabha Debates* (June 2, 1971), pp. 187-89; Vikram Chand Mahajan, *ibid.*, p. 198.
52. Somnath Chatterjee, *ibid.*, p. 192; S.M. Banerjee, *ibid.*, p. 201.
53. H.M. Patel, *ibid.*, p. 209; Virendra Agarwal, *ibid.*, p. 222.
54. N.L. Madan, *Congress Party and Social Change* (Delhi: B.R. Publishing Corp., 1984), p. 118; FICCI, *Correspondence...1972* (New Delhi: 1973), p. 171.
55. M. Das, *Fantasy of Coal Nationalisation* (Howrah: M. Das, 1975), pp. 12-20.
56. A.H. Hanson, *The Process of Planning* (London: Oxford University Press, 1966), pp. 465-68.
57. Satindra Singh, *Communists in Congress: Kumaramangalam's Thesis* (Delhi: D.K. Publishing House, 1973), pp. vii, 22, 43-44, 91.
58. CPI, *On the General Election of March 1971* (New Delhi: 1971), p. 9; *Political Report and Political Resolution: Adopted by Ninth Congress of the Communist Party of India* (New Delhi: 1971), pp. 126, 140, 144; and *Documents of the Tenth Congress of the Communist Party of India...1975* (New Delhi: 1976), p. 181. Emphasis in original.
59. Satindra Singh, pp. xx-xxiii.
60. S. Mohan Kumaramangalam, *Coal Industry in India: Nationalisation and Tasks Ahead* (New Delhi: Oxford and IBH Publishing Co., 1973), foreword.
61. K. Nayar, *India After Nehru*, p. 221.
62. Somnath Chatterjee, in *Lok Sabha Debates* (August 2, 1972), p. 183.
63. Kumaramangalam, *Coal Industry in India*, pp. 38-40, and in *Lok Sabha Debates* (December 10, 1971), pp. 32-39, (December 13, 1971), pp. 63-66.

64. M. Das, pp. 7-9, 20-37.
65. FICCI, *Correspondence...1972* (New Delhi: 1973), pp. 548-550; and Das. p. 47.
66. Kumaramangalam, *Coal Industry in India*, pp. 40-51, and in *Lok Sabha Debates* (March 15, 1973), pp. 212-21, 290-302.
67. Cited in Das, p. 11; see also Hanson, p. 468.
68. FICCI, *Correspondence...1973-74* (New Delhi: 1974), p. 98.
69. *Lok Sabha Debates* (March 15, 1973), p.213; Carras, p. 251.
70. Das, p.55.
71. *Lok Sabha Debates* (March 15, 1973), p.299.
72. Hindustan Steel Limited, *Statistics for Iron and Steel Industry* (Ranchi: 1970), p. 5.
73. William A. Johnson, *The Steel Industry of India* (Cambridge: Harvard University Press, 1966), pp. 79-80. See also Lok Sabha, Estimates Committee (1958-59), *Thirty-Third Report: Ministry of Steel, Mines and Fuel* (New Delhi: 1959), p. 7.
74. Johnson, p. 23.
75. Johnson, pp. 67-69, 81-83.
76. Hanson, p. 472; Johnson, pp. 35-36.
77. *Lok Sabha Debates* (August 21, 1972), pp. 333-343.
78. *Lok Sabha Debates* (August 30, 1974), pp. 213-16.
79. FICCI, *Correspondence...1972*, pp. 276-77.
80. M.R. Chaudhuri, *The Iron and Steel Industry of India: An Economic-Geographic Appraisal* (2nd. ed; Calcutta: Oxford and IBH Publishing Co., 1975), p. 98. See also Johnson, p. 167.
81. FICCI, *Correspondence...1972*, pp. 276-77.
82. *Lok Sabha Debates* (August 21, 1972), p. 343.
83. *Rajya Sabha Debates* (August 21, 1974), p. 196.
84. *Lok Sabha Debates* (August 30, 1974), p. 217.
85. Indrajit Gupta, in *Lok Sabha Debates* (August 19, 1976), pp. 181-82; Robin Sen, in *Lok Sabha Debates* (August 21, 1972), p. 346.
86. *Lok Sabha Debates* (August 19, 1976), pp. 167, 169.
87. Kumaramangalam, in *Lok Sabha Debates* (December 9, 1971), pp. 76-78.
88. *Lok Sabha Debates* (January 27, 1976), pp. 243-49.
89. *Lok Sabha Debates* (April 4, 1972), pp. 171-190, (August 30, 1972), pp. 229-237.
90. FICCI, *Proceeding of the 45th Annual Session...March 1972* (New Delhi: 1972), p.64.
91. FICCI, *Correspondence...1972*, p. 177.
92. FICCI, *Correspondence...1973-74* (New Delhi: 1974), p. 98; see also FICCI, *Proceedings of the 46th Annual Session...1973* (New Delhi: 1973), p. 41-46.
93. National Productivity Council, *Productivity Trends in Cotton Textile Industry in India* (New Delhi: 1976), p. 144.
94. R.D. Mohota, *Textile Industry and Modernisation* (Bombay: Current Book House, 1976), p. 162. See also Ashok V. Desai, "Technology and Market Structure under Government Regulation: A Case Study of Indian Textile Industry," *EPW*, (January 29, 1983), pp. 150-59.
95. Compare the duty payable by powerlooms of Rs. 25 to Rs. 150 (depending upon size of unit) per annum per shift with that payable by mills: Rs. 1876, Rs. 3276, Rs. 7700 and Rs. 11,032 for controlled higher medium, non-controlled higher medium, fine and superfine varieties of cloth. Mohota, p. 161.
96. Mohota, p. 162.
97. Madan Gaur, *The Textiles* (Bombay: PPSI, 1977), pp. 30, 63.
98. Zaidi, *Full Circle*, p. 192.
99. R.K. Hazari, "The Rents of Misdelivery," *EPW*, XX, no. 28 (July 13, 1985), p. 1177.
100. National Productivity Council, pp. 100-112; Gaur, p. 53.
101. Mohota, *Textile Industry and Modernisation*. It should be noted that, though coming from an industrialist family, Mohota is sympathetic to the social justice aims of the government in relation to the decentralized sector and agriculture.
102. Gaur, p. 80.
103. Minister of Foreign Trade, L.N. Mishra, in *Lok Sabha Debates* (December 15, 1972), pp. 253-57.
104. Gaur, p. 65.

105. Virendra Agarwal in *Lok Sabha Debates* (December 18, 1972), pp. 290-94, and Piloo Mody, in *ibid.*, (December 19, 1972), pp. 273-74. Agarwal declared: "In fact, today we have a sick government with a sick economic thinking."
106. *Lok Sabha Debates* (August 25, 1976), pp. 249-50, 287-289.
107. Roza Deshpande, in *Lok Sabha Debates* (November 26, 1974), p. 279; S.M. Banerjee, in *Lok Sabha Debates* (August 25, 1976), pp. 260-264.
108. *Lok Sabha Debates* (August 25, 1976), pp. 288-89.
109. *Eastern Economist*, vol. 59, no. 20 (November 17, 1972), p. 103.
110. Jyotirmoy Bosu, *Lok Sabha Debates* (December 19, 1972), p. 272.
111. R.N. Chopra, *Evolution of Food Policy in India* (New Delhi: Macmillan, 1981), p.5.
112. Chopra, pp. ix, 198.
113. *Ibid.*, p. 40
114. R.P. Noronha, *A Tale Told By An Idiot* (New Delhi: Vikas, 1976), pp. 52-60.
115. Cited in Chopra, p. 95.
116. D.R. Mankekar, *Lal Bahadur Shastri* (New Delhi: Publications Division, 1973), p. 141.
117. Chopra, pp. 10, 124.
118. Zaidi, *Great Upheaval*, pp. 354, 423.
119. AICC, *Congress Marches Ahead*, VII (1972), pp. 109, 114.
120. AICC, *Congress Marches Ahead*, VIII (1973), p. 363.
121. Benedict Costa, "Take-Over of Foodgrain Trade: Pros and Cons," *Illustrated Weekly of India*, vol. 94, no. 14 (April 8, 1973), pp. 30-33.
122. K. Nayar, *India After Nehru*, p. 219.
123. "Take-Over of Foodgrain Trade: An Alternative Institutional Framework." *EPW*, VIII, no. 27 (July 7, 1973), pp. 1181-83.
124. R.V. Murthy, "The Take-Over Menace," *Eastern Economist*, vol. 60, no. 11 (March 16, 1973), pp. 577-78; FICCI background paper on "Take-Over of Wholesale Trade in Foodgrains," *ibid.*, pp. 613 - 618; FICCI, *Proceedings of the 46th Annual Session...1973* (New Delhi: 1973), pp. 25-32; Chopra, p. 145. For theoretical and empirical support of the trader's position, see G. Parthasarathy, *Dilemmas of Marketable Surplus: The Indian Case* (Waltair, Visakhapatnam: Andhra University Press, 1979), pp. 59-73.
125. A Marxist scholar points out: Mrs. Gandhi as "a more successful practitioner of Bonapartist policies....achieved a far greater degree of independence of the conflicting social classes when she carried out.... the *nationalization* of banks, aboliton of privy purses and the like.... and at the same time was able to command the unstinted support of the ruling classes as well as mass adulation." Ajit Roy, "The Failure of Indira Gandhi," *EPW*, XIX (November 10, 1984), p. 1896.
126. "Government as Grain-Dealer," *Eastern Economist*, vol. 60, no. 12 (March 23, 1973), p 627.
127. Chopra, p. 137.
128. "Foodgrains Trade Take-Over: Who Killed Cock Robin?", *EPW*, VIII, no. 22 (June 2, 1973), pp. 968-69.
129. AICC, *Congress Marches Ahead*, VIII (1973), pp. 256-57.
130. Chopra, p. 147.
131. "And What About Rice Now?", *EPW*, VIII, no. 27 (July 7, 1973), p. 1180.
132. Cited in Costa, p. 33.
133. FICCI, *Proceedings of the Forty-Seventh Annual Session...1974* (New Delhi: 1974), p. 9.
134. FICCI, *Procedings of the Forty-Fifth Annual Session....1972* (New Delhi: 1972), p. 7, and *Proceedings of the Forty-Sixth Annual Session...1973* (New Delhi: 1973), p. 3.
135. Venkatasubbiah, pp. 135, 141, 154.
136. This is a recurrent theme in Marxist writing on the Indian bourgeoisie. See, for example, Amiya Kumar Bagchi, "Foreign Capital and Economic Development in India: A Schematic View," in Kathleen Gough and Hari P. Sharma (ed.), *Imperialism and Revolution in South Asia* (New York : Monthly Review Press, 1973), p. 67, and Paresh Chattopadhyay, "Some Trends in India's Economic Development," in *ibid.*, p. 126.
137. See, for example: (a) Amiya Kumar Bagchi, "Does Political Buccaneering Pay?" *Mainstream*, XXIII, no. 39 (May 25, 1985), pp. 26-31: "the irresponsibility and the callous indifference to the needs of the common man displayed by the budget figures ....What the budget signals is that there

is a group of Indian capitalists and politicians who, with the help of their foreign friends, are prepared to hijack the Indian economy, if the Indian people let them do so.... you can say good-bye to any idea of planning." (b) Prabhat Patnaik, "Political Economy of Indian 'Liberalization', "*Mainstream*, XXIV (September 21, 1985), pp. 19-26: "This move, spearheaded by certain sections of upstart big bourgeoisie, draws qualified support from the entrenched big bourgeoisie in the context of the crisis, and seriously threatens not only the economic position of the working people, but also that of large sections [of the] petty bourgeoisie and non-monopoly bourgeoisie.... The 'de-industrialising' impact of 'liberalization' has already figured in discussions of the new policy. Such 'de-industrialisation' would not only affect the workers, who would additionally face wage-freezes, restrictions of political rights and the whip of 'discipline' so that foreign capital is enticed into the country, but also the petty and non-monopoly bourgeoisie whose units would face closure. The pursuit of 'liberal' economic policy therefore necessarily entails a narrowing of the class-basis of the state.... the emergence to influence of new monopoly strata, associated with a narrowing of the class base of the state has authoritarian implications." (c) H.K. Paranjape, "New Lamps for Old! A Critique of the 'New Economic Policy'," *EPW*, XX, no. 36 (September 7, 1985), pp. 1513 - 1522: "The thrust of the new approach cannot but lead to making even the largest private sector concerns again and fully parts of private business empires, thus accelerating the greater concentration of economic power in the hands of some groups and families.... the new policy is also likely to create unemployment.... the new government is becoming one by the richer and better off strata of the population for their own benefit." (d) Report by a special correspondent on a conference held at the Indian Institute of Public Administration, New Delhi, "Current Economic Policy of Government and Alternatives," *EPW*, XX, no. 36 (September 7, 1985), pp. 1503-1507: "Participants... expressed grave concern on the manner in which the planning process has been grossly devalued, the economy increasingly left to the mercy of the anarchy of the market and government policy has been steering clear of the means to harness the potential of the working people." (e) Statement of Conference of 28 Economists in Calcutta, held at the invitation of the West Bengal government, 'Economists' Warning Against Policy Shift," *Mainstream*, XX IV, no. 8 (October 26, 1985), pp. 24-25: "The system of industrial licensing has been relaxed, MRTP limits raised and the public sector downgraded.... In our opinion, the combination of these policies, whose success greatly depends on private business enterprise, is certain to take the country further away from the *accepted national objectives* of growth with equity and self reliance.... these developments would result in a worsening of inequalities in the distribution of incomes" (emphasis added).

138. Asoka Mehta, "Growth of the Public Sector as the Dominant Sector," in Vakil (ed.), *Industrial Development of India: Policy and Problems*, p. 16.

Chapter VIII

Economic-Political Crisis and the Retreat from Radicalism (1974-1984)

The 1971 elections had culminated on a note of extraordinary hopes and great euphoria about the future. With an unambiguous political mandate, a shrewd and popular leader, and the assurance of political stability given the massive Congress majority in parliament, there were high expectations of significant economic progress. During its first three years the new administration, indeed, kept up the momentum of the earlier radical policy stance displayed during 1969-1971. It engaged in a spate of nationalization measures, the most massive of which was that relating to coal and the most ambitious and risky pertained to the wholesale wheat trade. However, by late 1973 and through 1974 India was in the throes of a grave economic crisis. This crisis was of momentous consequence in engendering a shift in overall economic policy and, accordingly, in the government's posture on the public sector. This latter was not immediately manifest because of the continued usage of radical rhetoric as a carryover from the previous phase. This chapter examines the change in economic policy over the decade 1974-1984 by focussing on (1) the initial economic crisis and its relationship to policy change as well as to the associated political crisis; (2) the nature of economic policy during the interregnum of the Janata Party; and (3) the persistence and carrying forward of the policy of economic liberalization after the restoration of Mrs. Gandhi to power in 1980.

1. Renewed Economic Crisis and Policy Change

Several factors had gone into the building up of the economic crisis in the early 1970s. First, there was the international crisis in 1971 relating to Bangladesh, which diverted national energies from economic and social advance in attempting to face up to the serious threat to national security from Pakistan, in turn backed by China and the United States. To cope with the influx of 10 million refugees constituted an enormous economic drain not only on its financial resources but also on its foodstocks. In addition, preparation for the impending outbreak of hostilities, the actual war, and then caring for a hundred thousand POWs for almost a year, added immensely to India's defence expenditures which, in the case of a developing country that lives on the margin in the best of times, had to be met through deficit financing. Second, but relatedly, compounding the economic burden was the suspension of economic aid by the United States and its discontinuance for several years thereafter. Moreover, the renunciation of concessional food imports from the United States -- in a nationalist response to American hostility towards India during the crisis -- made more difficult the management of the country's food economy. Henceforth, scarce foreign exchange would have to be spent on food imports in a period of high food prices on the international market, placing a great strain on the country's budget and balance of payments.

Third, very critically, the economy was in a state of stagnation during the early 1970s, so that per capita income remained below the 1970-71 level for the subsequent four years (see Table VIII.1). This overall economic stagnation was reflective of a production crisis where, similarly, agricultural production for the four years remained below the 1970-71 level while industrial production grew at a rate of only 3.5 per cent. Fourth, the economic crisis was further fed by the sharp jump in oil prices in October 1973, dictated by OPEC. The oil price increase fuelled inflation, added to the pressure on the country's balance of payments and, with its cascading effect on fertilizer prices, affected the consumption of fertilizer in agriculture and thus agricultural production.

Table VIII.1

Economic Performance 1970-1975 (Index Numbers)

	GNP	GNP per Capita	Agricultural Production	Industrial Production*	Wholesale Prices
1970-71	100.0	100.0	100.0	100.0	100.0
1971-72	101.4	99.1	97.0	104.3	108.2
1972-73	99.9	95.5	89.5	110.2	121.5
1973-74	105.1	98.2	96.5	112.0	158.0
1974-75	106.2	97.2	92.1	114.3	173.9

* Calendar Year, beginning 1970.

Source: India, *Economic Survey 1978-79* (New Delhi: 1979), Appendix.

Fifth, state policies and the public sector's poor performance aggravated the economic crisis. This was most vividly manifest in the area of foodgrains where the takeover of the wholesale wheat trade worsened food shortages and price increases. Nor was that all. A resolution of the AICC at New Delhi in July 1974 noted that "the shortfalls and inadequacies in the infrastructure support for industrial growth, particularly in power, coal and transport have had a crippling effect on production in numerous industries."[1] All three fields listed were largely in the public sector, with coal having been completely nationalized by January 1973. The coal supply situation deteriorated with nationalization and, combined with transport bottlenecks, worsened the supply of power for industry and agriculture. Furthermore, the restrictive measures against big business and the various nationalization measures did not inspire much confidence in the business community about government intentions. That community held government policy responsible for the industrial stagnation. J.R.D. Tata told the government in November 1972 that the country had been faced with an investment famine since the report of the Dutt Committee; a year earlier he had warned the government of the impractical nature of its policy which expected to obtain growth while constraining the large industrial houses that were responsible for 50 per cent of the country's industrial production. FICCI president Mangaldas in April 1973 called the recent spate of nationalization both "superfluous and self-defeating" and warned that it would prove a "deterrent to investment."[2]

Moreover, neglecting economic discipline, the government resorted to deficit financing in the implementation of populist policies to satisfy different sectors of the population: deficit financing "made it possible for the government to pay high prices to surplus farmers, to avoid introducing an agricultural tax and to allow black-market speculators to operate freely, while at the same time seeking to ameliorate poverty with inflationary doles and subsidies, much of which finds its way, not to the poor, but into the pockets of corrupt administrators."[3] The ruling party itself seemed to agree, confessing: "In part, the difficulties were aggravated because of pressures of competing demands of different sections of the population. In spite of our exhortations, it was generally ignored that the sectional interests of agricultural producers, manufacturers, traders, workers and consumers were dependent on and must be balanced with the larger national good."[4]

Sixth, the cumulative impact of the preceding factors was to fuel inflation enormously and therefore intensify public discontent. By April 1974, prices were 58 per cent higher than three years earlier, having increased at the rate of 10, 12 and 30 per cent successively over the three years. The agony of the public in the face of inflation of this intensity and of widespread shortages of essential goods, including food, burst forth into riots and violence, striking at the government in one fashion or another. Most social classes in the rural and urban areas became alienated from the government, which thus lost its legitimacy despite the massive majority in parliament. The inflation and shortages ravaged the poor everywhere and the working class and middle classes in the urban areas. Industrial unrest was widespread; compared to 4.9 million mandays lost in 1961, the number was 16.5, 20.5, 20.6 and 40.3 million in 1971, 1972, 1973, and 1974.[5] Business was alienated by the spate of nationalization, restrictions on the private sector, and a looming threat of further attack on private property. The rich and middle peasantry had already turned against the Congress party in the 1967 elections and now became antagonistic because of government intervention in the wheat trade and also because of fear of land reform which government rhetoric, if not practice, threatened. Besides, the intermediate strata of both rural and urban areas perceived as grim the future for their offspring, who had increasingly taken to education but found little prospect of employment in the face of what seemed like imminent economic collapse.[6] As inflation cut heavily into the real resources mobilized by the government, it reduced public investment -- somewhere around a fifth -- which then adversely affected production and employment. The general atmosphere of gloom and impending doom affected everyone, and reached its climax in 1974-75 which the *Economic Survey* characterized as "a year of unprecedented economic strains in the history of independent India."[7] The acute economic crisis and the resulting social disorder were soon transformed into a political crisis as the opposition naturally moved in to exploit the public discontent to mount a major offensive against the government.

The rising social disorder was spectacularly expressed in a mutiny by the provincial armed constabulary in Uttar Pradesh in mid-1973. Then in early 1974 there was a mass statewide revolt in Gujarat led by students, which was accompanied by large-scale rioting and violence, with the aim of overthrowing the state government but also of discrediting the central government. The student movement was successful in forcing the resignation of the state government and further in the dissolution of the state legislature. This served only to spark a similar movement in the state of Bihar, where Jayaprakash Narayan (J.P.) came to head a revolt against Mrs. Gandhi, spearheaded by cadres from the Jan Sangh and CPI(M). The J.P. movement soon acquired all-India dimensions, with the opposition parties rallying behind it in an effort to reverse the electoral verdict of 1971. While this movement was under way, the socialist leader George Fernandes masterminded a strike by some 2 million employees of the Indian railways in May 1974. Although critical of the exploitation of the discontent of the public by "either the right reactionaries or the left extremists masquerading as their friends for the moment," the Congress party's ire was mainly directed at "right reaction and vested interests" who allegedly aimed "to paralyse the entire apparatus of production and to plunge the national economy into deeper trouble. This is the historically familiar tactic of fascism and right reaction." These elements, in the party's view, "resorted to naked and systematic violence because the common people have in free elections repeatedly frustrated their attempts to capture power."[8]

Change of Economic Course

Despite its leftist rhetoric, however, the government itself was ready to distance itself from the leftist thrust that had characterized its policy in the preceding half-decade and, indeed, to reverse course in its economic policy. Mrs. Gandhi had apparently become disillusioned by the results of the earlier radical measures, especially after the fiasco in the wake of the takeover of the wholesale wheat trade. Late 1973 marked a turning point. The government now determined to jettison its earlier radical and populist posture of restrictions on the private sector and relentless expansion of the public sector, and to replace it with a growth-oriented strategy. Jha underlines: "It was part of a disenchantment, which can be traced back to the oil crisis, and the suspicion that grew in Mrs. Gandhi that she had been made use of by the communists and their fellow-travellers in the Congress to pass legislation whose effect was to disrupt production, without making society significantly more egalitarian."[9] The reference to the oil crisis here should be taken more as chronological rather than causal; the real underlying cause was the whole complex of radical policies pursued since 1969, reaching its culmination in the wholesale wheat trade takeover.

The new growth-oriented course was reflected initially in a deflationary policy package. First, the government in May 1974 ruthlessly crushed the railway strike which indicated its determination not to concede to workers demands for higher wages since that would, in the government's view, compound the economic crisis by intensifying inflation and depleting resources for public investment. This stern action came on top of the government's earlier crushing of strikes by employees of Indian Airlines and the Life Insurance Corporation—both vast far-flung public sector enterprises. Then in July 1974 the government issued three ordinances which shifted the focus of economic policy from the pursuit of social justice and socialism to discipline in the management of the economy and to orthodox measures to bring the economy under control. As the Congress party declared: "Inflation is the single biggest threat to progress with stability. Control of inflation has, accordingly, to be our first priority."[10] Toward that end, the government adopted a new policy package. Two of the ordinances placed limits on the distribution of dividends, on the one hand, and on the other impounded any increase in wages and half of any additional dearness allowances, which were now to be compulsorily deposited with the government. The Congress was apologetic over the constraints on wages and dearness allowances, stating that they should not be considered "an anti-working class measure," for inflation was ruinous for the working class and had to be dealt with "on a war footing". As if to balance the measures that hit the working class, the government carried out raids on the premises and residences of businessmen who were suspected of being blackmarketeers, smugglers and tax evaders.

Besides, the government applied restraint to its own expenditures and a squeeze on credit to the private sector, declaring "strict fiscal and monetary discipline has to be imposed on Government and private sectors." The government called for an earnest effort for increased production. Even earlier, in 1973, the government had relaxed somewhat its licensing guidelines—though such relaxation continued to be hedged by restrictions against the large industrial houses—to assure expanded production in respect of 54 industries. While business complained about the credit squeeze and limits on dividends, it was nonetheless appreciative of the government's actions, stating that "there was no other alternative for the government but to take measures to restrain consumption" and that "in the present inflationary situation, government were compelled to take unpleasant actions." Later, it characterized the measures as "timely bold and impressively unconventional."[11]

There can be no doubt that the severe economic crisis had induced recourse to economic orthodoxy and resort to the liberalization cure. The government could not very well go on the same path as before or reinforce it when that path, if it did not cause the crisis, did nothing to resolve it. Furthermore, importantly, there was the leadership's perception that the public mood was no longer hospitable in relation to the earlier path, for the public had come to associate its agony with precisely the socialist measures undertaken earlier. This was certainly the case with the takeover of the wholesale wheat trade and the nationalization of the coal industry, both of which had resulted in a deterioration of the existing situation in respect of higher prices and shortages. This deterioration was experienced directly by the public. Besides, the informed sections of the urban middle classes and the leaders of the prosperous peasantry had become alienated from the public sector because of its failure to perform and the huge losses it incurred, whose burden was then transformed into higher taxes on the population. Just as Mrs. Gandhi's turn to a radical path had been influenced by her perception of what the public mood wanted, so similarly her turning away from continuing that path was influenced by a changed political mood. Several years later one newspaper gave a good description of the changed political mood in an editorial, significantly titled "Not by Ideology Alone":

> A change of considerable significance is taking place in India. While the more dogmatic and less discerning among leftist intellectuals will dispute this view, the better informed among them will regret the change and describe it as being counter-revolutionary. But a change is a change....the emphasis has shifted from distributive justice to growth....more and more people have come to believe that their lot can improve only if the growth rate improves significantly. And fewer and fewer are willing to accept that expansion of the public sector is the answer to the twin problems of economic development and social justice. Indeed, the public sector is widely seen to have become a big drag on the economy.
>
> This disillusionment with the public sector has not led to greater social respect for businessmen....But faith in their capacity to produce had increased....As opposed to this, it is difficult to find an ordinary citizen who has something good to say about the performance of the State-controlled electricity board, railways, coal mines, telephones and so on. And if there is criticism of businessmen on account of their malpractices, it is more than matched by condemnation of bureaucrats and politicians on charges of corruption, inefficiency and violation of norms. In fact the prevalent view is that corrupt ministers, legislators and officials compel even honest businessmen to raise black money.
>
> Several factors account for the change in the atmosphere—the lacklustre performance of not only the public sector in our country but for most other centrally-controlled economies elsewhere and the decline in the standards of public life and administration, for example. Some social developments also deserve attention—the decline in the importance of the propertyless intelligentsia which inevitably tended to think in leftist terms because it could secure positions of influence for itself only through the political process and the state apparatus, the incorporation of a section of it into property-owning middle class and the rise of down-to-earth hard-headed members of upper peasantry in the country's political life.[12]

Ironically, such opinion came to be shared by some extreme leftists as well; a commentator in the radical journal *Frontier* wrote two years later:

> at a *certain point* market solutions are to be preferred as more egalitarian and better able to raise living standards from abysmal absolute levels, than the strangulation of corrupt State involvement....State control...can be more detrimental to the interests of the impoverished majority than impersonal market forces....In this situation market solutions which promote economic growth are inherently preferable to stagnation under State control....most people would prefer to have electricity than the socialism of the WBSEB! [13]

The change in the public mood was critical to policy change in 1974. It underlined once again the fundamental importance of the political system insofar as it required the government leadership to obtain eventually a revalidation of its political mandate at the ballot box. Given that requirement, it is unrealistic in the context of the change in the public mood to expect the leadership to have persisted in its earlier ideological course, indeed, to have intensified it, as some have done. [14]

Behind the more immediate phenomenon of change in the public mood there seemingly lay also more fundamental shifts in social stratification and mobilization. A keen political scientist calls our attention to these:

> the program of industrialization and agrarian modernization proceeded apace. Some social classes were becoming more clearly differentiated and were entering the political arena; these new classes tended to weight the political scales more and more toward the Right. The "green revolution" swelled the ranks of the richer farmers, while industrialization increased the numbers of big and small businessmen -- merchants, traders, industrialists, and bankers. As schools and newspapers proliferated, the government bureaucracy mushroomed; and as the judiciary expanded its role as arbiter of socioeconomic conflict, new classes of professionals and intelligentsia developed. All were well-organized and articulate, in sharp contrast to the still largely undifferentiated masses for whom the Congress also claimed to speak. [15]

The leadership could not close its eyes to the interests of these highly organized segments of the population, especially in the context of the serious economic crisis which affected them so directly.

Even if occasioned by the immediate economic crisis, the shift in the government's attitude to the management of the economy was also facilitated, and perhaps induced, by other factors. One of these pertained to the weakening of the Left within the Congress. The Left's eminence within the Congress party resulted from doing battle on Mrs. Gandhi's behalf, first against the Syndicate and then against the Grand Alliance of rightist groups. But its transparent arrogation of the Nehru legacy, its obvious thrust for domination within the party, and its shrill pressure for radical measures were perceived by other elements in the party as a Communist design to steer India on a course that would erode, indeed overthrow, democracy. These elements formed a Nehru Study Forum in 1972 to counter the Left's Congress Forum for Socialist Action. As the two forums quarrelled publicly, both were asked by Mrs. Gandhi a year later to disband; even though formally dissolved, they continued to function informally. The Congress Left also became divided within itself as one section, which had its antecedents in the Praja Socialist Party (PSP), attacked the ex-Communist faction for attempting to isolate the socialists. [16] The Left was further weakened by the death in May 1973 of its foremost theoretician and strategist, Kumaramangalam, and by the support extended by former PSP leaders to Jayaprakash Narayan as he mounted a political offensive against Mrs. Gandhi and her party, indeed, the political system. Thus the balance of power between the Left and others had changed.

Besides, Mrs. Gandhi -- subjected to repeated charges that she was a Communist, a crypto-Communist or a stooge of the Soviet Union -- must have been under pressure to demonstrate her independence of the Left. Moreover, one cannot discount the influence of a new political confidante who was emerging on the scene, her own son Sanjay. Himself active in establishing a major automotive plant, Sanjay was not enamoured by the public sector and favoured the private sector. He was also opposed to Communists, even of the CPI variety with whom Mrs. Gandhi was in tacit cooperation. Although the IMF and World Bank may have also given advice to the government for an orthodox stabilization package,[17] not much need be made of it. The real source of the policy package was the economic crisis itself. Indian economists were in the forefront of advocating harsh measures; 140 of them recommended a deflationary package that amounted to "a major surgical operation", compared to which the government's measures "were rather moderate."[18]

The harsh economic measures of the government had a dramatic impact on inflation by the end of September 1974, when prices began to decline; they were to assure price stability for the next year and a half even as inventories accumulated because of reduced demand. However, in the meantime, the associated political crisis had acquired its own momentum and it deepened as a result of Mrs. Gandhi's entanglement with the courts on the issue of electoral improprieties. Perceiving a political upheaval generated by the opposition forces led by J.P., Mrs. Gandhi imposed an emergency regime on the country. The roots of the emergency, however, lay -- as the prime minister herself recognized -- in "the prolonged economic crisis, which caused hardship to our people and rendered them vulnerable."[19] It is mistaken, however, to believe that "her reversal of economic strategy in 1974 contributed to the conditions that fueled the opposition engine and set her on a collision course with the forces against her" and then to attribute the political turmoil to her alleged ideological betrayal.[20] Quite the contrary, it was the discontent that originated in economic conditions brought about by the earlier policies inspired by ideology that engendered and fed into the movement against Mrs. Gandhi and her regime. Interestingly, even then "the urban workers, the landless laborers, and others of the backward classes stayed out of the action....the Congress was thought to be more in tune with the needs of the poor."[21] The J.P. movement was correctly perceived by them to be from the right, not from the left.

Economic Policy under the Emergency

The context of the economic and the related political crisis was critical to the declaration of emergency by Mrs. Gandhi on June 26, 1975 even though the immediate occasion was provided by the political turmoil surrounding the adverse court judgement against her. Although economic policy was not the motivation for the emergency, Mrs. Gandhi made it salient in her public posture after the declaration of emergency, either as a diversionary tactic or as a result of a genuine belief that the solution to India's crisis lay in a bureaucratic-authoritarian model, even if temporarily. She believed that the anti-government movement under J.P. "would have led to economic chaos and collapse," making the country "vulnerable to fissiparous tendencies and external danger."[22] To a considerable extent, the emergency can be said to constitute a continuity with, and intensification of, the rightward shift in economic policy that had been initiated in early 1974.

One public expression of the economic policy under the emergency was the 20-point programme announced by Mrs. Gandhi on July 1, 1975. This programme seemed to be designed to appeal to every important social class. Significantly, the first point pertained to control of inflation, with Mrs. Gandhi declaring that "the first and foremost challenge is on the price front" and that "this anti-inflation strategy has to be continued." In relation to the rural areas the programme was particularly addressed to the weaker sections of the population, whom it promised more rapid implementation of land reform, allocation of house-sites for the landless, abolition of bonded labour, liquidation of rural indebtedness, and upgrading of minimum agricultural wages. Although it had been difficult for the state in India to rigorously impose land ceilings against the interests of the rich and middle peasantry, the articulated programme should not be dismissed as idle rhetoric; rather it should be taken as aimed at what had been a loyal constituency of the Congress party, more particularly of Mrs. Gandhi. To the rich and middle peasantry, the major promise was in terms of expansion of irrigation. As for the middle classes, the programme included income tax relief by raising the exemption limit and the socialization of urban land to facilitate wider home ownership, while for the students it envisaged the provision of essential commodities, books and stationery at cheaper prices, and a new apprenticeship programme to increase employment. For the business community, it assured "liberalisation of investment procedures", "an accelerated power programme", and "strict economy in government expenditure". Significantly, for the workers it only promised "new schemes for workers' association in industry". More generally, it pledged to continue with the policy to assure "streamlined production, procurement and distribution of essential commodities."[23]

Apart from liberalization in investment procedures, the rightward shift was evident in the reduction of personal (though not corporate) taxes, in new investment incentives, and in the reduction by half of the annual bonus to workers and its abolition in case of loss-making enterprises. Besides, there was a more forthright positive appreciation of the possible contribution by the private sector. The AICC in a resolution in 1975 stated: "While the commanding heights of the economy must continue to rest with the Public Sector, the Congress recognizes the useful role of a socially conscious private sector in accelerating the development process. Recent changes in industrial licensing policies have been designed to facilitate this process."[24] But for the business community, the real significance of the emergency lay in more than this. It lay particularly in assuring labour peace through its barring strikes. This was a new boon for Indian busienss which had not been available to it hitherto. The emergency meant the conversion of the "soft-state" into a "hard-state" which, while against factory closures and lay-offs, wanted labour discipline for the higher purpose of increased production. Business was soon, however, to find itself as a result in a production glut.[25]

The new economic policy did not leave attitudes toward the public sector unaffected. Proceeding on the past analogy of radical measures of nationalization having been functional in the mobilization of political support, Congress president Barooah suggested further nationalization, specifically of the sugar and textile industries. His suggestions were immediately shot down.[26] Indeed, the day after the declaration of emergency, Mrs. Gandhi in a radio broadcast mentioned that "wild conjectures are circulating about impending nationalization of industries, etc. and drastic new controls", and emphatically declared "we have no such plans" and immediately added "our purpose is to increase production". Even before the emergency, Mrs. Gandhi had come to the view in respect of nationalization that "we should not do it unless it is a must. We have found that we are not really equipped to undertake such responsibility."[27] The new posture towards the public sector was reflected in the position of Sanjay Gandhi who, in an interview to an obscure journal in August 1975, not only let it be known that Communists were an anathema to him but also that he did not care much for the public sector. Holding the public sector to be inefficient, he wanted it to be allowed to die a natural death. He opposed both nationalization and controls; rather, he wanted a larger role for the private sector to assure economic advance, and asked for tax cuts.[28] Even Mrs. Gandhi warned public enterprise executives that the public sector would have to be judged on its performance in terms of the goods and services it provided and the surpluses it made available for capital accumulation, rather than the size of its investment.[29]

Despite the critical posture in relation to the public sector, two major measures of nationalization were undertaken during the period of the emergency. One related to oil, where the government effectively nationalized virtually the entire oil industry by a complete acquisition of Esso and Burmah Shell, thus gaining control of 95 per cent of the oil industry in India. This essentially represented a victory for Indian nationalism, which had endeavoured to remove foreign control over oil in the country after the government had signed what was subsequently regarded as a humiliating agreement in the 1950s with the oil multinationals for the construction of three oil refineries. But the present acquisitions were the culmination of actions and negotiations initiated long before the emergency. For example, 76 per cent of Esso shares had been acquired through agreement in March 1974.[30]

The other measure concerned the nationalization in 1976 of three major engineering firms in West Bengal, traditionally known for their high quality and profitability but which faced closure because of a financial crisis, allegedly resulting from mismanagement. The roots of their crisis lay in the mid-1960s when in a general state of recession the railways cut back on their orders for freight cars, on which these companies were heavily dependent. Some saw the more proximate origins of the crisis in "the traumatic experience of the United Front Government of West Bengal since 1967" which allegedly fostered a confrontation between capital and labour. Again, it is noteworthy that while the nationalization of these firms took place in 1976, that was more a formalization of an already accomplished takeover. One of the firms -- Braithwate and Company -- had been taken over in 1971 while the other two -- Burn Company and Indian Standard Wagon Company -- were similarly taken over in 1973 at which time the Ministry of Heavy Industry and Steel and Mines made "a definite policy announcement that our ultimate intention is to nationalise these two companies". Whatever their other differences with the government in 1976, both CPI and CPI(M) supported this measure of nationalization.[31]

The general shift in policy on the economy and the public sector was reflected in the changed configuration of influential advisers around the prime minister. Of course, her son Sanjay had emerged, as many charged, as an extra-constitutional centre of power. But the period also saw the decline of the influence of Siddhartha Shankar Ray, D.K. Barooah and Rajni Patel. Additionally, Jha underlines that "the fall of the radicals was accompanied by the rise of a new breed of henchmen" who were close to both Mrs. Gandhi and her son -- Bansi Lal, Mohammed Yunus, Yashpal Kapoor and R.K. Dhawan -- while "with Sanjay, there came into prominence a new breed of businessmen, all of whom enjoyed close personal ties with him."[32]

The changed stance on economic policy caused consternation in the CPI, while the CPI(M) and many leftists accused Mrs. Gandhi of a lurch to the right. Mrs. Gandhi vehemently denied any deviation from socialist policy, but equivocated: "sometimes due to circumstances and events beyond our control we have to take new steps, or we may have to make minor changes, not in the policy, not in the direction, not in the objectives but in the manner of attaining the objectives. So, there is no move to the right....We have always followed a particular road which some people have outlined as left-of-centre, and that is the road we intend to pursue. Of course, if circumstances demand that we have to do something more, if the people demand, then we will naturally have to do it. Our socialism is not a doctrinaire socialism."[33] Even so, politically Mrs. Gandhi had distanced herself from the CPI by the end of 1976 when she allowed, perhaps encouraged, an open attack on the CPI.

Jha maintains that the government during the emergency regime did not favour big business or monopoly bourgeoisie, rather it favoured the intermediate strata. But his view of the intermediate strata is so broad as to encompass almost all business; as he himself acknowledged, "the urban component of this class includes a nascent indigenous capitalist class of owner proprietors whom Marxists would call the national bourgeoisie."[34] Not surprisingly, the economic measures undertaken by the government during the emergency were highly welcome to big business. As its spokesman, FICCI president Harish Mahindra, placed on record: "how appreciative we are of the steps already taken by the government to refine policy measures and improve their operational efficiency. There is a positive and practical re-orientation in respect of industrial licensing, export and import regulations, price controls, direct and indirect taxation, etc....There is a better environment for accelerated growth." He believed that the government had now "come to place emphasis on results. And the results have so far been really good. The economy is on the move." Moreover, he thought that "the body-politic which almost became unworkable had come to life again," and wanted it ensured that "we do not ever slide back to the politics of turbulence and turmoil."[35] The FICCI noted that "the most important development is harmony in industrial relations," with the number of mandays lost being cut in half in 1975 over the preceding year. It was particularly satisfied with the 1976 budget, remarking that it had "many distinct features which mark it out as a watershed in the evolution of our fiscal policy....It is appreciated that Government have accepted the fundamental tax principle that high tax rates are self-defeating and that a larger tax revenue can be realised with reduced rates of taxation. From this view, the approach to personal taxation is realistic."[36] Of course, the FICCI wanted the principle to be extended to corporate taxation as well as other concessions to pull several industries out of recession. But there was no denying of its satisfaction at the new trend. On the other hand, the CPI—which had supported the declaration of emergency as an appropriate response to the allegedly reactionary aims of the J.P. movement—became disenchanted with it.

Among economists of India, especially from the Left, there has prevailed an orthodoxy that India's economic history since independence can be sharply divided in the mid-1960s, with the earlier period considered as being marked by economic advance and industrial growth, while the later period has been characterized as one of economic stagnation and industrial deceleration.[37] More recently, the eminent economist K.N. Raj has questioned this orthodoxy by demonstrating that since 1975 there has been an acceleration in the rate of economic and industrial growth.[38] If this is correct, then the credit for it would seem to belong to Mrs. Gandhi for launching India on a new course in economic policy in 1974. That new course should properly be seen as an endeavour to break the modernization stalemate that had come to characterize India since the early 1960s. Furthermore, Mrs. Gandhi at the end of the emergency left behind, whether by accident or design, a legacy of an unparalleled food buffer stock in India's history of some 18 million tons; an unprecedented pile of foreign exchange reserves of the order of nearly $3 billion; relative price stability compared to the hyper-inflation of the early 1970s; and a toned-up public sector. No earlier regime in India had been able to establish such a performance record.

However, the emergency had other consequences beyond initiating a period of economic growth and inducing discipline which was appreciated by many in the society. Press censorship and the suppression of civil liberties resulted in the isolation of the state from the public and consequently in its becoming increasingly arbitrary and capricious. In that condition, fascinated by the technocratic model and eager to resolve India's problems through a quick fix, the state resorted to drastic measures like massive sterilization of males, particularly in northern India, and slum clearance in Delhi. These resulted in the multiple alienation of precisely those groups which had been the bedrock of support for the Congress party in the past -- the poor, including the Harijans, and the minorities. The emergency regime accordingly became discredited and, when Mrs. Gandhi held new elections in early 1977, it was overthrown by a coalition of several parties hurriedly put together under the umbrella of the Janata party.

2. Janata and the Aversion Toward Large Industry, Public and Private

The Janata party took over power in 1977 in reaction, not to an economic crisis, but to a political crisis representing widespread public disapproval of the emergency. Economcally, even though India's deeprooted poverty persisted, the Janata government inherited from Mrs. Gandhi in immediate terms a situation that could only be described as an enviable one, combining as it did a huge food buffer stock and relatively large foreign exchange reserves. No government in post-independence had started out with or had come to possess such advantages. Moreover, in the course of the first year of Janata rule, the country also produced a record bumper food crop in India's history, but in no way could that achievement be ascribed to any policy of the Janata party. Briefly, the comfortable economic situation inherited by the Janata government allowed it a range of choice that had not come India's way since independence. The price situation was somewhat of a cause for concern but the foreign exchange reserves enabled the Janata government to import raw materials and commodities to remove scarcities and control prices.

While the Janata government faced no immediate economic crisis, which the previous government had had to contend with almost continuously since the mid-1960s, it had a built-in potential for political crisis and break-up. Indeed, a noteworthy aspect of Janata rule was precisely its very brevity, lasting only a little over two years. The Janata party was essentially a more consolidated and expanded version of the "Grand Alliance" of rightist groups in the 1971 elections and like that coalition, united by nothing but opposition to Mrs. Gandhi. But what had eluded it in 1971 came its way in 1977 as a result of the massive public alienation from the Congress in northern India. Paradoxically, while opposed to the Congress, the Janata party represented a reincarnation of it, not the least in that its upper level leadership had come from the Congress but also that it encompassed within itself the broad ideological spectrum from left to right and the diversity of India's social classes. Indeed, it has been suggested that "the Janata Party is 'Congress' without the liability of the Emergency."[39] Interestingly, the Janata party had a cooperative political arrangement with the CPI(M), corresponding to the one that the Congress had with the CPI. In the parliament, the party reflected the same domination of the intermediate strata or middle sectors that had been characterstic of the Congress. It is noteworthy that the agriculturists dominated the Janata caucus, with the modern professions of the middle class constituting the second largest group, while business and industry had only a negligible representation.

Notwithstanding the analogy with the Congress, the Janata party lacked the political coherence, howsoever feeble, that was characteristic of the Congress either as a function of value consensus or a dominant leader or both. Instead, the party consisted of half a dozen important groups representing different political interests and tendencies. The first of these, though small in numbers, was the Syndicate or Congress(O) which had seceded from the Congress party in 1969. Its key leader was Morarji Desai, who became the prime minister. Although it reflected the traditional value consensus of the Congress, it was more conservative in orientation. With its bitter hostility toward Mrs. Gandhi, deriving from its political clash with her, it had also moved away from the economic model of economic modernization and heavy industry identified with her father; instead, it had chosen to counterpoise against it the Gandhian model (after Mahatma Gandhi) with its emphasis on rural areas, agriculture, small scale and handicraft industry, and decentralization.

The same Gandhian model was also strongly advocated by the BLD group (Bharatiya Lok Dal) led by Charan Singh, which under its umbrella pursued most aggressively the interests of the middle-caste rich and middle peasantry, now possessed of a heightened class-consciousness. Although attacked as the partisan leader of the kulaks, Charan Singh offered the most coherently worked out alternative economic model, which opposed what is known in the development literature as "urban bias". He stridently championed the reduction of rural-urban disparities and favoured a massive shift in investment to the rural areas to aid both an agriculture-led economic growth and an employment-oriented works programme.[40] With a reputation for personal incorruptibility and possessed of enormous political ambition, he served as deputy prime minister and home minister and, after a break, as finance minister.

Another more recent recruit to the Gandhian economic philosophy of "small is beautiful" was the Jan Sangh, which provided the largest contingent for the Janata party, almost a third of the party's representation in the Lok Sabha. Representing the traditional middle class of shopkeepers and traders and the new middle class of small industrialists and white collar workers, this segment of the Janata favoured small industry, though of the modern type.

The socialists within the Janata party represented two strains, one that had come directly from the opposition and the other that had entered via the Congress party where it had been present in the form of the "Young Turks." Both, however, were inclined to accept Gandhian prescriptions because of their political ties to Jayaprakash Narayan. The latter had, after a communist phase, founded and led the Congress Socialist Party before independence and the Praja Socialist Party after independence, had later withdrawn but only intermittently from politics, and then moved on to a fuller conversion to Gandhism. More recently, he had led the movement against Mrs. Gandhi and emerged as the political and spiritual mentor of the Janata party only to be subsequently disappointed by its factional infighting. The outstanding representative of the socialists in the government was the radical labour leader, George Fernandes, who considered big business an abomination but headed the industry ministry. The faction from the Swatantra party also swore by Gandhian moral values, but economically it was an advocate of the classical laissez faire philosophy. It carried little weight within the party, but it provided the Janata government's first minister, H.M. Patel, a former civil servant.

Somewhat of an odd man out in this diverse group was the Congress for Democracy led by Jagjivan Ram, who had remained in Mrs. Gandhi's cabinet during the emergency but quit to oppose her only after she called the elections. Personally popular in his community of the former untouchables, Ram represented a moderate left-of-centre position. He favoured the Nehru model and was not enamoured of the emphasis on agriculture and rural areas which, he believed, was a cover for the furtherance of kulak sectional interests.

Despite the political heterogeneity within the Janata party, the party was united, at least initially, in its opposition to Mrs. Gandhi and sought to reverse the centralism and authoritarianism that it identified with her. In the economic arena, the Janata party attempted to foster a new alternative programme centered on a tripod of primacy for agriculture, rural development and small-scale industry.[41] It saw its distinctive economic programme as a way to avoid the evils of both capitalism and communism, much as the Congress Economic Committee had thought in 1948. The new economic thrust became manifest in two ways. One was a shift in resources to agriculture and rural development, especially with the budgets for 1978-79 and 1979-80.[42] That this was accompanied by the imposition of enhanced taxes in urban areas and increased subsidies to rural areas became a source of resentment in the urban areas. An important consequence of the shift in resources was the intensification of shortages at the end of Janata's rule in the critical sectors of power, coal and transport.[43] These shortages, in turn, were to define the economic agenda for the successor government.

The second way in which Janata's economic programme was distinctive related to the special position it accorded to small-scale industry. The Janata government largely accepted the basic structure established by the Industrial Policy Resolution of 1956 with its division of the industrial arena into the public sector and the private sector, with the latter further divided into large, small and cooperative sectors. In respect of the public sector, the Janata government's official position differed little from the previous government both in terms of intent and role:

> The public sector in India has today come of age. Apart from *socialising the means of production in strategic areas*, public sector provides *a countervailing power to the growth of large houses and large enterprises in the private sector*. There will be an expanding role for the public sector in several fields. Not only will it be the producer of important and strategic goods of basic nature, but it will also be used effectively as a stabilising force for maintaining essential supplies for the consumer.

But it was the promotion of cottage and small industries that constituted "the main thrust" of Janata's industrial policy. Its declared position was that "whatever can be produced by small and cottage industries must only be so produced."[44] Despite this proclamation, however, the policy's eventual beneficiary was the power-driven small-scale sector rather than the cottage or handicraft sector, given the stupendous task in organizing production in the cottage sector.[45] For the small-scale sector, the government in a spectacular move increased from 180 to 504 the products reserved for manufacture by it. In this fashion, the government sharply restricted the expansion of the large-scale sector; its policy statement explicitly declared furthermore that in respect of these products no expansion of capacity will henceforth be allowed in the large-scale sector. Whatever measures of liberalization of the industrial economy the government adopted were really oriented toward assisting the small-scale sector. The government also launched a programme to set up a District Industries Centre in each of India's districts to assist small-scale industries technically and economically. Janata leaders further threatened to dismantle large-scale production in respect of several industries, especially textiles, soap, and shoes. This new trend in government policy was, of course, not welcome to big business and it let its views be known though there was little it could do about it. Given its weak situation politically, it did not openly attack government policy, but it complained about the revived war of the sectors and the creation of a wall between the sectors.[46] There was also divergence over the conception of the desirable role of the small-scale sector; whereas the government conceived its role in terms of a direct producer of goods, big business wanted it to be ancillary to large-scale production.

Big business came under attack from Janata leaders, immediately after the new administration took power, for not having opposed Mrs. Gandhi's emergency regime. It was verbally abused by George Fernandes right in its own house: "...why do men who are supposed to be captains of industry and leaders in their trade, kowtow to those in authority? What is it that is missing in one's character that makes men behave like rats."[47] Without realizing it, Fernandes was with this statement essentially underlining the fact that big business had been only an object rather than a subject in relation to the state. No one had suggested that big business was opposed to Mrs. Gandhi; rather, business had been appreciative and supportive of the emergency and the economic policy that went with it. Consequently, her party's loss in the 1977 elections could not be taken as great testimony to the power of the bourgeoisie in the Indian political system. Interestingly, in a private interview in 1979 a counselor at the American Embassy raised the intriguing question: in what other country would a labour leader so opposed to business, such as Fernandes, become the minister in charge of industry?[48] Prime Minister Morarji Desai also talked down to big business in the fashion of Nehru, and chastised it for always finding fault with the government and others rather than looking at its own behaviour more critically.[49] More practically, the government refused to revise the definition of a large industrial house, which restricted the sphere of operation open to big business.

Beyond that, business generally was directly affected by government concessions to organized labour with the reversal of the predecessor regime's policies on bonuses, strikes, and impounding of additional dearness allowances.[50] On nationalization, business was baffled by contradictory statements from the government. While Desai swore by a mixed economy and promised to refrain from nationalization, other ministers, such as Fernandes and Biju Patnaik, threatened to "break up the Birla empire", to "demolish" the country's top 20 industrial houses, and to nationalize the Tata Iron and Steel Co.[51] In the event, the Janata bark was bigger than its bite, given the internal divisions within the party, and it basically pursued a policy of neither nationalization nor denationalization. Sick textile mills were not nationalized. However, business soon became disenchanted with the Janata regime, and complained about the deterioration in law and order, the state of high industrial unrest (often violent), the revival of conflict between the sectors, the threats of nationalization, the shortages of coal and power, the dismal state of the capital market, and the hostile treatment meted out to the private sector.[52] Business proved unresponsive in terms of new investment; the government's grant of easier credit, simplification of licensing procedures, and higher import quotas failed to mollify it.[53]

What is striking about the Janata party is not the hostility toward the corporate private sector, in which it certainly represented a continuity with the record of the Congress, but the simultaneous antagonism displayed toward the public sector on the part of key segments of the party. The government as a whole, no doubt, declared its intention to expand the role of the public sector. However, for the first time, important constituents of the government lumped the public sector with big business as suitable targets for attack. Charan Singh considered the public sector inefficient and a white elephant, and wanted it dismantled.[54] In this he was one with Sanjay Gandhi. During the Janata party rule, even that usually ardent defender of the public sector, the Bureau of Public Enterprises (functioning under Charan Singh as Finance Minister) advised the government to close down the Heavy Engineering Corporation and to sell Modern Bakeries to the private sector.[55] That was not acceptable, but earlier — soon after the Janata took over power — the government removed the 10 per cent price preference that public sector enterprises had in bidding for government contracts, and sought to place them on an equal footing with the private sector. The government was also willing to allow imports to provide competition to the public sector.[56] Furthermore, as government shifted resources to the rural areas there were fewer resources left for the public sector which then led to shortages in power, coal and transport. Critics were quick to attack the government for what they perceived as a deliberate design of de-industrialization of the country.

The Janata government finally collapsed under the weight of its factional divisions rather than of any defeat in elections. Personal ambition, no doubt, was involved, but so was conflict between the rural and urban segments of the intermediate strata. In the process, Charan Singh emerged as a successor prime minister with the support of other parties for his faction. One Marxist scholar, even though easily resorting to the standard formula of "the breakdown of national coalition of the bourgeoisie and the landlords/rich peasants", does underline the essential point of the drive of the rural intermediate strata "to control the centre of political power" and to obtain "an equality of power with the capitalist classes to run the state apparatus."[57] But support for Charan Singh proved unstable and he soon resigned to avoid a vote of non-confidence in parliament. The attempt of the kulaks under his leadership to capture the state thus ended in failure.

In the meantime, public discontent had mounted with the deterioration of law and order, the ugly display of party infighting, and the rise in prices of consumption goods under the impact of the 1979-80 budget (which was compounded by the drought of 1979). When new elections took place in early 1980, the electorate turned again to Mrs. Indira Gandhi as a provider of political stability, personal security and economic growth. She returned to power on the support of a political coalition that united the top and the bottom of the social hierarchy [58] plus some segments of the middle. Her victory importantly manifested thus the failure of the strategy to defeat her on the basis of the power of the rural intermediate strata alone -- as represented by Charan Singh and his party -- or of the urban intermediate strata alone, as represented by the other former constituents of the Janata party. It is significant at the same time of the ideological distance that Mrs. Gandhi herself had travelled and of her reading of the public mood that she did not promise, as she did through the radical slogan in 1971, to "remove poverty" but asked the electorate, through a more conservative and prosaic one, to "vote a government that works."[59] Although she drew her major support from the top and the bottom, her programme was thus middle of the road; it was a case of a centrist programme that largely lacked the support of classes in the centre.

3. The Return of Mrs. Gandhi: Toughening on Public Sector, Betting on Private Sector (1980-1984)

The return of Mrs. Gandhi to power with a massive mandate in 1980 carried further the logic of the policies that were initiated in 1974 and during the emergency. The hallmark of these policies was pragmatism and the shedding of ideology. This did not, however, imply the repudiation or neglect of policies aimed at relief for those at the bottom of the social hierarchy. That could not be, because Mrs. Gandhi's victory was, in large measure, a function of the electoral support of the poor. Not surprisingly, Mrs. Gandhi's new 20-point economic programme was committed, among other things, to a national rural employment programme, review and effective enforcement of minimum wage for agricultural labour, rehabilitation of bonded labour, accelerated programmes for the advancement of scheduled castes and tribes, allotment of house sites to rural families, and house-building schemes for the economically weaker sections.

Notwithstanding the electoral support from the top and the bottom of the social hierarchy, Congress membership in the parliament continued to reflect the dominance of the intermediate strata. But the ideological orientation of its parliamentary delegation and party organization had undergone fundamental change. With the departure of many of the Young Turks after the emergency, the various splits that took place in the Congress subsequently, and the alienation of the CPI from the Congress, there were few ardent leftists that remained in Mrs. Gandhi's Congress party. Her cabinet was bereft of any committed socialists.

Beyond the change in the party's ideological orientation, however, what underlay the pragmatism and the erosion of ideology in the posture of the new government was the economic situation that it inherited from the Janata party. Outstanding among its features was inflation over the preceding year of some 20 per cent. Of course, the public alienation that stemmed from it had been, in part, responsible for bringing Mrs. Gandhi back to power. But inflation was, in turn, a manifestation of serious deterioration in the economy, with the new Congress leadership characterizing it as an economy in shambles or as "a shattered and desperate economy."[59] There were serious shortages in essential commodities such as coal, steel and cement. There was also grave impairment, or even breakdown, of the infrastructure in respect of power and transport. The combination of infrastructural deterioration and commodity shortages created serious bottlenecks in the production of needed goods. Besides, after the drastic price hikes in oil by OPEC in 1979, the country was confronted with an enormous crisis in its balance of payments, with the oil import bill as a proportion of export exchange earnings jumping from 30.4 per cent for 1978-79 to 53.7 and 90.0 per cent for 1979-80 and 1980-81.[60]

The roots of the new and more open pragmatism thus lay in the compulsion of events, especially events originating in the international economy -- not, as traditionally understood, in the shape of the advanced capitalist countries and international financial institutions such as the IMF -- but in the form of OPEC, which necessitated going to the IMF for assistance in the first place. What the balance of payments crisis demonstrated was that India could not finance its energy needs: without boosting its exports, which it could not do without giving a more outward orientation to its largely inward-biased economy; without expanding production in an economy whose chief characteristic over two decades had been slow growth, even stagnation; without assuring quality for its goods through enhanced competition in an economy whose hallmark had been protection and control; without imparting efficiency to its high-cost economy; without technologically upgrading its productive apparatus which had been denied technological modernization through a restrictive policy on import of technology; and without augmenting its productive apparatus with new and large investments in an economic regime which barred large industrial houses from investing in many spheres of the economy.

The new imperative of expanded production, both to relieve internal shortages and to increase exports, resulted in new policy initiatives. However, if the old formula had been that the public sector was both an instrument of socialism and of faster economic development, the public sector was no longer perceived as the engine of growth, if it had ever been. Most striking for the leadership was the inability of the public sector to generate economic surpluses for new investment, which was attributed to its inefficiency, even as the public sector represented half the entire capital investment in the organized industrial sector. The record that the leadership had before it about the performance of the public sector over the pervious decade did not inspire much confidence in it as an instrument of growth, but rather engendered disillusionment (see Table VIII.2). The economic planners repeatedly and pointedly underlined the failure of the public sector to provide adequate surpluses for investment and the serious implications of this failure for the erosion of the resource base of the entire Indian fiscal system.[61] One Congress member of parliament asserted: "certainly 'socialism' has become a nasty word because we have unwittingly tried to identify the functioning through the instrumentality of public undertakings with the word 'socialism'."[62] Not only was the public sector not providing an economic return on the investment, it was further perceived to be a block to growth in the private sector by virtue of its pre-empting financial resources and its lack of performance in the critical areas of energy and transport. As against the record of the public sector, not only did the corporate private sector demonstrate profitability but also the ability to raise capital in the market, as made vivid by oversubscription of shares, especially those of FERA companies.

Table VIII.2

Financial Performance of Public Enterprises of the Central Government

Year	No. of Enterprises	Capital Employed (Rs. billion)	Profit before taxes (Rs. billion)
1970-71	87	36.06	.20
1971-72	93	40.89	.22
1972-73	101	47.56	.81
1973-74	114	53.76	1.49
1974-75	120	66.27	3.13
1975-76	121	88.24	3.07
1976-77	149	108.87	4.21
1977-78	155	121.30	1.60
1978-79	159	139.69	1.85
1979-80	169	161.82	2.25
1980-81	168	182.31	.39

Source: India, Ministry of Finance, Bureau of Public Enterprises, *Public Enterprises Survey 1980-81* (New Delhi: 1982), p.8.

At the same time, the leadership was less vulnerable if it turned away from the public sector to the private sector because its own disillusionment was matched by that of the intermediate strata that had supported it in the past. The perception of the middle classes especially was that the public sector had failed to deliver the goods what with their experience with shortages and rising prices of goods and services provided by the public sector. True, the public sector performed symbolic functions for the leadership in terms of a leftist image that was an advantage among the vast bulk of the population. More importantly, it performed substantive functions in terms of control over the direct management of about half the organized sector of the economy with the consequent political power and patronage that it vested in the political leadership and bureaucracy. But all that could be of little comfort to a political leadership if it ended up in economic stagnation. The ensuing shortages and inflation could only spell electoral disaster at the polls, with all that power and patronage evaporating into thin air for the incumbents. Political compulsions thus dictated the search for alternative growth mechanisms. Some began to see virtue in capitalism; Vice-President M. Hidayatullah criticized those who looked down on capitalism, holding that the country needed capitalists for development.[63] The government may also have been persuaded by the experience of the Green Revolution that what was necessary was the provision of an adequate infrastructure and appropriate incentives, and that the private sector will respond with increased growth.[64] If the leadership was somewhat disillusioned in relation to the public sector, it was disappointed with small scale industry as well. The protection given to that sector had not yielded the desired results, and shortages had often been the outcome. Besides, the new Congress leadership was not exclusively fascinated by small scale and village industries; even though recognizing their employment potential, its 1980 election manifesto underlined that "under this or any other pretext, it would not be proper to embark on a process of primitivization of the society."[65]

Tilting Toward the Private Sector

In the circumstance, the leadership perforce had to turn to the corporate private sector. But in doing so, it undertook no overthrow of the existing framework of the economy, but rather made changes at the margin to accommodate itself to its new compulsions. No denationalization of the public sector was contemplated; the public sector was here to stay. Even in terms of investment, the proportions between public and private sector investment remained titled in favour of the former. Besides, the public sector already had such a massive presence in the economy, having come into real possession of the commanding heights, that any change in the proportion of investment could have an effect on its relative status only over the long run. Moreover, the very fact that the economic infrastructure was in a state of serious crisis, and that its restoration to health and expansion was mandatory for both the public and private sectors, meant strengthening and reinforcement of the public sector since such infrastructure was largely a monopoly of the public sector. Nonetheless, the change was palpable: in terms of future investment the public sector was now seen primarily as an instrument for facilitating the expansion of the private sector. Investments envisaged for the public sector were designed to improve and expand the economic infrastructure. The Sixth Five Year Plan could thus be described as being basically a power, coal and transport plan. The plan outlays demonstrated the intent of assuring a more productive infrastructure (see Table VIII.3).

Table VIII.3

Sixth Five Year Plan: Public Sector Outlays

Sector	Rs. (billion)	Per cent
1. Agriculture	56.95	5.84
2. Rural Development	53.64	5.50
3. Special Area Programmes	14.80	1.52
4. Irrigation and Flood Control	121.60	12.47
5. Energy	265.35	27.22
6. Industry and Minerals	150.18	15.40
7. Transport	124.12	12.73
8. Communications and Information & Broadcasting	31.34	3.21
9. Science and Technology	8.65	0.89
10. Social Services	140.35	14.40
11. Other	8.02	0.82
Total	975.00	100.00

Source: India, Planning Commission, *Sixth Five Year Plan 1980-85* (New Delhi: 1981), pp. 57-58.

One could therefore say that the posture of an instrumental-socialist mixed economy was being reoriented toward a consummatory mixed economy, if not an instrumental-capitalist one. Even though no overthrow of the existing economic framework was undertaken, there were several areas of that framework in which policies underwent modification, or indeed reversal. Such modification was manifest in respect of (1) liberalization of the economy, which earlier had been marked by pervasive controls and restrictions, and (2) a tougher attitude toward the performance of the public sector and toward the take-over of sick units in the private sector.

Liberalization and competition "became the watchwords of the new regime in order to expand production and advance quality. No doubt, the IMF—which was approached by India for a massive loan of over $5 billion to cope with its impending balance of payments crisis— also favoured this course. But the impact of the IMF was to influence marginally the balance of opinion within the government in that direction, for the regime was oriented that way in any case. There was a time when government leaders had threatened penalties for enterprises that exceeded their licensed capacity. But now the regime adopted a more positive attitude, though still within the framework of bureaucratic control and monitoring, toward increased production in the private sector. In 1975, similarly compelled by the need to expand production, the government had already allowed automatic expansion of capacity by 5 per cent annually or 25 per cent over a five-year period in respect of 12 engineering industries out of a list of some 40 core industries divided into 19 categories; this expansion was over and above the normal 25 per cent expansion allowed. In 1980, the government made this facility available to the other remaining industries included in the list of core industries, such as chemicals, drugs, ceramics and cement.[67] This was an earnest of the government's new posture that "what is needed above all is a set of pragmatic policies which will remove the lingering constraints to industrial production and, at the same time act as catalysts for faster growth in the coming decades." More importantly, the industrial policy statement of 1980 was an indication of the government's determination not to set the small scale sector against the large-scale industrial sector as had been characteristic of the Janata government. Rather, it wished to see an "integrated industrial development" and endeavored "to reverse the trends of the last three years towards creating artificial divisions between small and large-scale industry under the misconception that these interests are essentially conflicting."

A more dramatic shift in industrial policy followed Mrs. Gandhi's declaration to make 1982 "the year of productivity". This declaration in itself reflected the dominant new-found pragmatic thrust in the government's economic policy. The economy had recovered from the crisis that the government had inherited, but the government was not eager to embrace a radical path. No doubt, in early 1982 it proclaimed a new 20-point programme, whose hallmark was to assure social justice, consistent with the support base of the Congress party. But the basic assumption of the government was that removal of poverty rested on expanded production. As one journal expressed it, "right now, the mood of the Government is not to be unduly bothered about theoretical issues - about distributive justice and socialism and such other relics from the bygone eras.... It is clear that the Government no longer cares for the nitty-gritty of ideology. As [Finance Minister] Mukherjee said: 'We're neither right nor left. You can still call us right in the non-ideological sense of the word, meaning correct'."[68] For her part, the prime minister felt that the enhanced strength that her government had provided to the economy and polity enabled the country now to afford greater liberalization.[69]

The new industrial policy was announced in April 1982 and was immediately hailed by the industry-controlled press as "economic pragmatism" and "dilution of dogma."[70] This policy incorporated two new measures: one was to accept the principle of automatic expansion of licensed capacity by one-third over the best production level in the preceding five years rather than just one-fourth as before; the other was to enlarge the list of core industries -- beyond the earlier 19 industry categories -- that would be open to large industrial houses and FERA companies.[71] It was the second measure, enlarging the scope of the industrial arena open to the large-scale sector, that was considered to mark a significant departure from the earlier posture of the government which had sought to bar the expansion of big business. Accordingly, this policy change marked for big business a new attitude toward the corporate private sector. Again, the government opened to the private sector areas of industrial activity, such as power and oil exploration, that were earlier closed to it.[72] The Tatas, for example, were allowed to install a 500 mw power plant in Bombay. Still again, to help the manufacturers generate resources for modernization and expansion by pre-emption of what was being cornered by middlemen into the black economy, the government decided on the abolition of the administered prices of pig iron; similarly, it decided on partial decontrol of cement.[73] All these measures reflected a realization on the part of government that it could not impose restrictions that inhibited the functioning of the private sector and at the same time expect industrial growth; they also reflected the realization of the futility of negatively identifying socialism with the imposition of controls on the private sector.

Furthermore, the government sought to create a more favourable climate for investment by business "with the conscious objective of creating an environment conducive to industrial dynamism"; as Finance Minister Mukherjee explained: "the annual budgets for the years 1980-85 have a distinct philosophy. Incentives were provided to encourage savings and to channel them into productive investment....In addition, excise tax relief on additional production was allowed. The Government has actively encouraged the corporate sector to mobilise the financial resources it needs for investment and modernisation directly from the public. This policy has been highly successful. The total amount of capital issued by the private corporate sector increased from a little over Rs. 300 crores in 1980-81 to Rs. 529 crores in 1981-82, and further to Rs. 809 crores in 1983-84. This is an expansion of 170 per cent in three years."[74] Meanwhile, the government also drastically simplified its procedures for approval of new investment proposals from the private sector. It also relaxed the regulations on conversion of loans to equity by public financial institutions.[75]

In addition, it adopted a more liberal policy on the import of raw materials, spare parts and technology. In its 1980 statement on industrial policy, the new government promised that it "will consider favourably the induction of advanced technology, and will permit creation of capacity large enough to make it competitive in world markets." Although the government's new technology policy statement of 1983 proclaimed the importance of development of indigenous technology, more significant was the rapid increase in the import of technology as evidenced in the number of technical collaboration proposals approved. As the industry minister explained: "Government believes in self-reliance; but not in technological isolation. Government believes in technological inter-dependence."[76] The greater openness to the import of technology stemmed from the government's appraisal and fear of increasing technological obsolescence that hampered the country's competing in world markets. The literal overthrow of past economic orthodoxy in respect of restrictions on import of technology and size of undertakings became especially manifest in respect of electronics and telecommunications, where earlier positions were reversed in 1983 to favour liberal import of technology, economic size units, and participation by big business.

The new trend in policy was welcomed by big business. The FICCI president characterized the 1982 industrial policy statement as "a bold and well-timed attempt", while the president of the Associated Chamber of Commerce and Industry held the measures to be in accord with what his organization had been advocating. However, representatives of small scale industries were critical of the new policy.[77] Similarly, left opinion was opposed to the shift in policy. For one left-oriented scholar reflecting this opinion, "the April 1982 changes in Industrial Policy are a culmination of this process" of "dilution and non-implementation of the 1956 policy" and that "presently the path which was rejected in the fifties seems to have been opted for with gusto, rather than choosing a path for completing the unfinished tasks."[78]

Tougher Posture Toward Public Sector

In relation to the tougher attitude toward the public sector there was first a new frankness and forthrightness, indeed brutal honesty -- never seen in the leaders at the very highest level of any Congress government before -- about the performance of the public sector. Soon after the new government entered office, Industry Minister Chanana in his statement on industrial policy confessed, in a strange reversal of cause and effect, that there had been "an erosion of faith in the public sector which has been reflected in its rather poor performance in recent years"; he attributed it though to the particular circumstance of the recent collapse of the Janata government and took it as "a gigantic task" for the new Government "to rehabilitate faith in the public sector." On the other hand, Finance Minister Venkataraman underlined that the inefficiency of the public sector had eroded its credibility as a driving force for economic growth and social justice. A few days later, Mrs. Gandhi characterized its performance as "a sad thing".[79] The finance minister returned to the theme the following year; he expressed distress at the poor capacity utilization in many public sector enterprises, and declared: "The Government placed great faith in the public sector. It has progressed in several directions but has not yet delivered results commensurate with the massive investments which have gone into it at great public sacrifice."[80]

Notwithstanding the new appraisal of the public sector, the government remained committed in its public pronouncements to the proposition that it must occupy the commanding heights of the economy. A more comprehensive position on the rationale and status of the public sector in the eyes of the government was by way of reiteration of a 1966 statement by Mrs. Gandhi: "We advocate a public sector for three reasons: to gain control of the commanding heights of the economy; to promote critical development in terms of social gain or strategic value rather than primarily on considerations of profit; and to provide commercial surpluses with which to finance further economic development."[81] The government had already achieved the first aim, but had been shaken in its faith in the tenability of the third and accordingly had become less persuaded of the advisability of the second. Its thrust for the expansion of the public sector had weakened except in the case of infrastructure. No longer did the public sector inspire the vision of a vanguard or harbinger of socialism. Instead, the concern in relation to it had become the more prosaic one of assuring greater efficiency of existing enterprises. Statements from the very top underlined the need for efficiency; typical was the warning by Industry Minister Tiwari: "I am, however, not prepared to show any indulgence to managerial inefficiency, lack of financial discipline and failure to motivate the work force."[82]

Mrs. Gandhi's new 20-point economic programme even came to incorporate as one of its points the "improvement in the working of the public sector enterprises", and in 1982 the Congress party's trade union front INTUC held a convention specifically on the subject. Interestingly, Mrs. Gandhi told delegates of their own share in the predicament of the public sector: "There is one difficulty. People say nationalise, but we nationalised and what happened. Labour wants this and that, with the result that nothing can run. They demand of us what they do not demand of private capitalists. Therefore it is difficult to operate the public sector efficiently or profitably."[83]

All this, however, did not mean that the government altogether refrained from nationalization of private sector enterprises. But no longer was the impulse ideological and generalized, rather it was instrumental and specific. Some enterprises were nationalized since they were already under government management because of industrial sickness; in some cases, even the decision to nationalize had been made by the predecessor government though it had not been implemented (Bengal Chemicals, National Company, Amritsar Oil Works). In many of these cases, the interests of labour were important in the initial takeover; they were important as well in the nationalization of British India Corporation with 8000 workers[84] and of several textile mills in Bombay which were subject to strikes and lockouts of nearly two years' duration in 1982 and 1983. In the last case the jobs, and therefore the votes, of 36,000 workers were at stake; Commerce Minister V.P. Singh underlined the government's desire, through the nationalization, "to end the misery of the workers."[85] Political and symbolic reasons underlay the nationalization of Maruti which had been identified with Mrs. Gandhi's younger son, Sanjay. Dadri Dalmia Cement was nationalized to restart a closed enterprise in order to meet the demand for cement. Oil India and Burmah Oil Company were nationalized in order to complete the process of exclusive government control in oil. Similarly, six large banks were nationalized to round off the earlier bank nationalization, giving government direct control over 90 per cent of bank deposits. Although business remained quiet about the various acts of nationalization, it protested the nationalization of banks -- which it thought was "tantamount to penalization of efficiency", for the performance of the private banks had been vastly superior to that of the nationalized banks[86] -- and that of the textile mills.

The government also adoped a tougher attitude toward its taking over of sick units. The government admitted that it had been a mistake to have been too indulgent in the past in this respect. Finance Minister Venkataraman openly wondered "what kind of socialism it is where the private sector takes the profits home and the government had to take over when it makes losses."[87] In its industrial policy statement of July 23, 1980, the government announced that future take-overs would be only "in exceptional cases". A month later, Industry Minister Chanana explained that "the Government of India would not take over sick units for the sake of taking over"; he indicated that in the first instance the government would seek the merger of a sick unit with a healthy unit, failing that the state government would be asked to take over, and only as a last resort would the central government take over "the factory in the interest of the public and also in the interests of the workers."[88] A member of the planning commission warned against "being blackmailed into taking over sick companies."[89] These proclamations represented reiteration of earlier similar pronouncements; but the fear of industrial unrest in the past had usually forced the government to take over sick enterprises. A major review in October 1981 resulted in new guidelines, which too were a reiteration of earlier policy; but, in what was considered a departure from previous practice, they provided for the return of sick units, not yet nationalized, to their original owners.[90]

Besides, the government underlined that it would resort to take-over "only in cases which would become viable within a reasonable time."[91] More bluntly put, it meant that the government would let some sick units die.

As Mrs. Gandhi's last term in office was coming to a close, the tougher attitude toward efficiency in the public sector was carried further. In late 1983, she sent a harshly-worded letter to chief executives of public enterprises about her dismay at the performance of the public sector; she warned that, without at least a 10 per cent return on investment, "we cannot afford the luxury of massive investment" in the public sector.[92] As part of a general economy drive, the government in 1984 reportedly decided to place a moratorium on monetary allocations to public sector enterprises and also not to provide any further allocations to enterprises that had suffered losses during 1983-84.[93] This represented a complementary counterpart to the government's new strategy of reliance on the private sector for growth. As an augury of government intentions in the future there was the significant appointment the same year of a committee headed by the prime minister's advisor on economic affairs, Arjun Sengupta, "to review the role of public enterprises in the context of the economic structure and overall planning" and, significantly, to recommend "an appropriate policy framework and simplified procedures for the closure of non-viable units."[94]

Summary and Conclusions

In 1974 economic policy in India underwent a dramatic shift which was masked by the continued use of the earlier radical rhetoric. Behind this shift there lay the perception on the part of the political leadership of a change in the mood of the public. It is immaterial that this public was basically constituted of the intermediate strata. The linkage of the perception of the public mood to this policy change once again underlines the critical impact of the nature of the political system for decisionmaking by government. No doubt, all governments must concern themselves with the public response to their policies, but this is an even more urgent consideration with those governments that must face the electorate to renew their term in office. In this instance, behind the change in the public mood lay the reality of persistent economic stagnation accompanied by severe inflation.

Technically, the government had the choice of persisting with the same path that it had adopted earlier and, indeed, intensifying its application on the consideration that India's economic situation required a more radical, deeper, and more comprehensive approach. This course was recommended by many, especially the CPI and radical economists—and continues to be so recommended—on the ground that a mixed economy in which the private sector continued to operate could not effectively cope with the country's economic problems. Rather, what the situation required, within their intellectual framework, was the elimination altogether of the corporate private sector, described as consisting of "monopolies".

However, the government instead decided to take the alternative route of changing course. Critical to the government's decision was its conclusion about the failure of the older complex of policies -- aggressive expansion of the public sector through nationalisation and massive state entrepreneurship, imposition of restrictions on big business, establishment of a regime of controls on the private sector more generally, and reliance on the small-scale sector and new entrepreneurs- to fill the economic space that had been denied to the corporate private sector. Given the immensity of the crisis and the perception that the roots of that crisis lay in the earlier complex of policies, the government resorted to a set of more orthodox policies which constituted a reversal of the earlier policies. More accurately, it was not so much a reversal as the initiation of moves which departed radically from reliance on the earlier policies; it did not constitute a reversal in the sense of an overthrow of the physical consequences of those policies in the form of the massive public sector that had already come into existence.

Whether the perceived evaluation of the failure of the public sector as an instrument of economic growth was justified or not -- leaving aside the question of its moral desirability as a road to socialism -- that perception was critical to the change of course. To the extent that that perception coincided with reality, it confirmed the observation of a Scandinavian scholar in another context that "the worst enemy of a socialist policy in any African country is bad economic performance."[95] Thus, it was not any design on the part of big business but the performance of the public sector itself that was to blame for turning away from the public sector. Of course, radical economists and communist spokesmen insistently maintain that the public sector cannot perform effectively as long as there exists alongside it a private sector, for the private sector aims to subvert it. Why the private sector should aim to do so is not explained, when it is also simultaneously alleged that it benefits from the public sector and that, indeed, the public sector was brought into being precisely for the advancement of the private sector. Nor is it explained why the private sector should be able to prevent the public sector from performing effectively when it could not prevent the more critical venture of creating the public sector, in considerable measure through the nationalization of the private sector. One suspects that it is really the relatively better performance of the private sector in terms of profitability that lies behind the recommendation of many radical scholars for the elimination altogether of the private sector. For, in that case there will be no basis for comparison of performance to the disadvantage of the public sector; in effect, there will then be no test at all of the effectiveness of the public sector.

In this light, the fact that Nehru fostered a mixed economy with a continuing, even if restricted, private sector was of momentous consequence, for it kept open the possibility of comparison between the two sectors and thus of rethinking on the future role of the private sector. It may be suggested that Nehru planned it that way, but there is no evidence of that just as there is none on the point that big business wanted it that way. Quite the contrary. To maintain that position is to confuse consequence with cause, or outcome with origins. Business was no more able to display any particular strength in relation to Mrs. Gandhi-- than it had been earlier in relation to Nehru-- to prevent her from reducing, when she was so minded, the role of the private sector through both nationalization and government controls.

Nor was Mrs. Gandhi's policy shift apparently tied to any substantial change in the class structure and state power. Her policy changed too quickly to have been responsive to structural changes in social stratification. Her economic policy during the post-Nehru period went through the following phases, representing quite sharp and substantial shifts:

1. 1966-1968 formed the "period of tutelage"[96] when Mrs. Gandhi deferred, on the one hand, to the Old Guard in party affairs and, on the other hand, to expert opinion of both technocratic-minded political leaders and high level bureaucrats. But policy based on the advice of the latter -- such as devaluation -- drove her into discord with the party leadership.
2. 1969-1973 constituted the radical phase in Mrs. Gandhi's policy posture. Its overwhelming feature is the spate of nationalization measures and the regime of restrictions in relation to the corporate sector.
3. 1974-1984 covered the retreat from radicalism, a more toughminded posture toward the public sector, a greater reliance on the private sector as the growth instrument and, accordingly, a relaxation of controls on the corporate sector.

What is significant is that over this entire period the Congress party continued to have the poor and the minorities as its support-base, and equally that the intermediate strata remained dominant in political representation in parliament, while the urban middle class continued to exercise hegemony in the executive. Business, whether favourably or unfavourablly treated, remained an object of decisions made by the state apparatus, not a subject and part of that apparatus. No doubt, the rich and middle peasantry had become more class conscious but its open emergence into state power at the centre had to await the victory of the Janata party. More significantly, even then the peasantry -- despite its numerical strength -- was marked by an incapacity to act on its own as demonstrated by its failure to capture state power after the collapse of the Janata government. The real critical element in the change in policy was the actual experience with radical or "socialist" policies on the part of essentially the same class and power structure. The adverse impact of these policies on the public led to changes in policy on the part of a government functioning within a democratic framework. In part, the new economic policies initiated in 1974 represented a correction in the course launched with bank nationalization in 1969. The new policies were attractive to big business but that class could do nothing to assure the continuation of Mrs. Gandhi's administration in power when she was confronted with a generalized opposition to the emergency regime.

What Mrs. Gandhi started in 1974 by way of liberalization of the economy was carried further after her return to power in 1980. Even though she made important changes in policy, however, "liberalization was still being looked upon more as an exception than as a ruling principle."[97] The logic of the course she initiated was then pushed still further by her son and successor, Rajiv Gandhi, after her death in 1984. In a marked policy shift, he came to rely more openly on the private sector as a growth mechanism, even if short of carrying the new course to its logical conclusion. His policy stance was particularly noteworthy not only in the greater liberalization, if not gradual dismantlement, of controls that he launched but especially in the reduction of corporate taxes to enable the private sector to generate resources for investment on its own.

With these changes, there had thus come into existence a philosophical reversal of the Nehru strategy in regard to the public sector. Nehru had made the public sector the focus of investment in an effort to supplant the private sector and make a transition to socialism. Now his grandson saw the private sector as the locus of further growth and investment and, as a result, thus attempted to create a situation where the private sector would come to supplant the public sector. The structure of the economy was now to be left to the determination of the market rather than that of socialist-minded planners. A transformation of a mixed economy from instrumental-socialist to instrumental-capitalist was patently under way, and the government was not averse to being seen as acting, even though still not proclaiming, as if it were building a capitalist system. This represented a reversion to the pattern that was becoming manifest in the mid-1960s before Mrs. Gandhi's turn to radicalism. Regardless of the escalation under Rajiv Gandhi in regard to reliance on the private sector, the roots of his policy must be properly seen as lying in the shift that was undertaken by Mrs. Gandhi in 1974. Notwithstanding the new direction, what is impressive nonetheless is that the cumulative weight of earlier policies had meanwhile made the public sector hegemonic in the Indian economy.

NOTES

1. AICC, *Congress Marches Ahead*, X (1975), p. 294.
2. FICCI, *Correspondence...1972* (New Delhi: 1973), p. 534; *Correspondence...1971* (New Delhi: 1972), p. 373; and *Proceedings of the Forty-Sixth Annual Session...1973* (New Delhi: 1973), p.4.
3. Zareer Masani, *Indira Gandhi: A Biography* (London: Hamish Hamilton, 1975), p. 305.
4. AICC, *Congress Marches Ahead*, XII (1976), p. 227.
5. S.D. Punekar (ed.), *Economic Revolution in India* (Bombay: Himalaya Publishing House, 1977), p. 4.
6. Myron Weiner, "Political Evolution -- Party Bureaucracy and Institutions," in John W. Mellor (ed.), *India: A Rising Middle Power* (Boulder, Colorado: Westview Press, 1979), p. 38.
7. Punekar (ed.), p. 10.
8. AICC, *Congress Marches Ahead*, X (1975), pp. 282-87.
9. Prem Shankar Jha, *India: A Political Economy of Stagnation* (Bombay: Oxford University Press, 1979), p. 176.
10. *Ibid.*, pp. 287-296.
11. FICCI, *Correspondence...1974-75* (New Delhi: 1975), pp. 12-13, 17; FICCI, *Proceedings of the Forty-Eighth Annual Session...1975* (New Delhi: 1975), p.3.
12. *Times of India*, February 22, 1981.
13. *Frontier*, XVI, no. 1 (August 20, 1983), pp. 5-9.
14. Mary C. Carras, *Indira Gandhi: In the Crucible of Leadership: A Political Biography* (Boston: Beacon Press, 1979), pp. 150-59.
15. Carras, p. 211.
16. Francine Frankel, *India's Political Economy, 1947-1977* (Princeton: Princeton University Press, 1978), p. 479.
17. Frankel, p. 515; Carras, pp. 169-73.
18. C.N. Vakil and others, *A Policy to Contain Inflation with Semibombla: Submitted to the Prime Minister on Behalf of 140 Economists* (Bombay: Commerce, 1974), p. 98; and C.N. Vakil and P.R. Brahmananda, *Memorandum on Inflation Reversal and Guaranteed Price Stability* (Bombay: Vora and Co., 1977), p. 4. The eminent economist V.K.R.V. Rao and his associates also advocated an orthodox anti-inflationary policy package; see V.K.R.V. Rao et al., *Inflation and India's Economic Crisis* (New Delhi: Vikas, 1973), See also Jha, *India*, pp. 144-53, who effectively demolishes the case for influence of international financial institutions.
19. A. Moin Zaidi, *Full Circle 1972-1975* (New Delhi: Michiko and Panjathan, 1975), p. 251.
20. Carras, p. 151.

21. Carras, p. 178.
22. Zaidi, *Full Circle*, p. 17. A comprehensive review of the various theories on the emergency is to be found in P.B. Mayer, "Congress (I), Emergency (I): Interpreting Indira Gandhi's India," *Journal of Commonwealth and Comparative Politics*, XXII, no. 2 (1984), pp. 128-50.
23. Zaidi, *Full Circle*, pp. 17-18, 125-26.
24. AICC, *Congress Marches Ahead*, XII, pp. 229-230.
25. Kuldip Nayar, *The Judgement* (New Delhi: Vikas, 1977), p.100.
26. *Ibid.*, p. 60.
27. Zaidi, *Full Circle*, p. 13; Khushwant Singh, *Indira Gandhi Returns* (New Delhi: Vision Books, 1979), p.66.
28. Nayar, *Judgement*, pp. 88-89; Jha, *India*, p. 170.
29. Zaidi, *Full Circle*, p. 254.
30. Statement by the Minister of Petroleum, *Lok Sabha Debates* (March 1, 1974), pp. 239-242, (January 16, 1976), pp. 125-132, and (August 23, 1976), p. 159.
31. *Lok Sabha Debates* (December 5, 1973), pp. 264-67; (August 23, 1976), pp. 255-60; (August 24, 1976), pp. 148-205.
32. Jha, pp. 171-73.
33. AICC, *Congress Marches Ahead*, XII, p. 104; see also *ibid.*, XIII (1976), pp. 79-80.
34. Jha, p. 147.
35. FICCI, *Proceedings of the Forty-Ninth Annual Session...1976* (New Delhi: 1976), pp. 2, 8.
36. FICCI, *Correspondence...1976-77* (New Delhi: 1977), pp. 36-37, 98-99.
37. See, among others, Prabhat Patnaik, "An Explanatory Hypothesis on the Indian Industrial Stagnation," in Amiya Kumar Bagchi and Nirmala Banerjee (ed.), *Change and Choice in Indian Industry* (Calcutta: K.P. Bagchi & Co., 1981), pp. 65-89; Deepak Nayyar, "Industrial Development in India: Growth or Stagnation," in *ibid.*, pp. 91-117; Ashutosh Varshney, "Political Economy of Slow Industrial Growth in India," *EPW*, XIX (September 1, 1984), pp. 1511-1517; Pranab Bardhan, *The Political Economy of Development in India* (Oxford, UK: Basil Blackwell, 1984); Isher Judge Ahluwalia, *Industrial Growth in India: Stagnation Since the Mid-Sixties* (Delhi: Oxford University Press, 1985); S.L. Shetty, "Structural Retrogression in the Indian Economy Since the Mid-Sixties," *EPW*, XIII, nos. 6-7 (February 1978), pp. 185-244.
38. See K.N. Raj, "Some Observations on Economic Growth in India Over the Period 1952-53 to 1982-83," *EPW*, XIX (October 13, 1984), pp. 1801-1804.
39. C.P. Bhambri, *The Janata Party: A Profile* (New Delhi: National, 1980), p.89.
40. Charan Singh, *Economic Nightmare of India* (New Delhi: National, 1981) and *India's Economic Policy: The Gandhian Blueprint* (New Delhi: Vikas, 1978).
41. C.N. Vakil, *Janata Economic Policy: Towards Gandhian Socialism* (New Delhi: Macmillan, 1979). See also Sachchidanand Sinha, *The Permanent Crisis in India: After Janata What?* (New Delhi: Heritage, 1978), ch.4.
42. Vakil, p. xxiv.
43. Vishnu Dutt, *Indira Gandhi: Promises to Keep* (New Delhi: National, 1980), pp. 119-120, 131.
44. India, Ministry of Industry, *Report 1977-78* (New Delhi: 1978), pp. 268-84. Emphasis added.
45. Jha, pp. 192-93.
46. FICCI, *Proceedings of the Fifty-Second Annual Session...1979* (New Delhi: 1979), pp. 2, 67, and *Proceedings of the Fifty-First Annual Session...1978*. (New Delhi: 1978), pp.2-3.
47. FICCI, *Proceedings of the Fiftieth Annual Session...1977* (New Delhi: 1977), p. xv.
48. Interview with Mr. Thomas Vrebalovich, Science Counselor, American Embassy, New Delhi, on October 17, 1979.
49. FICCI, *Proceedings...1978*, pp. 8-15.
50. Jha, pp. 188-89.
51. Cited in Marcus Franda, *India's Rural Development: An Assessment of Alternatives* (Bloomington: Indiana University Press, 1979), p. 99; FICCI, *Proceedings...1979*, pp. 4, 66-67; Vakil, pp. 112-13, 130.
52. FICCI, *Proceedings...1979*, pp. 2-4.
53. Franda, p. 99.
54. Charan Singh, *Economic Nightmare*, chs 10-11.

55. *Hindustan Times* (New Delhi), May 2, 1979.
56. Jha, p. 190.
57. Bhambri, pp. 118-19.
58. Myron Weiner, "Congress Restored: Continuities and Discontinuities in Indian Politics," *Asian Survey*, XXII, no. 4 (1982), pp. 339-55.
59. See speech by Commerce Minister Pranab Mukherjee, in FICCI, *Proceedings of the Fifty-Third Annual Session...1980* (New Delhi: 1980), p. 42.
60. India, *Economic Survey 1982-83* (New Delhi: 1983).
61. India, Planning Commission, *Sixth Five Year Plan 1980-85* (New Delhi: 1981), pp. 2, 67, 77, 260, 444.
62. V.B. Raju, in M.R. Virmani (ed.) *Public Enterprises -- From Nehru to Indira Gandhi: Objectives, Achievements & Prospects* (New Delhi: Centre for Public Sector Studies, 1981), p. 175.
63. *Times of India* (New Delhi), January 17, 1983.
64. C.T. Kurien, "The Budget and the Economy," *EPW*, XVII (March 20, 1982), p. 435-438.
65. Sharda Paul, *1980 General Elections* (New Delhi: Associated Publishing House, 1980), p. 198.
66. On the new-found importance of "liberalization" and "competition" in Mrs. Gandhi's economic thinking, see the editorial "The Need for Competition", *Times of India* (New Delhi), May 14, 1983.
67. See statement by Industry Minister Charanjit Chanana, in *Lok Sabha Debates* (July 23, 1980), pp. 367-382.
68. *India Today*, (February 15, 1982), pp, 62-71.
69. B.M. 'Prime Minister's Budget," *EPW*, XVII (March 6, 1982), pp. 350-51.
70. *Hindustan Times* (New Delhi), April 23, 1982, and *Statesman* (New Delhi), April 28, 1982.
71. Statement by Industry Minister N.D. Tiwari, in *Lok Sabha Debates* (April 21, 1982), pp. 484-85; Ministry of Finance, *Economic Survey 1982-83* (New Delhi: 1983), p. 30.
72. *Times of India* (New Delhi). August 19, 1982.
73. *Economic Survey 1982-83*, pp. 25, 30.
74. Pranab Mukherjee, *Beyond Survival: Emerging Dimensions of Indian Economy* (New Delhi: Vikas, 1984), pp. 58-59.
75. *Business India* (August 29-September 11, 1983), and Prem Shankar Jha, "A Return to Populism," *Times of India*, March 12, 1984.
76. Speech by Industry Minister N.D. Tiwari, in FICCI, *Proceedings of the Fifty-Fifth Annual Session...1982* (New Delhi: 1982), p. 31.
77. *The Hindu* (Madras), April 23, 1982.
78. Kamal Nayan Kabra, "Culmination of a Drift." *Patriot* (New Delhi), May 20, 1982.
79. *Lok Sabha Debates* (July 23, 1980), p. 370; *Times of India* (New Delhi), September 20, 1980; PTI story, in *Ceylon Daily News*, September 25, 1980.
80. *The Hindu* (Madras), May 12, 1981. Later, at its 77th annual session in December 1983 at Calcutta, the Congress stated in a resolution: "The large investments made in the previous plans in infrastructure, basic industry and mining have not yielded the expected returns." A. Moin Zaidi (ed.), *The Annual Register of Political Parties: 1983* (New Delhi: 1985), p. 402.
81. Mukherjee, *Beyond Survival*, p. 114.
82. *Hindustan Times* (New Delhi), November 6, 1982.
83. Speech in Hindi, in New Delhi, October 25, 1982.
84. The government's justification for nationalisation in this case was: "to prevent industial unrest and keep the Corporation going." See *Lok Sabha Debates* (August 27, 1981), pp. 307-311. For other cases, see *ibid*., (June 13, 1980), pp. 313-316; (July 3, 1980), pp. 193-196, 221-225; (November 18, 1980), pp. 409-412; (August 27, 1981), pp. 340-42; (September 7, 1981), pp. 395-97; (October 5, 1982), pp. 319-322.
85. *Business India*, No. 148 (November 7-20, 1983), pp. 54 - 62.
86. FICCI, *Correspondence and Relevant Documents...1980* (New Delhi: 1981), pp. 40-41.
87. *The Hindu* (Madras), September 5 and 14, 1981.
88. *Lok Sabha Debates*, (July 23, 1980), p. 379, (August 27, 1981), p. 359.
89. Mohd. Fazal, in Virmani (ed.), p. 187.
90. *The Hindu* (Madras), October 6, 1981.

91. *Economic Survey 1982-83*, p. 28.
92. B.M., "Public Sector and the 'Political Push'," *EPW*, XVIII, no. 50 (December 10, 1983), pp. 2098-99.
93. B.M., "Privatisation of Public Sector," *EPW*, XIX, no. 5 (February 4, 1984), pp. 193-194. See also editorial in *ibid.*, XIX, no. 8 (February 25, 1984), p. 317.
94. B.M., "Privatisation of Public Sector," *EPW*, XIX (1984), pp. 1858-59.
95. Knud Erik Svendsen, "Socialist Problems After the Arusha Declaration," *East Africa Journal* (Nairobi), May 1967, cited in William Tordoff and Ali A. Mazrui, "The Left and the Super-Left in Tanzania," *Journal of Modern African Studies*, X, no. 3 (1972), p. 439.
96. Dutt, p. 142.
97. L.K. Jha, "Liberalising the Economy," *Business India*, (May 6-19, 1985), pp. 97-98.

Chapter IX

The Hegemonic Position of the Public Sector

This study has so far examined the relative role of ideology and interest in the establishment and expansion of the public sector in India. The purpose of the present chapter is to evaluate the consequences of policy initiatives in the post-independence period for the status or position that the public sector has come to occupy in the Indian economy. The aim, however, is not to analyse different aspects of the role that the public sector plays in relation to the economy, society or polity, such as its contribution to capital accumulation, national integration and equality, or institution-building. It is rather to examine the size and nature of the public sector around the end of "the era of Mrs. Gandhi," that is, to determine the outcome of the endeavours of Nehru and Mrs. Gandhi to create and expand the public sector.

While differing on why the public sector was created and what interests it serves, many view the public sector in India as having achieved a hegemonic position in the economy. Asoka Mehta, a former head of the planning commission, maintained in the early 1970s that "the critical areas among 'the commanding heights' have thus been occupied. Strategically the Government has put itself in a controlling position." [1] A former high-ranking civil servant, H.V.R. Iengar, stated around the same time that "as time passed more and more emphasis was given to the public sector...on the ideological ground that the commanding heights of the economy should be in the hands of Government. The progress in this direction has been dramatic." [2] Another high-ranking civil servant, S.S. Khera, who worked closely with Nehru, asserted in the late 1970s that "the public sector has grown exponentially, and has assumed large proportions and a great variety of activities.... the public sector has come to assume its rightful place which the late Jawaharlal Nehru, India's first Prime Minister, had prescribed for it: it has come to occupy the commanding heights of the national economy." [3]

Similarly, Raj K. Nigam, an economist who was close to Mrs. Gandhi, observed in 1983 that "the growth of public sector enterprises has been enormous -- both in terms of investment and scope of activity. As a result, this sector has come to assume a dominant role in the national economy and its units are occupying many commanding heights in the economy." [4] Even as early as the mid-1960s, a public commission had noted that "the growth and expansion of the public sector has been phenomenal." [5] This position is shared by some on the Left as well. C. Rajeswara Rao, General Secretary of CPI, told an international seminar in 1976: "...the state sector occupies a very important place in our country's economy today. It is in a position to compete with the private sector. Moreover the state sector is a dominant factor in all the key and heavy industries like steel, heavy engineering, heavy electricals and oil. The monopolists and imperialists, who have been obstructing the development of the public sector through all devices, have failed miserably. The public sector has developed to giant proportions."[6] Even farther left, a critic notes in relation to the public sector that "the investment has gone largely into the capital intensive sectors like power, oil and heavy industry. Needless to say, these are the so-called 'commanding heights of the economy,' monopolised by state capital."[7] The term "commanding heights" repeatedly used in these assessments of the public sector refers largely to banking in the financial field; power, transport and communications in the area of infrastructure; and basic and heavy industry in the industrial arena. In the light of the above-cited comments, one could justifiably conclude that the public sector has come to occupy a position in the economy that was intended by its creators. The task remains to see whether the facts support the conclusion that the public sector occupies the hegemonic position in the economy that has been attributed to it.

However, the term "public sector" encompasses different things for different analysts. Here, several variants may be noted:

1. Economic enterprises of the *central government* that come under the jurisdiction of the Bureau of Public Enterprises, Ministry of Finance, Government of India. Known as *"non-departmental enterprises"*, they have been established as individual firms with varying degrees of autonomy for the purpose of providing specific goods and services. Often, the term public sector is applied to these enterprises only, numbering 228 in 1984.

   Strictly speaking, it is these enterprises that represent directly the consequences of the endeavours of Nehru and Mrs. Gandhi to create and expand the entrepreneurial role of government in the economy as a route to a socialist society. Given the overwhelming presence of the central government in the sector of non-departmental enterprises (see item 3 below), the central government's activity in this regard can be taken to be equivalent to that of the entire sector.

2. All economic enterprises of the *central* government, regardless of whether they are non-departmental enterprises or departmental enterprises. The latter consist of those enterprises that are managed directly by departments of the government without having been incorporated as separate firms. They largely comprise of public utilities, such as the railways, or strategic industries such as ordnance factories.

3. All economic enterprises of central, state and local governments, regardless of whether they are departmental or non-departmental.

   Although the public sector is often equated with the economic enterprises of the central government, economic enterprises at the state level play no inconsiderable part in the economy. One authority on the public sector maintains, quite accurately, that "in terms of investment, the State Governments public sector is nearly one-fourth of the Central Government public sector, but in numbers, the State Governments enterprises far exceed the total number of the Central Government public enterprises."[8] It may well be that different impulses have lain behind the creation and expansion of public enterprises at the state level than at the central level. On the other hand, the Congress party under the leadership of Nehru and Mrs. Gandhi was dominant not only at the central level but also in the states, especially during the Nehru era. Much of the investment at the state level has been in the realm of public utilities, such as electric power and transport.

The term *public sector* in this chapter will be used in reference to item 3 above, that is, all economic enterprises of the central, state and local governments. Items 1 and 2 will be treated as its sub-categories.

4. All government activity as distinguished from private activity. That will be referred to as the *government sector*, of which the public sector is taken to be a part. Often, the term public sector is used to refer to this category in discussions of the role of government in western developed economies. Although the government sector covers areas of activities that are not strictly economic -- such as defence, law and order, and social services -- yet a consideration of it is important to understanding the total context in which the public sector operates.

The Rapid Expansion of the Government Sector

The share of the government sector in the total economy has seen a rapid expansion since the beginning of planning. In 1980-81 the government sector had a share of over 20 per cent in the net domestic product while the private sector's share was less than 80 per cent. This share of the government sector needs to be appreciated in the background of a largely rural population, rather than an urban population living in an industrialized country; it seems somewhat meaningless to compare mechanically Third World and developed economies in this respect. What is more impressive is that the share almost tripled over three decades from 7.5 per cent in 1950-51 to 20.2 per cent in 1980-81. Within the government sector, the public sector saw an extremely rapid expansion. While the share of public administration and social services remained relatively stable, that of the public sector increased by five times from 3 per cent in 1950-51 to 15.1 per cent in 1980-81 (see Table IX.1).

Again, within the public sector, it was the segment of the non-departmental enterprises that was the site of unprecedented expansion. The expansion of the public sector was really co-extensive with the expansion of non-departmental enterprises. Whereas the share of departmental enterprises changed little, or indeed declined, that of non-departmental enterprises moved from literally noth-

Table IX.1

Net Domestic Product

(Rs. billion; at current prices)

	1950-51 Value	%	1960-61 Value	%	1970-71 Value	%	1980-81 Value	%
A. Government Sector	7.20	7.5	14.22	10.6	50.07	14.5	214.48	20.2
1. Public Admin. and Defence	4.30	4.5	5.38	4.0	16.35	4.7	54.15	5.1
2. Public Sector	2.90	3.0	8.84	6.6	33.72	9.8	160.33	15.1
a. Administrative Enterprises	-	-	1.97	1.5	7.66	2.2	29.27	2.8
b. Departmental Enterprises	-	-	5.22	3.9	13.30	3.9	39.11	3.7
c. Non-departmental Enterprises	-	-	1.65	1.2	12.76	3.7	91.85	8.6
B. Private Sector	88.30	92.5	119.13	89.4	295.12	85.5	847.61	79.8
Total	95.50	100	133.35	100	345.19	100	1062.09	100

Source: Centre for Monitoring Indian Economy, *Public Sector in the Indian Economy* (Bombay: 1983), p. 8.

ing in 1950-51 to 1.2 per cent in 1960-61, to 3.7 per cent in 1970-71, and then to 8.6 per cent in 1980-81. In absolute terms, the growth of non-departmental enterprises was massive. The real expansion of such enterprises, notwithstanding their initiation by Nehru, obviously occurred under Mrs. Gandhi. More narrowly, the greater expansion of non-departmental economic enterprises of the central government during Mrs. Gandhi's regime is manifest in Table IX.2. It is really after 1969 that there took place a phenomenal expansion in such enterprises.

Departmental enterprises represented activity that had been presaged in the colonial period, even if not in substantial measure. These enterprises related primarily to public utilities. On the other hand, non-departmental enterprises represented economic activity which the state undertook not only as a direct producer, especially in mining and manufacturing, but also intentionally with a

Table IX.2

Growth of Non-Departmental Public Enterprises
of the Central Government

Year*	Number of Enterprises	Total Investment (Rs. billion)
1951	5	0.29
1956	21	0.81
1961	48	9.53
1966	74	24.15
1969	85	39.02
1974	122	62.37
1979	176	156.02
1980	186	182.25
1981	185	211.02
1982	205	249.16
1983	209	300.38
1984	214	354.11

*Refers to either March 31 or April 1.

Source : India, Bureau of Public Enterprises, *Public Enterprises Survey 1983-84* (New Delhi: 1984), p. 166.

larger design in view for a transition to a different type of economic system. The change wrought in the nature of the public sector as a result of such state activity is palpable. Whereas the departmental enterprises segment with a share of 3.9 per cent was dominant over the non-departmental enterprises segment as late as 1960-61, with the latter having only a share of 1.2 per cent, by 1980-81 the situation had been reversed; the share of non-departmental enterprises at 8.6 per cent was far more than double that of departmental enterprises in 1980-81 (Table IX.1).

Apart from the share in the net domestic product, the rapid rise of the public sector is manifest in the growth rate of its sales or incomes as compared to that of the private sector in the organized portion (explained below) of the economy. Over the two decades from 1960-61 to 1980-81, the value of sales or income of the public sector increased 37-fold at an annual growth rate of 20 per cent compared to about 11 per cent for the private sector in the organized segment of the economy.[9] Especially noteworthy, moreover, is the change among the various components of the public sector in relation to sales or income (see Table IX.3). While in 1960-61 the Railways alone were far ahead of non-departmental non-financial enterprises, by 1980-81 they had been reduced to being only one-thirteenth the size of the latter. Even the other departmental enterprises as well as non-departmental financial enterprises had outperformed the Railways.

Table IX.3

Sales/Income of Public Sector Enterprises:
1960-61 to 1980-81

(Rs. billion; at current prices)

Year	Departmental Enterprises		Non-Departmental Enterprises		Total
	Railways	Other	Non-Financial	Financial	
1960-61	4.60	3.63	3.23	0.82	12.28
1965-66	7.34	6.57	15.69	1.84	31.44
1970-71	10.07	11.02	45.73	4.96	71.78
1975-76	17.75	23.44	164.97	15.50	221.66
1980-81	26.35	43.61	347.75	33.84	451.55

Source : Centre for Monitoring Indian Economy, *Public Sector in the Indian Economy* (Bombay : CME, 1981), p.24.

Employment as Indicator of Rapid Expansion

The really strategic position of the public sector in the economy can be seen by looking at its share in employment. Understandably, in an agricultural country and a backward economy the public sector, no matter how large, can cover only a small portion of total employment. Consequently, an adequate appreciation of its critical role requires distinguishing between the unorganized and organized sectors of the economy. The organized sector refers to the entire government sector and to all non-agricultural establishments in the private sector that employ 10 or more workers. It covers about 10 per cent of the labour force.

Of the nearly 25 million workers in the organized sector in 1984, between two-thirds and three-fourths (69.7 per cent) were in the government sector while less than one-third were in the private sector. The government sector, with far more than double the size of the private sector, dwarfed the latter (Table IX.4). Thus, it is clear that the government sector "overwhelmingly dominates the organized segment of the Indian economy."[10]

The share of the government sector in 1981 had increased by more than 10 per cent over 1971 and by 16 per cent over 1961. Correspondingly, the share of the private sector in 1981 had declined by 16 per cent from 1971 and by 22 per cent from 1961. This pattern reflects both the result of the series of nationalization measures during the 1970s and the faster growth of the public sector. Noteworthy, again, is the fact that much of the rise in the share of the public sector and the decline of the private sector took place after 1968, especially during 1969 to 1977 when Mrs. Gandhi was in full control of the government and Congress party; by contrast, the shares of the two sectors had remained relatively stable between 1961 and 1968. It is remarkable, indeed, that after 1970 there is consistently year after year a decline in the share of the private sector, and a corresponding rise in the share of the public sector. Looked at differently, employment in the government sector increased by nearly 44 per cent from 10.73 million in 1971 to 15.48 million in 1981. This is about double the rate of increase in population. In contrast, employment in the private sector increased by a bare 10 per cent in those 10 years.

Table IX.4

Employment in the Organised Sector of the Indian Economy

	Private Sector		Government Sector					
			Administration		Public Sector		Total	
	Million	%	Million	%	Million	%	Million	%
1956	-	-	3.03	-	2.21	-	5.23	-
1961	5.04	41.7	3.73	30.8	3.32	27.5	7.05	58.3
1968	6.52	40.0	5.23	32.0	4.57	28.0	9.80	60.0
1969	6.53	39.3	5.32	32.0	4.77	28.7	10.09	60.7
1970	6.70	39.3	5.47	32.0	4.90	28.7	10.37	60.7
1971	6.76	38.6	5.61	32.1	5.12	29.3	10.73	61.4
1972	6.77	37.4	5.86	32.4	5.45	30.2	11.31	62.6
1973	6.85	36.4	6.04	32.1	5.93	31.5	11.97	63.6
1974	6.79	35.2	6.23	32.3	6.25	32.5	12.48	64.8
1975	6.81	34.6	6.44	32.7	6.42	32.7	12.86	65.4
1976	6.84	33.9	6.64	32.9	6.68	33.2	13.32	66.1
1977	6.87	33.3	6.77	32.8	7.00	33.9	13.77	66.7
1978	7.04	33.1	6.92	32.6	7.28	34.3	14.20	66.9
1979	7.21	32.9	7.07	32.3	7.61	34.8	14.68	67.1
1980	7.23	32.4	7.22	32.4	7.86	35.2	15.08	67.6
1981	7.39	32.3	7.36	32.2	8.12	35.5	15.48	67.7
1982	7.55	32.1	7.55	32.1	8.40	35.8	15.95	67.9
1983	7.52	31.4	7.81	32.6	8.65	36.0	16.46	68.6
1984	7.34	30.3	7.98	33.0	8.89	36.7	16.87	69.7

Source : India, *Economic Survey 1969-70* (New Delhi: 1970), pp. 94-95, *Economic Survey 1981-82* (New Delhi: 1982), pp. 108-109, and *Economic Survey 1985-86* (New Delhi: 1986), pp.142-43.

[The figures for the private sector for 1961 refer to non-agricultural establishments employing 25 or more workers; after 1966 they include such establishments employing 10 or more workers.

The 1969 figures include transfer of staff to public sector after bank nationalization; this change is vivid in the change in figures for trade and commerce in public sector from 0.18 million in 1968 to 0.26 million in 1969, and a corresponding change for private sector from 0.35 million to 0.29 million. Similarly, such coal nationalization between 1972 and 1974 changed drastically the figures for mining and quarrying not only for those years but in subsequent years as well since such nationalization pre-empted an industry of vast expansion potential. Figures from 1971 to 1977 for the public sector in mining and quarrying are: 0.18, 0.26, 0.44, 0.61, 0.69, 0.72 and 0.76 million. Corresponding figures for the private sector, which demonstrate its contraction and stagnation are: 0.40, 0.35, 0.25, 0.13, 0.12, 0.13 and 0.13 million.]

Within the government sector, it is the public sector, comprising economic enterprises of various kinds, that has seen its share climb. Whereas the share of "administration and social services" in 1981 was only marginally different from that in 1971 (or, indeed, in 1968), though there were shifts in between, the public sector's share had increased by more than 20 per cent over 1971, from 29.3 per cent in 1971 to 35.5 per cent in 1981. In sheer numbers, while administration and social services increased their total employment by 31 per cent from 5.61 million to 7.36 million, employment in the public sector increased by almost double that rate at over 58 per cent from 5.12 million to 8.12 million. Employment in the public sector had by 1981 significantly outstripped that in administration (including defence) and social services. The ratio between the two was 52:48. In terms of employment in the organised sector of the economy, the private sector in 1981, with 7.39 million workers (same as in administration and social services), stood not just eclipsed by the government sector but also outstripped by the public sector; this change had occurred earlier in 1977. In other words, the economic muscle -- as reflected in manpower employed -- of the public sector alone exceeded that of the entire private sector in the organised segment of the economy. Truly, the public sector is an economic colossus.

Capital Accumulation as Indicator of Rapid Rise

The rise in the position of the government sector and its drive toward dominance over the economy is apparent in its role in capital formation. Whereas the ratio between the government sector and the private sector in gross domestic capital formation was 32:68 on the average for the five-year period 1950-55, it had changed to 46:54 for 1975-1980 (see Table IX.5); significantly, in the meantime, gross domestic capital formation as a proportion of GNP had increased from 9.9 per cent to 21.8 per cent for the respective five-year periods. Of course, the ratio for 1975-1980 was not the high point for the government sector; that was reached in the average for 1960-1965 (49:51) after which there was a decline and then a recovery. As a proportion of GNP, the government's share in gross domestic capital formation rose from 3.1 per cent on the average for 1950-1955 to 10.0 per cent for 1975-1980 which is the highest for any quinquennium since 1950; this represents a tripling of its share; on the other

Table IX.5

National Savings and Capital Formation by Government
and Private Sectors, 1950-51 to 1979-80

(Rs. billion; at current prices)

	Gross Domestic Saving				Gross Domestic Capital Formation			
	Rs.billion		% Share		Rs.billion		% Share	
	Govt	Priv	Govt	Priv	Govt	Priv	Govt	Priv
Average 1950-55	1.7	7.8	18	82	3.1	6.7	32	68
Average 1955-60	2.2	12.9	15	85	7.4	10.7	41	59
Average 1960-65	6.0	19.2	24	76	14.7	15.1	49	51
Average 1965-70	8.1	39.0	17	83	22.2	30.9	42	58
Average 1970-75	16.7	75.5	18	82	40.1	56.6	41	59
1975-76	33.4	115.1	22	78	76.8	70.5	52	48
1976-77	41.1	137.2	23	77	85.1	80.1	52	48
1977-78	40.6	154.5	21	79	74.5	105.9	41	59
1978-79	46.1	190.6	19	81	96.5	141.4	41	59
1979-80	47.9	193.1	20	80	117.8	129.0	48	52
Average 1975-80	41.8	158.1	21	79	90.1	105.4	46	54

Source : CMIE, *Public Sector in the Indian Economy* (Bombay: CMIE, 1983), pp. 14-15.

hand, the private sector's share increased from 6.8 per cent to 11.8 per cent, that is, it did not even double.[11]

As against the rapid strides made by the government sector in capital formation stands its much smaller role in gross domestic savings, the average for 1975-1980 being 21 per cent. Thus the government sector's capital formation is based on a draft on the community's savings, forced or otherwise (Table IX.5). Gross domestic savings as a proportion of GNP had increased from an average of 9.6 per cent for 1950-1955 to 22.3 per cent for 1975-1980. In relation to the formation of gross fixed assets, one estimate for the five-year period 1975-1980 is that, whereas large companies in the private sector relied on internal sources to the extent of 81 per cent and on external resources only up to 19 per cent for such purpose, almost the reverse was the case with non-departmental enterprises of the central government, the respective percentages being 26 and 74.[12] Regardless of the explanation or justification for the meager contribution of the public sector to savings, it is this phenomenon that became critical in the shift in government policy toward the public sector during and after the mid-1970s.

In Control of the "Commanding Heights"

The hegemonic position of the public sector emerges far more strikingly when it is examined, not just in the context of the total economy, but in relation to different sectors of the economy. For one thing, it is impressive that the reach of the public sector extends into literally every sector of the economy. Here, one is talking of the state not just in terms of regulation of the economy, but also as an active entrepreneur and manager. Werner Baer and his associates, in a discussion of Brazil, distinguish between (1) the state as regulator, a more traditional conception of the role of the state, and (2) the state as economic agent. They further differentiate, in respect of the former, between (a) fiscal function of the state revolving around government expenditures and the tax burden, and (b) direct regulation in relation to prices, production and foreign trade.[13] It is sufficient to note here that, outside the Communist countries, India has had one of the most extensive and stringent system of discretionary and non-discretionary controls in relation to the economy. As Myrdal pointed out in 1968, "the fact is that no major and, indeed, few minor business decisions can be taken except with the prior permission of the administrative authorities or at the risk of subsequent government disapproval. This implies that 'private' business in India is something entirely different from what it normally is in the Western countries."[14] Furthermore, a key economic adviser to the government notes, in regard to the system of controls, that "unfortunately many people, especially those who exercise control, think of it in purely negative terms -- an extension into the economic field of Section 144 of the Criminal Procedure Code."[15]

In respect of the state as economic agent, Baer and his associates distinguish between (a) the state as a banker and financier and (b) the state as a producer; one could add to this (c) the state as a trader. The state in India manifests itself in all these roles. But its position varies from sector to sector in terms of the degree of its domination.

A particularly useful indicator of the state's domination over the economy is the share of the government sector in different parts of the organised segment of the economy. This domination can be analysed from the perspective of both employment and production. A visual inspection of Table IX.6 illustrates dramatically the preponderant position of the state in the economy. In absolute numbers, the employment in state enterprises is simply mind-boggling, with nearly one and a half million in the Railways alone and close to two-thirds of a million in coalfields. But so is the reach of the public sector in various branches of the economy. Even in the field of agriculture and forestry, the public sector has a major presence with 36.2 per cent of employment. The state's role here basically revolves around the creation and maintenance of irrigation works, the control and upkeep of forest, and logging. Following the distribution of powers in the federal system, most of such activity is, of course, handled by the states.

As for the area of mining and quarrying, the state appears as a hegemon with over 85 per cent of employment. This sector, "which was earlier almost wholly dominated by the private sector, is now almost wholly dominated by the public

Table IX.6

Employment in the Organised Sector March 1977

Industry	Government Million	%	Private Million	%
1. Agriculture, hunting, forestry and fishing	.476	36.2	.838	63.8
2. Mining and quarrying	.757	85.3	.130	14.7
Coal	.643	96.8	.021	3.2
Others	.114	51.1	.109	48.9
3. Manufacturing	1.226	22.7	4.165	77.3
4. Electricity, gas and water	.563	94.2	.035	5.8
Generation/transmission of electricity	.154	91.1	.015	8.9
Distribution of electricity	.351	97.0	.011	3.0
5. Construction	1.009	92.4	.083	7.6
6. Wholesale and retail trade, and restaurants and hotels	.075	21.4	.275	78.6
Restaurants/eating places	.003	3.8	.076	96.2
Others	.072	26.6	.199	73.4
7. Transport, storage and Communications	2.467	97.2	.071	2.8
Railways	1.444	99.9	.002	0.1
Tram/bus services	.355	90.8	.036	9.2
Storage/warehousing	.054	100.0	-	-
Post, telegraphs, telephones	.468	100.0	-	-
Others	.147	81.7	.033	18.3
8. Financing, insurance, real estate and business services	.534	74.2	.186	25.8
Banking	.375	75.2	.124	24.8
Insurance	.060	100.0	-	-
Others	.099	61.5	.062	38.5
9. Government administration	3.292	100.0	-	-
10. Community, social and personal services	3.477	76.2	1.085	23.8
Total	13.876	66.9	6.868	33.1

Source: Adapted from CMIE, *Public Sector in the Indian Economy* (Bombay: CMIE, 1981), p.24.

sector."[16] This is a result both of nationalization of private enterprises and of reservation by the state of several spheres for exclusive development in the public sector.

The domination of the state in mining and quarrying that is manifest in the figures on employment is confirmed by the state's share in the value of production of all minerals. In 1980-81, the public sector had a share of 91.3 per cent in the total production value of Rs. 14.6 billion; this was in contrast to a share of 11.9 per cent in 1960-61 and 23.9 per cent in 1970-71.[17] This order of domination by the public sector in the area of raw materials has meant that the private sector has become dependent on the public sector for the supply of inputs.

Among fuel minerals, coal is literally a monopoly of the state with nearly 97 per cent of employment in the public sector; the only coal mines outside the control of the state are the captive ones of the Tata steel plant. In 1981-82, over 98 per cent of the country's production of 125 million tonnes of coal came from the public sector.[18] The state's share in lignite production of 6.7 million tonnes in 1982 was 92 per cent. The state has a monopoly in the production of crude oil (16.2 million tonnes in 1981-82) and natural gas (3.8 billion cubic metres). Interestingly, before 1960, all crude oil production and refining was in the private sector, more accurately, the foreign sector.

In non-fuel minerals, the state's share in the production of iron ore in 1981 (41.4 million tonnes) was 49 per cent; its share in the value of production of this mineral (Rs. 12.9 billion in 1981), however, was much higher at 62 per cent. In respect of bauxite, the state's share in 1981 was over one-fifth (22 per cent) of the total production of 19.2 million tonnes; this share is likely to increase significantly as projects under implementation are completed. Forty-four per cent of chromite ore (3.4 million tonnes) and 42 per cent of manganese ore (1.5 million tonnes) was produced in the public sector in 1981; the corresponding share of the public sector in the value of the production was 39 and 59 per cent. Besides, the production of several minerals was exclusively a monopoly of the public sector: copper ore, gold, diamonds, lead, zinc, phosphorite, fluorites, pyrites, sulphur, and rare earths. Indicative of the impact of nationalization of the early 1970s is that 94 per cent of the production of sillimanites in 1981 (over 12 thousand tonnes) took place in the public sector.[19]

In manufacturing, the share of the public sector in employment lay between one-fourth and one-fifth; by 1981 it stood at 24.1 per cent in respect of manufacturing in the factory sector.[20] Since this is the area which has been the location of entrepreneurship by the state the most and one where the Central government has been the most active since the mid-1950s, special attention is given to it in a separate section below.

The state is, again, overwhelmingly dominant in the sphere of electricity, gas and water. Over 90 per cent of the manpower employed in the generation and transmission of electricity belongs to the public sector. In 1978-79, 93 per cent of the country's electricity outside of captive power plants was generated in the public sector.[21] Although in more recent years the Central government has increasingly taken to the generation of electricity through mega-projects, primarily under the National Thermal Power Corporation and National Hydroelectric Power Corporation, electricity has largely been a function of the states. In the distribution of electricity, the role of the public sector is even more prominent, employing as it does 97 per cent of a much larger labour force.

In construction, too, with a work force of over 1 million constituting over 92 per cent of the labour force employed in that activity, the state has as significant a role as it does in electricity. Nor is it without a substantial role in trade and hotels, where it employs nearly 27 per cent of the work force. Especially significant is the enormous role of the state in import trade, with its share being 62 per cent in 1980-81 as it was in 1970-71. However, in 1974-75 and 1975-76, it was 76 per cent as a consequence of the increase in oil prices. The management of such a large proportion of the import trade by the state has, again, made for dependence of the private sector on the public sector for the supply of inputs. The state's share in export trade was 22 per cent in 1980-81.[22]

Nothing matches the domination of the state in transport, storage and communications; over 97 per cent of the labour force in this sector is employed by the state. In the railways, the state's share is 99.9 per cent while airlines, both domestic and international, are a state monopoly. Again, 100 per cent of the employment in storage and warehousing as also in posts, telegraphs and telephones is in the hands of the state; these facilities are basically operated by the central government. On the other hand, tram and bus services, where 91 per cent of the labour force is employed in the public sector, are provided by the states and local governments. In 1981-82, over 44 per cent of the passenger bus fleet belonged to the public sector; however, goods transport by trucks is basically in the private sector. The state's share in shipping in terms of tonnage in 1982 was 53.7 per cent.[23] In community and social services, the state dominates with over 76 per cent of the labour employed.

The state plays as well a dominant role -- one derived primarily from nationalization -- in banking, where it employs over 75 per cent of the labour force, and insurance where it has a monopoly. Particularly noteworthy here is the role of the state in the provision of credit. The pre-emption of resources by the state through heavy taxation in order to promote the public sector, together with the socialization of savings through the nationalization of insurance and banking, has led to an enormous dependence of the private sector on public financial institutions for resources for investment. The result has been a significant share being held by the public sector in the equity of private sector companies through direct purchase or conversion of loans. A study of 361 joint stock companies shows that in 1978 the public financial institutions held 25.7 per cent of their equity as a whole as against 33.8 per cent held by joint stock companies, 37.6 per cent by individuals, and 2.6 per cent by others.

For 35 new companies in the study, however, the level of participation by public financial institutions was much higher, standing at 42.6 per cent as against 24.8 per cent for the 326 old companies; the corresponding figures for 1965 were 24.9 and 17.9 per cent, underlining the substantial change that took place under Mrs. Gandhi. In 1959, the share of these institutions was only 6.6 per cent.[24]

As a consequence of this pattern, there is hardly a major private firm that does not have a substantial share in its equity from the public sector, and at times the share can be over 50 per cent. This is illustrated by an examination of the top twenty-five companies in the private sector (Table IX.7). Not even foreign firms are without a stake in their equity from public financial institutions. Especially significant is the case of two major private firms in north India that came under threat of a takeover by outside interests in the early 1980s - Escorts, a high profile engineering enterprise, and DCM, a textiles giant. The state's

Table IX.7

Share of Public Financial Institutions in Equity of Top Twenty-Five Private Indian Companies

Rank	Company	Affiliation	Equity	Year
1	Tata Engineering and Locomotive Co.	Tata	40.37	1986
2	Tata Iron and Steel Co.	Tata	42.25	1982
3	ITC	Foreign	34.80	1984
4	Hindustan Lever	Foreign	12.68	1985
5	DCM	Shri Ram	42.54	1982
6	Reliance Textile Industries	Ambani	15.66	1982
7	Associated Cement Companies	ACC	37.31	1982
8	Dunlop India	Foreign	33.82	1982
9	Ashok Leyland	Foreign	27.45	1982
10	Mahindra and Mahindra	Mahindra	45.30	1985
11	Southern Petro-Chemical Indust. Corp.	Chidambaram	41.60	1985
12	Hindustan Motors	Birla	22.39	1985
13	Voltas	Tata	42.12	1985
14	J.K. Synthetics	Singhania	18.84	1982
15	Ceat Tyres of India	Foreign	23.83	1984
16	Gwalior Rayon Silk Manufacturing	Birla	22.49	1985
17	Brooke Bond India	Foreign	24.80	1985
18	EID-Parry (India)	Parry	27.45	1985
19	Modi Rubber	Modi	41.40	1985
20	Escorts	Escorts	54.04	1980
21	Larsen and Toubro	L & T	40.12	1985
22	Indian Aluminium Company	For./Alcan	35.86	1985
23	Union Carbide India	Foreign	23.13	1982
24	Ballarpur Industries	Thapar	27.00	1980
25	Hindustan Aluminium Corporation	Birla	18.15	1985

"Foreign" refers to equity ownership of more than 25 per cent held abroad.

Source: Ranking is based on sales or income for 1982-83, as given in Indian Institute of Public Opinion, *Quarterly Economic Report*, XXVIII, no. 3 (1984), pp. XI-XVI. Information on equity is from S.K. Goyal, "Privatisation of Public Enterprises: Some Issues for Debate" (mimeo; New Delhi: Indian Institute of Public Administration, 1985), pp. 2-22, and personal communication from Dr. S.K. Goyal, dated July 7, 1986, based on data from the Corporate Information System, Indian Institute of Public Administration, New Delhi.

share in the equity of the former was 54 per cent and in that of the latter 42.5 per cent. The controversy surrounding the abortive takeover highlighted the critical role of the state in possible takeover threats to private firms.

Often, the very impressive participation by the public financial institutions has been taken to represent government support to the private sector, but it would seem that it is equally a manifestation of the enormous penetration of the state into the private sector. Regardless of how the state's direct presence in the private sector is used, it constitutes a tremendous source of power and control in the hands of the state in relation to the private sector; this is over and above all the other instruments of regulation that the state possesses. Whether the state uses its penetration for purposes of promotion or control, it is nonetheless, more narrowly, a beneficiary at the same time of the higher profitability of its investment in the private sector compared to that of the public sector.

Strictly in terms of direct ownership and management, the picture that emerges from a review of the share of the public sector in the chief areas of economic life is that in one area after another -- more specifically, mining (especially coal), electricity, construction, transport, storage, communications, banking and insurance -- the state is not only dominant but hegemonic, if not always the exclusive economic entrepreneur. Even in other areas -- irrigation, forestry, manufacturing, trade and hotels -- the state has fostered a substantial role for itself.

The State as Manufacturer

As in the economy as a whole, the state is present in literally every area of manufacturing. Even though its share in the number of factories is small, that is, 6 per cent in 1978-79, the state has a major presence in respect of fixed capital, where its share was 44 per cent.[25] In contrast with its major share in fixed capital, its share in the value of manufacturing output of factories (excluding electricity, gas, repair services and cold storage) was only 19 per cent. Whatever the reasons for the smaller share in output, and howsoever justified, it is this element of productivity that has been central to the attack on the public sector in economic policy in the mid-1970s and subsequent years. More in correspondence with its value of output, the share of the state in employment in manufacturing is 20 per cent while its wage bill is 26 per cent, indicating a labour class that is either more privileged or more skilled, or both.

No area of manufacturing is without the presence of the state, whether it is consumer goods, intermediate products or capital goods. In food manufacturing in the organized sector, the public sector through Modern Food Industries, with its factories in 13 major cities, produced 50 per cent of the country's 145 thousand tonnes of western-style bread in 1981-82. It also has a small share of the country's soft drinks market, the production for which expanded in the wake of the ouster of Coca Cola from India in 1977. In vanaspati or hydrogenated oil (shortening), the state with its three factories took 8 per cent of the market in 1981-82. Similarly, in sugar, the state with 42 sugar mills was responsible for a little over 7 per cent of the country's sugar production even as it employed about 17 per cent of the work force in the industry; over 52 per cent of the sugar production of 60 million tonnes in 1979-80 was from factories run by cooperatives.[26] The state has also entered into food processing with one maize mill and one oil mill.

In textiles, the state, through the National Textile Corporation and its subsidiaries, is the largest cotton textile producer in India. With some 120 textile mills, it has a share of about one-fourth of the country's textile production from mills in 1981-82. Its share in the value of cloth produced was 9.2 per cent, reflecting its concentration on production of cheaper cloth varieties. The state is also present in the production of jute textiles; its six mills produced 10 per cent of the country's jute textiles. It has a small share (2.4 per cent) in the production of woollen yarn. As for newsprint, the state has a monopoly in local production. The state also owns two units for the manufacture of rubber tyres and is as well a producer of leather goods, including footwear.

In the sphere of chemicals, while the state has only a minor share in production, in some products it is the sole producer (acetanilide, aniline) or controls over 80 per cent of the production (nitrobenzene, methanol); for some products its share is over 10 per cent (caustic soda, soap). In pesticides, the state is the sole producer of DDT in India. Nitrogenous fertilizer is a major field of activity of the state in the economy, with 52 per cent of the country's production of 3.1 million tonnes in 1981-82 being in the public sector. In the production of phosphate fertilizer, the state had a share of 31 per cent in the production of about one million tonnes. In both cases, the state's share in capacity was higher than its share in production. In drugs and pharmaceuticals, the state was responsible for over one-third of the country's production of bulk drugs in 1980-81, valued at Rs. 1.9 billion; in drug formulations, however, its share was much smaller at around 7 per cent of the Rs. 12 billion worth production. In many drugs, 100 per cent of the production has been in the public sector: for example, sulphaguanidine, sulphanilamide, vitamins B1 and B2, folic acid, primaquin, analgin, phenobarbitone, quinine salts.

All twelve oil refineries with a total refining capacity of 38 million tonnes and the entire production of petroleum and petroleum products are in the public sector. In petrochemicals, the Indian Petrochemicals Corporation Limited (IPCL) is the sole producer of DMT (used in the manufacture of polyester fibre), acrylonitrile, polybutadiene rubber, polypropylene, and linear alkyl benzene. In 1981-82, its share in the country's production of LD polyethelene was 73 per cent, acrylic fibres 71 per cent, ethylene glycol 67 per cent, xylenes 63 per cent, propylene 63 per cent, ethylene 54 per cent, and butadiene 48 per cent.

In non-metallic mineral products, the state's 14 cement factories produced over one-fifth of the country's 22.2 million tonnes of cement in 1982. In relation to basic metal industries, in the area of ferrous metal the country produced a total of some 9.1 million tonnes of steel in 1982-83, of which over 80 per cent was from major steel plants while nearly 20 per cent was from mini steel plants. Of the 7.3 million tonnes of production in the major steel plants, the share of the public sector stood at around 78 per cent. Almost all of the public sector's production came from the major steel plants at Bhilai, Durgapur, Rourkela, Bokaro and Burnpur, which are managed by the holding company, Steel Authority of India Limited (SAIL). The public sector also has a substantial share in heavy steel structurals.

Outside of aluminium and zinc, the public sector has a monopoly of the production in major non-ferrous metals (copper, lead, gold and silver). In aluminium, the state had a share of 17 per cent in the production of 208 thousand tonnes in 1981-82 even though its share in capacity stood at over 30 per cent. The public sector's share, however, will climb dramatically as the mega-projects under the National Aluminium Company (NALCO) are completed. In zinc, the state's share in the 58 thousand tonnes of production in 1981-82 was 81 per cent. Although the copper production of 27 thousand tonnes in 1981-82 was entirely in the public sector, that production met only about two-fifths of the country's requirements. The country's entire lead production of about 14 thousand tonnes in 1981-82 came from the public sector's Hindustan Zinc Limited (HZL) and met less than one-third of India's needs for that metal.

The next important sector in terms of manufacturing is the production of machinery, of which there are several kinds. In the area of non-electrical machinery, the public sector had a share of an estimated 18 per cent in the Rs. 40 billion worth of production in 1981-82.[27] Apart from an importance deriving from operating some of the more advanced segments of the industry, the public sector has a more dominant role in parts of it. In mining and allied machinery, the public sector's share in capacity in 1977-78 was 93 per cent even if in production it was lower at 70 per cent. The pre-eminent enterprise here is the Mining and Allied Machinery Corporation (MAMC), which supplies some 50 per cent of India's needs for such equipment. In construction machinery, the public sector controlled the entire capacity and production of crawler tractors in 1977-78. In industrial machinery, 80 per cent of the capacity for production of steel plant equipment was in the public sector in 1977-78, while the public sector's share in the production of machine tools, tractors, power tillers and road rollers was 45, 16, 28 and 72 per cent respectively. The state provided half of the Rs. 2.5 billion worth of production of machine tools in 1981-82; the bulk of its share came from the pre-eminent enterprise in the field, HMT.

In the area of electrical and electronic equipment and machinery as a whole, the public sector was responsible for some 39 per cent of the production valued at Rs. 36 billion in 1981-82. The entire capacity and production in hydro and steam turbines lies with the public sector. The key enterprise in this area is the renowned Bharat Heavy Electricals Limited (BHEL), which pioneered the field in India and ranks among "the ten largest manufacturers of power plant equipment in the world."[28] BHEL, with its production valued at Rs. 9 billion in 1981-82, meets over 80 per cent of the country's needs for power and transmission equipment. In 1981-82, the public sector's share in the production of transformers was over 60 per cent. The public sector also has a monopoly in respect of coaxial and telephone cables, telephones and teleprinters.

In electronics, about a third of the country's production is from the public sector.[29] While the public sector's role in consumer electronics is a minor one, the state has a dominant role in professional electronics. The public sector has about 15 per cent of the country's capacity for production of TV sets. Bharat Electronics Limited (BEL) in the public sector produces about 20 per cent of the country's output of TV tubes. In the field of telecommunication equipment, the state has had a monopoly in the production of telephones, teleprinters, telephone exchanges and transmission equipment. The public sector also has a major share in the production of control instruments.

Transport equipment is another area in which the public sector plays a predominant role, having a monopoly in the production of ships, aircraft, locomotives and railway coaches. In 1981-82 it had a share of about 52 per cent in India's production of nearly 18 thousand railway wagons. It even had a large share in the production of scooters; in 1981, the state produced 24.3 per cent of the country's output of 201.4 thousand scooters. In transport equipment as a whole, the public sector had a one-third share in the value of production of Rs. 38 billion in 1981-82; its share in value added was much higher at over 47 per cent.[30]

The area of transport equipment overlaps with defence equipment as in the case of ships and aircraft. Through its various enterprises, the public sector has a monopoly in the production of defence equipment. Defence production at about Rs. 18 billion in 1983-84 relates not only to equipment for the infantry but also highly sophisticated items such as tanks for the armoured divisions, frigates and submarines for the navy, and helicopters and jet aircraft (namely, Migs and Jaguars) for the airforce. India meets nearly 70 per cent of its requirements of defence equipment from its own industry.[31] But at the other end from this world of defence, the public sector plays an important role in consumer goods, taking a share of 64 per cent of the country's production of close to 5 million watches in 1979-80 as well as the entire output of 9.2 million square metres of photo film..

The state's reach in the economy overall and in manufacturing is thus staggering. Table IX.8 lists goods and services in the production of which the state has a share of between 95 and 100 per cent (unless otherwise noted). The number of such categories in which the state has such a dominant share is simply extraordinary. Here, again, through its dominant role, the public sector makes the private sector dependent on itself.

Although the state is present in almost every conceivable area of production of goods and services, the intensity of its investment varies from one area to another. Four-fifths of the investment in non-departmental enterprises of the central government is in enterprises producing and selling goods (see Table IX.9). The largest single industry group in terms of investment has been steel, claiming nearly one-sixth (16.15 per cent) as its share. Other major industry groups have been coal (11.49%), chemicals, fertilizers and pharmaceuticals (11.28%), petroleum (10.66%), minerals and metals (8.29%), power (7.09%), and heavy engineering (4.66%). At one time, the share of steel was far greater, amounting to as much as one-third, but subsequent attention to coal, oil, power and fertilizer has brought its share down considerably; still, it remains the most pre-eminent of different areas in total cumulative investment.

Table IX.8
Goods and Services with State Share of 95-100 Per cent

MINERAL PRODUCTION:

Fuels : coal, lignite (92%), natural gas, petroleum.
Metallic Minerals : copper ore, gold, lead and zinc concentrates, silver.
Non-metallic Minerals : diamonds, fluorites, phosphorite, pyrites, sillimanite (94%), sulphur.

MANUFACTURES:

Paper : newsprint.
Chemicals and Chemical Products : acetanilide, aniline, dimethyl terephthalate (DMT), methanol, metaamino phenol.
Pesticides: DDT
Fertilizers : ammonium sulphate nitrate, calcium ammonium nitrate, nitrophosphate.
Drugs and Pharmaceuticals : analgin, doxycycline, gentamycin, sulphaguanidine, sulphanilamide, quinine salts, vitamins B1 and B2, folic acid, antimalaria drug primaquin, phenobarbitone.
Petroleum Products : LPG, mobile gas, naphtha, kerosene, aviation turbine fuel, high speed diesel, light diesel oil, fuel oils, refinery throughput.
Non-metallic Mineral Products : opthalmic blanks and optical items.
Basic Metal Industries and Non-ferrous Metals : cadmium, copper, lead, gold, silver.
Ferrous Metals : pig iron.
Non-electrical Machinery : crawler tractors.
Electrical Equipment and Machinery : Hydro and steam turbines, coaxial cable, telephone cable, telephones, teleprinters.
Transport Equipment : shipbuilding, locomotives, railway coaches.
Miscellaneous : defence equipment, photo films.

OTHER AREAS:

Banking and Insurance : life insurance, general insurance, banking (91% of deposits).
Transport and Communications : railways, air transport, communications.
Commercial Energy : power generation (93%)

Source: Atul Sarma, "Public Enterprise: Policy Goals in India," *State Enterprise*, I, no. 3 (1983), pp. 252-77; Centre for Monitoring Indian Economy, *Public Sector in the Indian Economy* (Bombay: 1981 and 1983), chapters 4-9. The years for these items vary; the listing may be assumed to apply largely to 1980.

The State as Industrial Colossus

That the state in India is truly an industrial colossus there can be no doubt about. This is manifest not only in its dominance in key sectors of the economy, but also in its share of the total output of the 250 industrial giants of the country. The Centre for Monitoring Indian Economy estimates that in 1981 the total output of these 250 industrial giants was Rs. 458 billion (including excise duties). The share of the 58 public sector enterprises in the group was 55 per cent as against 12 per cent for 42 foreign enterprises, and 33 per cent for 150 other private sector enterprises.[32]

The position of the state as an industrial hegemon in the economy emerges even more starkly if one looks at it in the context of the 100 largest companies in 1982-3, ranked according to size of sales or income. What is impressive is that of the 100 largest companies -- which, of course, do not include departmental enterprises, such as the railways, posts and telegraphs, radio and television, and ordnance factories -- 44 are non-departmental enterprises of the central government, whose share in the aggregate income far exceeds their numbers, that is, 72.3 per cent (Table IX.10). The utter dominance by these state enterprises among the 100 companies stands out even more vividly when viewed from the perspective of paid-up capital and net assets. The state enterprises had 90.3 per cent of the aggregate paid-up capital -- with less than 10 per cent left over to the entire private sector -- and 77.4 per cent of the net assets. Given these dimensions, it is no surprise that, among the top 30 companies in the list, 22 companies are from the public sector while all top eight companies are state enterprises (see Table IX.11). The gigantic size of the public sector enterprises is apparent from the fact that the sales income of the first two enterprises -- Indian Oil Corporation (IOC) and Steel Authority of India Limited (SAIL) -- is nearly equal to that of the entire private sector in the sample. Even more significantly, the paid-up capital of just one public sector enterprise, SAIL, is more than three times that of the entire private sector put together among the group of 100 largest companies. Again, SAIL and ONGC (Oil and Natural Gas Commission) together have net assets greater than the aggregate net assets of the private sector. It is thus undeniable that the public sector bestrides the economy like a colossus.

This, of course, is in respect of only the non-departmental enterprises of just the central government among the largest 100 companies. The picture is not much different if the entire universe of publicly listed companies is taken into account. An estimate by the *Economic Times* for 1983-84 shows that a total of 93,294 private sector companies had an aggregate paid-up capital of Rs. 55.14 billion; on the other hand, 970 government-owned companies had a total paid-up capital of Rs. 164.15 billion, that is, about 75 per cent of the two sets of companies put together. This situation constitutes a dramatic reversal from the one prevailing in 1956-57 when the paid-up capital of the private sector companies amounted to Rs. 10.05 billion as against Rs. 0.73 billion for government-owned companies.[33] It should be noted that these enterprises of the government do not include departmental enterprises, such as the railways or electric power boards.

Table IX.9

Pattern of Investment in Non-Departmental Enterprises of
the Central Government (as at end of 1983-84)

Industry Group 1	Number of Enterprises 2	Share in Total Investment Rs. billion 3	% Share 4
1. Enterprises under construction	6	17.47	4.93
2. Enterprises producing and selling goods	146	281.16	79.40
a. Steel	6	57.17	16.15
b. Minerals and Metals	13	29.37	8.29
c. Coal	5	40.69	11.49
d. Power	2	25.11	7.09
e. Petroleum	12	37.74	10.66
f. Chemicals, Fertilizer, Pharmaceuticals	25	39.93	11.28
g. Heavy Engineering	14	16.52	4.66
h. Medium & Light Eng.	20	5.87	1.66
i. Transportation Equip.	12	11.79	3.33
j. Consumer Goods	14	8.01	2.26
k. Agro-based	10	0.36	0.10
l. Textiles	13	8.60	2.43
3. Service Enterprises	55	54.27	15.33
a. Trading & Marketing	19	8.23	2.33
b. Transportation	9	21.95	6.20
c. Contract & Construction	7	2.66	0.75
d. Industrial Development & Tech. Consultancy	11	0.94	0.27
e. Dev. of Small industries	1	0.48	0.14
f. Tourist	2	0.82	0.23
g. Financial	3	18.36	5.18
h. Other	3	0.83	0.23
4. Insurance Companies	7	0.12	0.34
Total	214	354.11	100.00

Source : India, Bureau of Public Enterprises, *Public Enterprises Survey 1983-84* (New Delhi: 1984), pp. 165-66.

Table IX.10
Data on 100 Largest Companies in India 1982-83 (Rs. Billion)

	Government Companies Value	%	Private Companies Value	%	Total
1. Number of companies	44	44	56	56	100
2. Paid-up capital	86.77	90.3	9.28	9.7	96.05
3. Net Assets	242.93	77.4	71.01	22.6	313.95
4. Sales/Main Income	339.43	72.3	130.24	27.7	469.68
5. Profit before Tax	17.66	73.4	6.38	26.5	24.04

Source : Indian Institute of Public Opinion, *Quarterly Economic Report*, XXVIII, no. 3 (1984), pp. I-XVI.

The Public Sector as a Parallel Economy

There prevails an argument among radical scholars that the public sector is simply a servant of the private sector, that it exists to provide subsidised inputs for enterprises in the private sector, enabling them to make easy and excessive profits. An adequate treatment of this argument would, no doubt, require another monograph. But, without pre-judging that issue, it should suffice here to note some aspects that would tend to raise questions about the logic of this position. In the first place, for most inputs provided to the private sector by the public sector, the prices are lower for imports, so that the private sector would benefit much more from importing the inputs rather than depending on the public sector for their supply.

Secondly, it is amazing how such a large public sector - whose dimensions have been sketched above - would have been necessary to subsidise such a relatively small private sector, at least in the sense of the organized segment of the economy. Nor should it be forgotten that it is the community that has provided the resources for investment in the public sector in the first place; the government has not created resources, it has simply mobilized them from the community.

Thirdly, even if inputs from public sector projects were to be made available but not at supposedly subsidised prices, it is difficult to imagine that that would necessarily mean a loss to private sector enterprises, for the increased input prices would simply be passed on to the consumers. Fourthly, it is extremely important to underline that in quite a few areas of production, the public sector exists alongside the private sector. Some of these areas would include:

Bicycles	Penicillin
Bulk drugs	Railway wagons
Cars	Scooters
Cement	Steel
Drug formulations	Television sets
Fertilizer	Textiles
Machine tools	

Since in these areas, the public and private sectors compete with each other, it is difficult to determine in what manner the contribution of the public sector is distinctive in subsidising the private sector. In one sense, it may be distinctive but in a manner quite different from what is commonly understood or implied. In order to cover higher costs of production in the public sector, the government often establishes higher prices and since these are laid down as uniform prices for entire sectors, such as steel, cement or fertilizer, the private sector tends to benefit more.

Notwithstanding the counter-arguments, there can be no doubt that there are linkages between the public sector and the private sector. But equally the argument can be made that the public sector, simply because it is so massive and far-flung, exists as a sector with strong linkages within itself. There is a strong inter-relatedness among enterprises of the public sector, with the result that the public sector exists as almost a parallel economy which often bypasses the organized private sector. Thus, for example, the steel industry provides steel to the defence industry, whose products are then inducted into the armed forces. The public sector's engineering consultants design the fertilizer plants, its construction enterprises may set up the plants, whose inputs come from state-owned oil refineries, which had earlier been provided oil by the Oil and Natural Gas Commission, which in turn not only pumped the oil but much earlier did the exploration and commissioning of the oil wells; the inputs for the fertilizer plants were also transported through pipelines constructed and owned by the state or through a railway network owned by the state as well. The fertilizer as output is then transported in good measure through the railways and taken to the consumer, often through state-sponsored co-operatives. The railways, in turn, are run on coal supplied by the nationalized coal mines, or on electricity from power plants owned by the state, which are fuelled by coal from the state's coal mines, or by diesel from the state's oil refineries. The locomotives for the running of the trains are manufactured by state enterprises while nearly 60 per cent of the freight cars are also produced by the state. To be sure, the goods and services are finally consumed by private individuals, at times mediated through private sector enterprises, but the ultimate consumers of economic goods and services are always private individuals, even in the most socialist of economies.

The inter-relatedness of the public sector has partly been a result of preferential treatment by public sector enterprises to each other in bidding for contracts, but largely it is a function of pre-emption of vast areas of economic life by

Table IX.11
Top Thirty Companies: 1982-83 (Rs. Billion)

Company	Paid-up Capital	Net Assets	Sales/Main Income
1. Indian Oil Corporation	1.23	12.78	97.86
2. Steel Authority of India Limited	32.38	57.62	30.54
3. Oil and Natural Gas Commission	3.43	41.45	23.76
4. Hindustan Petroleum Corporation	0.15	2.97	20.99
5. State Trading Corporation	0.15	1.65	18.32
6. Bharat Petroleum Corporation	0.15	1.90	17.24
7. Minerals and Metals Trading Corporation	0.35	1.62	12.05
8. Bharat Heavy Electricals Limited	2.03	9.72	11.79
*9. Tata Engineering and Locomotive Co.	0.42	4.03	8.53
*10. Tata Iron and Steel Company	0.83	5.47	7.58
11. Madras Refineries	0.13	1.21	7.14
12. Cochin Refineries	0.07	0.94	7.02
13. Air India	0.74	6.39	6.45
*14. ITC	0.27	2.16	6.07
15. Shipping Corporation of India	0.28	7.66	5.72
16. Central Coalfields	2.59	9.57	5.57
*17. Hindustan Lever	0.29	1.68	5.12
18. Western Coalfields	2.15	5.03	4.86
*19. DCM	0.22	1.89	4.59
20. IBP	0.03	0.17	4.49
21. Indian Airlines	0.50	5.42	4.39
22. Indian Petro-Chemical Corporation	1.86	4.52	4.31
23. Bharat Coking Coal	3.50	5.46	4.29
24. Eastern Coalfields	3.73	4.75	4.15
*25. Reliance Textile Industries	0.24	3.18	4.06
26. Oil India Limited	0.28	1.95	3.52
*27. Associated Cement Companies	0.33	2.18	3.48
*28. Dunlop India	0.16	0.95	3.40
29. Bharat Earth Movers Limited	0.16	2.06	3.35
30. National Fertilizers Limited	2.74	5.34	3.16

*Private sector

Source : Indian Institute of Public Opinion, *Quarterly Economic Report*, XXVIII, no. 3 (1984), pp. XI-XII.

the state. One decided implication of the inter-relatedness of the public sector is that operational efficiency, or the lack of it, has a cascading effect throughout the public sector.

Summary : Indian Economy as a Mutant

Such is the overwhelming presence of the public sector that it makes the Indian economy a new economic mutant in the shape of a "mixed economy". Certainly, the Indian economy cannot be characterized as a socialist economy, given the presence of sharp inequalities and of a large and dynamic private sector, even if hemmed in by restrictions and controls. On the other hand, where the public sector looms so large, it is difficult to take the economy to be simply a capitalist one. The Indian economy seems thus to represent a new amalgam where both public and private sectors are major forces in the economy and in national development. But, regardless of the future of the system, the shape of the new mutant extends beyond a mechanical mixture of the two sectors in the form of a mixed economy. It is not simply the existence of the state as an economic leviathan that puts into question the characterization of it as a mixed economy. Even in respect of what would be ordinarily considered the private sector, at least the organized segment of it, the public sector has an important presence not only in the investment and on the boards of directors but also through its framework of controls and licensing. Venkatasubbiah aptly comments:[34]

> These developments changed the structure of India's industrial economy. Functionally what began as two distinct sectors became hardly more than two sections of a single sector -- one wholly owned by the Government and the other so linked to it at crucial points that it (the private sector) cannot function without the public sector. The relative capitalization of the two sectors is immaterial. For this reason, private industry's stake in the health and survival of state enterprise is now equal to that of the state itself.
>
> Functionally and product-wise India's industrial economy is vertically integrated, with public enterprise at the top.... functionally there are no longer two sectors but two sections of a single sector....a structure of production dominated by the state.

It is often said that India has had as its aim the building of an independent capitalist system, that is, a capitalist system that ought not to be subordinate to the western developed capitalist economies, but coordinate with them. It may well be that the economic system that India has sought to construct would turn out to be an independent one. However, it is apparent that, if it is a capitalist system at all, it is a capitalist system of a special kind, given the domination of the public sector over the economy and the regulation of the private sector, especially by a state that has largely been independent of the bourgeoisie.

NOTES

1. Asoka Mehta, "Growth of the Public Sector as the Dominant Sector," in C.N. Vakil (ed.), *Industrial Development of India: Policy and Problems* (New Delhi: Orient Longman, 1973), pp. 15-27.
2. H.V.R. Iengar, "Role of the Private Sector," in *ibid.*, pp. 28-41.
3. S.S. Khera, *Government in Business*, (2nd ed.; New Delhi: National, 1977), p. vii.
4. Raj K. Nigam, "Public Sector for Public Welfare," (lecture; Patna, February 2, 1983), pp. 1-20.
5. India, Administrative Reforms Commission, *Report on Public Sector Undertakings* (New Delhi: 1967), p. 3.
6. Communist Party of India, *Role of State Sector in Developing Countries* (New Delhi: People's Publishing House, 1977), p. vi.
7. Dipanjan Rai Chaudhuri, "State Monopoly Capital," *Frontier*, XVI, nos. 8-10 (October 8-22, 1983), pp. 33-38.
8. Nigam, p. 12.
9. Centre for Monitoring Indian Economy, *Public Sector in the Indian Economy* (Bombay: CMIE, 1983), p. 5; hereafter referred to as CMIE (1983).
10. Centre for Monitoring Indian Economy, *Public Sector in Indian Economy* (Bombay: CMIE, 1981), pp. 14-15; hereafter referred to as CMIE (1981).
11. CMIE (1983), pp. 14-15.
12. CMIE (1981), pp. 159-60.
13. Werner Baer, Richard Newfarmer and Thomas Trebat, "On State Capitalism in Brazil: Some New Issues and Questions," *Inter-American Economic Affairs*, XXX, no. 1 (Summer 1976), 69-91.
14. Gunnar Myrdal, *Asian Drama : An Inquiry Into the Poverty of Nations* (New York: Pantheon, 1968), vol. II, p. 921.
15. L.K. Jha, *Shortages and High Prices: The Way Out* (New Delhi: Indian Book Company, 1976), p. 57.
16. CMIE (1981), p.43. The discussion in this and the following section draws largely on CMIE (1981) and CMIE (1983), chs. 4-9.
17. CMIE (1983), p.41.
18. CMIE (1983), p.55.
19. CMIE (1983), ch. 4.
20. CMIE (1983), p. 67.
21. CMIE (1983), p. 50.
22. CMIE (1983), pp. 127-28, and Atul Sarma, "Public Enterprise: Policy Goals in India," *State Enterprise*, I, no. 3 (1983), pp. 252-77.
23. CMIE (1983), pp. 137-38.
24. "Survey of Ownership of Shares in Joint-Stock Companies as at the End of December 1978," in Reserve Bank of India, *Bulletin*, vol. 37, no. 2 (February 1983), pp. 69-131.
25. CMIE (1983), p. 68.
26. CMIE (1983), p. 77
27. CMIE (1983), p. 105.
28. CMIE (1981), p. 96.
29. CMIE (1983), p. 113.
30. CMIE (1983), pp. 119-120.
31. CMIE (1983), p. 123.
32. CMIE (1983), p. 69.
33. *India Today* (May 15, 1985), p. 72.
34. H. Venkatasubbiah, *Enterprise and Economic Change: 50 Years of FICCI* (New Delhi: Vikas, 1977), p. 153.

Chapter X

Conclusions

The massive size of the public sector in India and its hegemony over the economy is patently obvious. Some parts of this public sector are, no doubt, a consequence of historical legacy, having been inherited from the colonial regime; these parts pertain primarily to public utilities. But, in the main, the public sector in India has been the result of policy in the post-independence period. The public sector, of course, covers (a) public utilities, which have seen a phenomenal expansion after independence. But it now also encompasses huge new areas of economic life that earlier fell outside the ownership by the state. These areas include, most distinctively and most massively since independene, (b) manufacturing, especially in the field of basic, heavy and strategic industries, and (c) banking and insurance. In all these areas, the state has come to own and control what are called the "commanding heights" of the economy.

It is a measure of the change that has taken place in respect of the public sector in India that it corresponds by and large to the aspirations of the Communist Party of India (CPI) in the mid-1960s. In its programme adopted in 1964, the CPI envisaged a comprehensive class coalition under its leadership, called the National Democratic Front, to achieve power with the goal of implementing its version of the transition to socialism. The aim of the resulting national democratic state would be, in the view of CPI, to "rapidly expand the scope of the state sector and make it the dominant sector in our national economy, by vigorously developing the key and heavy industries in the state sector and also by extending the sphere of nationalization to banks, general insurance, foreign trade, oil, coal and other mines, and plantations."[1] Surprisingly, with the exception of foreign trade and plantations, all that has come to pass, even though that has occurred under the auspices of the Congress party.

The route to the creation of the public sector has been both (i) entrepreneurship on the part of the state, and (ii) nationalization of existing private sector enterprises. Nationalization was especially a phenomenon of the regime under Mrs. Gandhi and, though neglected in treatments of the public sector, it has played a significant part in the expansion of the public sector. Both the aim of occupying the "commanding heights" as a route to socialism and the opportunity provided by "sinking sands" -- that is, sick enterprises in the private sector facing closure -- have been instrumental in nationalization.

The resultant public sector in India is truly overwhelming in size and scope. As such, it could not have come into existence in a mere fit of absent-mindedness

Conclusions

or as a result of happenstance. Important social and political forces would have had to exist to propel it into being. What are the factors that explain the creation and expansion of the public sector? Two general explanations exist in the social sciences: (1) the first explanation pertains to interests, whether of a single class or coalition of classes, while (2) the other explanation revolves around the impact of ideology and ideas. What is the potency of each explanation in relation to the public sector in India?

The public sector in India has come into being over a considerable period of time. Consequently, a diachronic or historical view of its creation and expansion is necessary. Here, it is manifest that critical to the creation of the public sector was the ideology-based choice in favour of an *instrumental-socialist* mixed economy under Nehru in the 1950s as it was to the subsequent expansion of the public sector under Mrs. Gandhi in the late 1960s and early 1970s. Perhaps the fostering of the public sector on the scale attempted in India may possibly have been undertaken by another regime committed to a capitalist economy or an instrumental-capitalist mixed economy. But to ignore the choice that was actually made in favour of the particular form of mixed economy -- that is, instrumental-socialist -- would respect neither the specificity of the Indian case nor the authenticity of what actually took place.

It is true that in other less developed countries, some -- but not necessarily all -- of the activities corresponding to the public sector in India may have been undertaken on the instrumental ground that the private sector did not have the capacity to take on huge capital-intensive projects or that the projected enterprises were in the nature of monopolies. The same considerations may have operated in the Indian case as well, but turn out to be largely and effectively irrelevant in terms of rationale. That is so because the ideological framework actually employed provided a much larger umbrella for the state's intervention in the economy, given the consummatory goal of the establishment of a socialist society. The state felt under no obligation to resort to instrumental arguments, given the legitimacy of the objective of a socialist order, ultimately if not immediately. To the extent that they were employed, such arguments represented rationalising and reinforcing elements to forestall and silence opposition, rather than motivating impulses. Consequently, the fact that a public sector of the same magnitude exists in India and some other country, say South Korea, in no way enlightens us on the underlying reasons for the creation of the public sector in the two countries. As Jones and Mason underline: "In sum, the exceptional similarity of the public-enterprise sectors in India and Korea is due to a constellation of factors -- history, ideology, interest-group struggle, and pragmatic responses to different economic conditions -- which act in offsetting directions and just happen to balance out in these two countries."[2]

However, potent as ideology is as an explanation in the Indian case, with Nehru and his daughter as its bearers, ideology by itself does not constitute a complete explanation for the installation and expansion of the public sector in India. Also important is the configuration of class interests that facilitated the implementation of policies issuing out of ideology. The most pertinent model relating classes to state power in India is that of Kalecki's "intermediate regime" where the intermediate strata control the state. What the empirical evidence demonstrates is that the intermediate strata were, indeed, dominant in the state in the 1950s as they have been in the period after that. However, the interests of the intermediate strata were not necessarily identified with any particular framework for the organization of the economy. There were sections of these strata that favoured the socialist framework or the socialist-instrumental mixed economy, but there were at the same time other sections that were oriented toward a capitalist framework. Thus, no tendency toward socialism or "state capitalism" necessarily inhered in the fact that the intermediate strata controlled state power, as Kalecki's "intermediate regime" model posits. To that extent, the model is overdeterministic. Notwithstanding that, to the extent that dominance by the intermediate strata excluded dominance by the bourgeoisie, the constellation of class interests was crucial insofar as it avoided automatic choice in favour of a capitalist order. A choice in favour of a full-fledged capitalism or an instrumental-capitalist mixed economy would certainly have followed had the bourgeoisie been dominant. To that extent, the configuration of class interests is certainly important. Still, the intermediate regime in India, because of its internal divisions, did not inherently carry any assurance for a choice in favour of a socialist or a socialist-instrumental mixed economy. Indeed, the basically conservative orientation of the bulk of the ruling Congress party seemed to tilt the choice in favour of a consummatory capitalist order or, at the most, a capitalist-instrumental mixed economy.

But, in reality, the choice was made in favour of an instrumental-socialist mixed economy. Fundamental to that choice was, firstly, the acceptance of Nehru as the sole dominant leader of both the Congress party and the Congress government at the centre and, secondly, the particular ideological orientation of Nehru. There were two important factors involved in the acceptance of Nehru as supreme leader of the party and government. One was intervention by fate which took away Nehru's political rival Patel, who had been more sympathetic to the bourgeoisie and the bourgeois framework for the economy; it is indicative of the basic ideological tendencies of the Congress party as a whole that, until his death, Patel controlled the party machine. The other factor flowed from the then recently-adopted constitutional framework, which required any party eager to hold state power to obtain a popular mandate from the newly-enfranchised population. This requirement underlines the importance -- beyond (1) ideology and (2) interests -- of (3) the structural context. Because of his mass popularity and charismatic appeal, deriving from a lifetime of sacrifice in the nationalist cause, Nehru was deemed indispensable to the Congress party in assuring victory at the polls at the time.

Given the Congress party's dependence on him for political power, Nehru was able to exploit his strategic position in party and government to push through policies that issued out of his ideological orientation, which he had evolved in terms of socialist model of development during the period before independence. This use of his critical position by Nehru is in line with what has been called the *"strategic contingencies theory,"* which refers to "the way in which particular participants in an organization can dominate, and influence structure, by their indispensability."[3] Nehru's model encompassed a "mixed economy" as a transitional phase to socialism. During that phase of transition to socialism, the state would expand the public sector not only absolutely but relatively so that, over a not indefinite period of time, the private sector would be reduced to a mere appendage of the economy while the principal means of production would be owned and managed by the state. Whatever little of the private sector would still be left would have, in addition, to function within the limits and constraints established by the state.

The state under Nehru, notwithstanding the structural requirement of an electoral mandate from society, had some degree of relative autonomy, primarily because of the low state of social mobilization and the consequent low political consciousness of the populace. Still, Nehru was not able to push through all the policies that he desired as part of his socialist pattern of society. This was so particularly when such policies would affect agrarian interests, precisely because of the enormous political power that the landed classes as one segment of the intermediate strata wielded in a political system based on adult franchise. But there is no doubt that his policies as such in relation to the public sector issued out of his ideology. They did not stem from any consideration on his part for the interests of the bourgeoisie or intermediate strata, or for notions of incapacity of the bourgeoisie to undertake massive capital-intensive projects. Most telling in this respect is the fact that his policies were opposed, even if fruitlessly, by the bourgeoisie while, on the other hand, they were consistently supported by the ideological opponents of the bourgeoisie, that is, the Communists and socialists. To attribute potency here to ideology in policy is, however, to make an empirical statement, not a value judgement.

Overall, then, Nehru's socialist-instrumental "mixed economy" model was rooted in ideology. Nehru thus seems to confirm Keynes' judgement that ideology is ultimately determining in social phenomenon: "The ideas of economists and political philosophers, both when they are right and when they are wrong, are more powerful than is commonly understood. Indeed, the world is ruled by little else." Nonetheless, there is no denying the relationship of class interests to the policies that Nehru implemented. Critical to that relationship was the exclusion of the bourgeoisie from state power, which allowed Nehru to push through his policies that circumscribed the bourgeoisie, without being blocked by that bourgeoisie. The fact that the intermediate strata were in power facilitated, even if it did not determine, that endeavour. Nehru, however, allowed for the continued existence of the bourgeoisie for the sake of utilizing all resources for economic development, even though in a circumscribed role. That left open the possibility that the bourgeoisie could become important in determining the direction that the mixed economy may eventually take, depending on the historical opportunities, especially if the performance of the public sector was not adequate.

Streeten has called attention to the fact that it is not only ideas and interests that are relevant to the analysis of policy but also praxis or experience.[4] This certainly seems to be the case after Nehru's death, when the experience of the working of the public sector (including the heavy industry sector) but more especially the experience of a major agricultural crisis made for rethinking about the relative role to be accorded to the public and private sectors. Working closely with the party leadership, Shastri seemed to move away from the ideology-based policy framework of Nehru to embrace a more pragmatic approach, which encompassed greater emphasis on agriculture, the market and the private sector. That pragmatic approach was continued initially by Mrs. Gandhi as head of the government, notwithstanding her own ideological preferences which were closer to her father's, if not more radical.

However, subsequently she shifted sharply to the left, a shift which underlines once again the importance of the structural context requiring electoral validation as a precondition for staying in power. The sharp decline of the Congress party at the polls in 1967, its defeat in many states where non-Congress governments then came to power, and the onset of revolutionary activity on the part of the Naxalites, led to the perception on the part of Mrs. Gandhi that a sharp turn to the left was essential if the Congress party was to survive electorally in the future. This perception coincided with her own known ideological orientation even as it issued out of the interest of electoral victory and therefore political survival; thus, both ideology and interest were intertwined, but no different from the situation with Nehru after Patel's death. Behind the leftward course lay the momentum of the movement for socialism -- initiated and nurtured by Nehru -- and the determination of a critical though not large group of ideologically-oriented socialists and former Communists, who vowed loyalty to the Nehru legacy. This group supported, perhaps pushed, Mrs. Gandhi on the leftward course. However, her perception of the need for a leftward shift arose just at the time that a conservative leadership had after Nehru's death entrenched itself in the Congress party. As the wielders of power in the party, this leadership was not tolerant of her spirit of independence in government, especially in more radical causes. In the conflict that ensued between the two sides, Mrs. Gandhi split the Congress party and assumed control of the breakaway party even as she retained control of the government. Her ideological posture of the time was manifest in the tacit alliance of her Congress party with the Communist party and the elevation of former communists to high office within her Congress party, while her opponents joined with the right-wing groups.

The split of the Congress party took place in the context of bank nationalization. After the split, Mrs. Gandhi's government moved to further curb and circumscribe the private corporate sector. Less than two years after the split, Mrs. Gandhi's approach seemed vindicated when she was returned to power in 1971 with a massive mandate on the basis of her credentials as a radical leader evidenced in bank nationalization and in her pledge to "remove poverty". There then followed a series of nationalization measures, most notably in coal and in wholesale wheat trade. However, experience again intervened to stem the leftward tide. In the mid-1970s, as the economy faced a serious crisis, the perception arose that the country's economic difficulties were related to the earlier policies of extensive nationalization undertaken, to the emphasis on the public sector, and to the restrictions on the private sector. Mrs. Gandhi then began to take a less hostile attitude to the private sector and, when she returned to power in 1980, she adopted an even more positive posture toward it and simultaneously a more toughminded approach to the public sector. Experience with policy was important in this change in the posture and policy of the government. But more important was the perception that the public mood had changed from an earlier endorsement of radical policies, centering on a vigorously expanding public sector, to a critical stance toward them in the belief that they had failed to deliver the goods. There thus lay great risk to the electoral prospects of the government in this changed public mood, and therefore the government changed course again, this time away from the radical leftist path.

Both Mrs. Gandhi's earlier turn to the left and then her turning away from it testify once again to the importance of the structural context — particularly the requirement of success at elections to continuance in power—to the evolution of policy in relation to the economy and the public sector. To that extent, policy change under Mrs. Gandhi shows the limits of the "intermediate regime" model. Concentrating on the social coalition in power, that model tends to ignore the structural context; in its emphasis on interests, it neglects the dynamics of the political process. The model is oblivious to the difference that the constitutional order makes to policy even when the same social coalition is in power. The policy change under Mrs. Gandhi is also a testimony to the limits of any mechanical application of the notion of relative autonomy of the state. Although that notion had some relevance in the 1950s, though not exclusively so even then, it had become increasingly questionable as greater social mobilization of the population took place; policy then had increasingly to take public opinion into account. The very constitutional order installed in the country had been instrumental in linking society and state. At the same time, threatened by an aroused public, the regime also resorted to increased coercion, as in the emergency, only to suffer unprecedented electoral punishment.

In the changed policy environment under Mrs. Gandhi in the mid-1970s and early 1980s, nationalization of private sector enterprises still took place but it was related more to specific class interests than motivated by a generalized ideological thrust. Significantly, the class interests served in this situation were those of labour, not those of capital.

Thus, over the historical period of the four decades after independence, both ideology and interest entered into the creation and expansion of the public sector. But the proportions in which they were present differed at different times over that period. Under Nehru, ideology played a particularly critical role in the state's entrance into manufacturing. Political factors were important in moving Mrs. Gandhi in a leftist ideological direction and subsequently in turning her away from it. Equally, before Nehru consolidated his power, political factors had blocked him from pursuing his ideology-based predilections. All through this period, with its shifts in one direction or another, the intermediate strata remained in control of the state. Though the relative balance of power between the constituent urban and rural components changed, and while the system confronted crises and instability, the "intermediate regime" persisted over an extended period, defying the assumption that such a regime is a temporary phenomenon. Thus the notion of the intermediate regime does not by itself assure a perference for the public sector insofar as the model concentrates only on the class dimensions. More importantly, one needs to incorporate into it the structural context which in the Indian case over the period under consideration happened to be a parliamentary system.[5] The requirement that the government must obtain the consent of the governed also makes important the consideration of experience with policy as it affects the governed. In brief, then, ideas, interests and institutions are all essential to a more adequate understanding of economic policy over an extended historical period rather than a single moment in time, when one or another factor seems more salient.

The liberalization of the economy from the mid-1970s onwards, especially after the death of Mrs. Gandhi, may be taken to suggest that the public sector was intended all along to perform a transitional role in economic development, not so much to advance socialism as to promote eventually the private sector and capitalism. But here one has to distinguish between causes and consequences; otherwise, in that manner, one could also hold that Mao Zedong intended all along to have his revolution facilitate ultimately a return to a market economy, or that similarly Stalin intended to create a ruling class of managers in the Soviet Union. In the Indian case, an inquiry into the origins of the public sector lays bare the fact that the fundamental impulses in the establishment of the public sector were ideological, even if the facilitating conditions were political. Nehru's ideology had as its ultimate goal, even if the route was that of "mixed economy", the ushering in of a socialist society rather than any hidden agenda to advance capitalism. If the country's mixed economy did not routinely advance to socialism on the backs of a relentlessly expanding public sector, it is because of the perceived failure (rightly or wrongly) of the public sector in performance.

On the other hand, just because some liberalization has occurred in the period after the mid-1970s does not necessarily assure a capitalist order. In a country where half the population lives below the poverty line, where nearly four-fifths of the population inhabits the rural areas, where because of these preceding elements there is hostility toward the rich, where the transformation to an industrial society still lies in the future, where the transfer of vast masses of human beings to urban areas in the context of development has yet to occur, where even the transition to a middle income country is still to take place -- where all these factors prevail, there is great potential for movement toward a more radical system than the one currently in place.

Here, the fact that the basic structure for that radical order is already in place as a result of the policies of Nehru and Indira Gandhi means that only marginal changes to that structure may be necessary to move it to a different economic order. There seems to be agreement on this point on a general plane between some Marxist and some liberal scholars. Thus the Marxist scholar Block points to the "tipping" phenomenon whereby "the growth of the state's role in the economy can reach a tipping point past which capitalists lose their capacity to resist further state intervention."[6] On the other hand, Barry underlines the theme of fear in twentieth century classical liberal thought that "piecemeal acts of intervention by government in a free economy and society will, if continued over an unspecified period of time, bring about the transformation of that society into a totalitarian regime in which all but the most trivial decisions affecting an individual are taken by the state."[7] Given the fact that the public sector already occupies a hegemonic position in the Indian economy suggests that it is not fanciful to envisage such a possibility. It is for that reason perhaps that radical scholars and leaders oppose, with a vehemence that amounts to making a fetish of the public sector, any changes in the basic structure in the direction of liberalization. The existence of the present system is therefore no guarantee against a possible socialist framework in the future, regardless of questions over its desirability, especially if it affects the political-institutional context that has been characteristic of India thus far.

On the other hand, the fact that acknowledged socialist countries, like China, have moved toward a more market-oriented economy perhaps strengthens the hands of those social and political forces that favour a capitalist order. But in the contention between opposing forces favouring different visions of the desirable social and economic organization, the outcome is likely to depend not only on the political configuration favouring the interests of the capitalist class but also on the relative performance of the public sector as well. Indeed, the political configuration itself would be intimately related to the public sector's performance.

NOTES

1. Communist Party of India, *The Programme of the Communist Party of India* (As Adopted...December 1964) (New Delhi: 1965), pp. 50-51.
2. Leroy P. Jones and Edward S. Mason, in Leroy P. Jones (ed.), *Public Enterprise in Less Developed Countries* (Cambridge, UK: Cambridge University Press, 1982), p. 23.
3. Christopher Ham and Michael Hill, *The Policy Process in the Modern Capitalist State* (Brighton, Sussex: Wheatsheaf Books, 1984), p. 120.
4. Paul Streeten, "A Problem to Every Solution," *Finance & Development*, June 1985, pp. 14-16.
5. The importance given to structure here, as part of the concern for alternative explanations for state policy, is in line with G. John Ikenberry, "The Irony of State Strength: Comparative Responses to the Oil Shocks in the 1970s," *International Organization*, vol. 40, No. 1 (Winter 1986). The article came to the attention of the present author after this chapter had been written, and therefore provides an independent confirmation of the importance of state structure in addition to, but not excluding, societal explanations.
6. Fred Block, "Beyond Relative Autonomy: State Managers as Historical Subjects," in Ralph Miliband and John Saville (ed.), *The Socialist Register 1980* (London: Merlin Press, 1980), p. 234.
7. Norman P. Barry, "Is There a `Road to Serfdom'?" *Government and Opposition*, XIX, No. 1 (Winter 1984), pp. 52-67.

Select Bibliography

Adhikari, G. *Communist Party and India's Path to National Regeneration and Socialism.* (New Delhi: Communist Party of India, 1964).
Ahluwalia, B.K. (ed.) *Jawaharlal Nehru: India's Man of Destiny.* (New Delhi: Newman Group of Publishers, 1978).
Ahluwalia, Isher Judge. *Industrial Growth in India: Stagnation Since the Mid-Sixties.* (Delhi: Oxford University Press, 1985).
Alavi, Hamza. "State and Class under Peripheral Capitalism," in Hamza Alavi and Teodor Shanin (eds.). *Introduction to the Sociology of "Development Societies".* (New York: Monthly Review Press, 1982).
Alavi, Hamza and Shanin, Teodor. *Introduction to the Sociology of "Developing Societies".* (New York: Monthly Review Press, 1982).
Alford, Robert R. and Friedland, Roger. *Powers of Theory: Capitalism, the State and Democracy.* (Cambridge, UK: Cambridge University Press, 1985).
Almond, Gabriel A. and Powell, G. Bingham.*Comparative Politics: A Developmental Approach.* (Boston: Little, Brown, 1966).
Amin, Samir et al.*Dynamics of Global Crisis.* (New York: Monthly Review Press, 1982).
Apter, David E.*The Politics of Modernization.* (Chicago: University of Chicago Press, 1966).
Arora, Satish K."Social Background of the Indian Cabinet,"*Economic and Political Weekly*, VII, Nos. 31-33, August 1972.
Avineri, Shlomo (ed.). *Karl Marx on Colonialism and Modernization.* (Garden City, N.Y.: Anchor Books, 1969).

Baer, Werner; Newfarmer, Richard; and Trebat, Thomas. "On State Capitalism in Brazil : Some New Issues and Questions," (*Inter-American Economic Affairs*, XXX No.1 summer 1976).
Bagchi, Amiya Kumar. *Change and Choice in Indian Industry.* (Calcutta: K.P. Bagchi & Co., 1981).
Bagchi, Amiya Kumar. "De-industrialization in India in the Nineteenth Century: Some Theoretical Implications," *Journal of Development Studies*, XII No.2, 1976.
Bagchi, Amiya Kumar. *Private Investment in India 1900-1930.* (Cambridge: Cambridge University Press, 1972).
Bardhan, Pranabh. *The Political Economy of Development in India.* (Oxford, UK: Basil Blackwell, 1984).
Barry, Norman P. "Is There a 'Road to Serfdom'?" *Government and Opposition.* XIX, No.1 (Winter 1984).
Baumol, William J. (ed.) *Public and Private Enterprise in a Mixed Economy.* (New York: St. Martin's Press, 1980).
Beling, Wilard A. and Totten, George O. (ed.)*Developing Nations: Quest for a Model.* (New York: Van Nostrand Reinhold Company, 1970).

Beteille, Andre. *Caste, Class, and Power.* (Berkeley: University of California Press, 1965).
Bettelheim, Charles.*Class Struggles in the USSR: First Period: 1917-1923.* (New York: Monthly Review Press, 1976), vol.1.
Bhagwati, Jagdish N. and Desai, Padma. *Indian Planning for Industrialization.* (London: Oxford University Press, 1970).
Bhambri, C.P.*The Janata Party: A Profile.* (New Delhi: National, 1980).
Bhatt, V.V.*Development Perspectives.* (Oxford, UK: Pergamon Press, 1980).
Bhuleshkar, Ashok V.*Growth of Indian Economy in Socialism.* (Bombay: Oxford and IBH Publishing Co., 1975).
Bhuleshkar, Ashok V.*Towards Socialist Transformation of Indian Economy.* (Bombay: Popular Prakashan, 1972).
Binder, Leonard et al.*Crises and Sequences in Political Development.* (Princeton: Princeton University Press, 1971).
Birla, Ghanshyam Das.*Bapu: A Unique Association.* (Bombay: Bharatiya Vidya Bhavan, 1977), vol. IV.
Birla, Ghanshyam Das. *In the Shadow of the Mahatma: A Personal Memoir.* (Bombay: Orient Longmans, 1953).
Block, Fred. "Beyond Relative Autonomy: State Managers as Historical Subjects," in Ralph Miliband and John Saville (ed.), *The Socialist Register 1980.* (London: Merlin Press, 1980), pp.227-42.
Branko, Horvat.*The Yugoslav Economic System.* (White Plains, N.Y.: International Arts and Science Press, 1976).
Braverman, Harry. *Labor and Monopoly Capital: The Degradation of Work in the Twentieth Century.* (New York: Monthly Review Press, 1974).
Brecher, Michael. *Nehru· A Political Biography.* (London: Oxford University Press, 1959).
Brecher, Michael. *Political Leadership in India: An Analysis of Elite Attitudes.* (New York: Praeger, 1969).
Brecher, Michael. *Succession in India.* (London: Oxford University Press, 1966).
Brewer, Anthony. *Marxist Theories of Imperialism: A Critical Survey* (London: Routledge & Kegan Paul, 1980).
Briggs, Asa. "The Welfare State in Historical Perspective," *European Journal of Sociology*, II, No.2, 1961.
Browett, John. "The Newly Industrializing Countries and Radical Theories of Development," *World Development*, vol.13, no.7 (1985).
Brus, Wlodzimierz. *Socialist Ownership and Political Systems.* London: Routledge and Kegan Paul, 1975).
Buchanan, James M. and Wagner, Richard E. *Democracy in Deficit: The Political Legacy of Lord Keynes.* (New York: Academic Press, 1977).

Carras, Mary C. *Indira Gandhi: In the Crucible of Leadership : A Political Biography.* (Boston: Beacon Press, 1979).
Centre for Monitoring Indian Economy.*Public Sector in the Indian Economy.* (Bombay: CMIE, 1983).

Chakravarty, S. "Nehru and the Public Sector," *Eastern Economist*, vol.75, No.22, Nov.28, 1980.

Chandidas, R.*India Votes*. (New York: Humanities Press, 1968).

Chandra, Bipan et al. "The Communists, the Congress and the Anti-Colonial Movement," *Economic and Political Weekly*, XIX, No.17 (April 28, 1984).

Chandra, Bipan. "The Indian Capitalist Class and British Imperialism," in R.S. Sharma (ed.), *Indian Society: Historical Probings in Memory of D.D. Kosambi*. (New Delhi: People's Publishing House, 1974), pp. 390-413.

Chandra, Bipan. "The Indian Capitalist Class...," Summary or Oral Presentation *Social Scientist*, XI, No.11 (1983).

Chandra, Bipan. "Jawaharlal Nehru and the Capitalist Class; 1936," *Economic and Political Weekly*, X, Nos. 33-35, August 1975.

Chandra, Bipan. "Peasantry and National Integration in Contemporary India," in K.N. Panikkar (ed.), *National and Left Movements in India*. (New Delhi: Vikas, 1980).

Chang, Sherman H.M.*The Marxian Theory of the State*. (Philadelphia: 1931).

Chattopadhyaya, Boudhayan. "The Ambivalence of Nehru: The Narodnik Utopia vs. The Liberal Utopia," in Centre for Social Research, *The Seminar on Nehru*. (Madras: Centre for Social Research, 1974), pp. 244-59.

Chaudhuri, Dipanjan Rai. "State Monopoly Capital," *Frontier*, XVI, Nos. 8-10 (Oct. 8-22, 1983).

Chaudhuri, M.R.*The Iron and Steel Industry of India: An Economic-Geography Appraisal*. (2nd edn.; Calcutta: Oxford and IBH Publishing Co., 1975).

Chilcote, Ronald H.*Theories of Comparative Politics: The Search for a Paradigm*. (Boulder, Colorado: Westview Press, 1981).

Chopra, R.N. *Evolution of Food Policy in India*. (New Delhi: Macmillan, 1981).

Christensen, Peter Moller. "Plan, Market or Cultural Revolution in China,"*Economic and Political Weekly*, XVIII, No.16-17 (April 16-23, 1983).

Cohen, Stephen S. *Modern Capitalist Planning: The French Model*. (Berkeley: University of California Press, 1977).

Collier, David (ed.). *The New Authoritarianism in Latin America*. (Princeton: Princeton University Press, 1979).

Communist Party of India. *Documents of the Ninth Congress of the Communist Party of India......1971*. (New Delhi: 1972).

Communist Party of India. *Proceedings of the Seventh Congress of the Communist Party of India: Bombay, 13-23 December 1964*. vol.3,*Discussions*. (New Delhi: 1965).

Communist Party of India. *The Programme of the Communist Party of India* (as adopted.... Dec. 1964). (New Delhi: 1965).

Communist Party of India. *Role of State Sector in Developing Countries*. (New Delhi: People's Publishing House, 1977).

Coombes, David. *State Enterprise: Business or Politics*? (London: George Allen and Unwin, 1971).

Crane, Robert I. (ed.). *Aspects of Political Mobilization in South Asia*. (Syracuse, N.Y.: Syracuse University, Maxwell School, 1976).

Crocker, Walter. *Nehru: A Contemporary's Estimate.* (New York: Oxford University Press, 1966).
Crouch, Colin (ed.). *State and Society in Contemporary Capitalism.* (New York: St. Martin's Press, 1979).

Dahl, Robert A. *Dilemmas of Pluralist Democracy.* (New Haven : Yale University Press, 1982).
Dahl, Robert A. *Who Governs? Democracy and Power in an American City.* (New Haven: Yale University Press, 1961).
Dange, S.A. *Crisis and Workers: Report to AITUC General Council, Bangalore Session, 14-18 January 1959.* (New Delhi: All India Trade Union Congress, 1959).
Dange, S.A. *General Report at Ernakulam: Silver Jubilee Session, All India Trade Union Congress, December 25 to 29, 1957.* (New Delhi: AITUC Publication, 1958).
Das, M.*Fantasy of Coal Nationalisation.* (Howrah: M. Das, 1975).
Das Gupta, Ranjit. "Nehru's Economic Thinking and India's Struggle for Economic Independence," in Centre for Social Research, *The Seminar on Nehru.* (Madras: Centre for Social Research, 1974).
Dasgupta, A. and Sengupta, N.K. *Government and Business in India.* (Calcutta: Allied Book Agency, 1978).
Desai, Ashok V. "Factors Underlying the Slow Growth of Indian Industry," *Economic and Political Weekly*, XVI, nos.10-12 (March 1981), pp. 381-392.
Dias, Clarence J. "Public Corporations in India," in International Legal Center, *Law and Public Enterprise in Asia.* (New York: Praeger, 1976).
Divekar, V.D. *Planning Process in Indian Polity.* (Bombay: Popular Prakashan, 1978).
Dore, R.P. "The Sociology of Development and Issues Surrounding Late Development,"*International Studies Quarterly*, XXVI, No.4 (1982).
Drieberg, Trevor. *Indira Gandhi: A Profile in Courage.* (New Delhi: Vikas, 1972).
Dunn, John. *The Politics of Socialism: An Essay in Political Theory.* (Cambridge, UK: Cambridge University Press, 1984).
Dupuy, Alex and Truchil, Barry. "Problems in the Theory of State Capitalism," *Theory and Society*, VIII, no.1 (1979).
Dutt, R.C. *Socialism of Jawaharlal Nehru.* (New Delhi: Abhinav Publications, 1981).
Dutt, R. Palme. *The Problem of India.* (New York: International Publishers, 1943).
Dutt, Vishnu. *Indira Gandhi: Promises to Keep.* (New Delhi: National, 1980).
Dutt, Ratna. "The Party Representative in Fourth Lok Sabha,"*Economic and Political Weekly*, IV (Annual number, January 1969).
Duvall, Raymond D.and Freeman, John R. "The State and Dependent Capitalism,"*International Studies Quarterly*, XXV, no.1 (March 1981), pp. 99-118.

Easton, David.*The Political System.* (New York: Alfred A. Knopf, 1971).

Echols, John M., III. "Does Socialism Mean Greater Equality? A Comparison of East and West Along Several Major Dimensions,"*American Journal of Political Science*, vol.25, no.1 (1981). pp. 1-26.
Engels, Frederick. "Preface," in Karl Marx, *The Communist Manifesto*. (Chicago: Henry Regnery Company, 1965).
Erdman, Howard L. *The Swatantra Party and Indian Conservatism*. (Cambridge, Great Britain: Cambridge University Press, 1967).
Evans, Peter. *Dependent Development: The Alliance of Multinational, State and Local Capital in Brazil*. (Princeton: Princeton University Press, 1978).

FICCI. *Proceedings of the Fifty-third Annual Session....1980*. (New Delhi: 1980).
FICCI. *Proceedings of the Fifty-fifth Annual Session....1982*. (New Delhi: 1982.
Fic, Victor M. *Peaceful Transition to Communism in India*. (Bombay: Nachiketa Publications, 1969).
Foster-Carter, Aidan. "Korea and Dependency Theory," *Monthly Review*, vol.37 (October 1985).
Franda, Marcus. *India's Rural Development: An Assessment of Alternatives*. (Bloomington: Indiana University Press, 1979).
Franda, Marcus. *Radical Politics in West Bengal*. (Cambridge, Mass: MIT Press, 1971).
Frankel, Francine. *India's Political Economy, 1947-1977*. (Princeton: Princeton University Press, 1978).

Galbraith, John Kenneth. *American Capitalism: The Concept of Countervailing Power*. (Boston: Houghton Mifflin, 1956).
Galbraith, John Kenneth. *Economics and the Public Purpose*. (New York: Houghton Mifflin, 1973).
Galbraith, John Kenneth. *The New Industrial State*. (Boston: Houghton Mifflin, 1978).
Galbraith, John Kenneth. *The Voice of the Poor: Essays in Economic and Political Persuasion*. (Cambridge, Mass: Harvard University Press, 1983).
Gandhi, Indira.*The Speeches and Reminiscences of Indira Gandhi..* (Calcutta: Rupa & Co., 1975).
Geithman, David T. *Fiscal Policies for Industrialization and Development in Latin America*. (Gainesville, Florida: University Press of Florida, 1974).
Gerschenkron, Alexander.*Economic Backwardness in Historical Perspective*. (Cambridge, Mass.: Belknap Press, 1966).
Ghai, Yash (ed.). *Law in the Political Economy of Public Enterprise: African Perspectives*. (Uppsala, Sweden: Scandinavian Institute of African Studies, 1977).
Ghosh, Atulya.*The Split in Indian National Congress*. (Calcutta: Jayanti, 1970).
Ghosh, D.N. *Banking Policy in India: An Evaluation*. (Bombay: Allied Publishers, 1979).
Gillis, Malcolm. "The Role of State Enterprises in Economic Development" *Social Research*, vol.47 (1980), 248-89.
Gold, David A.; Lo, Clarence, Y.H.; and Wright, Erik Olin. "Recent Develop-

ments in Marxist Theories of the Capitalist State," *Monthly Review*, XXVII, no.5 (Oct. 1975), pp.29-43, and no.6 (Nov. 1975), pp. 36-51.

Gopal, Sarvepalli. "The Formative Ideology of Jawaharlal Nehru," in K.N. Panikkar (ed.), *National and Left Movements in India*. (New Delhi: Vikas, 1980).

Gopal, Sarvepalli. *Jawaharlal Nehru: A Biography*: Volume One, 1889-1947. (Cambridge, Mass.: Harvard University Press, 1976).

Gopal, Sarvepalli. *Jawaharlal Nehru: A Biography*: Volume Two, 1947-1956. (Cambridge, Mass.: Harvard University Press, 1979).

Gopal, Sarvepalli. *Jawaharlal Nehru: A Biography*: Volume Three, 1956-1964. (London: Jonathan Cape, 1984).

Gordon, David M.; Edwards, Richard; and Reich, Michael. *Segmented Work, Divided Workers: The Historical Transformation of Labor in the United States*. (Cambridge, UK: Cambridge University Press, 1982).

Gough, Kathleen and Sharma, Hari P.(eds.). *Imperialism and Revolution in South Asia*. (New York: Monthly Review Press, 1973).

Granick, David. *Enterprise Guidance in Eastern Europe*. (Princeton: Princeton University Press, 1975).

Gregor, A. James. *Interpretations of Fascism*. (Morristown, N.J.: General Learning Press, 1974).

Gupta, Ram Chandra. *Lal Bahadur Shastri: The Man and His Ideas*. (Delhi : Sterling Publishers 1966).

Gupta, Vinod. *Anderson Papers: A Study of Nixon's Blackmail of India*. (Delhi: Indian School Supply Depot, 1972).

Ham, Christopher and Hill, Michael.*The Policy Process in the Modern Capitalist State*. (Brighton, Sussex: Wheatsheaf Books, 1984).

Hanson, F.R.*The Breakdown of Capitalism: A History of the Idea in Western Marxism, 1883-1983* (London: Routledge & Kegan Paul, 1985).

Hanson, A.H.*The Process of Planning*. (London: Oxford University Press, 1966).

Hanson, A.H. *Public Enterprise and Economic Development*. (2nd edn., London: Routledge & Kegan Paul, 1965).

Hardgrave, Robert L. *India: Government and Politics in a Developing Nation*. (New York: Harcourt, Brace & World, 1970).

Harris, Nigel. *India-China: Underdevelopment and Revolution*. (New Delhi: Vikas, 1974).

Hart, Henry. *Indira's India: A Political System Reappraised*. (Boulder, Colorado: Westview Press, 1976).

Hayward, Jack and Narkiewicz, Olga A. *Planning in Europe*. (London: Croom Helm, 1978).

Hazari, R.K. *Industrial Planning and Licensing Policy: Final Report*. (New Delhi: Planning Commission, 1967).

Heckscher, Eli F. *Mercantilism* (London: George Allen & Unwin, 1955), Vol.II.

Hirsch, Fred and Goldthorpe, John H. (eds.).*The Political Economy of Inflation*. (Cambridge: Harvard University Press, 1978).

Hirsch, Fred. *Social Limits to Growth* (Cambridge: Harvard University Press, 1976).
Hook, Sidney. *Marx and the Marxists* (New York: D. Van Nostrand, 1955).
Huntington, Samuel P. *Political Order in Changing Societies*. (New Haven: Yale University Press, 1968).

India, Administrative Reforms Commission. *Report on Public Sector Undertakings*. (New Delhi: 1967).
India, Ministry of Industrial Development and Company Affairs. *Report of the Industrial Licensing Policy Inquiry Committee*. (New Delhi: 1969).
India, Ministry of Industry. *Report 1977-78*. (New Delhi: 1978).
India, Planning Commission. *Report of the Committee on Distribution of Income and Levels of Living*. (New Delhi: 1964).
India, Planning Commission. *Sixth Five Year Plan 1980-85*. (New Delhi: 1981).
India (Republic). *Nationalisation of Banks: A Symposium*. (New Delhi: Publication Division, 1970).
India (Republic). *Report of the Monopolies Inquiry Commission*. (New Delhi: 1965).
Indian National Congress. *Resolutions on Economic Policy, Programme and Allied Matters* (1924-1969). (New Delhi: All India Congress Committee, 1969).
International Legal Central (ed.). *Law and Public Enterprise in Asia*. (New York: Praeger, 1976).

Jessop, Bob. "Recent Theories of the Capitalist State," *Cambridge Journal of Economics*, I, no.4 (1977), pp.353-373.
Jha, L.K. "Liberalising the Economy," *Business India*, (May 6-19, 1985).
Jha, L.K. *Shortages and High Prices: The Way Out*. (New Delhi: Indian Book Company, 1976).
Johnson, William A. *The Steel Industry of India*. (Cambridge: Harvard University Press, 1966).
Jha, Prem Shankar. *India: A Political Economy of Stagnation*. (Bombay: Oxford University Press, 1980).
Jones, Leroy P. and Sakong, Il. *Government, Business, and Entrepreneurship in Economic Development: The Korean Case*. (Cambridge, Mass.: Harvard University Press, 1980).
Jones, Leroy P. (ed.). *Public Enterprise in Less Developed Countries*. (Cambridge, UK: Cambridge University Press, 1982).
Joshi, P.C. "Reflections on Marxism and Social Revolution in India," in K.N. Panikkar (ed.) *National and Left Movements in India*. (New Delhi: Vikas, 1980).

Kabra, Kamal Nayan. "Culmination of Drift," *Patriot* (New Delhi), May 20, 1982.
Kabra, Kamal Nayan and Rao, Suresh Rama. *Public Sector Banking*. (New Delhi: People's Publishing House, 1970).

Karunakaran, K.P. *The Phenomenon of Nehru.* (New Delhi: Gitanjali Prakashan, 1979).
Katznelson, Ira. "Considerations on Social Democracy in the United States," *Comparative Politics,* XI (October 1978).
Kaushik, Susheela. *Election in India: Its Social Basis.* (Calcutta: K.P. Bagchi & Co., 1982).
Kautsky, John H. (ed.). *Political Change in Underdeveloped Countries: Nationalism and Communism.* (New York: John Wiley, 1962).
Kautsky, John H. *The Political Consequences of Modernization.* (New York: John Wiley, 1972).
Khera, S.S. *Government in Business.* (New Delhi: National, 1977).
Kidron, Michael. *Foreign Investments in India.* (London: Oxford University Press, 1965).
Kissinger, Henry. *White House Years.* (Boston: Little, Brown, 1979).
Kochanek, Stanley A. *Business and Politics in India.* (Berkeley: University of California Press, 1974).
Kochanek, Stanley A. *The Congress Party of India: The Dynamics of One-Party Democracy.* (Princeton: Princeton University Press, 1968).
Kohli, Atul. "Parliamentary Communism and Agrarian Reform: The Evidence from India's Bengal,"*Asian Survey,* XXIII, no.7 (July 1983), pp. 783-807.
Kothari, Rajni. "The Congress System,"*Asian Survey,* IV, no.12 (1964), 1161-73.
Kothari, Rajni. *Politics in India.* (Boston: Little, Brown, 1970).
Krasner, Stephen D. *Defending the National Interest: Raw Materials and U.S. Foreign Policy.* (Princeton: Princeton University Press, 1978).
Kuipers, S.K. and Lanjouw, G.J. (eds.). *Prospects of Economic Growth.* (Amsterdam: North-Holland, 1980).
Kumar, Dharma (ed.). *The Cambridge Economic History of India.* (Cambridge, UK: Cambridge University Press, 1983), II.
Kurian, K. Mathew (ed.). *India—State and Society: A Marxian Approach.* (Bombay: Orient Longmans, 1975).
Kumar, Narendra (ed.).*Bank Nationalisation in India: A Symposium.* (Bombay: Lalvani Publishing House, 1969).
Kumaramangalam, S. Mohan. *Coal Industry in India: Nationalisation and Tasks Ahead* (New Delhi: Oxford and IBH Publishing Co., 1973).
Kurien, C.T. "The Budget and the Economy," *Economic and Political Weekly,* XVII, (March 20, 1982).
Kurien, C.T. *Indian Economic Crisis: A Diagnostic Study* (Bombay: Asia Publishing House, 1969).

Lamont, Douglas F. *Foreign State Enterprises.* (New York: Basic Books, 1979).
Lane, David. *The End of Social Inequality? Class, Status and Power under State Socialism.* (London: George Allen & Unwin, 1982).
Lange, Oskar and Taylor, Fred M. *On the Economic Theory of Socialism.* (New York: McGraw Hill, 1964).
LaPalombara, Joseph L. (ed.). *Bureaucracy and Political Development.* (Princeton: Princeton University Press, 1963).

Select Bibliography

Lee, Edmund. "Economic Reform in Post-Mao China: An Insider's View," *Bulletin of Concerned Asian Scholars*, XV, no.1 (1983).

Levine, Andrew and Wright, Erik Olin. "Rationality and Class Struggle,"*New Left Review*, no.123 (1980).

Levine, Herbert S. and Bergson, Abram (ed.). *The Soviet Economy: Toward the Year 2000*. (London: George Allen & Unwin, 1983).

Lindberg, Leon N. et al. (ed.).*Stress and Contradiction in Modern Capitalism: Public Policy and the Theory of the State*. (Lexington, Mass.: Lexington Books, 1975).

Lindblom, Charles E. *Politics and Markets: The World's Political-Economic Systems* (New York: Basic Books, 1977).

List, Friedrich. *The National System of Political Economy*. (New York: Augustus M. Kelley, 1966).

Madan, N.L. *Congress Party and Social Change*. (Delhi: B.R. Publishing Corpn., 1984).

Maddison, Angus. *Class Structure and Economic Growth: India and Pakistan Since the Moghuls*. (London: George Allen & Unwin, 1971).

Malaviya, H.D. *Socialist Ideology of Congress: A Study in its Evolution*. (New Delhi: 1966).

Mandel, Ernest. "On the Nature of the Soviet State,"*New Left Review*, no.108 (1978).

Mankekar, D.R. *Lal Bahadur Shastri*. (New Delhi: Publications Division, 1973).

Manley, John F. "Neo-Pluralism: A Class Analysis of Pluralism I and Pluralism II," *American Political Science Review*, vol.77, no.2 (June 1983), pp. 368-83.

Markovits, Claude. *Indian Business and Nationalist Politics 1931-1934*. (Cambridge: Cambridge University Press, 1985).

Marx, Karl. *Capital*. (Moscow: Progress Publishers, 1954), Vol. I.

Marx, Karl and Engels, Frederick. *Manifesto of the Communist Party*. (Moscow: Progress Publishers, 1975).

Masani, Zarrer. *Indira Gandhi: A Biography* (London: Hamish Hamilton, 1975).

Mehta, V.R. *Ideology, Modernization and Politics in India*. (New Delhi: Manohar, 1983).

Mellor, John W. (ed.). *India: A Rising Middle Power*. (Boulder, Colorado: Westview Press, 1979).

Miliband, Ralph. "Poulantzas and the Capitalist State," *New Left Review*, no.82 (Nov-Dec.1973), pp.83-92.

Miliband, Ralph and Saville, John (eds.). *The Socialist Register 1980*. (London: Merlin Press, 1980).

Miliband, Ralph. *The State in Capitalist Society*. (New York: Basic Books, 1969).

Miliband, Ralph. "State Power and Class Interests,"*New Left Review*, no.138 (1983), pp.57-68.

Mills, C. Wright. *The Power Elite*. (New York: Oxford University Press, 1957).

Mishra, Girish. *Public Sector in Indian Economy*. (New Delhi: Communist Party Publication, 1975).

Mohota, R.D.*Textile Industry and Modernisation.* (Bombay: Current Book House, 1976).
Moore, Barrington. *The Social Origins of Democracy and Dictatorship: Lord and Peasant in the Making of the Modern World.* (Boston: Beacon Press, 1966).
Moraes, Frank *Jawaharlal Nehru: A Biography.* (New York: MacMillan, 1956).
Morris-Jones, W.H. *The Government and Politics of India.* (London: Hutchinson University Library, 1964).
Morris-Jones, W.H. *Parliament in India.* (Philadelphia, Penn.: University of Pennsylvania Press, 1957).
Mukherjee, Aditya. "Indian Capitalist Class and the Public Sector 1930-1947," *Economic and Political Weekly*, XI, no.3 (January 17, 1976).
Mukherjee, Pranab. *Beyond Survival: Emerging Dimensions of Indian Economy.* (New Delhi: Vikas, 1984).
Mullins, Willard A. "On the Concept of Ideology in Political Science," *American Political Science Review*, vol.66, no.2 (June 1972), pp. 498-510.
Myrdal, Gunnar. *Asian Drama* (New York: Pantheon, 1968), vol.II.
McCully, Bruce. *English Education and the Origins of Indian Nationalism.* (New York: Columbia University Press, 1940).

Namboodiripad, E.M.S. *Economics and Politics of India's Socialist Pattern.* (New Delhi People's Publishing House, 1966).
Namboodiripad, E.M.S. *Indian Planning in Crisis.* (New Delhi: National Book Centre, 1982).
Nanda, B.R. (ed.). *Socialism in India,* (Delhi: Vikas, 1972).
Nayar, Baldev Raj. "Business Attitudes Toward Economic Planning in India," *Asian Survey*, XI, no.9 (1971), pp. 850-65.
Nayar, Baldev Raj. *India's Quest for Technological Independence.* vol.1 - *Policy Foundation and Policy Change.* (New Delhi: Lancers Publishers, 1983).
Nayar, Baldev Raj. *India's Quest for Technological Independence.* vol.2 - *The Results of Policy.* (New Delhi, Lancers Publishers, 1983).
Nayar, Baldev Raj. *The Modernization Imperative and Indian Planning.* (New Delhi: Vikas, 1972).
Nayar, Baldev Raj. "Political Mainsprings of Economic Planning in the New Nations: The Modernization Imperative versus Social Mobilization," *Comparative Politics*, VI, no.3 (April 1974).
Nayar, Baldev Raj. *Violence and Crime in India: A Quantitative Study.* (New Delhi: MacMillan, 1975).
Nayar, Kuldip. *India After Nehru.* (New Delhi: Vikas, 1975).
Nayar, Kuldip. *India: The Critical Years.* (New Delhi: Vikas, 1971).
Nayar, Kuldip. *The Judgement.* (New Delhi: Vikas, 1977)
Nehru, Jawaharlal. *An Autobiography.* (London: John Lane The Bodley Head, 1936).
Nehru, Jawaharlal. *The Discovery of India.* (New York: John Day Company, 1946).
Nehru, Jawaharlal. *Planning and Development: Speeches of Jawaharlal Nehru,*

1952-56. (New Delhi: Publications Division, New Delhi).
Nehru, Jawaharlal. *Speeches:* volume 1, Sept. 46-May 49. (New Delhi: Publications Division, 1977).
Nehru, Jawaharlal. *Speeches:* volume 3, March 1953 - August 1957. (New Delhi: Publications Division, 1958.
Nehru, Jawaharlal. *Speeches:* volume 5, March 1963 - May 1964. (New Delhi: Publications Division, 1968).
Nehru, Jawaharlal. *Toward Freedom.* (New York: John Day Company, 1942).
Nigam, Raj K. *Public Sector for Public Welfare.* (Lecture delivered in Patna)(New Delhi: Documentation Centre, 1983).
Nordlinger, Eric A. *On the Autonomy of the Democratic State.* (Cambridge, Mass.: Harvard University Press, 1981).
Norman, Dorothy (ed.). *Nehru: The First Sixty Years.* (New York: John Day Company, 1965).
Noronha, R.P. *A Tale Told by an Idiot.* (New Delhi: Vikas, 1976).
Nove, Alec. *Political Economy and Soviet Socialism.* (London: George Allen & Unwin, 1979).
Nove, Alec. *The Soviet Economic System.* (London: George Allen & Unwin, 1977).

O'Connor, James. *The Fiscal Crisis of the State.* (New York: St. Martin's Press, 1973).
Organski, A.F.K. *The Stages of Political Development.* (New York: Alfred A. Knopf, 1965).
Overstreet, Gene D. and Windmiller, Marshall. *Communism in India.* (Berkeley: University of California Press, 1960).

Panandiker, V.A. Pai and Sud, Arun. *Changing Political Representation in India.* (New Delhi: Uppal Publishing House, 1983).
Panikkar, K.N. (ed.) *National and Left Movements in India.* (New Delhi: Vikas, 1980).
Pantham, Thomas. "Elites, Classes and the Distortions of Economic Transition in India," in Sachchidananda and A.K. Lal (eds.), *Elite and Development.* (New Delhi: Concept Publishing Company, 1980), pp. 71-96.
Parthasarathy, G. *Dilemmas of Marketable Surplus: The Indian Case.* (Waltair, Visakhapatnam: Andhra University Press, 1979).
Patankar, Bharat and Omvedt, Gail. "The Bourgeois State in Post-Colonial Formations,"*Insurgent Sociologist*, IX, no.4 (Spring 1980), pp.23-38.
Paul, Sharda. *1980 General Elections.* (New Delhi: Associated Publishing House, 1980).
Peter, Flora and Heidenheimer, Arnold J. (eds.). *The Development of Welfare States in Europe and America* (New Brunswick, N.J.: Transaction Books, 1981).
Petras, James. *State Capitalism and the Third World: Critical Perspectives on Imperialism and Social Class in the Third World.* (New York: Monthly Review Press, 1978).

Poulantzas, Nicos. "On Social Classes,"*New Left Review*, no.78 (1973).
Poulantzas, Nicos. *Political Power and the Social Classes*. (London: New Left Books, 1973).
Poulantzas, Nicos. "The Problem of the Capitalist State," *New Left Review*, no.58 (Nov-Dec. 1969), pp. 67-78.
Provizer, Norman W. (ed.). *Analyzing the Third World*. (Cambridge, Mass.: Schenkman Publishing Company, 1978).
Punekar, S.D. (ed.). *Economic Revolution in India*. (Bombay: Himalaya Publishing House, 1977).

Rahman, M.M. *The Congress Crisis*. (New Delhi: Associated Publishing House, 1970).
Raj, K.N. "The Politics and Economics of Intermediate Regimes," *Economic and Political Weekly*, VIII, no.27 (July 7, 1973), pp. 1189-98.
Ralf Dahrendorf. *Class and Class Conflict in Industrial Society*. (Stanford, Calif.: Stanford University Press, 1959).
Rao, M.B. (ed.). *Documents of the History of the Communist Party of India*. (New Delhi: People's Publishing House, 1976). vol.VII (1948-1950).
Rao, R.P. *The Congress Splits* (Bombay: Lalvani Publishing House. 1971).
Rao, V.K.R.V. et al. *Inflation and India's Economic Crisis* (New Delhi: Vikas, 1973).
Rao, V.K.R.V. and Joshi, P.C. "Some Fundamental Aspects of Socialist Transformation in India," *Man & Development*, IV, no.2 (1982).
Rao, V.K.R.V. "Some Fundamental Aspects of Socialist Change in India," in Ashok V. Bhuleshkar (ed.),*Growth of Indian Economy in Socialism*. (Bombay: Oxford and IBH Publishing Co., 1975), pp. 485-501.
Ray, Baren. *India: Nature of Society and Present Crisis*. (New Delhi: Intellectual Book Corner, 1973).
Ray, Rajat K. *Industrialization in India: Growth and Conflict in the Private Corporate Sector 1914-47*. (Delhi: Oxford University Press, 1979).
Reich, Michael. *Segmented Work, Divided Workers: The Historical Transformation of Labor in the United States*. (Cambridge, UK: Cambridge University Press, 1982).
Reisman, D.A. *Adam Smith's Sociological Economics*. (London: Croom Helm, 1976).
Robinson, Joan. "Introduction," to Michal Kalecki,*Essays on Developing Economies*. (Hassocks, Sussex: Harvester Press, 1976),
Roemer, John E. *A General Theory of Exploitation and Class*. (Cambridge, Mass.: Harvard University Press, 1982).
Rosen, George. *Democracy and Economic Change in India*. (Berkeley: University of California Press, 1966).
Rosenthal, Donald B. *The Limited Elite: Politics and Government in Two Indian Cities*. (Chicago: University of Chicago Press, 1970).
Rostow, W.W. *Politics and the Stages of Growth*. (Cambridge: Cambridge University Press, 1971).
Roxborough, Ian. *Theories of Underdevelopment*. (Atlantic Highlands, N.J.:

Humanities Press, 1979).
Roy, Ajit. *Monopoly Capitalism in India.* (Calcutta: Naya Prokash, 1976).
Roy, Ajit. "Sharers in Indian State Power," in K. Mathew Kurian (ed.), *India — State and Society: A Marxian Approach* (Bombay: Orient Longmans, 1975), pp. 129-43.
Rudolph, Lloyd I. and Susanne H. *Agrarian Power and Agricultural Productivity in South Asia.* (Delhi: Oxford University Press, 1984).
Rustow, Dankwart A. *A World of Nations: Problems of Political Modernization.* (Washington, D.C.: Brookings Institution, 1967).

Sachchidananda and Lal, A.K. *Elite and Development.* (New Delhi: Concept Publishing Co., 1980).
Sachs, Ignacy. *Problems of Public Sector in Underdeveloped Economies.* (Bombay: Asia Publishing House, 1964).
Scase, Richard (ed.).*The State in Western Europe.* (New York: St. Martin's Press, 1980).
Schatz, Sayre P. "Socializing Adaptation: A Perspective on World Capitalism," *World Development*, XI, no.1 (1983).
Schumpeter, Joseph A. *Capitalism, Socialism and Democracy.* (London: George Allen & Unwin, 1950).
Scott, Peter Dale. "Peace, Power and Revolution,"*Alternatives*, IX (1983-84).
Sen, Anupam. *The State, Industrialization and Class Formations in India: A Neo-Marxist Perspective on Colonialism, Underdevelopment and Development.* (London: Routledge & Kegan Paul, 1982).
Sen, Asok. "Bureaucracy and Social Hegemony," in *Essays in Honour of Prof. S.C. Sarkar.* (New Delhi: People's Publishing House, 1976), pp. 667-85.
Sen, Mohit (ed.). *Documents of the History of the Communist Party of India.* (New Delhi: People's Publishing House, 1977), vol.VIII (1951-56).
Shah, K.T. (ed.). *Industrial Finance* (Report of the Sub-Committee). (Bombay: Vora & Co., 1948).
Shah, K.T. *National Planning, Principles & Administration.* (Bombay: Vora & Co., 1948).
Sharma, R.S. (ed.). *Indian Society: Historical Probings in Memory of D.D. Kosambi.* (New Delhi: People's Publishing House, 1974).
Sharma, Harish C. *Nationalisation of Banks in India.* (Agra: Sahitya Bhawan, 1970).
Sharpe, Myron E. *John Kenneth Galbraith and the Lower Economics.* (White Plains, N.Y.: International Arts and Sciences Press, 1974).
Shastri, Lal Bahadur. *Selected Speeches of Lal Bahadur Shastri* (June 11, 1964 to January 10, 1966). (New Delhi: Publications Division, 1974).
Shastri, Sunil and Bhalla, Chander M. (eds.) . *Lal Bahadur Shastri: Commemoration Volume.* (New Delhi: 1970).
Shepherd, William G. *Public Enterprise: Economic Analysis of Theory and Practice.* (Lexington, Mass.: Lexington Books, 1976).
Shepsle, Kenneth and Weingast, Barry R. "Political Solutions to Market Problems,"*American Political Science Review*, vol.78, (1984), 417-33.

Shirokov, G.K. *Industrialization of India.* (Moscow: Progress Publishers, 1973).
Shonfield, Andrew. *In Defence of the Mixed Economy.* (Oxford: Oxford University Press, 1984).
Shonfield, Andrew. *Modern Capitalism: The Changing Balance of Public and Private Power.* (London: Oxford University Press, 1965).
Shonfield, Andrew. *The Use of Public Power.* (Oxford: Oxford University Press, 1982).
Singh, Charan. *Economic Nightmare of India.* (New Delhi: National 1981).
Singh, Charan. *India's Economic Policy: The Gandhian Blueprint.* (New Delhi: Vikas, 1978).
Singh, Khushwant. *Indira Gandhi Returns.* (New Delhi: Vision Books, 1979).
Singh, Satindra. *Communists in Congress: Kumaramangalam's Thesis.* (Delhi: D.K. Publishing House, 1973).
Sinha, Sachchidanand. *The Permanent Crisis in India: After Janata What ?* (New Delhi: Heritage, 1978).
Skidelsky, Robert (ed.). *The End of the Keynesian Era: Essays on the Disintegration of the Keynesian Political Economy.* (New York: Holmes & Meir Publishers, 1977).
Skocpol, Theda. *States and Social Revolutions.* (Cambridge: Cambridge University Press, 1966).
Smith, Adam. *An Inquiry into the Nature and Causes of the Wealth of Nations.* (London: Methuen & Co., 1925).
Sobhan, Rehman and Ahmed, Muzaffer. *Public Enterprise in an Intermediate Regime: A Study in the Political Economy of Bangladesh.* (Dacca: Bangladesh Institute of Development Studies, 1980).
Solo, Robert A. *The Political Authority and the Market System.* (Cincinnati: South-Western Publishing Co., 1974).
Sood, P. *Indira Gandhi and the Constitution.* (New Delhi: Marwah Publications, 1985).
Stepan, Alfred. *The State and Society: Peru in Comparative Perspective.* (Princeton: Princeton University Press, 1978).
Stephens, John D. *The Transition from Capitalism to Socialism* (Atlantic Highlands, N.J.: Humanities Press, 1980).
Streeten, Paul. "A Problem to Every Solution," *Finance and Development,* June 1985.
Subramaniam, V. *The Managerial Class of India* (New Delhi: All India Management Association, 1971).
Suntharalingam, R. *Indian Nationalism: An Historical Analysis.* (New Delhi: Vikas, 1983).
Szymanski, Albert. *The Capitalist State and the Politics of Class* (Cambridge, Mass.: Winthrop Publishers, 1978).

Tandon, B.C. *Management of Public Enterprises.* (Allahabad: Chaitanya Publishing House, 1978).
Thakurdas, Purshotamdas et al. *Memorandum Outlining a Plan of Economic Development for India.* (London: Penguin Books, 1945).

Thapar, Romila (ed.). *Situating Indian History.* (New Delhi: Oxford University Press, forthcoming).
Therborn, Goran. "The Prospects of Labour and the Transformation of Advanced Capitalism," *New Left Review,* no.145 (1984), 5-38.
Therborn, Goran. *What Does the Ruling Class Do When It Rules?* (London: New Left Books, 1978).
Tomlinson, B.R. *The Political Economy of the Raj 1914-1947: The Economics of Decolonization in India.* (London: MacMillan, 1979).
Trivedi, M.L. *Government and Business.* (Bombay: Multi-Tech. Publishing Co., 1980).
Truman, David B. *The Governmental Process: Political Interests and Public Opinion* (New York: Knopf, 1951).
Tucker, Robert. *The Marxian Revolutionary Idea.* (New York: W.W. Norton, 1969).
Tupper, Alan. "The State in Business,"*Canadian Public Administration,* XX, no.1 (1981).

Vakil, C.N. (ed.). *Industrial Development of India: Policy and Problems.* (New Delhi: Orient Longmans, 1973).
Vakil, C.N. *Janata Economic Policy: Towards Gandhian Socialism.* (New Delhi: MacMillan, 1979).
Vakil, C.N. and Brahmananda, P. R. *Memorandum on Inflation Reversal and Guaranteed Price Stability.* (Bombay: Vora and Co., 1977).
Vakil, C.N. and others. *A Policy to Contain Inflation with Semibombla: Submitted to the Prime Minister on Behalf of 140 Economists.* (Bombay: Commerce, 1974).
Venkatasubbiah, H. *Enterprise and Economic Change: 50 Years of FICCI.* (New Delhi: Vikas, 1977).
Virmani, M.R. (ed.). *Public Enterprises from Nehru to Indira Gandhi: Objectives, Achievements and Prospects.* (New Delhi: Centre for Public Sector Studies, 1981).

Weiner, Myron. "Congress Restored: Continuities and Discontinuities in Indian Politics," *Asian Survey,* XXII, no.4 (1982).
Weiner, Myron. *Party Building in a New Nation: The Indian National Congress.* (Chicago: University of Chicago Press, 1967).
Wilson, A. Jeyaratnam and Dalton, Dennis (eds.). *The States of South Asia.* (London: C. Hurst & Company, 1982).
Wolfe, Alan. "Has Social Democracy a Future?"*Comparative Politics,* XI (October 1978).

Zaidi, A. Moin. *Full Circle, 1972-1975* (New Delhi: Michiko and Panjathan, 1975).
Zaidi, A. Moin. *The Great Upheaval 1969-1972* . (New Delhi: Orientalia, 1972).

Index

Adhikari, G. 86
agitations. *See* social disorder
agriculture :
    favoured by Janata Party, 341
    favoured by Shastri, 255
    fostered by Indira Gandhi, 260, 275-77
    neglect by Nehru, 248, 254-55
    public sector position in, 369, 374-75
agricultural bourgeoisie, 104-107, 116
Agricultural Prices Commission, 255
Ahmed, Muzaffer, *See* Sobhan
aircraft industry, 377
airlines nationalization, 193
AICC Economic Programme Committee, 181, 183, 283, 341
All India Congress Committee. *See* Congress party
All India Rural Credit Survey, 283
All India Trade Union Congress, 148, 214, 242-43
Almond, Gabriel 28, 63, 67
Alavi, Hamza 37, 41, 45-46, 119
Amin, Samir 21
"Andhra thesis", 78-79, 80
Arora, Satish 112
Asiatic mode of production, 73
Associated Chambers of Commerce, 223, 350
Avadi resolution, 194, 195, 299

Baer, Werner 369
Bagchi, Amiya Kumar 132, 326n
Bangladesh crisis, 258, 295, 296, 297, 318, 328
banking:
    contribution of private sector to, 284-85
    dominant role of public sector in, 372
    nationalisation of, 181, 195, 250, 251, 263, 269, 271, 272, 282-89, 292, 355
    social control over, 261, 284-86
Baran, Paul A. 45
Bardhan, Pranab 115, 119, 120, 123, 207
Barooah, D.K. 336, 338
Barry, Norman P. 392
Basic industries, 134, 135, 193, 195, 197, 199, 201, 202, 222, 233, 234, 237-38, 248, 251, 254, 301, 358n, 386
Bernstein, Eduard 4
Bettelheim, Charles 15, 95
Bharat Heavy Electricals Limited (BHEL) 376
Bharatiya Jan Sangh 93, 117, 249, 262, 272, 273, 288, 289, 293, 294, 298, 331, 341
Bhatt, V.V. 207
Bhuvaneshar resolution, 195, 235, 250-51, 284, 299
big business :
    as target of policy, 198, 287-88, 289-92, 297, 320, 329, 330, 342-43, 355
    role of, 78-79, 85-86, 88-89, 95, 96-99, 104-105, 112, 115, 120, 122, 123, 236-43, 247n, 320-22, 327n, 338
    See also business
Birla, G.D. 86, 162, 163, 164, 172n, 228, 232
Birlas, 232, 270, 288, 291, 292-93, 304, 307, 318
Block, Fred 39, 392
Bolsheviks, 3, 26
Bombay Plan, *See* Tata-Birla Plan
Bonapartist state, 35, 46
Bose, Subhash Chandra, 188
Bourgeois-landlord coalition :
    CPI concept of, 77-79, 236, 239, 344
    CPI(M) concept of, 96-97, 102
    CPI(M-L) concept of, 107-109
    collaboration with imperialism by, 77-79, 80, 96-99, 107-110
    lack of social hegemony of, 114
bourgeoisie:
    as agent of industrialization, 21
    in Soviet Union, 15
    the state as, 15-16, 61n
    weakness of, 110, 114
    See also big business, business
Brecher, Michael 147, 170n, 228, 229, 257
Brenner, Robert 21
Brus, Wlodzimierz 15-19
bureaucracy, 114-115, 133, 219
"bureaucratic-authoritarian" model, 41-42, 335, 339
business :
    advance of, 131-32, 168

alienation of, 234-35, 249, 251, 262, 292-93, 308-309, 330
attack on, 264, 289-92, 342
comprador role of, 107-109, 110, 161, 179-80
CPI position on, 71-96, 236-43, 353, 354
CPI(M) position on, 96-104
divisions in, 231-32, 234, 288, 292-93
"dual role" of, 83, 87, 89, 92, 97, 99, 105, 107, 239-40, 301
negative image of, 133, 162, 179, 224, 246n, 306, 333, 392
non-comprador role of, 97, 105, 109-110, 180
opposition to economic power of, 201, 217-18, 263, 284
opposition to big business by, 85-86, 88, 115, 123, 288
opposition to imperialism by, 82-84, 100, 123, 161-62, 244
opposition to Nehru model by, 220-35, 236, 244, 249,
participation in politics by, 231-32, 234-35, 262, 263-64, 284, 320, 324n, 340, 384
posture on government policy under Indira Gandhi, 262, 288, 291, 292-93, 296, 301, 305, 308-309, 317-18, 320, 330, 338, 343, 349, 350-51, 355
posture on government policy under Janata Party, 342
posture on government policy under Nehru, 180-82, 196, 201, 220-35, 236, 244, 249, 251, 389
posture on government policy under Shastri, 253
posture on Nehru, 150, 153, 154, 162-64
role in nationalist movement, 73, 137-43, 161-62, 170n, 172n, 173n, 223
role in planning before independence, 156, 164-67
role of, 104-107, 175-80, 236-43, 290, 308-309, 317, 320-22, 327n, 354, 384, 389
weakness of, 111, 162, 168, 178-80, 189, 193, 208, 216, 223-24, 227, 228, 246n, 288, 291, 292, 301, 306, 308, 309, 320-21, 342-43, 354, 355, 389
See also big business

Calcutta Electric Supply Company, 222
capitalism:
  Adam Smith on, 4-6
  attitudes toward state intervention in, 161
  attraction of, 10-11, 393
  breakdown in, 9, 53n
  crisis in 1970s, 11
  economic behaviour in, 160-61
  its lack of appeal, 13-14, 16
  Marx's view of, 2-4
  modification of laissez faire, 6-14
  persistence of, 7

prerequisite for socialism, 13
  relationship to democracy, 12
capitalist class. *See* business
caste, role of, 113, 133, 137, 138, 139, 178
cement industry, 350, 352
Chakravarty, S. 207
Chanana, Charanjit 351, 352
Chandra, Pipan 141-42, 160, 162, 163, 164, 170n
Charan Singh 106, 340-41, 343, 344
Chattopadhyaya, Boudhayan 140
Chattopadhyaya, D.P. 313, 314
Chaudhuri, S. (Finance Minister) 260
Chavan, Y.B. 297
chemical industry, 375
Chilcote, Ronald H. 119
China, 19, 26-28, 143, 191-92, 248, 328, 393
class:
  and the state in Third World, 41-50, 61n
  bipolarization, 2, 7, 33
  changing balance of power, 175-78, 287, 334
  concept in Marxism, 3, 119
  decomposition of, 7
  in CPI analyses, 71-96
  in India, 120-121, 174-80, 208-209, 213-20, 293-95, 299, 335, 344-45, 354-55, 389
  in nationalist movement, 139-43, 168
  weakness of classes, 121, 123
  See also big business, business, intermediate strata, new middle class, peasantry
coal industry:
  management by private sector of, 302-303
  nationalization of, 298-303, 308-309, 329
  public sector role in, 371
Colombo Plan, 225
colonial rule:
  impact of, 130-35, 168, 202
  laissez faire policy under, 131
  planning under, 134, 168
Comintern, 73-74
"commanding heights", 202, 206, 207, 217, 244, 261, 274, 284, 286, 287, 316, 320, 336, 347, 351, 360, 361, 386
communications, 372
communism, 3, 4, 15-20, 26
Communist Party of India:
  analysis of the Indian state, 71-96, 120-23, 220, 221, 236-43, 244
  appraisal of Nehru, 77, 78, 140, 159, 160, 237
  appraisal of Patel, 77, 80
  as co-sharer in state power, 121
  as serving interests of intermediate strata, 241, 245, 247n
  collaboration with Congress government by, 81, 91, 214, 239, 241, 263, 287, 289, 293, 299-301, 335, 340, 390
  collaboration with imperialism by, 74, 141
  criticism of Shastri by, 256
  critiques of position of, 94-96, 239, 241-42,

Index

244
  divisions within, 84-87, 241
  electoral performance of, 262, 293
  perspective on efficacy of mass pressure, 88-89
  perspective on parliamentary system, 88-89, 93
  position on government policy, 74-87, 236-43, 287, 298, 303, 306, 308, 313-14, 321-22, 323n, 337, 338-39, 389
  position on imperialism, 74, 140-41, 237-40
  position on public sector, 84-87, 88-94, 236-43, 244-45, 252, 354, 386
  position on the Nehru model, 236-43
  posture on Congress party, 73-76, 91, 140-43, 236-37, 239-40, 249
  relationship with Mrs. Indira Gandhi, 261, 262, 282, 287, 345
  role in nationalist movement, 73-74, 140-41, 142, 160
  shift to constitutionalism, 79-87, 236
  similarity with views of Nehru, 239, 241, 243, 244, 321
  social basis of, 236
  strategy of infiltration into Congress party, 263, 289, 298-301, 321, 331, 334
  strategy of unity and struggle, 240-43, 300-301
Communist Party of India (Marxist):
  analysis of the Indian state, 96-107, 120-23, 221
  appraisal of public sector, 97-104
  as co-sharer in state power, 97, 121
  compromising with peasantry, 106-107
  conflict with CPI, 289
  critiques of CPI(m) position, 104-107
  electoral performance of, 262, 293
  perspective on government policy, 95, 97-99, 288, 298, 301, 303, 306, 308, 313-14, 322, 331, 337, 338
  perspective on mass pressure, 97
  perspective on parliamentary system, 97, 100-102
Communist Party of India (Marxist-Leninist):
  analysis of the Indian state, 107-109, 120, 123
  appraisal of public sector, 108
  critiques of position of, 109-110
  posture on Congress party, 107
concentration of economic power, 13, 181, 189, 194, 196, 198, 203, 284, 287, 290, 296, 297, 317, 327n
Congress of Oppressed Nationalities, 147
Congress party:
  acceptance of socialism by, 192-95, 250-52
  appraisal by Marxists of, 71-115, 188-89, 220
  conflict in Indira Gandhi period, 261-75, 277-78
  conflict in Nehru period, 182, 186-88, 250, 388
  divisions within, 241, 261, 284-86, 334-35, 390
  economic policy, 181, 192-95, 250-52, 261, 268-74, 283, 284, 295-97, 317, 332, 336-37, 345, 386
  electoral performance of, 262, 263, 264, 267-68, 293, 390
  failure in implementing socialist policies, 267
  ideological orientation of, 181-82, 185-86, 188-89, 192-95, 207-209, 214-20, 250-52, 253, 263-75, 277-78, 284-86, 289-301, 321, 344-45, 349, 386
  leadership interests of, 215-20
  left wing, 183, 186, 187, 188, 208, 253, 263ff, 284, 286, 289-301, 334, 335, 337-38
  nature of, 63-71, 175-78, 187, 188-89, 248, 252, 274, 280-81n, 299
  right wing, 184, 186, 187, 188, 189, 253, 390
  shift of power in, 251-52, 262, 263-64, 272-74, 289, 293-95, 298-301, 334, 335, 337-38
  social basis of, 175-80, 209, 214-20, 251-52, 293-95, 299, 335, 344-45, 353, 355
  split in, 262, 263, 267, 273, 277, 390
  status in political system of, 63-71, 175, 187, 188, 248-49, 262, 293, 339, 340, 344, 362
  weaknesses of, 66
Congress Forum for Socialist Action, 263, 301, 334
Congress Socialist Forum, 257
Congress Socialist Party, 138, 153, 341
"Congress system", 64, 66, 91
Congress (O) *See* Syndicate
consummatory 'mixed economy", 129
controls on private sector, 113, 116, 117, 133, 165-67, 168, 181, 201, 219, 224-35, 248, 251, 289-92, 311, 316, 320, 321, 331, 342, 343, 355, 369, 374, 384, 390
  relaxation of, 332, 338, 348-50, 354, 355
cooperative farming, 195, 215, 233, 249, 251
copper industry, nationalization of, 307-309
corporation, the rise of, 7-8
corruption, 67-68
countervailing power
  of labour, 7-8
  of public sector, 197, 214-20, 234, 235, 238-39, 241, 243, 244, 250, 288, 292, 320, 342
criminalization of politics, 67-68

Dahl, Robert 12
Dahrendorf, Ralf 7
Das Gupta, Ranjit 170n
Dasgupta, Biplab 95-96, 109-110, 120
DCM (Delhi Cloth Mills), 372
defence expenditures, 248, 328

deficit financing, 328, 330
deindustrialization, 131, 344
dependency theory, 20-22, 44-45, 57n
Desai, Ashok 117-118, 209, 256, 262
Desai, Morarji 250, 253, 266, 268, 269, 270, 272, 277, 285, 286, 340, 343
Deshmukh, C.D. 283
devaluation, 260, 261, 284, 318, 319, 355
development:
 induced and spontaneous, 1
 Indian model of, 63-65
 See also state
Dhar, D.P. 301, 317
Dhawan, R.K. 338
Dhebar, U.N. 188
dialectic (the) in Marxism, 2
 in Marxism, 2
diffusionism, 20-22
displacement-effect hypothesis, 53n
Djilas, Milovan 16
Dravida Munnetra Kazhagam (DMK), 282
Dutt, R.C. 137
Dutt, S. 291
Duvall, Raymond D. 44-45

economic crisis:
 under Indira Gandhi, 260-261, 263, 328-35, 354
 under Nehru, 180-81, 260, 274
 under Shastri, 254, 260, 274
economic policy:
 shift in course of, 187-88, 193-95, 231, 319-29, 328, 331-32, 344, 353-56, 388, 390, 391
 under Indira Gandhi, 260-61, 263, 272, 282-322, 331-39, 344-56, 356n, 386
 under Rajiv Gandhi, 326-327n, 355-56
 under Janata Party, 339-44
 under Nehru, 174-209, 263, 316
 under Shastri, 253-56, 316
economic liberalization:
 under Indira Gandhi, 117, 260, 284, 291, 319-20, 331-39, 342, 348-51, 353, 355, 358n, 392
 under Rajiv Gandhi, 326-327n, 355-56
 under Janata Party, 342, 343
economic planning *See* Five Year Plans
economic stagnation, 115-118, 261, 290, 329-30, 339, 353, 357n
elections:
 by-elections, 249-50
 fifth, 293-95, 320, 324n, 331
 first, 188
 fourth, 261, 263, 264, 288, 293, 330, 390
 second, 249,
 seventh, 344
 sixth, 339, 343
 third, 249

electricals, 376-77
elites:
 role in Indian development, 63-65, 136
 ruralization of, 175-78
Emergency, 259, 335-39, 357n, 391
employment, 131-32
Engels, Frederick 33, 40
Ernakulam, AITUC session at, 242-43

fascism, 26, 45, 67, 147, 151, 152, 331
Federation of Indian Chambers of Commerce and Industry. *See* FICCI
Fernandes, George 331, 341, 342, 343
FICCI, 223-35
Five Year Plans:
 Fifth, 250
 First, 190-91, 192, 196, 197, 226
 Fourth, 260
 Second, 83, 191, 192, 195-98, 228-31, 236-43, 263, 304
 Sixth, 347-48
 Third, 197-98, 232-34, 249, 263
Food Corporation of India, 297, 317
foreign enterprise, 199, 223
Forum for Free Enterprise, 223
Forward Bloc, 188
Frank, Andre Gunder 20, 21
Frankel, Francine 280n
Freeman, John R. 44-45

Gadgil, D.R. 185, 197
Galbraith, John Kenneth 11, 13
Gandhi, Mrs. Indira:
 alliance with the Left, 264, 271, 272, 289
 American view of, 258
 as representative of intermediate strata, 117
 attack by the Left on, 260-61
 attitude to public sector, 117, 271, 351-53
 attitude towards Soviet Union, 258
 attitude toward United States, 258, 318, 328-29
 authoritarianism as characteristic of, 259, 293, 326n, 341
 background of, 257
 charge of being communist, 262, 273, 334
 commitment to democracy, 259
 commitment to self-reliance, 258
 commitment to national interest, 258, 264
 commitment to socialism, 258-60, 264-75, 277-78, 281n, 338
 comparison with Nehru, 275-77, 322
 conflict with Syndicate, 263-75, 277-78
 criticism of, 67-70
 criticism of Shastri by, 256, 257, 260
 disenchantment with nationalization, 319, 336-37, 352-53
 economic policy under, 260-61, 263, 272, 282-322, 331-39, 344-56

Index

evaluation of achievements of, 275-77, 339
ideological orientation of, 256-60, 262, 264-75, 277-78, 280n, 281n, 331, 334-35, 338, 344, 390
Lohia's view of, 274
minority government of, 273-74, 282
Morarji Desai's view of, 262, 270, 274
nationalization thrust under, 263, 282-322, 352
phases in economic policy of, 354-55
political calculations in split with Syndicate, 264-67
political coup by, 274
policy moderation under, 266, 278
popular appeal of, 256, 266, 272-73, 293
posture on communism, 265-66
pragmatic orientation of, 259, 331-32, 344
radical phase of, 263-74, 282-322, 355, 390
reformist strategy of, 259, 265-66, 322
relationship with CPI, 262, 282, 289, 293, 335, 390
relationship with the poor, 293, 295, 336, 344-45
retreat from radicalism, 300, 314, 319-20, 328, 331-32, 335-37, 344, 353, 355-56, 390-91
selection as PM, 256, 262
state power under, 122
threat to political survival of, 266, 267, 272, 390
See also Congress party, elections
Gandhi, Mohandas Karamchand (Mahatma) 135, 137-38, 315, 340
Gandhi, Rajiv 326-327n, 355-56
Gandhi, Sanjay 335, 337-38, 352
Gandhism, 144, 167, 341
Ganesh, K.R. 301
general insurance, 297-98, 372
Gerschenkron, Alexander 25, 128
Ghosh, Ajoy 79, 82
Ghosh, Atulya 262, 280n
Gillis, Malcolm 128
Giri, V.V. 272
Gopal, S. 170n
Government and business, 223-235, 249, 251, 253, 262, 292-93, 308-309, 330, 343
Grand Alliance, 293-94, 334, 340
Gupta, Bhupesh 257, 299

Haksar, P.N. 317
Hangen, Welles 257
Hanson, A.H. 121, 183, 186, 190, 199, 216, 222-23, 227, 231, 298, 304
Harris, Nigel 143, 222
Hasan, Nurul 301
Hayek, F. A. 17
Hazari, R.K. 291
heavy industry, *see* basic industries

Hirschman, Albert O. 26
Hoselitz, Bert 20
Huan Xiang, 26
Hungary, economic reform in, 18-19

ideology:
  decline of, 319-20, 331-35, 344-45, 349, 391
  hegemony of socialist, 321-22
  modification of, 20, 27, 28, 51
  nature of, 1
  role of, 1-2, 20, 25-26, 51-53, 92-93, 107, 121, 123, 128
  under Indira Gandhi, 256-60, 262, 263-75, 277-78, 282-322, 331-35, 344-45, 360, 387-92
  under Nehru, 92-93, 107, 121, 123, 185-86, 188-89, 192-95, 200-201, 204, 206, 206-209, 210n, 211n, 213, 216-20, 226-35, 236, 247n, 387-92
  under Shastri, 253-56
  See also Marxism, Marxist analyses, socialism
Iengar, H.V.R. 360
Ikenberry, G. John 393n
Imperial Bank of India, 195, 283
imperialism, identification with capitalism 14
India-China War, 87, 235, 249, 263, 264, 284, 320
India-Pakistan war, 254, 256, 260
Indian Airlines, 332
Indian Fiscal Commission, 131
Indian Industrial Commission, 131
Indian Iron and Steel Company
  mismanagement of, 305, 309
  nationalization of, 303-307, 309
Indian National Congress. *See* Congress party, nationalist movement
Indian National Trade Union Congress, 214, 351
Indian Oil Corporation, 379
Indo-Soviet Treaty, 295
industrial licensing policy, 191, 291-92, 327n, 331-32, 336, 338, 343, 348-50, 384
Industrial Licensing Policy Inquiry Committee, 291, 301, 330
Industrial Policy Resolution of 1948, 182-85, 190, 194, 225, 303
Industrial Policy Resolution of 1956, 198-201, 227-28, 291, 298, 303, 342, 350
Industries (Development and Regulation) Act of 1951, 191, 226, 312
industry:
  advance of, 131-32
  state's role in, 369-84
industrialization:
  differences in early and late, 24-25
  difficulties in late, 24-25
  goal of rapid, 193
  military factors in 23, 24

Soviet, 25-26
See also modernization
inflation, 254, 260, 261, 263, 330, 332, 336, 339, 353
infrastructure, state of, 329, 341, 344, 345, 347, 358n
insurance. See general insurance, life insurance
instrumental-capitalist "mixed economy", 129, 134, 157, 166, 168, 184, 185, 191, 207, 208, 348, 356, 387, 388
instrumental-socialist "mixed economy", 129, 157, 164, 168, 190, 197, 199, 207, 213, 236, 244, 289, 321, 348, 356, 387, 388
"instrumentalist" approach to the state, 35-41, 43
intellectuals, See new middle class
interests, role of, 2, 20, 23, 24, 26, 30, 53, 123, 160-61, 188, 213-23, 231, 259, 266, 295, 313, 317, 322, 332-33, 336, 344-45, 353-55, 387-92
"intermediate regime":
  in India, 102, 112-118, 120-22, 123, 142, 168, 178-80, 198, 208-209, 214-20, 289, 321-22, 388-92
Kalecki's concept of, 46-48, 49, 130, 387
critique of concept, 102, 388, 391-92
intermediate strata, 46-48, 50, 112-118, 120, 122, 123, 168, 175-80, 189, 198, 208-209, 214-20, 241, 245, 247n, 287-89, 292, 321-22, 323n, 324n, 338, 340, 344, 346-47, 353, 355, 388-92
alienation of, 261, 263, 330, 332-33, 345
ideological orientation of, 168, 178-80, 189, 208-209, 214-20, 241, 388
See also new middle class, peasantry
International Monetary Fund, 335, 345, 348, 356n
international system, role in development, 24

Jagjivan Ram, 274, 341
Jan Sangh, See Bharatiya Jan Sangh
Janata Party:
  economic policy of, 339-44
  divisions in, 340-41
  nature of, 340-41
  relationship with CPI(M), 340
Jha, L.K. 260
Jha, Prem Shankar 115-117, 120, 338, 356n
Johnson, Harry 20
Jones, Leroy P. 128, 387
Joshi, P. C. (Communist leader), 75, 82
Joshi, P.C. (Marxist scholar), 73, 155, 157, 159, 170n

Kaldor, Nicholas 13
Kalecki, Michal 46-48, 49, 112, 120, 130, 213, 387, 388
Kamaraj Plan, 250
Kamaraj, K. 252, 256, 262, 266, 267
Kanoria, S.S. 246n, 320

Kapoor, Yashpal 338
Karachi resolution, 150, 151, 224
Karunakaran, K.P. 188-89, 220, 221-22
Kautsky, John 48-49, 63
Kautsky, Karl 4
Keynes, John Maynard 9, 11, 12, 389
Khadilkar, R.K. 301
Khera, S.S. 360
Kidron, Michael 197
Kissinger, Henry 258
Kochanek, Stanley 112, 138, 176-77, 179, 199, 224, 227, 246n
Kohli, Atul 122
Kondratieff cycle, 11
Kothari, Rajni: 91, 123, 124n
  criticism of Mrs. Gandhi by, 67-71
  critique of, 66-67, 70-71
  on the Indian state, 63-71
Krishnamachari, T.T. 284
Kumaramangalam, S. Mohan 141, 257, 294, 298-309, 317, 334
Kurien, C.T. 165, 194, 207

labour:
  alienation of, 330
  as beneficiary of public sector, 89, 93, 101, 103, 374
  as factor in nationalization, 303, 305, 306-307, 313-314, 322, 337, 352, 358n
  as privileged group, 213-214
  as ruling class, 118, 121
  concessions to, 34, 89, 343, 351-52
  countervailing power of, 7-8
  divisions in, 54n
  government opposition to, 331-32, 336
  role in nationalist movement, 138
labour aristocracy 118
Laclau, Ernesto 21
Lal, Bansi 338
Lamont, Douglas F. 129
land reforms, 47, 102, 157, 179, 195, 215, 233, 248, 251, 255, 330, 389
Lange, Oskar, 11, 17
Left, the, 251, 252, 262, 263ff, 284, 286, 289, 290, 292, 293-301, 307, 333, 334-35, 353, 360, 390
legitimacy, 38-39, 59n, 107, 128, 204, 209, 321
Lenin, V.I.:
  on the state, 33, 42
  revision of Marxism, 14
liberal-democratic model, 31-33, 37, 52, 63-67, 85, 89, 103
liberalism:
  common assumptions with Marxism, 21-22, 24
  impact on Nehru, 144, 157, 159, 204, 206
  theory of the state in, 31-33
liberalization. See economic liberalization

Index

life insurance, 195, 227, 283, 292, 372
Life Insurance Corporation, 305, 332
Lindblom, Charles, 12
List, Friedrich 22-24
Lok Dal, 106, 107

machinery, 376-77
Maddison, Angus 245
Mahalanobis, P.C. 195-96, 290, 291
Mahindra, Harish 338
Malaviya, K.D. 303, 305, 306
Mangaldas 320, 330
market
  failure of, 8-9, 27-28, 128, 222
  in Adam Smith, 4-6
Maruti, 352
Marx, Karl
  endorsement of imperialism, 22
  industrialized world as model, 1
  position on public ownership, 2-4
  on Asiatic mode of production, 73, 110
  on revolution, 3, 7
  on the state, 35, 40, 46
Marxism:
  appeal of, 34
  common assumptions with liberalism, 21-22, 24
  concern with state power, 118-119
  lack of original work in India on, 72-73
Marxist analyses, assessment of, 71-73, 119, 140-43 160-65, 167, 188-89, 207, 209, 220-23, 247n, 313, 320-21, 339, 353, 354, 381-82, 388, 392-93
Marxist models of the state, 33-50
Marxist models of the Indian state, 71-115, 119-123
Masani, Zareer 269
Mason, Edward S. 387
Matthai, John 167
Mayer, P.B. 357n
McClelland, David 20
Mehta, Asoka 316, 322, 360
Mendlowitz, Saul 67
Menon, Krishna 249, 261
Mensheviks, 3
metal industries, 375-76
middle class, as ruling class in Kalecki, 46-48. *See also* new middle class
middle peasantry. *See* peasantry
middle sectors *See* intermediate strata
Miliband, Ralph 10, 34, 35, 39, 43, 50
military, role in Third World, 41, 42, 49
mining industry, 369-71
minorities, role of, 293, 324n, 339, 355
Mishra, D.P. 187
Mitra, Ashok 117
mixed economy:
  as a new mutant, 384

  in India, 184, 185, 190, 191, 197, 199, 204, 206, 207, 210n, 213ff, 253, 354, 384
  rise of, 7-10
  types of, 128-30
  See also instrumental-capitalist "mixed economy", instrumental-socialist "mixed economy"
modernization, 1, 24, 28, 44, 51, 134, 137
"modernization imperative", 24, 51
Mohota, R.D. 325n
Monopolies and Restrictive Trade Practices Act, 1969 (MRTP), 290, 327n
Monopolies Inquiry Commission 179, 235, 290, 291
Morris-Jones, W.H. 275, 293, 324n
Mukherjee, Pranab 349, 350
multi-class coalition, 94, 121, 139, 140
Myrdal, Gunnar 213, 220-21, 369

Namboodiripad, E.M.S. 150, 221, 247n
Naoroji, Dadabhai 137
Narayan, Jayaprakash 192, 211n, 275, 319, 331, 334, 335, 341
National Democratic Front, 85, 90, 91, 289, 386
national goals, typologies of, 28-29, 51-52, 58n
national independence as a goal, 28, 51, 195, 199, 237-40, 244
National Planning Committee, 156, 164, 283
national power
  as factor in development 23
  relationship to public sector, 26
national security
  as factor in development, 24
National Textile Corporation, 312, 313, 375
nationalism in the Third World, 48
nationalist movement: 64
  comparison with Chinese communist movement, 143
  development into mass movement, 137-38, 146
  Gandhi's role in, 135, 137-38
  ideological differences in, 147, 148-53, 162, 168
  ideological orientation of, 149-53, 167, 168
  ideological trends in, 135-43, 147-53, 155, 159, 167
  Marxist perspective on, 73-75, 140-43
  mobilization of peasantry in, 136-38
  phases in growth of, 136-38
  position of the Left in, 138, 142, 148, 149, 150, 151, 153, 160, 163-64
  position of the Right in, 152-53, 162, 163, 164, 167, 168
  Resolution on Fundamental Rights and Economic Policy, 150, 151, 165
  Social background of leadership of, 136-38, 321
  social basis of, 135-43

weakness of socialism in, 139, 188, 207-208
nationalization: 181, 182, 184, 195, 200, 202, 203, 206, 224, 227, 247n, 263, 264, 282-89, 298-332, 336-37, 343, 353, 354, 355, 365
  threat of backdoor, 290, 291-92, 304, 305
Nayar, Baldev Raj 112-113
Naxalite movement, 264, 289, 390
Nehru, Jawaharlal:
  adaptation of socialist vision by, 151, 154-60, 202-206
  as alleged leader of bourgeoisie, 188, 20 247n
  as leader of the Left, 148
  as target of business hostility, 153, 163-64
  assessment of CPI perspective by, 140
  assessment of ideology of, 143-45
  attitude toward private sector, 155, 193, 201, 202-206, 207, 217, 239, 241, 243, 244, 247n
  commitment to Congress party, 153, 159
  commitment to democracy, 144, 157, 159, 204, 206, 250-51
  commitment to socialism, 136, 143-60, 189, 191-95, 200-201, 202-208, 214, 216, 250-51, 281n
  critical role in policy shift to socialism, 189, 209, 251, 388
  development of socialist vision of, 143-60
  differences with Gandhi, 147, 148-49, 151, 155, 160
  economic policies of, 180-209, 219-20, 248, 354, 389-90
  evaluation of achievements of, 275-77
  failure in land reform, 195, 215, 248, 389
  fear of disruptive forces, 158, 185
  hostility to private profit, 146, 155, 193
  ideological development of, 143-60
  ideological position of, 143-60, 164, 167-68, 388
  impact of Gandhi on, 144, 146-47, 148, 153, 159
  impact of Marxism on, 144-45, 147, 148-49, 151-52, 155
  impact of Soviet Union on, 147, 151
  imprisonment of, 146, 150-51, 156
  model for transition to socialism of, 154-60, 197, 202-206, 355-56, 388, 392
  moderateness of, 148, 153, 156, 158
  opposition to fascism, 147, 151, 152
  opposition to private economic power, 217, 238-39, 241, 243, 244
  phases in ideological development of, 145-60, 283
  policy moderation of, 180-85, 266
  political coup by, 185-88, 208, 266
  popular appeal of, 148, 153, 164, 168, 182, 187, 208, 248, 388
  position on communism, 147, 151
  position on violence, 147, 148-49, 159
  position in nationalist movement, 135-36, 143-44, 147-48, 151
  quest for distinctive Indian approach, 155-56, 202-204
  rejection of capitalism by, 155, 201, 205
  reorienting nationalist movement towards socialism, 136, 138, 148, 149-54
  restraint in nationalization, 203, 205-206, 282, 284, 303, 316, 322
  role as planner before independence, 154, 156-59, 164, 167
  role of mixed economy in ideology of, 157, 159
  social background of, 113, 143, 145-46, 321
  socialist model of, 156-57, 167-68, 354, 355-56
  State power under, 122, 389
  views on technology, 203, 322
  weaknesses of, 164
Nehru model :
  breakdown of consensus over, 250, 270-71
  consensus over, 219-20, 244, 248, 252, 327
  retreat from, 355-56
Nehru Study Forum, 334
new middle class:
  alienation of, 261, 330, 332-33
  decline of, 263
  hegemony of, 139, 162, 168, 295, 321-22, 355
  role in India of, 102, 103, 111, 112-118, 160, 162, 175-80, 189, 198, 208-209, 214-20, 236, 287, 293-95, 333-34, 336, 340, 341, 344, 347
  role in nationalist movement of, 136-43, 168, 170n
  development of, 7, 133, 136
  ideological orientation of, 138-39, 143, 178-79
  role in Third World, 48
Nigam, Raj K. 360
Nijalingappa, S. 269, 270, 272, 273

O'Connor, James 38, 161, 209
O'Donell, Guillermo 41-42
Oil and Natural Gas Commission (ONGC), 379, 382
oil industry, nationalization of, 337, 352, 375
Old Guard. See Syndicate
Omvedt, Gail 43-44
OPEC, 329, 345
Open Door policy, 20, 22
Organski, A.F.K. 28

Palghat, CPI Fourth Party Congress at, 82, 239-41, 299
Pantham, Thomas 114-115, 119, 120, 123
Paradigms, disintegration of, 20

Index

Paranjape, H.K. 291, 327n
Patankar, Bharat 43-44
Patel, H.M. 341
Patel, Rajni 299, 301, 338
Patel, Sardar Vallabhbhai 77, 80, 143, 148, 150, 153, 168, 182, 186, 187, 188, 208, 220, 223, 266, 388, 390
Patil, S.K. 262
Patnaik, Biju 343
Patnaik, Prabhat 126n, 327n
Peasantry:
   alienation of, 261, 330, 332-33
   as ruling class in Kalecki, 46-48
   new assertiveness of, 263, 344
   ideological orientation of, 179, 220, 248
   role in nationalist movement, 138-43, 168, 170n
   role in politics of, 104, 106-107, 111, 112-118, 175-80, 208-209, 215, 248, 263, 287, 293, 294-95, 317, 324n, 333, 336, 340, 341, 355, 389
People's Democratic Front 80, 99
Petras, James 49
petty bourgeoisie, role of, 102, 103, 111, 220, 287
planning (economic):
   in capitalist countries, 9
   in communist countries, 14-20
   in India, 134, 154-59, 164-67, 216-217
   in Marxism, 2-4
   See also Five Year Plans
Planning Commission, 189, 254
pluralism. *See* liberalism
political system:
   changes in, 252, 263
   implications of, 93, 100, 101, 105-106, 111, 130, 175, 205, 208, 259, 266, 277, 289, 295, 297, 307, 313, 332-34, 336, 344-45, 353-55, 388, 389, 390-92
   nature of, 174-75, 353
   poor role in politics of the, 293, 295, 313, 336, 339, 344-45, 355
Poulantzas Nicos 15, 35-37, 43, 218, 245n
poverty, 62, 132, 137, 146, 149
power (energy), 371
Praja Socialist Party, 188, 192, 215, 262, 334, 341
Prasad, Rajendra 143, 153, 168
praxis (experience), importance of, 355, 389-91, 392
private sector:
   dependence on public sector, 306, 372-74, 377
   penetration by the state of, 292
   performance of, 263, 284-85
   status in economic policy under Indira Gandhi, 260-61, 263, 268-69, 289-92, 295-96, 320, 331-32, 336, 347-51, 353, 355, 365-66, 390-91
   status in economic policy under Rajiv Gandhi, 326-327n, 355-56
   status in economic policy under Janata Party, 342
   status in economic policy under Nehru, 181-84, 185, 190-91, 194, 195-201, 202-206, 223-35, 251, 354, 390
   status in economic policy under Shastri, 253, 263-64, 390
privy purses issue, 292
proletariat, role of, 3
Pryke, Richard 30
public distribution system, 276, 293ff, 317, 318, 319
public financial institutions, 292, 302, 305-306, 308, 313, 372-74
   factors in the development of, 26-30, 51-53, 57-58n, 92-94, 103-104, 123, 128-30, 160, 178-80, 193-95, 214-20, 236-45, 282-89, 298-322, 337, 352, 358n
   goal of national independence in, 81-87, 89-96, 100, 237-41, 244
   hegemonic position of, 194, 196, 197, 200, 202, 206, 356, 360-84, 386, 393
   ideological role of, 103, 204, 347
   impact on economic growth, 205
   in capitalist countries, 9-10, 20
   in communist countries, 14-20
   in intermediate regimes, 47
   opposition to, 222-35, 244-45, 245n, 249, 270-71
   performance of, 118, 205, 263, 305, 308, 322, 329, 332, 333, 339, 343, 346-47, 351, 354, 358n, 368, 374, 389-90, 392
   public control of, 93, 100, 101
   role in development, 26-30
   role in economy, 360-84
   role in transtion to socialism, 197, 202, 206, 392-93
   role of ideology in, 26-27, 51-53, 57n, 92-93, 123, 128, 130, 191-98, 200-201, 204, 206-209
   role of interests in, 20, 130, 213-20, 235-45
   role of nationalization in expansion of, 282-89, 298-322, 322n, 353, 365-66, 386, 390
   role of state entrepreneurship in expansion of, 207, 282, 361, 369-79, 386
   status in economic policy under Indira Gandhi, 260-61, 282-322, 346-47, 348, 351-53, 355, 363, 372, 387-92
   status in economic policy under Nehru, 182-84, 190-91, 195-201, 364, 387-92
   status in economic policy under Shastri, 253-56
   types of, 361-62, 384

Rahman, M.M. 280n
railways, 331, 332, 369, 372

Raj, K.N. 112, 119, 120, 339
Rajagopalachari, C. 143, 168
Ram, Charat 320
Ram, Shri 224
Ranade, M.G. 137
Ranadive, B.T. 76
Rao, C. Rajeshwar 78-79, 360-61
Rao, R.P. 280n
Rao, V.K.R.V. 129-30, 356n
Ray, Baren 141
Ray, Siddhartha Shankar 338
Reddy, K.V. Raghunath 301
Reddy, Sanjiiva 272
refractories industry, nationalization of, 307
Reisman, D.A. 6
Reserve Bank of India, 283, 285
revolution, 3, 7, 218, 265
rich peasantry, See peasantry
riots See social disorder
Robinson, Joan 48
Rosen, George 211n
Rosenstein-Rodan, Paul 26
Rosenthal, Donald B. 175
Rostow, W.W. 20, 26, 28
Roy, Ajit 95, 104-107, 109-110, 120, 126n, 161-62, 326n
Rudra, Ashok 247n
Rural Banking Enquiry Committee, 383
Rustow, Dankwart 28

Sachs, Ignacy 129, 164, 197, 228
Sakong, 11, 128
Samyukta Socialist Party, 262, 294
Sathyamurthy, T.V. 126n
Satindra Singh, 301
savings, socialization of, 292, 302, 305, 372
scheduled castes, relationship to Congress party, 139, 293, 295, 339
Schumpeter, Joseph 10, 11
sectoral demarcation of economic activities:
    business position on, 224-35, 292-93, 342, 343
    government policy on, 182-84, 190-91, 195-201, 289-92, 332, 342, 347, 348-50, 355-56, 390
Sen Anupam 110-111, 119, 123, 178, 391
Sen, Asok 114-15, 119, 120, 123
Sen, Mohit 104
Sengupta, Arjun 353
Shah, K.T. 158, 184
Sharma, Shanker Dayal 296
Shastri, Lal Bahadur (Prime Minister), 253-56
shipping, 372
Shirokov, G.K. 215
Shonfield, Andrew 11
Shroff, A..D. 166
Singh, V.P. 352
Singhania, Lakshmipat 227

"sinking sands" theory, 307, 386
Sino-Indian war. See India-China war
Skocpol, Theda 39
small-scale industry, 110, 116-117, 118, 181, 184, 193, 195, 200, 215, 229, 233, 238, 240, 242, 250, 287, 309-310, 341, 342, 347, 349, 350, 354
Smith, Adam:
    critique by List of, 22-23
    invisible hand in, 3, 5, 8, 12
    position on market and state, 4-6
Sobhan, Rehman (and Muzaffer Ahmed) 178, 179, 209
social disorder, 261, 314, 319, 330-31, 343
socialism:
    absence of democracy under, 16, 17
    advance toward, 188, 192-98
    appeal of, 13-14, 20, 193-94, 216, 218-219
    economic inefficiency under, 17
    hegemony of ideology of, 321-22
    in Marx, 3, 13
    lack of appeal in developed countries, 10, 16
    legitimizing role of, 107, 110, 123, 387
    modification of, 14-20, 26-28, 393
    opposition to, 162-63, 220-35, 249
    performance as test of, 319, 353-54, 389-91, 392
    policy measures for, 195-201
    practice in communist countries, 14-20
    premature installation of, 13, 205
    rationale for, 192-95, 196
    retreat from, 185-86, 328-56
    role of, 128-29
    See also ideology, Marxism, Marxist analyses
socialistic pattern of society, 192-95
socialists, 244, 252
Sola, Robert A. 210n
Soviet Union:
    appeal of to Third World, 14
    obstacles to economic reform in, 19, 56n
    state capitalism in, 15-18, 56n
Stalin, Joseph 14
state:
    Adam Smith's view of, 4-6
    as bourgeoisie 61n
    as entrepreneur, 26-28, 128, 207, 282, 361, 369-79, 386
    autonomy of, 32, 35-41, 110-111, 115, 245n, 278, 308, 309, 318, 320, 326n, 336, 374, 384, 389, 391
    Bonapartist, 35
    bureaucratic, 41-42
    classes and, 41-50
    in the Third World, 32, 34, 41-50, 52-53
    intervention by the, 6, 7, 9, 10, 12, 55n, 128, 165, 387
    limited capacity of, 190, 203, 205-206, 224, 282, 284, 303, 315, 322, 337

Index

models of the Indian, 62-123, 278
models of the, 30-50, 52-53
multi-class state, 94, 121
penetration of private sector by, 306, 372-74
relationship to society, 30-50
role of in development, 1, 8-9, 20-30, 51-53, 369
theories of the state, 30-50
withering away of, 3-4
See also intermediate regime, intermediate strata
State Bank of India, 284
state capitalism:
  in Soviet Union, 15, 18, 56n
  in India, 86, 89, 92, 98, 112, 114, 123, 236, 247n
  in intermediate regimes, 47
  CPI on, 86, 89
state ownership. *See* public sector
state trading, 227, 233, 251
State Trading Corporation, 227
Steel Authority of India Limited (SAIL), 376, 379
steel industry, 303-307, 371
Stepan, Alfred 40
strategic contingencies theory, 388
structural context, importance of, 208, 259, 266, 332-34, 353, 388, 390-92, 393n
structural-functional approach, 63, 65
"structuralist" approach to state, 35-41, 43
Subramaniam, C. 260
Swatantra Party, 223, 234-35, 249, 262, 272, 273, 288, 289, 293, 294, 297, 341
Syndicate, 252, 253, 256, 260ff, 286, 289, 293, 294, 296, 334, 340, 355

Tandon, Purshottamdas 186-87, 220, 266, 272
Tata, J.R.D. 86, 296, 330
Tata-Birla Plan, 86, 164-67
Tata Electric Company, 222
Tata Iron and Steel Company, 303, 306, 343
Tatas, 231, 349
taxation, 197, 229, 251, 336, 338, 355
technological dependence, 25
technology policy, 350
teleology, in industrialization, 24
Ten Point Programme, 268-69, 267, 298
textile industry:
  crisis in private sector, 309-312
  government policy responsibility for crisis in, 309-11

nationalization of, 309-14, 352
public sector role in, 375
Therbon, Goran 42
Thorner, Daniel 185
Tilak, Bal Gangadhar 145
Tiwari, N.D. 351
Toennies, Ferdinand 53n
Toynbee, Arnold 41
transport, 372
"triple alliance," 42, 44
Trojan Horse, 245
trusteeship theory, 138, 139, 167
Twenty Point Programme, 336, 344-45, 349, 351
"two-pillar policy" of AITUC, 243

Unit Trust of India, 305-306
United States, relations with, 254, 258, 260, 261, 276, 315, 318, 328-29
USS Enterprise, 258

Vakil, C.N. 127n
values, consummatory and instrumental, 1, 20, 26, 51-53, 129-30, 194, 204, 211n, 282, 301, 305, 307, 352, 387, 388
Venkataraman, R. 351, 352
Venkatasubbiah, H. 216-217, 228, 234
violence, *See* social disorder
von Mises, Ludwig 17

Wagner's Law, 53n
Wallerstein, Immanuel 21
wars, impact on economy, 131, 132, 190
Warren, Bill 21
Weiner, Myron 112, 175, 324n
welfare, relationship to public sector, 26
welfare state, rise of, 7-8, 54n
wholesale wheat trade:
  critical impact on shift in economic policy, 319, 331
  nationalization of, 314-20, 330
  retreat from nationalization of, 319
World Bank, 335, 356n

Yadav, Chandrajit 301, 307
Yugoslavia, economic reform in, 18-19
Yunus, Mohammed 338

Zakir Hussain 268
"zero-sum" game in capitalism, Marxist assumption of, 2, 33, 241